Eco-evolutionary Dynamics

Eco-evolutionary Dynamics

ANDREW P. HENDRY

PRINCETON UNIVERSITY PRESS
Princeton and Oxford

Copyright © 2017 by Princeton University Press

Published by Princeton University Press, 41 William Street, Princeton, New Jersey 08540

In the United Kingdom: Princeton University Press, 6 Oxford Street, Woodstock, Oxfordshire OX20 1TR

press.princeton.edu

Cover photo: A medium ground finch (*Geospiza fortis*) interacts with the fruit and seed of *Castela galapageia* on Santa Cruz Island, Galapagos. Courtesy of the author

All Rights Reserved

ISBN 978-0-691-14543-3

British Library Cataloging-in-Publication Data is available

This book has been composed in Minion Pro

Printed on acid-free paper. ∞

Printed in the United States of America

10 9 8 7 6 5 4 3 2 1

For my family,

who successfully and productively distract me from eco-evolutionary dynamics.

Contents

Preface

It seems appropriate to start by repeating the preface I wrote in 2008 for the book proposal. It gives an idea of my motivation and goals before I started writing the book.

*   *   *

Research initiatives in ecology and evolution have periodically dated but never married. When one looks at long enough time scales, the value of such a union is obvious and undisputed. That is, different ecological conditions have clearly led to dramatic evolutionary changes: some plants perform well in high-light environments and others in low-light environments, some insects are faithful pollinators and others are nectar robbers, and some birds are very good at eating seeds whereas others are very good at eating insects. It is also clear that large evolutionary changes have dramatic ecological consequences: tall trees cause low-light environments beneath them, specialized pollinators and nectar robbers influence seed set in many plants, and specialized bird species can reduce seed or insect abundance. Such examples of reciprocal interactions between ecology and evolution on long time scales are so numerous and unequivocal as to remove the need for lengthy justification. *My goal in the present book will be to consider eco-evolutionary dynamics on contemporary time scales, ranging from just a few generations to several hundred generations.*

Perhaps the most dramatic recent progress in eco-evolutionary dynamics has been made in the area of "contemporary evolution," also called "rapid evolution." Indeed, this body of work now provides many of our iconic examples of how natural selection drives evolutionary change: toxic mine tailings cause the evolution of plant resistance to heavy metals, industrial pollution causes shifts in the frequency of melanic moths, droughts that alter food resources drive the evolution of finch beaks, and changing predator pressures altered the evolutionary trajectories of guppy populations. Following this acceptance that ecology drives contemporary evolution within populations, increasing attention has been focused on how ecology can drive the origin of new species. Examples of the contemporary onset of such "ecological speciation" come from plants adapted to different soils, insect populations adapted to different host plants, birds adapted to different food types, and fish adapted to different foraging environments in lakes. *One of my primary goals in the present book will be to provide an integrated understanding of the promoting and constraining forces that determine how ecological changes drive contemporary evolution and speciation.*

Ecology thus drives evolutionary dynamics in contemporary time, but does evolution drive ecological dynamics on similar time scales? A venerable tradition here has been to assume that evolution takes place over such extended periods of time that its impact on contemporary ecological dynamics can be safely ignored. Recent work, however, has challenged this assumption from the viewpoint of population dynamics, community structure, and even ecosystem function. The few existing examples are tantalizing. Phenotypic and genetic variation within populations contributes to population dynamics in feral sheep, salmon, and butterflies. The evolution of one species drives population dynamics in coexisting species: changing virulence of an introduced pathogen (myxoma virus) alters population dynamics of introduced rabbits and changing genetic compositions of green algae cause oscillations and extinctions of rotifers. Evolutionary changes in one species can also influence communities and food webs: plants with different levels of genetic variation have different arthropod communities and planktivorous fish with different (recently evolved) foraging traits can alter zooplankton communities. Such effects may also cascade to ecosystem function. For example, changes in the frequency of cottonwood tree genotypes, owing to selective harvesting by beavers, alters decomposition rates in streams. *Another goal of my book will be to consider the conditions under which evolutionary change might influence ecological dynamics and to suggest key experiments and synthetic efforts that will inform those conditions.*

Existing studies have thus focused on how ecology drives evolution and how evolution can also drive ecology—but a full understanding won't be achieved simply by looking at each pathway individually. Instead, ecology and evolution will interact and feed back on each other in a number of important and nontrivial ways. Very few studies have yet travelled far down this road. As one tantalizing example, changes in fish foraging traits influence zooplankton communities, which then alter selection on foraging traits, which then further influence zooplankton communities. *My book will investigate areas where such interactions seem particularly likely and important.*

To summarize, my book will be guided by several major goals. I first hope to summarize existing knowledge on eco-evolutionary dynamics, focusing particularly on natural populations in contemporary time. I further hope to integrate disparate fields and approaches to ecological and evolutionary dynamics, thus fostering conversation among researchers who might otherwise have continued to work separately. I also hope to identify key areas of ambiguity and weakness in the existing body of work, thereby stimulating novel approaches and experiments. Most importantly, however, I feel that this book is mainly a long advertisement for the value and sheer intrigue of studying eco-evolutionary dynamics in natural populations.

* * *

Having now finished the book seven years later, I find that I mostly stayed true to these objectives and, in so doing, found myself trying to straddle two goals. The first was to present each topic with enough simplicity and clarity that it could be understood and appreciated by nonspecialists. The second goal was to address key questions and controversies that captivate and bedevil specialists in each topic. It quickly became apparent that I had one foot in each of two boats that were inexorably drifting apart. Instead of giving up on one boat in favor of the other, I have gamely and perhaps stubbornly attempted to straddle the middle, which could turn out not to be the high ground. That is, my presentation of any topic might be too technical for nonspecialists but not technical and detailed enough for specialists. Oh well, this is the book that I would want to read and so it is the book that I have written. I hope it is also useful to you.

In trying to cover so much research in so many allied fields, it was clear early on that I could not possibly intellectually master even a fraction of the ideas and the literature. As a result, I have relied heavily on the help and advice of many individuals who have suggested articles, read sections of the book, provided data, and engaged in formative discussions and arguments. The listing of these individuals does not imply their endorsement of the arguments I have made in the book—in fact, I know a number of folks I here thank strongly disagree with some of my conclusions.

In first-name alphabetical order, thanks to: Adam Siepielski, Åke Brännström, André De Roos, Andrew Gonzalez, Andrew Storfer, Aneil Agrawal, Anthony Merante, Antoine Paccard, Anurag Agrawal, Ayco Tack, Ben Haller, Ben Phillips, Ben Sheldon, Blake Matthews, Brent Emerson, Cameron Ghalambor, Carl Walters, Charles Davis, Chelsea Chisholm, Constantinos Yanniris, Cristiaan Both, Dan Bolnick, Daniel Berner, Dan Flynn, Doug Futuyma, Dany Garant, Daniel Ortiz-Barrientos, Daniel Sol, David Stern, Derek Roff, Elizabeth Nyboer, Eric Palkovacs, Fanie Pelletier, Graham Bell, Gregor Fussmann, Gregor Rolshausen, Gregory Crutsinger, Hans Pörtner, Ian Wang, Ilkka Hanski, Jason Hoeksema, Jean-Sébastien Moore, Jennifer Schweitzer, Jenny Boughman, Jocelyn Lin, Joe Bailey, Joe Hereford, John Paul, Jonathan Davies, Josef Uyeda, Josh Van Buskirk, Josianne Lachapelle, Judy Myers, Kevin Pauwels, Katie Peichel, Kiyoko Gotanda, Lennart Persson, Lima Kayello, Loeske Kruuk, Luc De Meester, Luis Fernando De León, Lynn Govaert, Marc Cadotte, Marc Johnson, Marcel Visser, Marissa Baskett, Mark Urban, Mark Vellend, Martin Lind, Matthew Osmond, Matthew Wund, Michael Becker, Mike Kinnison, Nelson Hairston Jr., Oscar Gaggiotti, Patrik Nosil, Rana El-Sabaawi, Rees Kassen, Robert Colautti, Roger Thorpe, Ron Bassar, Rowan Barrett, Sam Yeaman, Samantha Forde, Sarah Huber, Sara Jackrel, Scott Carroll, Sinead Collins, Sonia Sultan, Stephanie Carlson, Steve Johnson, Steven Franks, Takehito Yoshida, Thomas Ezard, Thomas Hansen, Thomas Near, Thomas Quinn, Thomas Reed, Thomas Smith, Travis Ingram, Vincent Fugère, Yael Kisel, several working groups funded by the Québec Centre for Biodiversity Science (QCBS), and members of the DIVERSITAS (and then Future Earth) core project bioGENESIS, especially cochairs Michael Donoghue, Tet Yahara, Dan Faith, Luc De Meester, and Felix Forest. *I apologize to folks whose help I have forgotten owing to a hard drive failure that eliminated many of my older emails.*

Special thanks are due to those individuals, acknowledged in the figure captions, who provided data for reanalysis or redrafting figures. In addition, Eric Palkovacs and his graduate class at the University of California Santa Cruz read the book in its entirety, as did my graduate class at McGill. Eric Palkovacs, Marc Johnson, Blake Matthews, Joe Bailey, and Jenn Schweitzer were my go-to folks for help with the literature on community and ecosystem consequences of evolutionary change. Katie Peichel, Patrik Nosil, and Rowan Barrett played the same role for the genetics chapter. Thomas Hansen was particularly critical—helpfully so—of my chapters 3 and 4. Nelson Hairston Jr. and three other reviewers provided extensive comments during the review stage. Mike Kinnison was particularly fundamental in my early thinking about eco-evolutionary dynamics. Colin Garroway, Craig Benkman, Felipe Dargent, and Taylor Ward read parts of the "in press" book and pointing out a number of errors that needed fixing. At Princeton University Press, the various stages of book preparation and production were shepherded by Alison

Kalett, Mark Bellis, Joe Pastore, Betsy Blumenthal, Jaime Estrada, Alexandria Leonard, Quinn Fusting, and David Campbell. The references were cleaned up considerably by Melisa Veillette and Fiona Beaty. The book proofs were read in their entirety by Alice Hendry (thanks, Mom!) and Lesley Fleming, who caught a number of errors. Also, I am particularly grateful to Caroline Leblond, who kept my lab running while I had my nose buried in this book.

What does one say having finished a book that is 12 chapters long, has more than 1500 references, and took 7 years to write? Only that I hope you will find the book useful and perhaps even inspiring. I also hope that you will see that eco-evolutionary dynamics as a field has so many unknowns that plenty of opportunities exist for future work. If I write another edition of this book 10 years hence, I expect that it—especially the evo-to-eco chapters—will look very different. You could be the reason.

Chapter 1

Introduction and Conceptual Framework

Ecology and evolution[1] are so closely intertwined as to be inseparable. This reality is obvious on long timescales given that different species are clearly adapted to different environments and have different effects on those environments (Darwin 1859). Yet, traditionally, evolutionary and ecological processes have been thought to play out on such different time scales that evolution could be safely ignored when considering contemporary ecological dynamics (Slobodkin 1961). However, the past few decades have seen a shift away from this "evolution as stage—ecology as play" perspective toward the realization that substantial evolutionary change can occur on very short time scales, such as only a few generations (reviews: Hendry and Kinnison 1999, Reznick and Ghalambor 2001, Carroll et al. 2007). If contemporary evolution can be this rapid, and if the traits of organisms[2] influence their environment, it follows that evolution will need to be considered in the context of contemporary ecological dynamics. This point is not a new one (e.g., Chitty 1952, Levins 1968, Pimentel 1968, Antonovics 1976, Krebs 1978, Thompson 1998) but the growing realization of its importance is crystalizing into a new synthesis that seeks to integrate ecology and evolution into a single dynamic framework (Fussmann et al. 2007, Kinnison and Hairston Jr. 2007, Haloin and Strauss 2008, Hughes et al. 2008, Pelletier et al. 2009, Post and Palkovacs 2009, Schoener 2011, Genung et al. 2011, Matthews et al. 2011b, 2014, Strauss 2014, Duckworth and Aguillon 2015).

According to Web of Science and Google Scholar, the earliest use of the term "eco-evolutionary" was Kruckeberg (1969) and the first use of the term "eco-evolutionary dynamics" was Oloriz et al. (1991); yet modern usage really began with a 2007 special issue of *Functional Ecology* (Fussmann et al. 2007, Carroll et al. 2007, Kinnison and Hairston Jr. 2007). To illustrate, Web of Science tallies 18 articles that used "eco-evolutionary" in the title, abstract, or keywords prior to 2007 and 445 articles since that time (as of Apr. 8, 2016). Consistent with that modern usage, I here define eco-evolutionary dynamics as *interactions between ecology and evolution that play out on contemporary time scales*, with

---

[1] The Merriam-Webster definition will suffice for "ecology" (relations between a group of living things and their environment) whereas I offer my own definition for "evolution" (changes in the genetic composition—usually allele frequencies at particular places in the DNA sequence—of a population).

[2] When I refer to an "organism," I mean the collection of individuals that make up a population or species, not a single "individual" within that population or species.

"contemporary" intended to encompass time scales on the order of years to centuries (or one to hundreds of generations). These interactions can work in either direction. In one, ecological changes lead to contemporary evolution (eco-to-evo), such as the ongoing adaptation of populations to changing environments. In the other direction, contemporary evolution can lead to ecological changes (evo-to-eco), such as when trait change in a focal species alters its population dynamics, influences the structure of its community, or alters processes in its ecosystem. Moreover, these interactions can feedback to influence one another: that is, ecological change can cause evolutionary change that then alters ecological change (Haloin and Strauss 2008, Strauss et al. 2008, Post and Palkovacs 2009, Genung et al. 2011, Strauss 2014). In this first chapter, I provide an overall conceptual framework for studying eco-evolutionary dynamics, and I explain how the rest of the book fits into that framework.

The style of this first chapter differs from those that follow. In this first chapter, I provide a very simple and general introduction that builds a framework on which to hang the more detailed deliberations that will follow later. I have therefore here written with a minimum of jargon, citations, and footnotes; and I have provided boxes that outline simple and clear examples. This writing style is intended to provide a stand-alone introduction accessible to all evolutionary biologists and ecologists, as opposed to only those already well versed in the topic. Rest assured, the subsequent chapters will be awash in enough jargon, citations, footnotes, and details to be of interest even to specialists.

Key elements of the book: phenotypes of real organisms in nature

When studying eco-evolutionary dynamics, one might focus on genotypes or phenotypes. My focus will be squarely on the latter: for two key reasons. First, selection acts directly on phenotypes rather than on genotypes. Genotypes are affected by selection only indirectly through their association with phenotypes that influence fitness. Understanding the role of ecology in shaping evolution therefore requires a phenotypic perspective. Second, the ecological effects of organisms are driven by their phenotypes rather than by their genotypes. Genotypes will have ecological effects only indirectly through their influence on phenotypes that have ecological effects. In some cases, eco-evolutionary dynamics might be similar at the genetic and phenotypic levels, most obviously so when a key functional trait is mainly determined by a single gene. However, this situation will be rare because most traits are polygenic and are also influenced by environmental (plastic) effects, topics considered at depth in later chapters. These two properties muddy (in interesting ways) the genotype-phenotype map and dictate that studies of eco-evolutionary dynamics should have, as their focus, organismal phenotypes. This focus does not mean that genotypes should be ignored and, indeed, genotypes are explicitly considered at many junctures in this book—but the central focus must be on phenotypes.[3]

---

[3]Here are some definitions of related terms as they will apply throughout the book. "Phenotypes" are physical characteristics, whether physiology, morphology, behavior, or life history, expressed by organisms in ways that can interact with the environment. "Traits" are phenotypes that can be defined and studied at least partly independently of other such traits, such as metabolic rate, body size, aggression, or fecundity. (Of course, such traits will often be correlated with each other.) Adaptive traits are traits whose particular values (e.g., larger or smaller) influence the fitness (survival and reproduction—more about this later) of organisms. Such traits are often called functional traits in the plant literature.

Eco-evolutionary dynamics can be studied in theory or in real organisms. Theoretical studies, such as those employing analytical (symbolic) math or computer simulations, are critical for helping to delineate the various possibilities that arise from an explicit set of assumptions. Theory also can help to formalize conceptual frameworks, develop analytical tools, and evaluate predictive structures for the study of real organisms. For these reasons, theory will make frequent appearances in the book, typically as a means of setting up expectations and for helping to interpret the results of empirical studies. In the end, however, theory is only a guide to the possible—it can't tell us what actually happens; and so an understanding of eco-evolutionary dynamics requires the study of real organisms.

Eco-evolutionary studies with real organisms could proceed in the laboratory or in nature. Advantages of the laboratory are manifold: populations can be genetically manipulated, environments can be carefully controlled, replicates and controls can be numerous, and small organisms with very short generation times (e.g., microbes) allow the long-term tracking of dynamics (Bell 2008, Kassen 2014). These properties dictate that eco-evolutionary studies in the laboratory are elegant and informative, yet only in a limited sense. That is, such studies tell us what happens when we impose a particular artificial environment on a particular artificial population and, hence, they cannot tell us what will actually happen for real populations in nature. Understanding eco-evolutionary dynamics as they play out in the natural world instead requires the study of natural populations in natural environments. I will therefore focus to the extent possible on natural contexts, although I certainly refer to laboratory studies when necessary.

The study of real populations in real environments is usually considered to be compromised in several respects. For instance, such studies have difficulty isolating a particular ecological or evolutionary effect because it might be confounded, or obscured, by all sorts of other effects that exist in the messy natural world. To me, this suggested weakness is actually a major strength because we obviously want to know the importance of a particular effect within the context of all other effects that also might be important. By contrast, it seems of limited value to isolate and evaluate a particular effect in a controlled situation if that effect is largely irrelevant in natural contexts. Moreover, elucidating causal effects and their interactions is possible even in nature through experimental manipulations (Reznick and Ghalambor 2005). However, the limitations of studying real populations and real environments are certainly real and important: replication and controls are harder to implement, experimental manipulations are less precise, and ethical and logistical concerns prevent some experiments. Yet such studies ultimately will be the key to developing a robust understanding of eco-evolutionary dynamics.

Conceptual framework and book outline

My primary goal in this first chapter is to provide a conceptual framework for eco-evolutionary dynamics. The framework will be presented in three parts. The first part (*eco-to-evo*) outlines how ecological change influences evolutionary change, and thereby amounts to a review and recasting of the classic field of evolutionary ecology. The second part (*evo-to-eco*) outlines how evolutionary change influences ecological change, and thereby amounts to the set of effects that have crystallized and driven the emergence of eco-evolutionary dynamics as a term and as a research field. The third part (*underpinnings*) considers the genetic and plastic basis of eco-evolutionary dynamics, which can apply with equal relevance to our understanding of both preceding parts. Within each of these parts, important components of the framework will be presented sequentially and their correspondence to the various chapters will be explained. In the current presentation, I will only rarely refer to specific empirical results because those results will be discussed in detail in the chapters that follow. Rather, I will provide a series of linked examples drawn from a single empirical system: Darwin's finches on Galápagos (Grant 1999, Grant and Grant 2008). This choice of system doesn't imply that Darwin's finches provide the best illustration of every concept, but rather that they are suitable for explaining how different components of the conceptual framework fit together for a single well-known study system.

PART 1: ECO-TO-EVO

The eco-to-evo side of an eco-evolutionary framework obviously starts with ecology. By "ecology" in this context, I mean any combination of biotic or abiotic features of the environment that can impose selection on the phenotypes of some focal organism. In the context of a single population, I will generally refer to ecological *change*. In the context of multiple populations, I will generally refer to ecological *differences*. Either term might be used when generalizing to both contexts.

A single population in a stable environment should be characterized by phenotypes that are reasonably well adapted for that environment. Stated another way, the distribution of phenotypes in a population should correspond reasonably well to the phenotypes that provide high fitness (survival and reproductive success): that is, the distribution of phenotypes should be close to a fitness peak on the "adaptive landscape" (fig. 1.1). In this scenario, an obvious eco-to-evo driver is ecological change that shifts the fitness peak away from the phenotypic distribution. (A similar effect arises if the phenotypes shift away from the peak, such as through gene flow—see below.) This shift imposes selection on the population by increasing fitness variation among individuals with different phenotypes (Endler 1986, Bell 2008). If the phenotypic variation is heritable (passed on from parents to offspring), the next generation should see a phenotypic shift in the direction favored by selection: that is, toward the fitness peak. Under the right conditions, the phenotypic distribution should eventually approach the new peak and directional selection should disappear. In reality, peaks will be constantly shifting and populations might have difficulty adapting owing to genetic or other constraints as will be considered in detail later. Box 1 provides an illustrative example of directional selection and adaptation in Darwin's finches.

Selection is thus the engine that drives eco-evolutionary dynamics, and so the more detailed chapters of the book must begin there. Chapter 2 (Selection) starts with a description of how the mechanism works and how it is studied in natural populations. It then draws on recent meta-analyses to answer fundamental questions about selection in nature, such as how strong and consistent it is, how often it is stabilizing (disfavoring extreme individuals) or disruptive (favoring extreme individuals), what types of traits (e.g., life history or morphology) are under the strongest selection, and how selection differs when fitness is indexed as mating success (sexual selection) or survival/fecundity (natural selection).

The expected outcome of selection is adaptive phenotypic change (fig. 1.1), which should then shape eco-evolutionary dynamics. Chapter 3 (Adaptation) first outlines how to conceptualize and predict adaptive evolution based on information about selection

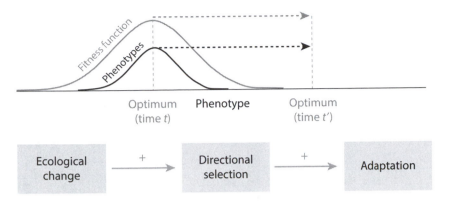

Fig 1.1. Graphical representation of ecological change, directional selection, and adaptation in a single population. Before ecological change (time *t*), the frequency distribution of phenotypes (lower curve) in a well-adapted population is centered near the phenotypic value that maximizes fitness (optimum). This optimum corresponds to the peak of the "fitness function" (upper curve) that relates phenotypes (*x*-axis) to fitness (*y*-axis). Ecological change occurring to time *t'* (upper dashed line) shifts the optimum phenotype to a new location, which imposes directional selection on the population, which should thus evolve toward the new optimum (lower dashed line). In reality, the fitness function is likely much wider than the phenotypic distribution. Also, this depiction assumes no constraints on evolution

Box 1

An example of natural selection and adaptation in Darwin's finches on the small Galápagos island of Daphne Major (Boag and Grant 1981, Grant and Grant 1995, 2003). Conditions during 1976–1977 caused a drought that prevented reproduction by most plants. During this period, the resident population of medium ground finches (*Geospiza fortis*) rapidly depleted available seeds from the environment and many individuals starved to death, resulting in a population size decrease of about 85%. The depletion of seeds was nonrandom because all the finches can consume small/soft seeds whereas only finches with large beaks can consume large/hard seeds. As the drought progressed, the seed distribution therefore became increasingly biased toward larger/harder seeds, and the mortality of *G. fortis* became size-selective. Birds with larger beaks were more likely to survive, resulting in directional selection for larger beaks. When the rains commenced in 1978, the finches that had survived to breed were those whose beaks were larger (on average) than the population before the drought. Beak size is highly heritable (large-beaked parents produce large-beaked offspring) and, hence, the generation of birds produced after the drought had larger beak sizes than the generation of birds produced before the drought. Ecological change caused directional selection that led to adaptive evolution.

(Continued)

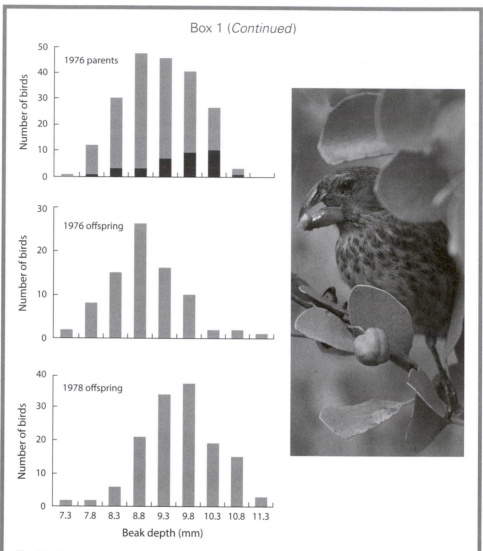

Fig B.1. Natural selection and evolution in *Geospiza fortis* (photo from Santa Cruz by A. Hendry) on Daphne Major. The top panel shows the frequency distribution of beak size in the breeding population in 1976 and the subset (black bars) of those birds that survived through the drought. The lower two panels show the frequency distributions of fully grown offspring hatched the year before the drought (middle panel) and the year after the drought (lower panel).

and genetic variation. It then introduces and explains adaptive landscapes (*x*-axis = mean phenotype; *y*-axis = mean population fitness), a concept that has proven useful in guiding our understanding of evolution. Finally, it reviews empirical data to answer fundamental questions about adaptation in nature, including to what extent short- and long-term evolution is predictable, how fast is phenotypic change, to what extent is adaptation constrained by genetic variation, and how well adapted natural populations are to their local environments.

Moving beyond selection and adaptation *within* populations, eco-evolutionary dynamics will be shaped by biological *diversity*: that is, different populations and species have different effects on their environment. This diversity arises when a single population splits into multiple populations that begin to evolve independently and could ultimately become separate species, which are then the roots of even the most highly divergent evolutionary lineages. Stated plainly, biological diversity at all levels has its initial origins in population divergence. Thus, the next step in developing a conceptual framework for eco-evolutionary dynamics is to expand our discussion from the evolution of single populations into the realm of population divergence.

The stage for population divergence is set when different groups of individuals from a common ancestral population start to experience different environments. These environmental differences could result from any number of factors, such as different abiotic conditions (temperature, pH, moisture, oxygen), different predators or parasites, different competitors, or different resources. Faced with this heterogeneity, selection will favor different phenotypes in the different groups, leading to *divergent (or disruptive) selection*. If the traits under selection are heritable, the expected outcome is adaptive divergence among the groups (now populations) that improves the fitness of each in its local environment (fig. 1.2) (Schluter 2000a). Box 2 provides an illustrative example from Darwin's finches.

Chapter 4 (Adaptive Divergence) focuses squarely on this process. It starts by explaining how the adaptive landscape concept can be extended from a single population in a single environment to multiple populations in multiple environments. Specifically, different environments produce different fitness peaks and divergent selection then drives different populations toward those different peaks (fig. 1.2). The chapter then outlines methods for inferring adaptive divergence with respect to both phenotypes (*x*-axis of the

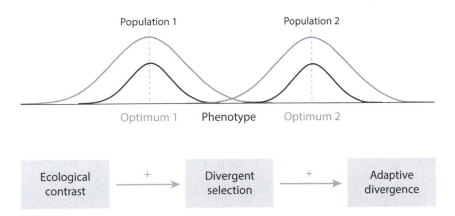

Fig 1.2. Graphical representation of an ecological contrast experienced by two populations in different environments, the divergent selection that this contrast imposes, and the adaptive divergence that is expected to result. Divergent selection occurs because the two environments have fitness functions (upper curves) with different optima. Adaptive divergence generates phenotypic differences (lower curves), such that each population becomes at least reasonably well adapted to its local optimum. As in figure 1.1, this depiction assumes no constraints on adaptive evolution

Box 2

The previous box described how one population of finches adapted to the seed resources that were locally available. Across populations, we would therefore expect average beak size to be correlated with the average size/hardness of available seeds. That is, adaptive divergence should take place as a result of differences in seed size/hardness distributions. To test this expectation, Schluter and Grant (1984) quantified the seed distributions on 15 Galápagos islands. They used these distributions to predict the mean beak sizes that would be expected for three seed-eating species: the small ground finch (*Geospiza fuliginosa*), the medium ground finch (*Geospiza fortis*), and the large ground finch (*Geospiza magnirostris*). These predictions (of what amount to adaptive landscapes) were then compared to observed beak sizes for the same species. Results for three of the islands are shown below and more complete results are provided in chapter 4. In many (although not all) cases, observed beak sizes closely matched predicted beak sizes. Adaptive divergence occurred in response to ecological differences.

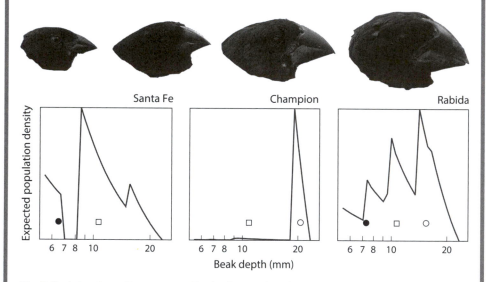

Fig B.2. Adaptive divergence. Beak size varies dramatically among populations and species of Darwin's finches in a manner that matches local food resources. The pictures at top show *Geospiza fuliginosa* (left), *Geospiza magnirostris* (right), and *Geospiza fortis* morphs with either small (center-left) or large (center-right) beak sizes (photos by A. Hendry). The figures at bottom show that mean beak sizes of the species present on each island (filled circle is *G. fuliginosa*, open box is *G. fortis*, and open circle is *G. magnirostris*) typically match the beak sizes expected based on the available seed types. Those expectations are shown as peaks depicting the expected density of finches of a given beak size. The data are from Schluter and Grant (1984), wherein further details are provided adaptive landscape) and fitness (*y*-axis). The chapter then turns to a review of empirical data informing several key questions about adaptive divergence in nature, including how prevalent and strong it is, how many peaks adaptive landscapes have (ruggedness), how many of the peaks are or are not occupied by existing populations (empty niches), how predictable it is (parallel and convergent evolution), and what is the role of sexual selection in modifying adaptive divergence.

The above description might give the impression of populations inevitably evolving the phenotypes best suited for their local environments. The reality, however, is that many factors can constrain adaptation well short of optimality. Some of those factors are considered in chapter 4, but I here wish to draw special attention to the role of dispersal, which can take place for individuals, gametes, or propagules (eggs, seeds, or spores). If the dispersers successfully reproduce, the resulting genetic exchange (gene flow) can disallow independent evolution of the recipient populations. When this gene flow is high and occurs among populations in different environments, adaptive divergence can be strongly hampered (Lenormand 2002, Garant et al. 2007). The expected outcome is a balance between divergent selection pushing populations apart and gene flow pulling them together, such that adaptive divergence will occur but not to the degree expected in the absence of gene flow (fig. 1.3). Box 3 provides an example of how gene flow hampers adaptive divergence in Darwin's finches.

Chapter 5 (Gene Flow) starts by outlining empirical methods for quantifying gene flow and inferring its role in adaptive divergence. An important point made therein

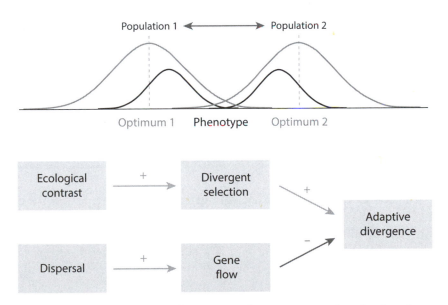

Fig 1.3. Graphical representation of how gene flow can constrain adaptive divergence. Dispersal between the populations (double-headed arrow) is expected to cause gene flow that prevents their independent evolution, and thus reduces adaptive divergence. The outcome is a balance between selection and gene flow: the populations show adaptive divergence (lower curves) but not as much as would be expected in the absence of gene flow (compare to fig. 1.2)

Box 3

As described in the previous box, multiple Darwin's finch species are present on many islands, and divergent selection has caused differences in their beak size: that is, they manifest adaptive divergence. A number of these species are very closely related and remain reproductively compatible, such that hybridization and introgression are not uncommon (Grant et al. 2005). Work on Daphne Major shows how this gene flow can reduce adaptive divergence between species. Starting in the 1990s, *G. fortis* showed increasing hybridization with *Geospiza scandens*, its pointier-beaked congener that mainly feeds on the pollen, nectar, and seeds of *Opuntia* cactus. The result was increasing introgression of *G. fortis* genes into *G. scandens*, which decreased differences between the species in both neutral genetic markers and adaptive phenotypic traits (Grant et al. 2004). Strikingly, the beaks of *G. scandens* became less pointed, thus starting to converge on the blunter beak shape of *G. fortis*. Increased gene flow reduced adaptive divergence.

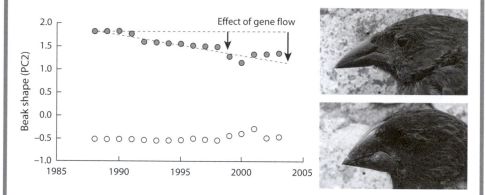

Fig B.3. Gene flow influences the beak shape of *Geospiza scandens* (top picture and dark circles in the figure) and *Geospiza fortis* (bottom picture and light circles in the figure) on Daphne Minor. The data are from Grant et al. (2004) and the photos from Santa Cruz are by A. Hendry and J. Podos. Gene flow is greater, and therefore its effects stronger, into *G. scandens* than into *G. fortis*

is that gene flow can sometimes *aid* adaptation, such as when it enhances the genetic variation on which selection acts. The key questions addressed with empirical data are therefore divided into the potential negative versus positive effects. On the negative side, questions include to what extent gene flow constrains adaptive divergence among environments, and how the resulting maladaptation might cause population declines and limit species' ranges. On the positive side, questions include whether gene flow has a special benefit in the case of antagonistic coevolution, and whether it can save (rescue) populations that would otherwise go extinct. Some of these questions begin to invoke evolutionary effects on demography, and thus grade into the evo-to-eco side of the story that we will take up later.

I earlier argued that divergent selection causing adaptive divergence is at the roots of evolving biological diversity, yet I have now just argued that gene flow hinders adaptive divergence and should thereby constrain that evolution. The exuberant diversity of life makes clear that this potential impasse is often broken—but how? One obvious solution is the presence of physical barriers (mountains, rivers, oceans, deserts) that eliminate dispersal among populations, but the more interesting situation occurs when dispersal remains possible. A likely solution in this case is *ecological speciation*, whereby adaptive divergence causes the evolution of reproductive barriers that reduce gene flow (Schluter 2000a, Nosil 2012). This process starts because populations can begin to adapt to different environments even in the presence of some gene flow (fig. 1.3). This initial divergence will increase the fitness of residents relative to dispersers, which will reduce gene flow. This reduction in gene flow allows further adaptive divergence, which further reduces gene flow, which allows further adaptive divergence—and so on until adaptive divergence is high and gene flow is low (fig. 1.4). Under the right conditions, the populations can become so divergent and reproductively isolated as to be considered separate species. Box 4 provides an example from Darwin's finches as to how this scenario might play out.

Chapter 6 (Ecological Speciation) starts by discussing how populations in different environments can fall at different stages along a continuum of progress toward ecological speciation. It then outlines how this variation can be used to infer ecological speciation through either of two general approaches: (1) integrated signatures of reproductive isolation based on measures of gene flow, and (2) confirmation of the ecological basis of reproductive barriers. The first two questions about ecological speciation in nature are

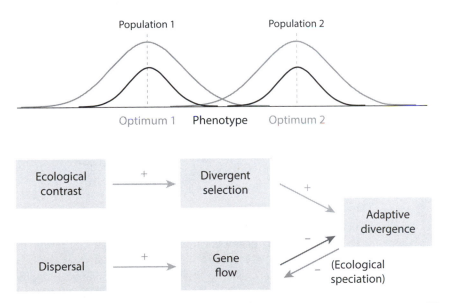

Fig 1.4. Graphical representation of how ecological speciation is expected to modify the balance between selection and gene flow that was previously illustrated in figure 1.3. Increasing adaptive divergence is expected to promote the evolution of reproductive barriers that reduce gene flow. This reduction in gene flow should allow increased adaptive divergence, potentially generating a positive feedback loop through which gene flow decreases to the very low levels characteristic of separate species

Box 4

As described in Box 2, adaptation to different seed distributions causes beak size divergence among populations and species of Darwin's finches. This divergence contributes to several reproductive barriers that influence ecological speciation. First, divergence in beak size (and bite force) causes divergence of the songs that males sing (Podos 2001, Herrel et al. 2009). Given that songs and beaks influence breeding behavior in Darwin's finches (Grant 1999, Grant and Grant 2008), populations diverging in beak size can begin to show positive assortative mating. Second, hybrids will show reduced survival if their intermediate beak sizes fall into valleys between the fitness peaks to which their parents were adapted (Grant and Grant 1993, 1996). To test these expectations, it helps to study populations in the early stages of speciation, such as *G. fortis* that are bimodal for beak size (Hendry et al. 2006). Studies of one such population confirmed the above expectations: large and small beak size morphs of *G. fortis* have different diets (De León et al. 2011), sing different songs (Huber and Podos 2006), respond most strongly to the songs

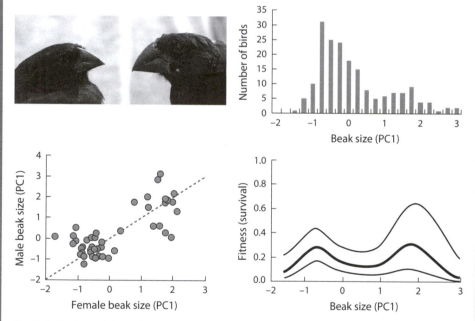

Fig B.4. Ecology drives diversification. The top left panel shows mature male *Geospiza fortis* morphs with small and large beak sizes: both birds were caught in the same mist net at the same time (photo: A. Hendry). The top right panel shows that beak sizes of this species at El Garrapatero fall into small and large modes, with relatively few intermediates (data from Hendry et al. 2009b). The bottom left panel shows that male-female pairing is assortative by beak size (data from Huber et al. 2007, provided by S. Huber). The bottom right panel shows that disruptive selection acts between the two modes (data from Hendry et al. 2009b). In all cases, beak size is PC1 from measurements of beak length, depth, and width. In the bottom right panel, the curves are a cubic spline with confidence intervals (see Hendry et al. 2009b for details)

Box 4 (*Continued*)

of the same morph (Podos 2010), show assortative mating by beak size (Huber et al. 2007), have higher survival than birds with intermediate beak sizes (Hendry et al. 2009b), and manifest genetic differences indicative of reduced gene flow (De León et al. 2010). Ecological differences within a species are driving the evolution of reproductive barriers that reduce gene flow.

---

these: When speciation occurs, how often is it ecological? And when ecological differences exist, how often do they cause speciation? Other questions consider the rapidity of ecological speciation (rapid speciation), at what point progress toward ecological speciation becomes irreversible (speciation reversal), to what extent ecological speciation is driven by competitive (adaptive speciation) or reproductive (reinforcement) interactions, and how many traits (magic traits) and selective pressures (dimensionality) are involved.

For this eco-to-evo part of eco-evolutionary dynamics, it would certainly be possible to add additional effects and processes, with more boxes and arrows; yet the above outline seems sufficient to lay the groundwork. First, selection (chapter 2) is the primary force driving the contemporary adaptation of populations (chapter 3). Second, divergent selection is the primary force driving adaptive divergence (chapter 4). Third, dispersal among populations in different environments can constrain adaptive divergence (chapter 5). Fourth, adaptive divergence can lead to reproductive barriers that reduce gene flow and thereby cause ecological speciation (chapter 6). Stated simply: ecology drives evolution! It is now time to consider the reverse: evolution driving ecology.

PART 2: EVO-TO-ECO

One way to outline the evo-to-eco side of eco-evolutionary dynamics would be to add arrows in figure 1.1 from adaptation back to ecological change (the single population case), and in figures 1.2–1.4 from adaptive divergence back to the ecological contrast (the multiple population case). As an example, the evolution of finch beaks will influence the seeds they consume and should therefore alter seed and plant distributions. In the end, however, presenting the key ideas will work more effectively through a different, although complementary, set of boxes and arrows. In particular, I now wish to specify interactions among different levels of biological variation: genes, phenotypes, populations (population dynamics), communities (community structure), and ecosystems (ecosystem function). Some of these levels refer to a particular focal organism: genes, phenotypes, and population dynamics (e.g., numbers, rates of increase, age structure, stability). Other levels refer to composite variables external to the focal organism: community structure (e.g., number and diversity of species, food web length, resistance to invasion) and ecosystem function (e.g., productivity, biomass, decomposition rates, nutrient fluxes).

How might change at each of these levels of variation influence change at the other levels? We can start from the simple recognition that interactions are sure to occur among the population dynamics of a focal organism, the structure of the community in which it is embedded, and the functions that exist in the encompassing ecosystem. For instance, changes in the abundance of a predator will influence the structure of prey communities and vice versa (populations ↔ communities), changes in the abundance of an important herbivore will influence ecosystem productivity and vice versa (populations ↔ ecosystems), and changes in a community of herbivores will influence ecosystem productivity and vice versa (communities ↔ ecosystems). Thus, we can start by drawing three boxes (populations, communities, ecosystems) and connecting them all with arrows that go both ways.

We can next recognize that the properties of populations, communities, and ecosystems could each influence selection on phenotypic traits. For example, beak size in a Darwin's finch population will be influenced by the number of individuals (population dynamics influences seed availability and therefore selection), the community of seeds and other finches (community structure influences seed availability and therefore selection), and soil moisture and nutrients (ecosystem function influences plant reproduction and therefore selection). We can depict such effects with arrows connecting each of the ecological levels to the phenotypes of a focal organism. Then, if those phenotypes are to evolve in response to selection, we need arrows from phenotypes to genes (or, more generally, genomes) and back again. Alternatively, variation at the three ecological levels could plastically influence organismal phenotypes without causing genetic change. Combining all these effects (fig. 1.5), we have an alternative way of presenting the eco-to-evo sequence originally depicted in figure 1.1.

This new representation of eco-to-evo effects acting on phenotypes within a population can be extended to represent eco-to-evo effects acting on phenotypic divergence among populations. First, we need to recognize that dispersal influences not only gene flow but also population dynamics: for example, immigration can help to maintain population size, such as in the case of "source-sink" dynamics. Second, gene flow influences genetic divergence among populations, which then influences phenotypic divergence—as explained above. Finally, adaptive divergence can reduce gene flow through ecological

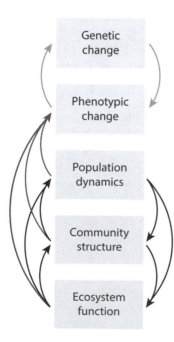

Fig 1.5. Graphical representation of the eco-to-evo side of eco-evolutionary dynamics for a population of a focal organism (top three boxes) in relation to composite aspects of its environment (bottom two boxes). This depiction is akin to that shown in figure 1.1, with the addition of separating three levels of ecological variation: population dynamics, community structure, and ecosystem function. Arrows indicate that these three levels can influence each other, as well as the phenotypes—and thereby genotypes—of the focal organism

speciation—also as explained above. Combining these effects (fig. 1.6), we have an alternative way of presenting the eco-to-evo effects originally depicted in figure 1.4.

To simplify presentation of the ideas that follow, I now fuse the within-population (fig. 1.5) and between-population (fig. 1.6) perspectives by simply referring to genes, phenotypes, populations, communities, and ecosystems. This concatenation means that references to a given level could refer to variation within a population or to variation among populations, with the latter implicitly including (but no longer explicitly represented with boxes and arrows) potential effects of dispersal and gene flow. To this new way of representing eco-to-evo effects, we can now add the reverse (evo-to-eco) side of the story (fig. 1.7). Specifically, phenotypes can influence population dynamics, community structure, and ecosystem function. I now discuss each of these potential effects in turn.

The phenotypes of a focal organism should have strong effects on its population dynamics. In particular, the mismatch between a population's current phenotypes and the phenotypes that would maximize fitness will influence population growth rate. Specifically, better-adapted populations (smaller mismatch) should have higher mean fitness, faster population growth, and perhaps larger population size (fig. 1.8). Thus, factors that change the mismatch should shape population dynamics. For instance, environmental change that increases the mismatch should precipitate population declines and extirpations. However, contemporary adaptation should decrease the mismatch and thereby promote population increases and range expansion: a process sometimes called "evolutionary rescue" (Gomulkiewicz and Holt 1995, Carlson et al. 2014). An example of adaptation influencing population dynamics in Darwin's finches is presented in Box 5.

Chapter 7 (Population Dynamics) starts with a more detailed outline of the various possibilities, including complexities that move beyond the above simplified scenario. It then evaluates various methods for inferring how phenotypes/genotypes influence population dynamics, including extensions of the year-by-year tracking approach illustrated

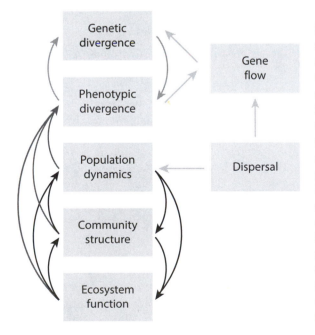

Fig 1.6. Graphical representation of the eco-to-evo side of eco-evolutionary dynamics taking into consideration multiple populations of a focal organism (akin to fig. 1.4). Extending the effects described in figure 1.5, ecological differences (at the population, community, or ecosystem levels) can influence, through selection or plasticity, phenotypic differences between populations of the focal species, which can cause genetic differences. In addition, dispersal between populations will influence population dynamics and gene flow, with the latter then influencing genetic divergence. Finally, phenotypic divergence can reduce gene flow through ecological speciation

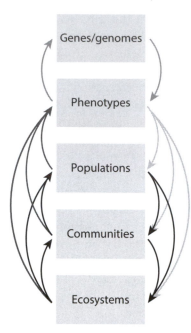

Fig 1.7. Graphical representation of a complete framework for considering eco-evolutionary dynamics. This representation fuses the within-population (fig. 1.5) and between-population (fig. 1.6) perspectives for eco-to-evo effects, and also adds the evo-to-eco realization that phenotypes can influence each ecological level. This representation also makes clear that feedbacks are expected between phenotypes and ecological variables. The effects of dispersal and gene flow (e.g., fig. 1.6) are implicit, but not explicit, in this representation

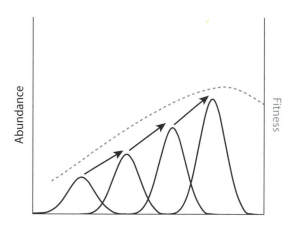

Fig 1.8. Graphical representation of how adaptation that reduces the mismatch between current phenotypes and optimal phenotypes can increase population size. The dashed curve shows mean fitness (*y*-axis at right) for populations with different phenotypes (*x*-axis) and the solid curves show the expected abundance of individuals (*y*-axis at left) with a given phenotype. The population starts far from the optimum where fitness is low and so too is population size. Adaptive evolution shifts (dashed arrows are time steps) the phenotypic distribution toward the new optimum, which thus increases fitness and so too population size. Nuances to this process are discussed in chapter 7

in Box 5. The key questions then provide an empirical assessment of the effects outlined above, starting with how maladaptation resulting from environmental change might decrease individual fitness and contribute to population declines, range contractions, and extirpations. The following questions consider the extent to which contemporary

Box 5

As noted in Box 1, the population size of Darwin's finches varies dramatically with rainfall, and mortality during the stressful drought periods is influenced

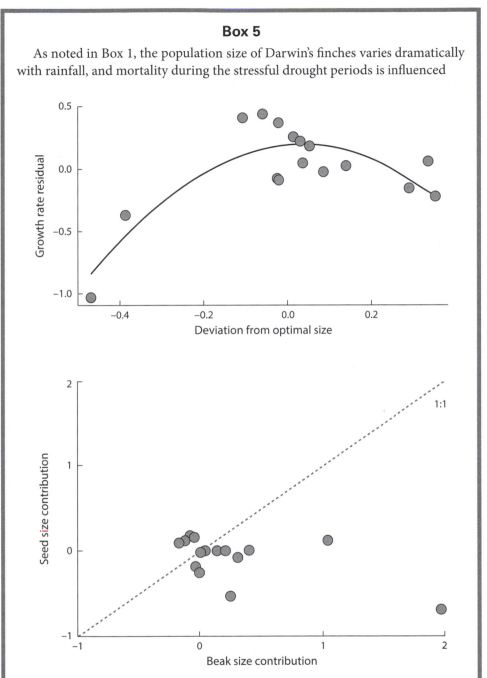

Fig B.5. Evolution influences population dynamics in *Geospiza fortis* on Daphne Major. The top panel shows the relationship between an estimate of the deviation of mean beak size from the optimal beak size (*x*-axis) and annual population growth rate "corrected" for effects of rainfall (growth rate residual, *y*-axis). The bottom panel shows the relative contributions to population growth rate of interannual changes in beak size (*x*-axis) and interannual changes in seed size (*y*-axis). The data are from Hairston Jr. et al. (2005), wherein more details are provided

(Continued)

> ## Box 5 (*Continued*)
>
> by beak size. Changes in population size from one generation to the next therefore should be related to how well existing beak sizes match the optimal beak sizes set by the seed distribution. In years when beaks are well adapted, population sizes should grow. In years when beaks are poorly adapted, population sizes should decrease. Hairston Jr. et al. (2005) tested these expectations by relating interannual changes in population size for *Geopsiza fortis* on Daphne Major to the degree of beak size adaptation, based on data from Grant and Grant (2002). The analysis showed that population growth rate was maximal at the estimated "optimal" beak size, and was lower if average beak sizes were larger or smaller. Hairston Jr. et al. (2005) further showed that this effect of variation in beak size on population growth rate was approximately twice as large (on average) as was the effect of environmentally driven variation in seed size. Evolution had an important influence on population growth.

evolution then helps to recover individual fitness and population size, which might then make the difference between persistence versus extirpation and range expansion versus contraction. A final question asks how phenotypic variation within populations and species influences population dynamics.

All discussion up to this point has considered effects on the properties (genotypes, phenotypes, population dynamics) of some focal organism, whereas I now transition to effects on composite ecological variables—starting at the community level. Phenotypic change in a focal organism could influence community structure though two basic routes, which I will refer to as "direct" and "indirect".[4] Through the direct route, the phenotypes of individuals could alter their per capita effects, depicted as the arrow from phenotypes to communities in figure 1.7. As an example, the foraging traits of a predator will influence the prey types it can consume and will therefore shape the prey community (Post and Palkovacs 2009). Box 6 provides a putative example for Darwin's finches. In this route, phenotypic change in the focal organism will alter community structure even if the abundance of that organism remains constant. Through the indirect route, the above-described influence of phenotypes on population dynamics could cascade to influence community structure, depicted as the arrows from phenotypes to populations to communities in figure 1.7. In this case, phenotypic change in the focal organism will alter community structure even if the per capita effects of that organism remain constant. Importantly, direct and indirect effects can act in the same direction, thus "reinforcing" or "amplifying" the total effect, or in opposite or "opposing" directions, thus "offsetting" or "canceling" the total effect.

Chapter 8 (Community Structure) starts with an outline of mathematical approaches for evaluating how genotypes/phenotypes might alter community structure, which then points to predictions about when such effects should be strongest in nature. The chapter

---

[4]Alternative terms, such as trait-mediated versus density-mediated, might seem similar but they would not suffice in the present context. As will be made clear later, density-mediated effects could themselves be trait-mediated. In addition, other sorts of indirect pathway beyond "density mediated" will be considered.

Box 6

Darwin's finches deplete seeds from the environment in relation to their beak size. The evolution of finch beaks should therefore influence the seed distribution,

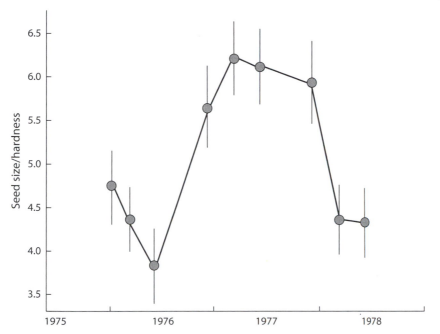

Fig B.6. Effects of finches on seed community structure. The graph shows the average size/hardness of seeds on Daphne Major from before to after the 1976–1977 drought. The data points are means and standard errors in fifty 1 m² quadrats sampled through time (from Boag and Grant 1981, wherein the details appear). The photograph shows some food types (fruits and seeds) available for consumption by *Geospiza* ground finches (on Santa Cruz). The top row shows *Cordia lutea*, *Scutia spicata*, *Tournefortia pubescens*, and *Bastardia viscosa*. The middle row shows *Portulaca oleracea*, *Cryptocarpus pyriformis*, *Tournefortia psilostachya*, and *Commicarpus tuberosus*. The bottom row shows *Vallesia glabra* and *Castela galapageia*

(Continued)

Box 6 (*Continued*)

which should then influence plant communities, which should then further influence finch beak sizes. This full eco-evolutionary feedback has been suggested (Post and Palkovacs 2009) but not formally demonstrated. However, the feedback can be inferred indirectly based on changes attending the colonization of Daphne Major in 1982 by the large ground finch, *Geospiza magnirostris*. As outlined in Box 1, drought conditions in 1976–1977 led to *G. fortis* depleting small/soft seeds from the environment. The remaining large/hard seeds led to selection on *G. fortis* for larger beaks. Another major drought occurred in 2003–2004 and selection was the opposite: *G. fortis* with smaller beaks were more successful than those with larger beaks (Grant and Grant 2006). It seems that, during the 2003–2004 drought, the larger-beaked *G. magnirostris* depleted the larger/harder seeds that *G. fortis* had used during the 1976–1977 drought. In short, the distribution of *Geospiza* beak sizes altered the seed distribution, which then fed-back to influence selection on the finches. Although the causal change in the beak size distribution here resulted from the addition of a new species, the situation remains a good proxy for variation within a species because *G. fortis* and *G. magnirostris* are very closely related and differ only in body/beak size. It remains to be determined to what extent a finch-induced change in the seed distribution alters plant communities.

then summarizes common approaches for empirical work, which might be broadly classed as (1) the effects of genotypes/phenotypes within and among populations, and (2) the year-by-year correspondence between phenotypic change and community change. The first two key questions that follow summarize the current state of knowledge for two classic applications of evolutionary thinking to community theory: predator-prey interactions and competition. The next question considers the importance of intraspecific genetic diversity for community structure, which echoes and extends the intense interest surrounding the effects of interspecific diversity (Loreau et al. 2001, Hooper et al. 2005). Other key questions relate to the relative strength of phenotypic/genetic effects, the time frames over which such effects play out, and whether they are direct or indirect in the sense described above.

The effects of genotypes/phenotypes on ecosystem function are a logical extension of the effects on community structure. For instance, such effects can be direct (when phenotypes differ in their per capita effects on ecosystem variables) or they can be indirect through several pathways: phenotypes to populations to communities to ecosystems, phenotypes to populations to ecosystems, or phenotypes to communities to ecosystems (fig. 1.7). Box 7 provides a concrete example by first suggesting how such effects might work in Darwin's finches and then showing how they actually do work in cottonwood trees (*Populus* spp.).

Chapter 9 (Ecosystem Function) first explains how the mathematical frameworks, empirical methods, and predictions introduced for community structure in chapter 8 can be extended to ecosystem function. Also outlined is an alternative conceptual framework (biological stoichiometry) for evaluating eco-evolutionary dynamics at

Box 7

No information exists on how the beaks of Darwin's finches influence ecosystem function, although such effects do seem likely. In particular, finch-induced changes in the seed community (Box 6) should have cascading consequences for ecosystem variables, such as decomposition rates, primary productivity, and nutrient cycling. Given that we don't know anything more for finches, we must switch to a different empirical system. Beavers (*Castor canadensis*) prefer to eat cottonwood genotypes with low levels of condensed tannin (Bailey et al. 2004). Cottonwood *(Populus)* stands subject to beaver activity thus become biased toward high-tannin genotypes. Condensed tannins influence many ecological properties, including soil microbial activity, and so a likely outcome of selection by beavers is reduced decomposition and nitrogen mineralization (Schweitzer et al. 2004). Thus, selection imposed by one species (beavers) can change the genetic composition of another species (cottonwoods), which alters ecosystem processes. Genetic variation within species, and presumably its evolution, influences ecosystem function.

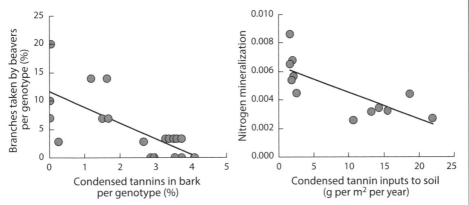

Fig B.7. Effects of selection by beavers (*Castor canadensis*) on ecosystem process. The left panel shows that beavers preferentially harvest cottonwood (*Populus*) genotypes that have low levels of condensed tannin (each data point is a cottonwood clone). The right panel shows that higher inputs of condensed tannin into the soil decrease rates of nitrogen mineralization (the data points are individual trees). The data were digitized from Whitham et al. (2006)

the ecosystem level (Elser 2006, Matthews et al. 2011b, Jeyasingh et al. 2014). Many of the key questions addressed in this chapter echo those first considered in the previous chapter: what is the importance of intraspecific diversity, what is the relative strength of the various effects, on what time scales do the effects play out, and to what extent are the effects direct or indirect? Also considered are some more synthetic questions: do the effects of genotypes decrease toward higher levels of complexity (from phenotypes to communities to ecosystems), and to what extent are feedbacks evident (traits influence ecosystems which then influence traits)?

At the end of these three evo-to-eco chapters, I hope that readers have an appreciation for the theoretical perspectives, empirical approaches, and key questions relating to this side of eco-evolutionary dynamics. Given that much less work has been done on this evo-to-eco side of the equation than on the eco-to-evo side, the general conclusions I draw will be increasingly tentative and haphazard. The message that I hope comes across is that much more work needs to be done in this area, and that many opportunities exist for investigators to make novel and important contributions to the field's development.

PART 3: UNDERPINNINGS

As will be repeatedly emphasized throughout the book, phenotypes are the nexus of eco-evolutionary dynamics—because they (as opposed to genotypes) are influenced by, and have influences on, other organisms and the environment. Genes do not have these properties except indirectly through their association with phenotypes. All of the above chapters therefore focus—with a few exceptions—on phenotypic changes within populations and phenotypic differences among populations. Yet we can't ignore the sources of phenotypic variation because the role of phenotypes in shaping eco-evolutionary dynamics will depend on how they are influenced by genes and the environment. Thus, the final part of the book will consider in more detail the two major contributors to phenotypic variation: genetic variation and phenotypic plasticity.

Evolution occurs when allele frequencies change across generations within populations. When these evolutionary changes differ among populations, genetic divergence takes place. In some cases, a single gene might explain most of the phenotypic variation in a particular trait, such as wrinkled versus smooth peas, or *Cepaea* color patterns, or human blood types, or sickle cell anemia (Bell 2008). In such cases, phenotypic variation should closely mirror underlying genetic variation. Most traits, however, are influenced by many genes of small-to-modest effect and also by the environment, necessitating a quantitative genetic approach (Roff 1997, Lynch and Walsh 1998). In its simplest form, this approach asks how much of the variation within or among populations has a genetic (as opposed to environmental) basis, and to what extent this basis stems from the additive effects of alleles and genes (additive), interactions among alleles at a locus (dominance), and interactions among genes (epistasis). The resulting determinants of genetic variation influence phenotypic responses to selection and thereby contribute to eco-evolutionary dynamics.

Chapter 10 (Genetics and Genomics) first outlines common empirical methods for studying the genetics of adaptation: quantitative genetics, quantitative trait locus (QTL) linkage mapping, association mapping, genome scans, gene expression, and candidate genes. The key questions then address various aspects of adaptation, speciation, and eco-evolutionary dynamics. First, how much additive genetic variation exists in fitness-related traits, with the answer informing evolutionary potential? Second, to what extent does nonadditive genetic variation (dominance and epistasis) influence phenotypic variation? Third, how many loci are involved in adaptation and how large are their effects, with the extremes being "many-small" versus "few-large"? Fourth, to what extent does the adaptation of independent populations to similar environments involve parallel/convergent genetic changes? Fifth, is adaptation to changing environments driven mainly by new mutations or by standing genetic variation, and what do the resulting "adaptive walks" look like (how many steps and in what order)? Finally, to what extent are the ecological effects of individuals, considered in the context of extended phenotypes, transmitted among generations—the so-called community heritability (Shuster et al. 2006)?

The other principal driver of phenotypic variation, and therefore eco-evolutionary dynamics, is phenotypic plasticity. Specifically, the environmental conditions experienced by an individual can cause developmental or behavioral changes in phenotype without any genetic change (Schlichting and Pigliucci 1998, West-Eberhard 2003). Some readers might wonder whether plasticity belongs in a book on eco-*evolutionary* dynamics, yet I can see several reasons why its inclusion is essential. First, the plasticity expressed by an individual has often evolved as a result of past selection, and so plasticity can be adaptive and can have a genetic basis. In essence, one can think of such plasticity as a current manifestation of past genetic change. Second, plasticity can evolve on contemporary time scales, and so phenotypic changes that accompany environmental change might partly reflect the *evolution* of plasticity. Third, plasticity modifies selection on genotypes, and thereby influences genetic responses to ecological change and ecological responses to genetic change. For all of these reasons, plasticity needs to be an integral part of any discussion of eco-evolutionary dynamics.

Chapter 11 (Plasticity) first outlines in more detail the nature of plasticity and how it can be studied, focusing in particular on the "reaction norm" approach. The subsequent key questions first evaluate whether or not plasticity is typically adaptive, with the main alternative being maladaptive physiological responses to stress. The next question informs the costs and limits to plasticity, without which any environment-phenotype mismatch could be easily bridged. The subsequent questions consider when adaptive plasticity should be strongest, such as when environments are variable in space or time, when gene flow is high, and when reliable cues exist. Also considered are alternative hypotheses for how genetic change and plasticity interact: that is, plasticity might enhance or constrain genetic evolution and ecological speciation. A final question considers how rapidly plasticity can evolve when populations experience new environments.

At the close of these two "underpinning" chapters, I hope that readers will agree that the phenotypic focus adopted in the book is valuable and appropriate. At the same time, I hope it remains clear that a phenotypic perspective on eco-evolutionary dynamics does not preclude rigorous investigations into its genetic and plastic basis.

What are eco-evolutionary dynamics—and what are they not?

Now that I have outlined a conceptual framework for eco-evolutionary dynamics, it is time to return to a consideration of just what they are and what they are not. I earlier defined eco-evolutionary dynamics as interactions between ecology and evolution that play out over contemporary time scales, such as decades or centuries. These interactions might be classified into five different categories.

1. Ecological change influences evolutionary change—but not vice versa.
2. Evolutionary change influences ecological change—but not vice versa.
3. Ecological change influences evolutionary change, which then influences ecological change—with the upstream ecological driver (e.g., population density) being different from the downstream ecological response (e.g., nutrient cycling).
4. Evolutionary change influences ecological change, which then influences evolutionary change—with the upstream evolutionary driver (a particular trait) being different from the downstream evolutionary response (a different trait).

5. Ecological and evolutionary change reciprocally influence each other through the same traits and ecological variables, inclusive of situations where intermediate traits or variables are involved. For example, a change in ecological variable A could cause a change in trait 1, which could then directly influence ecological variable B, which could influence ecological variable A.

Each of the five categories represents eco-evolutionary dynamics as long as the interactions occur in contemporary time. Further, categories 3 and 4 will be considered eco-evolutionary *feedbacks* in the broad sense and category 5 will be considered eco-evolutionary *feedbacks* in the narrow sense. These feedbacks can be positive (versus negative), such as when an increase in the level of ecological variable A causes the evolution of trait 1 in a manner that further increases (versus decreases) the level of ecological variable A. Positive feedbacks can reinforce (or "accelerate" or "enhance" or "exaggerate") eco-evolutionary dynamics, whereas negative feedbacks can oppose (or "dampen" or "slow") those dynamics. Note that all of the above designations focus on *change* as a driver of dynamics, whereas I will repeatedly emphasize that *a lack of change* as a driver of stability can just as easily (and perhaps even more importantly) be the result of cryptic eco-evolutionary dynamics, a phenomenon that might be called "eco-evolutionary stability."

The above five options might seem so inclusive as to dictate that all changes, whether ecological or evolutionary, fall under the umbrella of eco-evolutionary dynamics. It is therefore useful to also suggest some scenarios that would *not* be considered eco-evolutionary dynamics.

1. Evolutionary changes that are not the result of ecological changes, such as many of those caused by genetic drift or genomic interactions unrelated to the ecological environment.
2. Ecological changes that are not the result of evolutionary changes, such as those resulting from geological forces such as volcanic activity or continental drift.
3. Evolutionary changes that do not cause ecological changes, such as in traits that have little influence on fitness and the environment (classically, bristle number in *Drosophila*).
4. Ecological changes that do not cause evolutionary changes in the focal organism under study.

In addition, we might not invoke eco-evolutionary dynamics if ecology and evolution are interacting on such long time scales as to be largely unchanging in contemporary time. (Although a full eco-evolutionary view of life will ultimately require integration across all time scales.) The extent to which these various alternatives occur in nature is an open empirical question. Perhaps most ecological and evolutionary change is eco-evolutionary, or perhaps not.

Some additional explanations and clarifications are helpful. First, every environmental change probably drives evolutionary change in at least some organism, whereas the above arguments are intended to apply to a particular focal organism. By this I mean that some species will not evolve in response to some ecological changes, even if other species are strongly affected. Second, I have phrased the above discussion as a yes-or-no proposition (eco-evolutionary dynamics are or are not occurring), which is not the real question of interest. Instead, we should be more concerned with quantifying *rates* of eco-evolutionary dynamics and the *strength* of their effects. That is, eco-evolutionary effects may be fast or slow and strong or weak, or anything in between.

Limitations and scope

In attempting to provide a unified and comprehensive framework for studying eco-evolutionary dynamics, the book will end up covering a lot of ground. This broad scope made it impossible to go into great detail on any particular topic. I have instead tried to extract the most relevant considerations, the most critical questions, and the most informative empirical studies. For the same reason, I have not provided a detailed review of the many antecedents to eco-evolutionary dynamics or to alternative conceptual frameworks for its study. Instead, my goal was to integrate everything together, which led to the above framework that gives me the broadest possible scope for discussing interactions among genes, phenotypes, populations, communities, and ecosystems. More detailed work exists for particular subsets of these interactions, and I will reference these other efforts as the book unfolds.

I should also point out that this book does not represent, depict, or espouse a particular *hypothesis* or *theory*, such as the Ecological Theory of Adaptive Radiation (Schluter 2000a) or the Geographical Mosaic Theory of Coevolution (Thompson 2005). (Although I do suggest, in the final summary chapter 12, an emerging "Española-Isabela Hypothesis" for the eco-to-evo part of eco-evolutionary dynamics.) That is, I am not trying to marshal the evidence in support of a particular view of the world among potential alternative views. I am instead trying to review the evidence and tie together a series of disparate fields and subfields into a somewhat unified whole. No one disputes that ecology and evolution influence each other, but we lack a general conceptual framework and comprehensive empirical assessment of how these interactions play out in nature. That is what I am trying to achieve.

Chapter 2

Selection

This chapter focuses on natural selection, the primary force generating evolutionary change (Endler 1986, Schluter 2000a, Bell 2008).[1] Natural selection is also often the ultimate cause of nongenetic changes in phenotypes, because plasticity evolves in response to selection and is often adaptive (see chapter 11). Given that selection is the primary cause of phenotypic change, and that phenotypic change is the main driver of eco-evolutionary dynamics, selection is the foundation of all those dynamics. We must therefore ground our investigation of eco-evolutionary dynamics in an understanding of selection on phenotypes.

The present chapter examines selection within "populations," which are considered throughout the book to be conspecific groups of individuals within which interbreeding is common (close to panmixia) but among which interbreeding (and therefore gene flow) is restricted.[2] Subsequent chapters take up the consequences of this selection for evolution within populations (chapter 3) and among populations (chapter 4). In the first section of this chapter, I consider how to measure selection in natural populations. I will go into considerable detail because the accurate quantification of selection can be critical to understanding eco-evolutionary dynamics. Even though the level of detail might seem excessive, the reality is that many more nuances and details are not considered and instead appear in the various cited works, starting with Endler's (1986) classic book. In the second major section of the chapter, I consider key questions about selection in nature, and I try to answer those questions through the evaluation of empirical data. The specific questions were chosen based on long-standing interest or current controversies, as well as their relevance to eco-evolutionary dynamics. My treatment of each question is necessarily cursory and my answers are certainly not

---

[1] This phenotypic approach, justified in chapter 1, dictates that this chapter emphasizes selection on phenotypes, not genotypes—although a huge literature also exists on the latter topic (Bell 2008). In addition, natural selection is here generally considered to include sexual selection, until the very end of the chapter.

[2] Different populations are usually found in different locations, such that interbreeding can be restricted owing to geographical separation. However, different populations can also be fully sympatric, in which case interbreeding is usually restricted by intrinsic reproductive barriers. If interbreeding is sufficiently low, different populations are typically considered different species. I do not here specify a level of gene flow that would allow one to call groups separate populations (e.g., Waples and Gaggiotti 2006) or separate species, as it is often counterproductive to place artificial thresholds on a continuum of divergence and gene flow (Hendry et al. 2000a).

agreed upon by everyone, as will be the case in all chapters. The "answers" thus should be viewed as my own reading of the data rather than an attempt to always represent a consensus opinion.

Quantifying and visualizing selection

Selection is the nonrandom fitness of individuals with respect to phenotype or geno-type, with the focus here being squarely on the former. The most direct and informative way to estimate this (phenotypic) selection is to relate the fitness of individuals[3] to their phenotypes. To illustrate how this can be done, I will use data for great tits (*Parus major*) from Wytham Woods, UK. The data consist of measurements of two phenotypic traits for females (date of clutch initiation—the date the first egg is laid, and clutch size—the number of eggs) and a fitness estimate for those females (the number of recruits they produce; that is, offspring from the clutch that were recorded as breeding adults in future years). The analyses I perform with these data are intended to illustrate the visualization and quantification of selection, not to reveal anything specific about great tits in Wytham Woods. For much more complete analyses and interpretations of selection in this population see Garant et al. (2007) and Charmantier et al. (2008).

The first step in any selection analysis should be the visual inspection of relationships between traits and fitness. These relationships are shown in figure 2.1 for 256 Wytham Woods great tit females breeding in 1999. Among these females, dates of clutch initiation varied by more than 40 days, clutch sizes varied by more than 10 eggs, and most clutches produced no recruits. Some clutches, however, yielded one recruit and some yielded two or even three recruits. The question is whether this measure of female fitness is associated in some nonrandom way with date of clutch initiation or clutch size, which would be evidence of selection. Initial inspection of the data suggests that high-fitness clutches are more common at earlier dates of clutch initiation and at intermediate clutch sizes, but we need a formal way of making such inferences.

To formally infer the relationship between a trait and fitness (i.e., the "individual fitness function"), it is best to begin without assuming any particular a priori parametric shape, as would be the case in a linear or quadratic model. The most common route to this unconstrained individual fitness function is nonparametric cubic splines (Schluter 1988, Schluter and Nychka 1994, Gimenez et al. 2006). The trick is to decide just how smooth the spline function should be. Here one strives to illustrate the important and relevant bumps and dips in the fitness function but not the unimportant or irrelevant ones. The choice of a smoothing parameter can be informed by various measures of fit (Schluter 1988) but is often somewhat subjective. The splines I settled on for the Wytham Woods data (figure 2.1) broadly confirm the above assertions that selection seems to favor earlier dates of clutch initiation and intermediate clutch sizes, although the latter effect is pretty weak and (as we will see) not significant. The splines also suggest that linear and quadratic relationships might reasonably approximate these associations, but that some additional nuances are present. For example, any date of clutch initiation after May 5 (day 35) leads to zero fitness.

---

[3] According to Orr (2009), this is "individual fitness," as opposed to fitness as a summary statistic for a given genotype. Further discussion of "individual fitness" is deferred until later in the chapter.

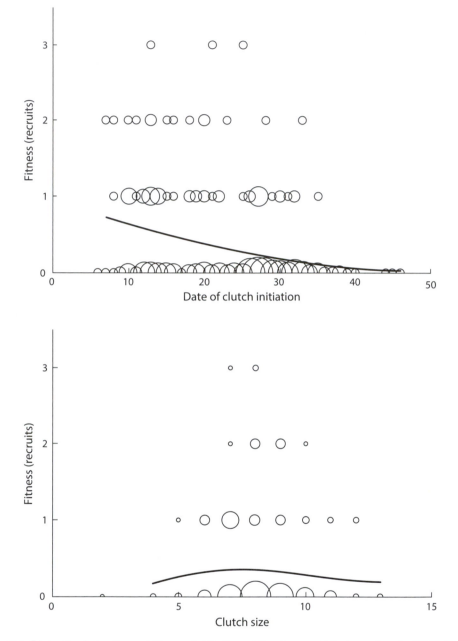

Fig 2.1. Phenotypic traits and fitness in Wytham Woods great tits breeding in 1999. Date of clutch initiation is the number of days after April 1 and clutch size is the number of eggs laid. Fitness is the number of birds in a given clutch that recruited into the population the following year. The size of bubbles is proportional to the number of clutches with each combination of trait and fitness. The curves are cubic spline estimates of relationships between each trait and fitness. The smoothing parameter (GCV) for the splines is 2 for date of clutch initiation and 0 for clutch size. The data were provided by B. Sheldon

Nonparametric fitness functions are useful visual representations of selection that facilitate the identification and interpretation of patterns of selection acting on individuals. Evolution, by contrast, acts at the level of the population. We therefore need a bridge between estimates of selection on individuals and equations for predicting the resulting evolutionary change of populations. This bridge is obtained by fitting parametric equations to individual fitness functions, which then yields "selection coefficients." Parametric functions can also be fit to cubic splines to obtain coefficients (Morrissey and Sakrejda 2013). The next chapter outlines the evolutionary equations, whereas the present chapter focuses on the selection coefficients themselves.

Evolutionary changes in the *mean* value of a trait will be related to directional selection acting on those traits, which can be estimated as the slope coefficient of a simple linear regression of relative fitness (absolute fitness divided by mean fitness) on phenotypes (Lande and Arnold 1983). This slope coefficient is termed[4] the "linear selection differential," abbreviated S_i, where subscript i refers to the ith trait. For the data shown in figure 2.1, the linear selection coefficients are $S_{date\ of\ clutch\ initiation} = -0.066$ ($P < 0.001$) and $S_{clutch\ size} = -0.062$ ($P = 0.447$). These coefficients correspond to the above interpretations that directional selection favors earlier dates of clutch initiation but does not seem to act on clutch size.

Evolutionary changes in the *variance* of a trait will be related to both directional selection and nondirectional selection. The nondirectional effects are typically considered by adding a squared term (i.e., a new variable that represents the squared deviation of that individual's phenotype from the mean of all individuals in the analysis) to the above-described linear regression that already includes the unsquared term (i.e., the actual phenotype of each individual) (Lande and Arnold 1983). The partial regression coefficient for the squared term (multiplied by two, Stinchcombe et al. 2008) is termed the "quadratic selection differential" (C_i). If this coefficient is negative, selection tends to disfavor extreme individuals, and so is expected to reduce the variance. This result can be interpreted as "stabilizing selection" if the fitness function really does disfavor extreme individuals on both sides of the mean (Mitchell-Olds and Shaw 1987, Schluter 1988). If the coefficient is positive, selection tends to favor extreme individuals, and so is expected to increase the variance. This result can be interpreted as "disruptive selection" if it really does favor extreme individuals on both sides of the mean. For the data shown in figure 2.1, the quadratic selection differentials are $C_{date\ of\ clutch\ initiation} = 0.0001$ ($P = 0.777$) and $C_{clutch\ size} = -0.029$ ($P = 0.326$). These coefficients suggest that significant quadratic selection is not present, even for clutch size—where it had seemed possible (although weak) under visual examination.

Selection differentials, both linear and quadratic, can thus be estimated by relating individual relative fitness to individual phenotypes. These same differentials can be estimated by measuring changes in the first (mean) and second (variance) moments of the distribution of phenotypes from before to after selection—but before reproduction (Endler 1986, Brodie III et al. 1995). This assumes, of course, that these shifts are caused by selection rather than plasticity, growth, or other nonselective changes. Although this moment-based approach appears simpler, the individual-fitness approach is more

---

[4]I have tried to adopt reasonably common terms for types of selection coefficients but be aware that different authors can use different terms (Brodie III et al. 1995, Matsumura et al. 2012).

flexible because it allows for a number of extensions. Among these is the potential to better separate direct selection (acting specifically on a particular trait) from indirect selection (acting through correlations with other traits). This separation is commonly achieved by adding multiple traits into the above regression models. In such cases, the partial regression coefficients for (1) each trait represent "univariate linear selection gradients" (β_i), (2) each trait deviation squared (again multiplied by 2) represent "univariate quadratic selection gradients" (λ_{ii}), and (3) the cross-product between the deviations for each pair of traits (i.e., multiply the deviation from the mean for one trait by the deviation from the mean for the other trait in the same individual) represent "bivariate quadratic selection gradients" (λ_{ij}, where the subscripts i and j refer to the two traits). These bivariate coefficients indicate selection on trait combinations, which will be explained later by reference to real data. Table 2.1 shows these values in the "not standardized" column for the data in figure 2.1. These coefficients fit directly into multivariate equations ($\Delta\mathbf{Z}=\mathbf{G}\boldsymbol{\beta}$) predicting evolutionary change in phenotypic traits (chapter 3) and, for this reason, into some equations for predicting ecological change (chapter 8).

The coefficients for each term in this multiple regression model represent the effect of that variable on fitness while statistically holding constant the effects of all other variables included in the model (Lande and Arnold 1983), with the usual caveats about interpreting partial regression coefficients in multiple regression models. Selection gradients are thus interpreted as capturing "direct selection" acting on a trait, as opposed to the above selection differentials that capture "total selection" acting on a trait (i.e., including "indirect selection" acting through other traits). Of course, one needs to remember that selection gradients can only remove indirect selection acting through traits that are measured and included in the model—that is, indirect selection on unmeasured correlated traits still could be a part of "direct" selection estimated as above (Walker 2014).

What do these gradients suggest about selection in the data set under consideration? First, date of clutch initiation appears to be under negative selection (favoring earlier laying) regardless of whether or not clutch size is included in the analysis. Second, although the quadratic selection *differential* for clutch size was far from significant, the univariate quadratic selection *gradient* is much closer to significance. This possible difference tells us that clutch size could potentially be subject to stabilizing selection once we remove the correlated effects of date of clutch initiation. Third, the nonsignificant bivariate quadratic selection gradient suggests that selection does not strongly favor particular combinations of date of clutch initiation and clutch size. For the moment, then, we can

Table 2.1. Selection coefficients estimated for the Wytham Woods great tit data shown in figure 2.1. See the text for details.

	Not standardized	*P value*	*Variance standardized*
$\beta_{\text{clutch initiation}}$	-0.071	< 0.001	-0.598
$\beta_{\text{clutch size}}$	-0.130	0.102	-0.198
$\lambda_{\text{clutch initiation}}$	-0.0001	0.795	-0.030
$\lambda_{\text{clutch size}}$	-0.050	0.094	-0.114
$\lambda_{\text{clutch initiation x clutch size}}$	-0.016	0.120	-0.206

simply note that combinations of these traits might matter for selection—and so we should investigate these interactions more carefully. This requires a return to visual inspection of the individual fitness function—but this time from a multitrait perspective. This can be achieved through parametric approaches based on quadratic approximations (Phillips and Arnold 1989) or through nonparametric methods (Schluter and Nychka 1994, Calsbeek 2012).

A bivariate cubic spline for date of clutch initiation and clutch size is shown in figure 2.2. Coupled with the above analyses, this representation allows more refined interpretations. First, early laying is favored overall—but perhaps most obviously so at intermediate clutch sizes. Second, stabilizing selection on clutch size is closer to significance when date of clutch initiation is added to the model because different intermediate clutch sizes are favored at different dates of clutch initiation (a ridge of high fitness). Third, correlational selection is not statistically strong because selection is best approximated by a ridge, not a saddle. That is, high fitness individuals aren't only those that breed late and have

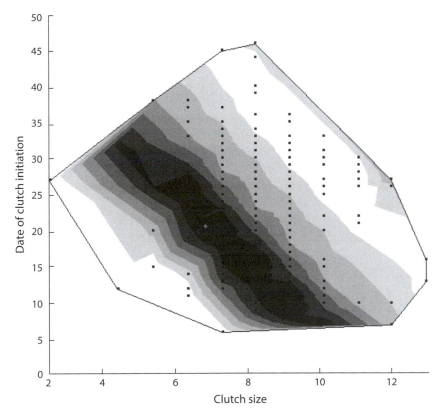

Fig 2.2. Bivariate cubic spline estimates from projection pursuit regression (Schluter and Nychka 1994) of relationships between phenotypic traits and fitness in Wytham Woods great tits in 1999. Darker contours indicate higher fitness, with the number of recruits shown in 0.1 unit gradations (the darkest contour corresponds to an expectation of 0.6 recruits per clutch). This plot is the bivariate equivalent of the univariate relationships shown in figure 2.1. The smoothing parameter is −4. The data were provided by B. Sheldon

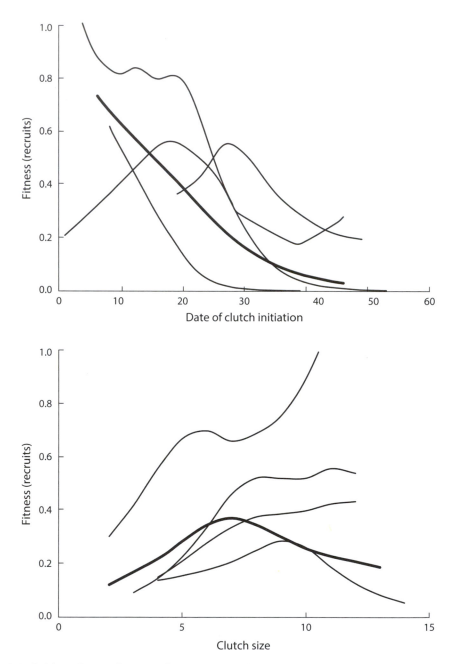

Fig 2.3. Cubic spline estimates of relationships between phenotypic traits and fitness in Wytham Woods great tits breeding in five years (1996, 1998, 1999, 2002, and 2007). The specific years were selected to show some of the variation and the smoothing parameters are as in figure 2.1. The thicker lines show the estimated splines for 1999 corresponding to figure 2.1, although the y-axis scale has changed for ease of interpretation. The data were provided by B. Sheldon

small clutches or that breed early and have large clutches (which a bivariate quadratic function is attempting to fit)—but also individuals that are intermediate for both date of clutch initiation and clutch size.

I have illustrated selection by reference to only a single year of Wytham Woods data. However, a robust picture of selection will often require data for multiple generations. The reason is that selection can sometimes, perhaps often, vary through time—as will be considered in more depth later in the chapter. For now, I simply illustrate the value of temporal replication by reference to five years of data for Wytham Woods (fig. 2.3). Using univariate cubic splines (for simplicity), we can see that temporal replication allows some generalization: selection consistently disfavors very late dates of clutch initiation and very small clutches. The graph also suggests some nuances: selection only sometimes favors early breeding and only sometimes disfavors large clutches.

STANDARDIZATION

Selection differentials and gradients can be calculated in the original units of measurement—as I have done above. Doing so is not a problem for predicting evolutionary change, but standardization is needed if coefficients are to be compared objectively across traits and studies (Endler 1986, Kingsolver et al. 2001). Two main standardizations have been suggested. The first standardization is by units of phenotypic standard deviation, yielding "variance-standardized" selection coefficients. This standardization can be accomplished for linear coefficients by dividing selection differentials and multiplying selection gradients by the phenotypic standard deviation before selection (Lande and Arnold 1983). For quadratic coefficients, the above division or multiplication is by the trait variance or the trait-by-trait covariance, rather than the standard deviation (Lande and Arnold 1983). The same standardizations can be accomplished at the outset by standardizing all traits in the sample before selection to a mean of zero and a standard deviation of unity—and then performing the above regression analyses. The variance-standardized coefficients for the Wytham Woods selection gradients are shown in the last column of Table 2.1.

The second standardization uses trait means rather than standard deviations (Hereford et al. 2004). These "mean-standardized" selection coefficients reflect the increase in relative fitness for a proportional change in the mean value of the trait. That is, a mean-standardized gradient of 0.54—the average value from the database of Hereford et al. (2004)—would mean that a doubling of the trait value would increase fitness by 54%. In addition, mean-standardized selection on fitness itself is expected to be unity. By this comparison, a value of 54% means that selection on the average measured trait is 54% as strong as selection on fitness itself (Hereford et al. 2004, Matsumura et al. 2012). These mean-standardized selection coefficients are similar to elasticities (van Tienderen 2000, Hereford et al. 2004), which are commonly used in demographic studies (Caswell 2001, Matsumura et al. 2012). I do not calculate mean-standardized coefficients for the Wytham Woods data because doing so requires ratio scale (Hereford et al. 2004), whereas date of clutch initiation is on an interval scale.

COMPLICATIONS AND CAVEATS

The calculation of selection coefficients is traditionally based on linear regression, which has a number of limitations. First, it assumes that all traits directly influence fitness, whereas one trait might influence another that then influences fitness. A suggested solution to this problem is to use path models (Kingsolver and Schemske 1991, Scheiner et al. 2000), although caveats remain (Petraitis et al. 1996). Second, regressions and path analysis have a number of potentially limiting assumptions, most obviously homoscedasticity (equal variances) and a normal distribution of residuals (Mitchell-Olds and Shaw 1987). To better meet these assumptions, data transformations (e.g., logarithmic) are sometimes performed before analysis (Lande and Arnold 1983). A particularly obvious and common deviation from normality is dichotomous fitness values, such as survival (score 1) or death (score 0).[5] Analyses of such data therefore often use logistic regression, and the resulting coefficients can be converted to their linear equivalents (Janzen and Stern 1998). An additional concern is that failure to observe an individual at the end of a selection episode is typically assumed to indicate its death, whereas that individual might instead be present but not observed or it might have moved elsewhere. At best, violation of this assumption will introduce noise into the selection estimate. At worst, it can lead to bias, such as when individuals with particular phenotypes are more or less likely to be observed or are more or less likely to disperse (Letcher et al. 2005). (An argument can be made that emigration is not a problem in the context of selection estimates because emigrants represent selective losses to a local population even if they don't die.) Various mark-recapture methods have been proposed to better deal with these problems (Kingsolver and Smith 1995, Gimenez et al. 2009). Taking the problem a step further, individuals with particular phenotypes might die before they are ever measured: the so-called "invisible fraction" (Hadfield 2008, Mojica and Kelly 2010). Cognizance of these possibilities can improve study design and make clear the appropriate caveats in interpretation.

An extremely difficult problem in measuring selection is how to best estimate fitness, a topic about which much has been written (e.g., Benton and Grant 2000, Roff 2002, Brommer et al. 2004, Orr 2009). As fitness itself is essentially impossible to measure (it is even exceptionally difficult to define), investigators work with fitness surrogates. These surrogates vary widely from simple trait measurements (e.g., gonad mass, body size, foraging rate, condition, or growth rate) to more direct fitness components (e.g., survival or mating success) to even more inclusive fitness surrogates (e.g., life time reproductive success). A first point is that using simple trait measurements is generally not a good idea—because they are far removed from fitness, as well as for the reasons below. Indeed, many studies have estimated selection *on* gonad mass, body size, foraging rate, condition, or growth rate—and this selection is not consistently positive. In my own studies, for example, viability selection generally does not favor large size, fast growth, or high condition (Carlson et al. 2004, DiBattista et al. 2007). A second point is that direct fitness components are a reasonable starting point for most studies that lack the intensive multigeneration monitoring data that are typically necessary for more inclusive measures. However, care should be taken to remember that different fitness components (e.g., reproduction versus survival: Roff 2002, p. 129) can—although not necessarily (Kingsolver and Diamond 2011)—trade off with each other, particularly at different

---

[5]Errors are here expected to be binomially distributed. Count data, such as the number of recruits used in the above Wytham Woods analysis, are expected to have Poisson-distributed errors, which I ignored to simplify the presentation. Approaches for dealing with complicated distributional problems in estimates of selection are discussed by Shaw and Geyer (2010).

stages of an organism's life cycle (Schluter et al. 1991, Rollinson and Rowe 2015). Thus, maximal information can be gained with methods that incorporate multiple selection episodes in a life cycle and that assess multiple fitness components (Arnold and Wade 1984a, 1984b, Wade and Kalisz 1989, Shaw and Geyer 2010). Finally, I would argue that the most inclusive fitness surrogates, particularly lifetime reproductive success (e.g., Brommer et al. 2004), are presumably best when evolutionary prediction is the goal.

Many other issues surround the estimation of selection, of which I can here only list a few. First, attaining statistical significance for typical strengths of selection typically requires very large sample sizes (Kingsolver et al. 2001, Hersch and Phillips 2004). In the Wytham Woods data, for example, sample sizes for the five years were always greater than 300 birds—and yet apparent selection is only occasionally statistically significant. Second, estimates of selection on phenotypes will not necessarily translate directly into evolutionary responses (Merilä et al. 2001b, Morrissey et al. 2010). This apparent disconnect will be discussed at length in chapter 3, but I here raise one point of particular relevance to the estimation of selection. In particular, phenotypes and fitness might be associated with each other because of environmental, rather than genetic, effects (Fritz and Price 1988, Schluter et al. 1991, Rausher 1992, van Tienderen and de Jong 1994, Stinchcombe et al. 2002). For example, some individuals might have larger traits and higher fitness simply because they are in good locations or have good parents. For this reason, a number of methods have been developed for estimating selection specifically on the genetic component of a phenotypic trait (Rausher 1992, Stinchcombe et al. 2002, Hadfield et al. 2010, Morrissey et al. 2010, Milot et al. 2011), although these cannot be implemented in many study systems.

Finally, it is important to remember that measuring selection does not reveal *why* that selection is occurring (Endler 1986, Wade and Kalisz 1990, MacColl 2011). Determining the specific cause of selection, such as competition, predation, parasitism, or temperature, requires additional methods. The causes of selection are usually a part of studies of adaptation, and so this topic is considered at greater length in the next two chapters.

Selection in nature

A large number of studies have now estimated selection in natural populations—and the resulting linear and quadratic coefficients have been collated in several meta-analyses. Endler (1986) reviewed selection coefficients available at the time—but his analysis was limited given the only recent publication of regression methods for estimating multivariate selection (Lande and Arnold 1983). Kingsolver et al. (2001) compiled more than 2500 variance-standardized selection coefficients from 63 studies. Hereford et al. (2004) compiled 580 mean-standardized selection coefficients from 38 studies. Siepielski et al. (2009) compiled 5519 variance-standardized selection coefficients from 89 studies where estimates were available for the same population in at least two different years. A variety of other studies have since used these databases for further analysis (Hoekstra et al. 2001, Hersch and Phillips 2004, Kingsolver and Pfennig 2004, Knapczyk and Conner 2007, Siepielski et al. 2009, 2011, Kingsolver and Diamond 2011, Kingsolver et al. 2012, Siepielski et al. 2013). I will use these various compilations in an attempt to answer several general questions about the strength and form of selection in nature: (Q1) how strong is selection?, (Q2) to what extent is selection on a trait direct versus indirect (acting through correlated traits)?, (Q3) what types of traits (e.g., morphology vs. life history vs. physiology) are under the strongest selection?, (Q4) to what extent is nonlinear selection typically stabilizing (against extreme phenotypes) versus disruptive (favoring extreme phenotypes)?, (Q5) how temporally variable is selection?, and (Q6) does the strength of natural selection (acting on viability) differ from sexual selection (acting on mating success)?

QUESTION 1: HOW STRONG IS SELECTION IN NATURAL POPULATIONS?

If eco-evolutionary dynamics commonly play out on contemporary time scales, then trait change is expected on similar time scales. Given that such change is likely driven by selection (Endler 1986, Schluter 2000a, Bell 2008), eco-evolutionary dynamics will depend on the strength of natural selection. A literal reading of Darwin gives the impression that selection must be weak: for example, "natural selection will always act very slowly, often only at long intervals of time, and generally on only a very few of the inhabitants of the same region at the same time" (Darwin 1859, p. 108). This general view prevailed for the next century, notwithstanding a few very early studies in which selection was recorded or inferred for natural populations (e.g., Bumpus 1899, Weldon 1901, di Cesnola 1907). Perceptions began to shift in the 1960s with the emergence of studies showing phenotypic change occurring on short time scales (see chapter 3), which implied that selection must be important on those same time scales.

Now that a large number of studies have used the Lande and Arnold (1983) approach to estimate selection, we can begin to ask questions about its typical strength and prevalence in nature. Interestingly, similar compilations of selection coefficients by different authors have led them to assert different conclusions. Endler (1986, p. 222) noted that "strong selection is not rare and may even be common." This conclusion was based largely on the observation that some studies document quite strong selection. While not disputing the assertion that selection is sometimes strong, Kingsolver et al. (2001, p. 253) emphasized that "directional selection on most traits and in most systems is quite weak." This conclusion was based largely on the observation that most estimates of selection were nonsignificant and clustered around zero. Finally, Hereford et al. (2004, p. 2140) used mean-standardized coefficients to argue for "extremely strong selection overall." This conclusion was based largely on the calculation that ". . . selection [on traits] is on average 54% as strong as selection on fitness." Many other authors have also weighed in on the debate on one side or the other.

What should we make of this seemingly fundamental disagreement? One important point is that "strong" or "weak" are generally subjective characterizations—and so one person's "strong" can be another person's "weak." Indeed, Hereford et al. (2004) argued that selection was strong based on average coefficients that were weaker than those Kingsolver et al. (2001) used to argue that selection was weak. Conclusions thus depend on the frame of reference, which I discuss further below. Another important point is that the distribution of selection coefficients is so broad that it can be misleading to infer a "typical" strength of selection. It is instead more appropriate to consider the entire distribution of coefficients (fig. 2.4). This distribution is characterized by a large number of observations at the low end of the range—essentially no selection at all—and a steadily decreasing number of observations toward higher values. This basic pattern holds regardless of whether one looks at selection differentials, selection gradients, linear selection, univariate quadratic selection, or bivariate quadratic selection, and whether one considers variance or mean standardizations. It also holds in every meta-analysis thus far conducted (Endler 1986, Hoekstra et al. 2001, Hersch and Phillips 2004, Kingsolver and Pfennig 2004, Knapczyk and Conner 2007, Siepielski et al. 2009, 2011, Kingsolver and Diamond 2011, Kingsolver et al. 2012, Siepielski et al. 2013). Interestingly, meta-analyses of selection at the genomic level yield similar patterns (Thurman and Barrett 2016). Thus, if we assume that selection varies from weaker to stronger, the former is clearly much more common than the latter.

As the above quotations indicate, evolutionary biologists often move beyond the relative categorizations of "weaker" versus "stronger" to the more absolute categorizations of "weak" versus "strong." On the one hand, I could argue that it is unhelpful to convert an obviously continuous pattern of variation into a dichotomous characterization. On the other hand, the practice is common—and so we may as well consider how one might make such inferences. Three possible frames of reference come to mind: (1) How much morphological change could be accomplished by the observed selection?, (2) How strong is selection on traits relative to selection on fitness?, and (3) How much of the variation in fitness can be explained by the measured traits?

In considering the first frame of reference, even selection coefficients at the low end of the distribution could lead to large phenotypic changes given enough time. This

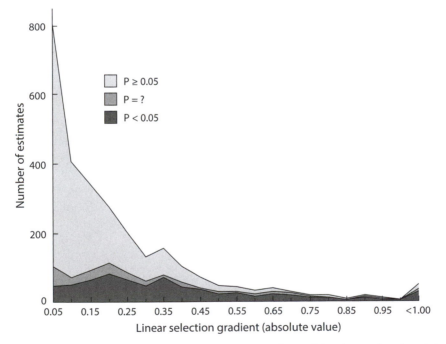

Fig 2.4. Frequency distribution of linear selection gradients (absolute values) estimated for phenotypic traits in natural populations. Significance levels are shown with different shading, including instances where the original study did not indicate significance (P = ?). The data are from Kingsolver and Diamond (2011), who compiled the Kingsolver et al. (2001) and Siepielski et al. (2009) databases

observation might seem unhelpful because then all selection is strong, and so the designation has no meaning. Or perhaps that is the point—any selection should be considered strong because of what it could accomplish if sustained (although selection is unlikely to be consistent over long time frames—see below). Considering the second frame of reference (Hereford et al. 2004), the average strength of selection on individual traits is estimated to be 56% as strong as selection on fitness itself (28% after correcting for sampling error—more about this below). This value seems unrealistically large and suggests biases in the estimation of selection (Hereford et al. 2004). An alternative interpretation of this average mean-standardized coefficient is that a doubling of the trait value would increase fitness by 56% (or 28%). To me, this could be interpreted as weak given that selection would be extremely unlikely to double a trait value on a reasonable time frame. Take as an exemplar the classic example of intense selection on Darwin's finches during droughts (Table 2 in Grant and Grant 1995). I calculate that the average shift in trait values owing to selection is 2.5%, which would correspond to a bias-corrected improvement in mean fitness of only 0.7%. Considering that this is one of the strongest selection episodes documented in nature, we might conclude that selection is generally weak.

The third frame of reference is a standard one in science: how much of the variation in a response variable (here fitness) can be explained by variation in a predictor variable (here the trait). As would be expected from the large spread in selection coefficients, a large spread is also found in the extent to which variation in fitness can be explained by variation in traits (fig. 2.5). Based on the data compiled by Hereford et al. (2004), 85% of the total gradients show an explanatory power of $r^2 < 0.125$ and only 8% show an explanatory power of $r^2 > 0.25$.[6] Based on the data compiled by Hersch and Phillips (2004), the corresponding percentages for selection differentials are 96% and 1%. I would argue, once again, that these values suggest most selection is on the "weaker" side of things; but how can we decide what constitutes "weak" versus "strong"? One approach is to compare the above explained variance to that seen in other ecological or evolutionary phenomena. Jennions and Møller (2002) performed a meta-analysis of meta-analyses in ecology and evolution and found that the median value for variance in a particular response variable explained by a particular predictor variable was 3.24%. The corresponding values from the above compilations of selection are less than 1%. By this criterion, then, selection on single traits seems weaker than other single-factor causal relationships in ecology and evolution. Peek et al. (2003) took a different approach by asking how much of the variance in ecological and evolutionary responses is *not* explained by all the predictor variables included in a study. They found that, on average, about half of the variance could be explained. It would be interesting to do a similar analysis for selection.

It is also important to note that selection coefficients are probably often upwardly biased. First, when selection estimates are converted to their absolute values, random sampling error will inflate regression coefficients if sample sizes are small (Hersch and Phillips

---

[6]The values in this section were calculated based on data provided by J. Hereford and E. Hersch that corresponded to the analyses in Hereford et al. (2004, fig 2) and Hersch and Phillips (2004, fig. 3). The calculations exclude cases where the variance explained was greater than unity—as this "could only be due to errors in calculating selection estimates" (Hereford et al. 2004, p. 2137). As another comment, the estimates in this section are based on linear selection, whereas the variance in fitness explained by all selection (e.g., including stabilizing selection) is likely much higher (e.g., Johnson et al. 2014).

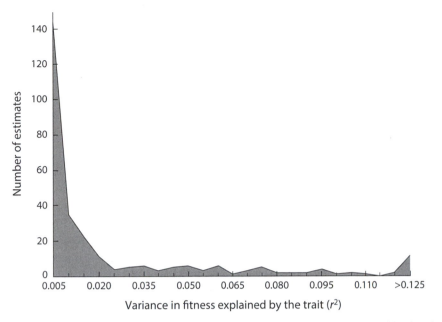

Fig 2.5. Frequency distribution of the variance in fitness explained by individual traits in studies of phenotypic selection in natural populations. I excluded nine estimates that exceeded 1.0 as they were likely errors in estimation. The data are from Hersch and Phillips (2004) and were provided by E. Hersch

2004, Hereford et al. 2004), although the magnitude of this bias might not be strong for selection (Knapczyk and Conner 2007). Correction for this effect deflates the median variance-standardized selection coefficient by 33% and the median mean-standardized selection coefficient by 52% (0.56 to 0.28 as noted above; Hereford et al. 2004). Second, investigators often target traits that they expect to be under strong selection, whereas the "average" trait might be under weaker selection. Third, studies might be more likely to be published if they have at least some significant selection estimates (Hersch and Phillips 2004, but see Knapczyk and Conner 2007). Fourth, estimates of directional selection from only a portion of the life cycle likely will be inflated relative to that over the entire life cycle (Schluter et al. 1991). Indeed, selection based on viability is weaker when estimated over longer time periods (Hoekstra et al. 2001). Fifth, estimates of selection based on only one fitness component are likely to be inflated relative to more inclusive fitness surrogates (Hereford et al. 2004), although other authors have not found this pattern (Kingsolver and Diamond 2011).

The most reasonable conclusions seem to be that (1) **selection in most natural populations on most phenotypic traits is quite weak most of the time (in the sense that it explains little of the variation in fitness and would make only a minor fitness improvement in a single generation), but (2) selection in some populations on some traits can be quite strong some of the time.** The first conclusion makes sense because most populations are probably reasonably well adapted for their local conditions (their phenotypes reside in the general vicinity of a local fitness peak) or they would not persist

(Hendry and Gonzalez 2008). Under such conditions, current selection is expected to be weak (Haller and Hendry 2014). In addition, a number of other factors, including density dependence, serve to flatten the fitness surface in the vicinity of an adaptive peak (Haller and Hendry 2014) as explained in detail in chapter 3. In essence, then, the fitness landscape experienced by most populations is probably relatively flat. The second conclusion makes sense because selection can be reasonably strong when phenotypes are displaced from local adaptive peaks. This displacement can occur for a variety of reasons, including colonization of a new environment (e.g., Clegg et al. 2008), strong environmental change (Grant and Grant 1995, Brown and Brown 2011, Husby et al. 2011), and strong constraints owing to gene flow (e.g., Bolnick and Nosil 2007, but see Rolshausen et al. 2015a).

QUESTION 2: WHAT IS THE RELATIVE IMPORTANCE OF DIRECT VERSUS INDIRECT SELECTION?

A classic question in evolutionary biology is the extent to which traits can be isolated from each other and considered individually with respect to selection and evolution, as opposed to being evaluated as an integrated, multivariate complex of traits (Olson and Miller 1958, Gould and Lewontin 1979, Walsh and Blows 2009). Of course, the answer will not be an either-or proposition; instead, some traits (or sets of traits) will act nearly independently, some will act as closely integrated units, and the rest will fall somewhere in between. One way to consider this range is to evaluate "phenotypic integration" among traits (Olson and Miller 1958, Cheverud 1982, Pigliucci 2003). An example is shown in figure 2.6 for correlations among four morphological traits and three pigmentation traits in isopod (*Asellus aquaticus*) populations from two habitats (reed and stonewort) in two lakes (Krankesjön and Tåkern) in Sweden (Eroukhmanoff and Svensson 2009). In this example, phenotypic integration tends to be higher within a given trait type (morphology vs. pigmentation) and within stonewort as opposed to reed isopods. Another way to consider variation in trait associations is from the context of "modularity," which identifies integrated units that are genetically and developmentally decoupled from other such units (Wagner and Altenberg 1996, Hansen 2003). For instance, the morphological and pigmentation traits of isopods form two (mostly) separate modules in reed habitats but not in stonewort habitats (fig. 2.6).

Correlations among traits could have their foundation in environmental effects (plasticity) or genetic effects (pleiotropy or linkage). In subsequent chapters, I will take up the question of how genetic correlations among traits might influence their evolution. Here I ask how *selection* is influenced by *phenotypic* correlations among traits. Two alternative extremes can be used to bracket the possibilities. At one extreme, total selection acting on a given trait might be the result of direct selection on that trait alone. For such traits, the selection coefficient should be the same whether (gradients) or not (differentials) other traits are included in the analysis. At the other extreme, total selection on a given trait might be entirely the result of indirect selection acting through other traits—with no direct selection on the focal trait itself. In this case, the selection on a given trait would vanish when the other traits were added to the analysis.

As might be expected, the answer lies somewhere between the extremes. First, selection differentials (simple linear regression) and selection gradients (multiple linear regression) are strongly correlated with each other when considered across all traits and studies (Kingsolver et al. 2001, Kingsolver and Diamond 2011; fig. 2.7). Second,

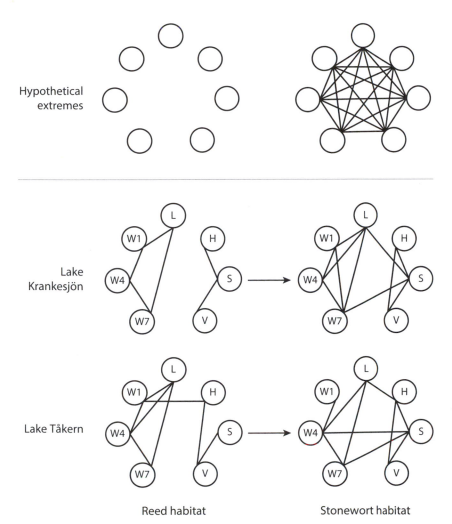

Lake Krankesjön

Lake Tåkern

Reed habitat   Stonewort habitat

Fig 2.6. Conditional dependence graphs representing correlations among traits, both hypo-thetical (top two graphs) and real (bottom four graphs). The top graphs show hypothetical extreme possibilities for the case of seven traits, with the left graph showing the case where all traits are uncorrelated with each other and the right graph showing the case where all traits are correlated with each other. The bottom four graphs show actual correlations for isopods in each of two habitats in each of two lakes. The traits are body length (L); body widths at the first, fourth, and seventh segments (W1, W4, W7); and pigmentation expressed as color (H), saturation (S), and brightness (V). The arrows represent the direction of evolu-tion of the correlation matrix within each lake: that is, the stonewort habitats are relatively new and so have only recently been colonized by isopods. The graphs are redrawn from Eroukhmanoff and Svensson (2009)

the association is approximately 1:1 for traits not related to body size ("nonsize traits"). These patterns might suggest that direct selection on traits is usually not compromised by indirect selection acting through other traits, but caveats remain. For instance, traits that are related to body size often show smaller differentials than gradients, suggesting that direct selection on size-related traits is partly offset by opposing indirect selection

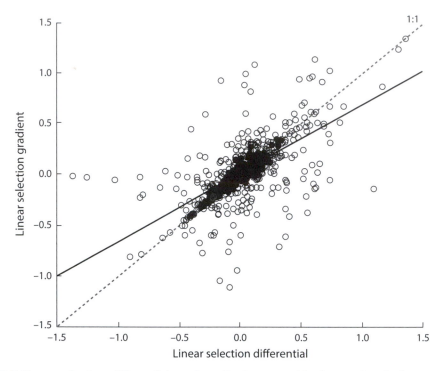

Fig 2.7. Linear selection differentials and gradients are positively correlated—but not perfectly so—in studies of phenotypic selection in natural populations. Seven outlier data points are not shown but are included in the estimation of the best-fit line (solid line). Each data point represents selection on a particular trait in a particular population during a particular selection episode. The data are from Kingsolver and Diamond (2011), who compiled the Kingsolver et al. (2001) and Siepielski et al. (2009) databases

acting through other traits (Kingsolver and Diamond 2011, Rollinson and Rowe 2015). In addition, even for nonsize traits, many individual comparisons deviate dramatically from the 1:1 association. To illustrate, I regressed gradients on differentials across all relevant data points from Kingsolver et al. (2011): $N = 749$, $F = 326.04$, $r^2 = 0.304$. That is, knowing the differentials for all nonsize traits in the database would allow one to predict only 30.4% of the variance in gradients among those same traits. Thus, indirect selection makes a substantial contribution to total selection in many instances. In addition, the role of indirect selection is certainly underestimated. The reason is that the effects of indirect selection are only removed for those traits that are actually included in the analysis (Walker 2014)—and any given study can't measure all potentially relevant traits. Overall, **I suggest that we still don't have a good idea of the extent to which traits are isolated versus integrated in the context of selection.**

QUESTION 3: WHAT TYPES OF TRAITS ARE UNDER THE STRONGEST SELECTION?

Another classic debate in evolutionary biology is whether or not different types of traits (e.g., life history, morphology, behavior, or physiology) have different levels of genetic variation (Mousseau and Roff 1987, Roff and Mousseau 1987, Houle 1992, Stirling et al. 2002, Hansen et al. 2011). This question originally derives from the expectation that genetic variation will be removed by most forms of selection (Fisher 1930, Merilä and Sheldon 1999), and that different types of traits should be under different forms and strengths of selection. I will return in chapter 10 to the question of whether or not different types of traits have different levels of genetic variation. Here I consider the putative foundation of such differences: that is, do different types of traits typically experience different strengths of selection? A common expectation here is that life history traits are more closely related to fitness than are morphological traits, and so the former should be under stronger selection. This expectation can be evaluated by taking a database of selection estimates, assigning traits to categories representing the different trait types, and testing whether typical strengths of selection differ among those types. The assignment of traits to different categories has been variable among studies (Mousseau and Roff 1987, Houle 1992, Kinnison and Hendry 2001, Stirling et al. 2002), and I here simply use the categorization applied in a recent compilation (Kingsolver et al. 2012). In that categorization, life history includes traits such as age at maturity, fecundity, various aspects of phenology, habitat preference, flower number, clutch size, nectar production rate, and life span. Morphology includes traits related to body size or condition factor, linear measurements of external characters (e.g., limbs, spines, floral traits), or counts of meristic characters.

Is the expectation that life history traits are under stronger selection than morphological traits supported by the data? The simple answer appears to be no. For example, Kingsolver et al. (2001) found that linear selection was, on average, stronger on morphological traits than on life history traits, and that no differences were evident with respect to quadratic selection.[7] If the traits are partitioned into sub-categories, however, some interesting patterns emerge. In particular, selection on phenology (timing of life history events such as reproduction) is somewhat skewed toward negative values and selection on size-related traits is somewhat skewed toward positive values (fig. 2.8). Although these differences are small and inconsistent (some phenology is under positive selection and some size traits are under negative selection), it is nevertheless tempting to speculate as to potential causes. First, negative selection on phenology is not surprising given that climate warming in many parts of the world is causing shifts toward earlier timing of spring phenology (Parmesan and Yohe 2003, Root et al. 2003). Second, positive selection on body size was argued by Kingsolver and Pfennig (2004) to be consistent with Cope's rule, which states that species within fossil lineages tend to evolve toward larger body size through time (Cope 1887, Alroy 1998). **Thus, although the original hypothesis about life history versus morphology does not seem to hold, some different types of traits do seem to differ in typical strengths of selection.**

Some conceptual problems attend the idea of comparing strengths of selection among trait types. The main problem is the one argued in Question 1 (above): evolution should bring most traits reasonably close to their adaptive optima—and so total selection should not be stronger on any particular type of trait. Stated another way, if selection is stronger on one trait than another, evolution should improve adaptation for that trait and thereby reduce selection on it: selection "erases its traces." **Thus, the finding that selection is**

---

[7] Also, the analysis of Kinnison and Hendry (2001) found that life history and morphological traits did not differ in their variance-standardized rates of phenotypic change in contemporary natural populations.

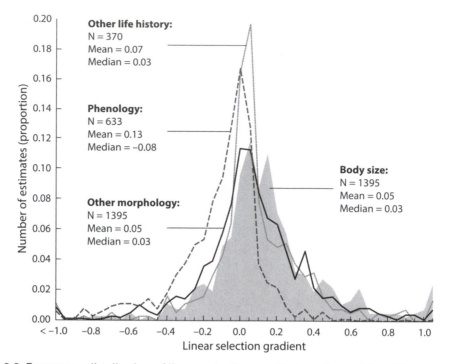

Fig 2.8. Frequency distribution of linear selection gradients estimated for different types of phenotypic traits in natural populations. The distributions are standardized across traits by using proportions (as opposed to counts). The data are from Kingsolver and Diamond (2011), who compiled the Kingsolver et al. (2001) and Siepielski et al. (2009) databases

stronger on one type of trait than another suggests at least one of the following: (1) maladaptive mutational inputs differ among traits, (2) selection has recently changed in a somewhat coherent manner for a particular type of trait—as was suggested above for phenology, or (3) biases in the estimation of selection differ among traits. I would argue on several grounds that body size, in particular, suffers from a positive bias in the estimation of selection. First, body size is not generally evolving to be larger in contemporary populations (Meiri et al. 2009, Gotanda et al. 2015), as would be predicted if selection were typically positive. Second, indirect selection acting through other traits will often act in opposition to direct selection on body size (see above and Rollinson and Rowe 2015)—and a number of these traits will be missed in most studies of selection. Third, many examples exist where larger and faster individuals show higher mortality rates or manifest other problems that would compromise fitness (Arendt 1997, Blanckenhorn 2000, Wikelski and Romero 2003, DiBattista et al. 2007, Carlson et al. 2008).

QUESTION 4: HOW COMMON IS STABILIZING OR DISRUPTIVE SELECTION?

The traditional view of adaptation is that selection pushes populations to locations in phenotypic space that represent local fitness peaks (Schluter 2000a, Arnold et al. 2001, Haller and Hendry 2014). This process will be discussed at length in the next two chapters but, for now, I simply consider what it means for expected patterns of selection. One prediction is that the distribution of phenotypes within a population will straddle a fitness optimum. If so, individuals with intermediate phenotypes should have the highest fitness and those deviating to either side should have lower fitness: that is, stabilizing selection should predominate. An alternative prediction stems from the idea that selection depends on competition for resources within populations (Dieckmann and Doebeli 1999, Dieckmann et al. 2004, Rueffler et al. 2006). In such cases, intermediate individuals may experience the strongest competition, leading extreme individuals to have the highest fitness (Bolnick 2001, Rueffler et al. 2006, Bolnick and Lau 2008). That is, disruptive selection should predominate. This dichotomy is an empirical problem that can be informed by reference to selection databases (Kingsolver et al. 2001, Kingsolver and Diamond 2011). In the following paragraphs, I first discuss the overall strength of quadratic coefficients (regardless of whether they are positive or negative) and then the extent to which those coefficients are consistent with stabilizing versus disruptive selection.

Most quadratic selection coefficients measured in nature are nonsignificant and very close to zero. As noted above for linear selection, a number of biases exist that suggest these values are upwardly biased, meaning that quadratic selection is even weaker than it appears. However, two opposing arguments exist that these values might be downwardly biased. One argument is that regression coefficients for quadratic terms need to be doubled in order to obtain quadratic selection coefficients (Lande and Arnold 1983, Phillips and Arnold 1989), whereas many studies in the existing databases failed to implement this adjustment (Stinchcombe et al. 2008). Another argument is that stabilizing selection will be most evident only in multivariate trait space, rather than in the above single-trait quadratic coefficients. For example, Blows and Brooks (2003) reanalyzed the (few) relevant studies in the Kingsolver et al. (2001) database in a multivariate context and suggested that the average strength of quadratic selection had been underestimated by a factor of at least 1.5. "Therefore, although nonlinear selection may well be generally weak on the individual traits included in any particular selection analysis as reported in many empirical studies, it may be much stronger on at least one of the composite traits that may be the actual target of nonlinear selection in many situations." (Blows and Brooks 2003, p. 818).

Regardless of any bias, inferring whether quadratic selection is "strong" or "weak" is subjective, as explained earlier with respect to linear selection. I therefore prefer the more objective criterion of how much variation in fitness is explained. Although these data are not readily available for quadratic coefficients—as they were for linear coefficients (fig. 2.5)—it is nevertheless clear that variance explained is even weaker here. For instance, sample sizes of 500–1000 are necessary for the typical quadratic coefficients in nature to be found statistically significant (Kingsolver et al. 2001). I would therefore argue that quadratic selection is very weak relative to many other ecological and evolutionary effects. Reassuringly, this result is to be expected if populations are relatively close to adaptive optima (Estes and Arnold 2007, Haller and Hendry 2014). The reason is that individuals with genes that lead to large phenotypic deviations from adaptive optima will be continually removed from populations, leaving genes/phenotypes that are reasonably well adapted.

Whatever its strength, quadratic selection appears to be disruptive as often as it is stabilizing (fig. 2.9). This apparent equivalence has been considered surprising under the above expectation that selection optimizes phenotypes around local fitness optima (Kingsolver et al. 2001, Estes and Arnold 2007), but it is not surprising under the expectation that intraspecific competition is a major contributor to fitness variation within populations (Rueffler et al. 2006, Bolnick and Lau 2008). In truth, both processes are likely acting in many populations at the same time, and might thereby offset each other, leading to no obvious quadratic selection in most cases (Haller and Hendry 2014). It is nevertheless clear that strong stabilizing or disruptive selection is sometimes present in nature (fig. 2.9). Assuming these cases do not reflect biases associated with poor fitness surrogates or fitness tradeoffs, it is worth considering the biological circumstances that contribute to their emergence.

Strong disruptive selection implies that individuals with intermediate phenotypes have low fitness relative to individuals with extreme phenotypes. This situation can occur owing to either of three main effects (Doebeli 1996, Thibert-Plante and Hendry 2011a). First, disruptive *sexual* selection can arise through alternative mating tactics, such as fighting versus sneaking (Gross 1996). Second, a population might use two reasonably distinct resources and the phenotypes best suited for those resources are different: that is, performance tradeoffs exist in adaptation to the different resources. Third, intraspecific competition might be strong for shared resources; such that individuals with phenotypes specialized for underexploited (extreme) resources have higher fitness. Any of these

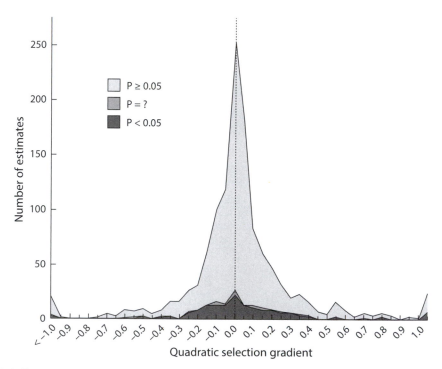

Fig 2.9. Frequency distribution of quadratic selection gradients estimated for phenotypic traits in natural populations. Significance levels are shown with different shading, including instances where the original study did not indicate significance (P = ?). The data come from Kingsolver and Diamond (2011), who compiled the Kingsolver et al. (2001) and Siepielski et al. (2009) databases

effects could easily occur episodically in any given population but it is hard to maintain them through time—because populations under disruptive selection should evolve so as to either specialize on only a subset of resources or to split into reproductively isolated species that use different resources (Rueffler et al. 2006). Instead, persistent disruptive selection requires the continual production of intermediate individuals. This situation is most likely to occur when assortative mating is absent or incomplete. One example comes from a bimodal population of Darwin's finches (*Geospiza fortis*), where selection disfavors birds with intermediate beak sizes (fig. 2.10). In this system, birds generally mate assortatively by beak size (Huber et al. 2007), but they still break this rule often enough to produce intermediates, which then suffer lower viability owing to a scarcity of appropriate resources. As another example, selection appears to sometimes disfavor intermediate foraging phenotypes in lacustrine threespine stickleback (*Gasterosteus aculeatus*), probably owing to intraspecific competition and the presence of reasonably distinct benthic and limnetic resources (Bolnick and Lau 2008). This disruptive selection might be maintained because assortative mating by phenotype is either weak (Snowberg and Bolnick 2008) or absent in most of these populations.

By reverse analogy to the above, strong *stabilizing* selection requires low fitness of extreme individuals on both sides of the phenotypic distribution coupled with their continual production. The first part of the requirement simply states that particular phenotypes should be best suited for particular conditions (i.e., a fitness peak must be present), which seems inevitable (see chapter 3, Adaptation). The second part of the requirement can be met if development is noisy (Hansen et al. 2006), if mutation rates are high (Burt 1995), or if gene flow is present from populations/species farther out on the phenotypic distribution. The above-mentioned Darwin's finch result provides an example here too: selection not only disfavors intermediate individuals but also extreme individuals (fig. 2.10). Stated another way, stabilizing selection is present around each of two beak size modes—and therefore is also disruptive between them (fig. 2.10). The continual production of extreme individuals in this case probably results from hybridization with even smaller (*Geospiza fuliginosa*) and larger (*Geospiza magnirostris*) congeners (Grant and Grant 2009, De León et al. 2010). Similar patterns have been documented for other polymorphic bird populations (e.g., Smith 1993).

Quadratic selection is usually absent or weak within populations, albeit with some informative exceptions. Moreover, no obvious tendency is seen for quadratic selection to be more often stabilizing or disruptive. These conclusions are consistent with the above idea that most populations are reasonably well adapted for local fitness optima and that competition flattens fitness surfaces in the vicinity of those optima (Estes and Arnold 2007, Haller and Hendry 2014).

QUESTION 5: HOW VARIABLE IS SELECTION?

Selection can vary in space, such as among populations in different environments, or through time, such as through the life cycle or across seasons or years. Spatial variation is considered in chapter 4, whereas I here focus on interannual differences in the strength and direction of selection within populations. This variation, and its autocorrelation, has an important bearing on several key questions in evolutionary biology (Siepielski et al. 2009, 2011, Bell 2010, Chevin and Haller 2014). First, greater variation in selection is, under some conditions, expected to maintain greater genetic variation within populations

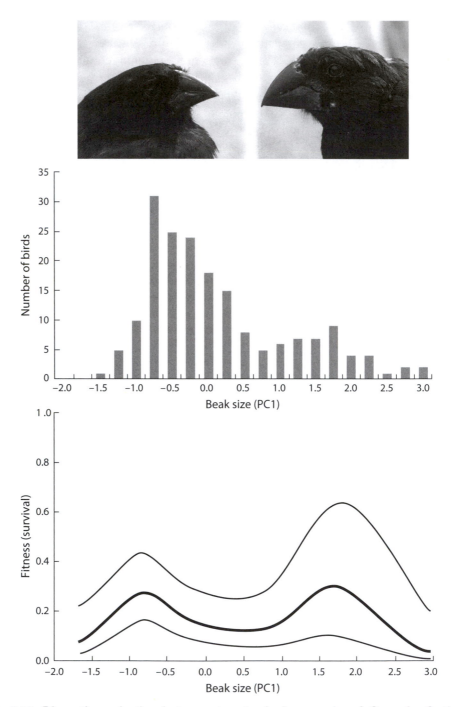

Fig 2.10. Disruptive selection between two beak size morphs of *Geospiza fortis* at El Garrapatero, Santa Cruz, Galápagos. The top panel shows examples of the two morphs (photo by A. Hendry). The middle panel shows the number of birds of different beak sizes (PC1 of beak length, depth, and width) captured in 2005. The bottom panel shows estimated selection (cubic spline with 95% bootstrap confidence interval) based on survival (*y*-axis value of 1) versus disappearance (*y*-axis value of 0) from 2005 to 2006. The data are from Hendry et al. (2009)

(Sasaki and Ellner 1997, Bergland et al. 2014). This increase in genetic variation can enhance the ability of populations to respond to future evolutionary change (chapter 10) but it can also make them less well adapted for prevailing conditions (Hansen et al. 2006). Second, variation in selection implies variation in optimal trait values: that is, movement of the adaptive peak (Chevin and Haller 2014, see also chapter 3). Such variation would suggest that mean trait values within populations are rarely at the adaptive optima at any given point in time (although making correct inferences can be difficult: Chevin and Haller 2014). That is, phenotypes rarely attain a fitness peak but must always (meta-phorically) chase it around in phenotypic space. Third, and from a practical perspective, increasing temporal variation means that we must measure selection across multiple, perhaps many, years if we are to accurately predict evolution.

A number of specific instances are known where selection indisputably fluctuates through time. One likely cause is temporal variation in environmental conditions, such as rainfall (Grant and Grant 2002, Carlson and Quinn 2007), temperature (Johnson 2011, Bergland et al. 2014), predation (Reimchen and Nosil 2002, Losos et al. 2006, Cunningham et al. 2013), snowfall (Karell et al. 2011), and competition (Grant and Grant 2006, Calsbeek and Cox 2010). Another likely cause is frequency-dependent selection (selection on a phenotype or genotype depends on its abundance relative to other phe-notypes or genotypes) resulting from factors such as predation (Olendorf et al. 2006), parasitism (Dybdahl and Lively 1998, Decaestecker et al. 2007), sexual conflict (Svensson et al. 2005, Le Rouzic et al. 2015), competition for mating opportunities (Sinervo and Lively 1996), or competition for resources. A good example of the last of these drivers is that selection in scale eating cichlids (*Perissodis eccentricus*) favors left-handed jaw curvature when right-handers are more common but favors right-hand jaw curvature when left-handers are more common (Hori 1993).

A number of specific examples of temporal variation in selection linked to temporal variation in causal effects are thus well established—but just how common and strong is temporal variation overall? Several studies have sought to inform this question through meta-analyses of temporally replicated selection estimates (Siepielski et al. 2009, 2011, Kingsolver and Diamond 2011, Morrisey and Hadfield 2012). In the first analysis, Siepiel-ski et al. (2009) reported three main findings. First, temporal variation in the magnitude of selection (standard deviation across time in the absolute value of coefficients) on a given trait in a given population is often as large as the average selection coefficient for that trait in that population (fig. 2.11). Second, the *direction* of selection changes through time in a substantial fraction of the studies. Third, the qualitative shape of the fitness function often varies dramatically among years—as can be seen in the Wytham Woods data presented above (fig. 2.3). The apparent amount of temporal variation is, however, almost certainly an overestimate of the real variation. The reason is that sam-pling error will cause selection coefficient *estimates* to differ among samples even if the true selection coefficient *parameter* in the population does not change (Siepielski et al. 2009). Accounting for this sampling error suggests that temporal variation in the actual underlying selection coefficients is low—that is, the signal of any real temporal variation cannot be separated from the noise in the estimates (Morrissey and Hadfield 2012). In addition, other analyses suggest that temporal variation has little influence on the expected cumulative selection experienced by a population (Kingsolver and Diamond 2011, Morrissey and Hadfield 2012).

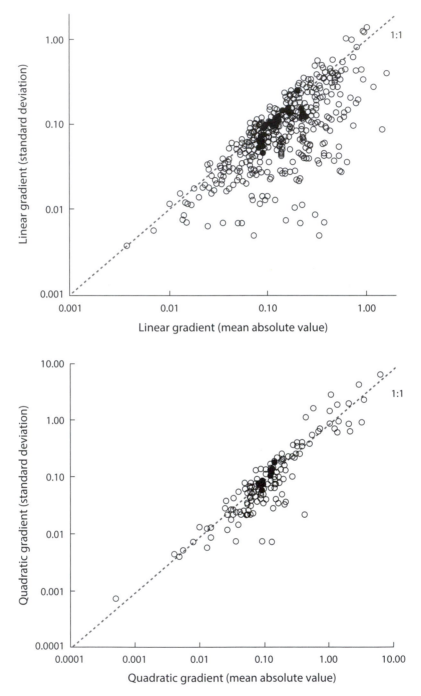

Fig 2.11. Selection coefficients (top panel: linear gradients; bottom panel: quadratic gradients) are highly variable in that the standard deviation among temporal replicates is often as high as the across-replicate mean. Each data point represents selection on a particular trait in a particular population measured across multiple time periods. The data are from Siepielski et al. (2009) and were compiled here by K. Gotanda

Given the difficulties in estimating selection with high reliability (see above), some authors (Gotanda and Hendry 2014, Chevin and Haller 2014) have argued that important complementary insights can be gained by the analysis of spatiotemporal variation in the means of adaptive traits, which are much easier to estimate and track through time. In an analysis of male guppy (*Poecilia reticulata*) color, Gotanda and Hendry (2014) found that spatial variation was very consistent through time, suggesting that temporal variation was relatively minor in comparison. Although this result suggests that temporal variation in selection is much weaker than spatial variation, the guppy system may be exceptional in that regard (Gotanda and Hendry 2014). Additional formal analyses of spatiotemporal variation in adaptive traits—ideally along with selection coefficients (Chevin and Haller 2014)—would be valuable.

Although selection acting on a population clearly varies dramatically through time in a number of instances, it remains uncertain just how typical this phenomenon is in natural populations. That is, selection might or might not be temporally variable in any given instance. Regardless, it cannot be disputed that temporal variation in selection *estimates*, even if caused by sampling error, is quite large relative to the average value. This observation highlights the importance of large sample sizes and temporally replicated estimates—as illustrated for the Wytham Woods data described earlier. And, again, it reinforces the point that selection is quite weak in most instances. **That is, most populations are relatively near their fitness peaks (see also** chapter 3) **and so selection on them will usually be very weak—and therefore selection estimates will be seemingly variable in relation to their overall magnitude.**

QUESTION 6: DO NATURAL AND SEXUAL SELECTION QUALITATIVELY DIFFER?

Following Darwin (1871), natural selection in the broad sense, as thus far discussed, can also be divided into natural selection in the narrow sense versus sexual selection. This distinction is usually drawn based on whether fitness is measured as survival or fecundity (natural selection) versus mating success (sexual selection). It is important to now make this separation because natural and sexual selection might (1) respond differently to environmental differences (eco-to-evo), and (2) have different effects on traits that then have ecological influences (evo-to-eco). Three expectations might be advanced for how these two types of selection might differ from each other. First, sexual selection can be stronger than natural selection because it can be more open-ended. For instance, females might often prefer males having the most extreme values in the population, regardless of the overall trait mean (Ryan and Keddy-Hector 1992). (The same arguments would also often apply to male choice of females and to intrasexual competition for mates.) Second, sexual selection might be more temporally consistent—for the reason just noted. That is, natural selection might vary from year to year depending on the environmental conditions, whereas females might always prefer more ornamented males. Third, natural and sexual selection might offset each other, such as when conspicuousness increases mating success but also increases susceptibility to predators (Endler 1980, Zuk and Kolluru 1998) or when investment into breeding reduces energy reserves for subsequent survival (Hunt et al. 2004). Indeed, sexual selection was the original explanation for why males evolved exaggerated traits that would seem costly to survival (Darwin 1859).

Meta-analyses of linear selection have repeatedly found that sexual selection based on mating success is generally stronger than natural selection based on viability (Hoekstra et al. 2001, Siepielski et al. 2011, Kingsolver et al. 2012). Although this difference is consistent with the first expectation above, a number of biases might also act in this direction; for instance, selection estimates based on mating success are usually over shorter periods of time and the traits might be more likely to have been chosen owing to an a priori expectation of selection (Hoekstra et al. 2001, Siepielski et al. 2011, Kingsolver et al. 2012). Inconsistent with the same expectation, however, is the observation that selection estimates based on fecundity are just as high as those based on mating success. So what really remains to be explained is why selection based on survival is lower than selection based on the other fitness components.

With respect to temporal variation, Siepielski et al. (2011) found that reversals in *direction* were less likely for sexual selection and for natural selection based on fecundity than for natural selection based on survival. Much of this difference probably can be attributed to the fact that coefficients based on survival are smaller (see above) and so sampling error is more likely to lead to apparent variation in direction. However, sexual selection gradients, although not differentials, were more variable in *magnitude* than were natural selection gradients based on either fitness component. For example, sexual selection can vary dramatically between years depending on factors such as temporal variation in the traits that reliably signal male quality (Chaine and Lyon 2008) and in the frequency of different female types (Gosden and Svensson 2008, Le Rouzic et al. 2015).

The possibility that natural and sexual selection act in opposition to each other has now been examined in several meta-analyses. Jennions et al. (2001) found that males with greater trait expression had, on average, *higher* survival, although the association was weak. In addition, Kingsolver and Diamond (2011) compared selection coefficients for the same traits in the same studies based on different fitness components: mating success, fecundity, and survival. They found no apparent association (i.e., no trade-off) among any of these fitness components. In short, although opposing effects of natural versus sexual selection are found in some specific studies—no evidence exists that this trade-off is general. One possibility is that survival, fecundity, and mating success are condition-dependent, and so males in high condition do well with respect to multiple fitness components (Jennions et al. 2001). Another possibility is that selection favors breaking the trade-off, such that "private" mating signals often evolve to increase conspicuousness to conspecifics while maintaining crypsis to heterospecifics such as predators (Stoddard 1999, Cummings et al. 2003, Millar and Hendry 2012).

Sexual selection differs in some respects from natural selection, the former being stronger (although a bias cannot be ruled out) and less variable (when the latter is based on viability). However, although sexual and natural selection do sometimes trade off with each other, this does not appear to be a general phenomenon, perhaps because of condition dependence or private signaling.

Conclusions, significance, and implications

The most important conclusion of this chapter is that selection is usually weak on most traits in most populations in nature. The reason is that most populations are reasonably well adapted for their local environments and, hence, maladaptive phenotypes are relatively rare. This phenomenon also explains why selection estimates are temporally variable, why this temporal variability is difficult to separate from sampling error, and why quadratic selection is weak and as often disruptive as stabilizing. (Another reason for this last observation is that competition flattens fitness surfaces.) Current selection in nature is thus a poor indicator of the selection that got a population to its current phenotypes: that is, selection erases its traces (Haller and Hendry 2014). An important implication for studies of eco-evolutionary dynamics is that evolutionary processes and their ecological effects can be difficult to study in well-adapted populations where little change is currently taking place.

Beyond the above generalization, specific instances of strong directional, stabilizing, disruptive, and temporally variable selection are well known. These instances are usually the result of changing environments, with a particularly general example being directional selection on phenology—likely as a result of climate change. Strong selection can also result from trade-offs among fitness components, such as natural and sexual selection, or from strong gene flow that drags populations off local adaptive peaks. For these reasons, the pragmatic route to documenting eco-evolutionary dynamics might be to study instances of environmental change, while also bearing in mind that such instances might not be representative of the typical situation.

Chapter 3

Adaptation

The previous chapter detailed how selection acts on phenotypes, and how this process can be studied in natural populations. The outcome of this selection is expected to be adaptation[1] that improves the ability of individuals to survive and reproduce in their local environments (eco-to-evo). The precision and accuracy of this adaptation will thus influence the fitness of individuals and the mean fitness of populations, the latter potentially altering population size, growth, or persistence (evo-to-eco: CHAPTER 7). Given that phenotypic traits and population dynamics then could influence both community and ecosystem variables (evo-to-eco: chapters 8 and 9), the study of adaptation is crucial to the study of eco-evolutionary dynamics.

It might seem awkward to separate the consideration of selection (previous chapter) from the consideration of adaptation (present chapter). Apart from the pragmatic benefit of avoiding overlong chapters, this separation has several conceptual merits. First, selection can occur without any adaptation resulting—for reasons that will be discussed below. Second, adaptation can be studied without measuring selection. That is, an investigator can sometimes infer that trait changes were driven by selection (i.e., adaptation) even if selection itself is not measured. Third, selection and adaptation are expected to be decoupled in reasonably stable environments, simply because selection "erases its traces," as explained above.

I first outline a common conceptual framework for linking selection to adaptation: the breeder's equation and its extensions. This outline will be somewhat detailed because, as in the study of selection, adaptation is fundamental to many of the following topics. I then use this conceptual framework to consider several key questions regarding the rate of phenotypic change and the extent to which it is adaptive. As usual, the proffered answers are my own reading and will not necessarily jibe with the opinions of others.

---

[1]The term "adaptation" as a process will be used to mean phenotypic change that improves fitness and has a genetic basis. "Adaptive change" will be used to mean phenotypic change that improves fitness regardless of whether or not it has a genetic basis. "Adaptation" or "adaptive" as a state (or condition) will be used for any trait that improves fitness in a given environment relative to plausible alternative traits in that environment.

Conceptualizing and predicting adaptation

The approach taken here falls under the umbrella of evolutionary quantitative genetics, where traits are considered as means and (co)variances (Falconer and Mackay 1996, Roff 1997, 2007, Lynch and Walsh 1998). This approach ignores genetic particulars, such as which genes are involved, and instead considers statistical associations between the phenotypes of relatives. Conveniently, quantitative genetics is the appropriate starting point for a study of eco-evolutionary dynamics, owing to the central importance of phenotypes as noted earlier. In the following sections, I first consider the evolution of a single trait (i.e., the breeder's equation) and then transition to the consideration of multiple correlated traits (i.e., the "Lande equation"). I also discuss an alternative, but related, approach based on the "Price equation."

UNIVARIATE ADAPTATION

A simple way to consider phenotypic evolution is by analogy with linear regression. For a sexual population, simply plot the mean "trait value" (mean phenotype) of parents on the x-axis (average value of the two parents, called the "midparent" value) and the mean trait value of their offspring on the y-axis (average value across all offspring for a given pair of parents). The regression line through these points gives the average trait value of offspring with respect to the average trait value of their parents. The relationship thus can be used to predict how a change in the mean phenotype of parents along the x-axis (owing to selection) should change the mean phenotype of the population by the next generation along the y-axis (i.e., the difference in mean phenotype of offspring, measured at maturity, from one generation to the next).

$$slope = \frac{\text{evolutionary response measured on offspring}}{\text{selection on parents}} = \frac{R}{S}.$$

Rearranging this equation gives the standard breeder's equation: $R = (slope)(S)$. Thus, to predict an evolutionary response (R), we need to know the strength of selection acting on the parents (S) and the slope of the relationship between parent and offspring phenotypes. This slope is called the **heritability** of the trait, and it is abbreviated h^2.

$$slope = h^2 = \frac{COV_{(MeanOffspringxMidparent)}}{V_{(Midparent)}}$$

This equation allows expansion to a more general formulation of heritability. Starting with the denominator, if the phenotypic variance for the trait is the same in both sexes, the phenotypic variance of the midparent values ($V_{(Midparent)}$) is half of the total phenotypic variance in all the parents (V_P). A number of factors can contribute to this variance: $V_P = V_A + V_D + V_I + V_E$, where V_A is the additive genetic variance, V_D is the dominance variance, V_I is the epistatic variance, and V_E is the environmental variance.

Turning to the numerator (covariance between parents and offspring), we need to consider the "breeding value": the trait value of a parent expressed as the mean (or "expected") trait value of its offspring (Falconer 1989, p. 117). By extension, the breeding value of an actual pair of parents, and therefore the mean (and expected) breeding value of their offspring, is the average of the two parental breeding values. Thus, a regression of mean offspring trait values on mid-parent trait values is a regression of parental breeding values against mean parental phenotypes.

A key point about breeding values is that they are determined by V_A (for a detailed explanation see Falconer 1989, p. 115–123). Specifically, the covariance between mid-parent values and mean offspring values is $1/2V_A$ and this parameter is theoretically—but not always in practice—independent of V_D, V_I, and V_E (Roff 1997, p. 30, and Falconer 1989, p. 150). Thus, we now know that:

$$slope = h^2 = \frac{COV_{(Offspring \times Midparent)}}{V_{(Midparent)}} = \frac{\frac{1}{2}V_A}{\frac{1}{2}V_P} = \frac{V_A}{V_P}.$$

Heritability thus generalizes to the proportion of the total phenotypic variance that has an additive genetic basis. Any means of estimating V_A can thus provide an estimate of heritability, and therefore be used to predict evolutionary responses using $R = h^2S$. Many methods exist to estimate V_A and some of the more common ones are parent-offspring regressions (as above) and half-sib mating designs (Falconer and Mackay 1996, Roff 1997, Lynch and Walsh 1998). Heritabilities can also be inferred for natural populations when pedigrees—or at least familial associations—are known. This approach is the "animal model;" which, in essence, exploits the different kinds of crosses that take place in natural populations (Kruuk 2004, Hadfield et al. 2010, Milot et al. 2011).

The approach just described is the classic theoretical framework for predicting the evolution of a trait given knowledge of selection and heritability ($R = h^2S$). It is important to remember, however, that this equation requires a large number of assumptions that are unlikely to be met in many natural populations (Merilä et al. 2001b, Morrissey et al. 2010). In the next section, I first discuss a way to deal with one of these assumptions (independence among traits) before moving on to additional issues.

MULTIVARIATE ADAPTATION

Traits do not evolve independently of each other, and so phenotypic evolution might not be well predicted by the above univariate approach. Assume, for illustration, that running speed and body size are phenotypically correlated, with speed being under positive selection. In this scenario, slower individuals are more likely to die, and so the survivors will be faster on average than the mean of the population before selection. The survivors will also be larger than the mean before selection—not because size has a direct influence on fitness but because size is correlated with speed: remove the slower individuals and you also—coincidentally—remove the smaller individuals. Stated in accordance with the concepts in chapter 2, speed is under direct selection and size is under indirect selection.

Selection on speed has thus shifted size within a generation, but how will this correlated shift translate to evolution across generations? Imagine first that size and speed are correlated simply because both are influenced by foraging opportunity, which is strictly environmentally determined. In this case, the phenotypic correlation between speed and size has no genetic basis: faster individuals do not have genetically larger body sizes. As a result, the mean *breeding value* for size is unaffected by selection on speed, and so no evolution of size is expected. Imagine next that the correlation between size and speed is determined by the same set of purely additive genes and is not influenced by the environment. The resulting covariance between breeding values for the two traits is called the "additive genetic covariance." In this case, selection on speed will cause a shift in breeding values for size (as well as for speed) and will therefore lead to the evolution of larger size, despite the fact that size was not directly under selection.

The problem of correlated traits requires a multivariate solution, which first involves rearranging the breeder's equation into an alternative form. Remembering that $h^2 = V_A/V_P$ and (by convention) relabeling R as ΔZ and V_A as G, the breeder's equation becomes $\Delta Z = G\beta$, where β is the selection differential and G is the additive genetic variance. This alternative formulation can be easily extended to multiple traits (Lande 1976, 1979) in which the βs become selection "gradients." For instance, when considering the evolution of a trait ("trait 1"), we can consider not only direct selection acting on that trait (β_1) and its additive genetic variance (G_{11}), but also indirect selection acting through another trait ("trait 2", β_2) and the additive genetic covariance between the two traits (G_{12}). The expected evolutionary response of trait 1 will be the sum of evolutionary effects through these two avenues of selection: $\Delta Z_1 = G_{11}\beta_1 + G_{12}\beta_2$. Similarly, the evolutionary response of trait 2 is expected to be $\Delta Z_2 = G_{22}\beta_2 + G_{21}\beta_1$. These equations allow us to predict how direct and indirect selection will influence evolutionary changes for both traits (Table 3.1). The final step is to generalize from the above two-trait equation to any number of traits:

$$\Delta Z_1 = G_{11}\beta_1 + \sum_{j=1}^{n} G_{1j}\beta_j \,,$$

where the subscript "j" is for each of the remaining n traits. An analogous equation exists for each of the other traits. Thus, the evolutionary response of each trait is a function of direct selection acting on each trait, the additive genetic variances for those traits, and the additive genetic covariances between traits. This system of equations is easier to write in matrix form, with the evolutionary responses and selection gradients shown as vectors and the additive genetic variances and covariances shown as a matrix (called the **G** matrix). Here is an example for three traits:

$$\begin{bmatrix} \Delta Z_1 \\ \Delta Z_2 \\ \Delta Z_3 \end{bmatrix} = \begin{bmatrix} G_{11} & G_{12} & G_{13} \\ G_{21} & G_{22} & G_{12} \\ G_{31} & G_{32} & G_{33} \end{bmatrix} \begin{bmatrix} \beta_1 \\ \beta_2 \\ \beta_3 \end{bmatrix},$$

The rules of matrix multiplication provide three equations—one for each trait. Note also that G_{ij} and G_{ji} are equivalent. The expansion to more traits is straightforward and is usually written more simply as $\Delta \mathbf{Z} = \mathbf{G}\boldsymbol{\beta}$, with the bold type signifying vectors and matrices.

ADAPTIVE LANDSCAPE

The above equations are related to a predictive evolutionary framework termed the "adaptive landscape" (Lande 1979, Fear and Price 1998, Schluter 2000a, Arnold et al. 2001). The adaptive landscape is a curve (for one trait) or an n-dimensional surface

Table 3.1. Examples provided here use the Lande equation to highlight interesting peculiarities of multivariate evolution, such as how one trait can evolve to be smaller and one larger even though both are under positive (+ve) direct selection (Scenario 9) or how both traits can evolve to be larger even though one is under negative (-ve) direct selection (Scenario 8).

	Scenario:	β_1	β_2	G_{11}	G_{22}	$G_{12} = G_{21}$	ΔZ_1	ΔZ_2
1	Selection (+ve) on trait 1 but absent for 2. Reasonable G for both traits. No additive genetic covariance.	1.0	0	0.5	0.5	0	0.5	0
2	Selection (+ve) on trait 1 but absent for 2. Reasonable G for both traits. Positive additive genetic covariance.	1.0	0	0.5	0.5	0.25	0.5	0.25
3	Selection (+ve) on trait 1 but absent for 2. Reasonable G for both traits. Negative additive genetic covariance.	1.0	0	0.5	0.5	−0.25	0.5	−0.25
4	Selection (+ve) on both traits. Reasonable G for both traits. Positive covariance.	1.0	1.0	0.5	0.5	0.25	0.75	0.75
5	Selection (+ve) on both traits. Reasonable G for both traits. Negative covariance.	1.0	1.0	0.5	0.5	−0.25	0.25	0.25
6	Opposite selection: +ve on 1, −ve on 2. Reasonable G for both traits. Positive covariance.	1.0	−1.0	0.5	0.5	0.25	0.25	−0.25
7	Opposite selection: +ve on 1, −ve on 2. Reasonable G for both traits. Negative covariance.	1.0	−1.0	0.5	0.5	−0.25	0.75	−0.75
8	Opposite selection: strong +ve on 1, weak −ve on 2. Reasonable G for both traits. Positive covariance.	1.0	−0.25	0.5	0.5	0.25	0.38	0.13
9	Selection (+ve) on both traits, but stronger on 1 than 2. Reasonable G for both traits. Negative covariance.	1.0	0.25	0.5	0.5	−0.25	0.44	−0.13

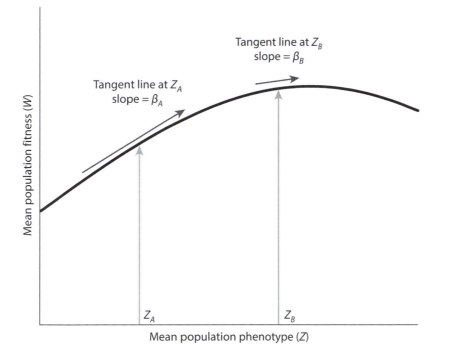

Fig 3.1. The curved line shows a hypothetical adaptive landscape with a single peak. Also shown are hypothetical selection gradients (β_A and β_B) for populations with mean phenotypes at two values (Z_A and Z_B). Selection gradients are tangent to the adaptive landscape and are generally expected to be stronger farther away from the fitness peak, as represented here by different lengths of the tangent lines. Although this illustration is for a single trait, the concept generalizes to any number of traits

(for n traits) of mean population fitness in relation to mean population phenotype.[2] That is, mean population fitness is depicted for a range of possible mean population phenotypes (assuming a particular variance), and a simple hypothetical example for a single trait is shown in figure 3.1.

Adaptive landscapes can be used to predict how the mean phenotype and mean fitness of a population will change through time. This change is determined by the direction and steepness (a vector) of the function or surface at a given location on the landscape. This vector is tangent to the landscape at that location (fig. 3.1) or, equivalently, is the first derivative of the function evaluated at that location. It points in the immediate direction of maximum steepness on the surface and has a length determined by that steepness. Conveniently, this vector can be estimated as the multivariate selection gradient (**β**) acting on the population (Lande 1976, 1979, Lande and Arnold 1983), in the manner described in the previous chapter.

---

[2]The term "adaptive landscape" has been used for a variety of related concepts (review: Schluter 2000). I here focus on Simpson's (1944, 1953) phenotype-based adaptive landscape, which was mathematically formalized by Lande (1976, 1979). This adaptive landscape is most relevant to the present book because phenotypes are critical for understanding of eco-evolutionary dynamics. Also, note that the adaptive landscape is different from the *individual fitness landscape*, which is a curve or surface for individual fitness versus individual trait values (see chapter 2).

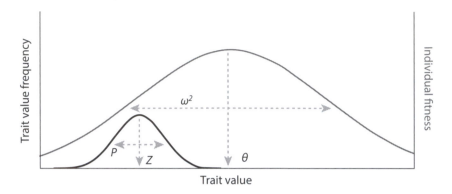

Fig 3.2. Illustration of the parameters used for estimating a selection gradient assuming a Gaussian function: $\beta = -\left(Z - \theta\right) / \left(P + \omega^2\right)$. The lower curve represents a phenotypic distribution having a given mean (Z) and variance (P). The upper curve represents an individual fitness surface having an optimum (θ) and a width (ω^2)

Selection will change as a population evolves across an adaptive landscape. Figure 3.1 provides an example by reference to a population with different mean trait values at two times (Z_A and Z_B). The tangent lines to the adaptive landscape at these locations in phenotype space will be the direction and strength of selection acting on the trait (β_A and β_B). The lines representing the βs both point to the right, showing that selection favors larger trait values at both locations/times. However, this selection is stronger at Z_A than at Z_B, as depicted with different lengths of the tangent lines.

If β is constantly changing as the mean phenotype moves across an adaptive landscape, how does one model evolution with a single equation? One solution is to employ an equation that relates β to the distance of the mean population phenotype from the optimum, with an example being

$$\beta = \frac{-\left(Z - \theta\right)}{\left(P + \omega^2\right)}$$

where Z is the mean trait value, θ is the optimal trait value, P is the phenotypic variance, and ω^2 is the strength of stabilizing selection around the optimum (Via and Lande 1985, Arnold et al. 2001) (fig. 3.2). (This equation assumes, among other things, a single optimum on an unchanging adaptive landscape and that phenotypes and stabilizing selection are Gaussian.) As ω^2 increases, the width of the fitness function around the optimum increases and the strength of stabilizing selection around the optimum decreases.

A critical distinction is that the fitness curve shown in figure 3.2 is no longer the adaptive landscape but rather the individual fitness function. What the above equation then does is to weight the individual fitness function by the phenotypic distribution in the population, thereby generating the corresponding selection gradient (β) that the population mean phenotype would experience on the adaptive landscape. Using the above equation for β in $\Delta Z = G\beta$, we can model the expected evolutionary trajectory

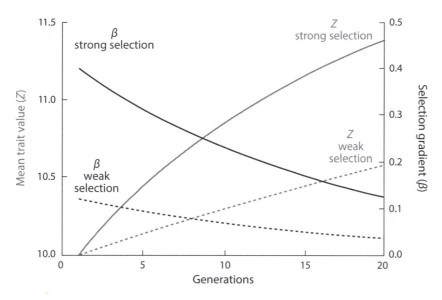

Fig 3.3. Illustration of evolutionary trajectories and selection gradients through the course of adaptation to a new optimum. The curves were produced using $\Delta Z = G\beta$, where $\beta = -\left(Z - \theta\right) / \left(P + \omega^2\right)$. In both cases illustrated (weak and strong selection), the starting phenotypic trait mean (Z) is 10, the optimum (θ) is 12, the phenotypic variance (P) is 1, and the additive genetic variance (G) is 0.3 (and therefore so too is the heritability). The difference between the two cases is that stabilizing selection (width of the fitness function, ω^2) is 4 in the strong case and 16 in the weak case

of a population that starts some distance from the optimum (fig. 3.3). This exercise confirms the expectation (fig. 3.1) that the strength of directional selection decreases as the population nears an optimum. As a result, the trajectory of evolution follows an asymptotic pattern—rapid at first but ever slower as the optimum is approached. It also confirms the expectation that stronger stabilizing selection around the optimum predicts a more rapid approach of mean phenotype to the optimum. Also made clear is the expectation that evolution is faster when the trait has more additive genetic variance (relative to its phenotypic variance) (fig. 3.4).

Beyond adaptive landscapes

The adaptive landscape is a simple and elegant construct that helps us understand the process of adaptation, but it rests on many restrictive assumptions, particularly in its mathematical formalization (Lande 1976, 1979, Fear and Price 1998, Arnold et al. 2001). These assumptions are so unrealistic at face value that some skeptics have called for the constructs complete abandonment. One criticism is that adaptive landscapes are assumed to be constant through time—but this is not problematic because temporal variation is easily incorporated (Simpson 1953b, Arnold et al. 2001, Estes and Arnold 2007). Another criticism is the absence of intraspecific interactions, which are known to fundamentally alter evolutionary trends. For instance, as a population evolves toward a fitness peak, increasing numbers of individuals compete for

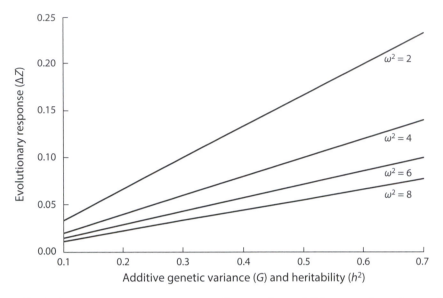

Fig 3.4. Evolutionary responses are stronger/faster when additive genetic variances (and heritabilities) are higher and when stabilizing selection is stronger. The equations used to generate the predictions are those shown in figure 3.3. In all cases, the phenotypic trait mean (Z) is 10, the phenotypic optimum (θ) is 11, and the phenotypic variance (P) is 1 (and so the heritability is the same as G). This figure is redrawn from Stockwell et al. (2003)

the resources represented by that peak. This competition effectively decreases the height of the peak (Dieckmann and Doebeli 1999, Dieckmann et al. 2004, Rueffler et al. 2006, Haller and Hendry 2014). Real adaptive landscapes are therefore like a *Sphagnum* bog (Rosenzweig 1978) or like "memory foam"—with peaks sinking ever lower as more individuals become positioned near the apex. This effect should flatten adaptive landscapes and more evenly distribute fitness among phenotypes (Haller and Hendry 2014). Similarly, traditional adaptive landscape theory largely ignores frequency dependence, and yet we know that the fitness of a given phenotype depends on the distribution of other phenotypes in the population, most obviously in game theoretic contexts (Gross 1996, Sinervo and Lively 1996, Sinervo et al. 2000, Svensson et al. 2005, Le Rouzic et al. 2015). Both of these effects are examples of the many ways in which ecological and evolutionary change interact with each other, which is—of course—the overall theme of this book; hence these and other mechanisms for such dynamics will be repeatedly revisited.

I would argue that a real adaptive landscape does exist for a given taxon at a given point in time, and this landscape implicitly includes all of the above factors that influence individual and mean population fitness. However, this "real" landscape is not easily captured by the current theory, and it is contingent not only on properties of the environment but also on properties of the populations, both of which will change through time. Indeed, this fundamental realization of interactions between environments and populations is at the heart of eco-evolutionary dynamics. I will continue to use the adaptive landscape concept, often in this less formal "real" interpretation, as a useful construct for relating phenotypes to environments.

ALTERNATIVES

The above exposition shows how the Lande equation ($\Delta \mathbf{Z} = \mathbf{G}\boldsymbol{\beta}$) can be used to predict the evolution of a population, and its success in doing so will be empirically evaluated later in this chapter. However, I first need to outline an important limitation used to argue for an alternative predictive framework (Morrissey et al. 2010, 2012). Specifically, the Lande equation assumes that the measured phenotypic association between traits and fitness reflects an underlying additive genetic association. An alternative is that other factors have generated a noncausal correlation between traits and fitness, in which case evolution will not proceed as predicted. One such disconnect can arise owing to phenotypic correlations among traits, where a focal trait that is not causally related to fitness is phenotypically (but not genetically) correlated with another trait that is causally related to fitness. Although the Lande equation is specifically designed to address such multivariate situations, it remains impossible for empiricists to measure all potentially correlated traits. Another disconnect is that environmental variation among individuals can influence both traits and fitness (Rausher 1992, Stinchcombe et al. 2002), such as when certain individuals occupy better habitats and so have larger traits and higher fitness—even though the larger traits were not the reason for the higher fitness. Stated another way, a major assumption of the Lande equation is that a phenotypic correlation between traits and fitness reflects sole causation; that is, genetically based trait variation is the *cause* of the fitness variation (Morrissey et al. 2010).

An approach suggested to obviate the above limitations is based on the "Robertson-Price Identity" (Robertson 1966, Price 1970):

$$\Delta Z = COV(z, w).$$

where $COV(z, w)$ is the additive genetic covariance between a trait z and fitness. The above causation problem is here solved because selection is estimated directly on the additive genetic component of the trait. (The Robertson-Price identity gives the same prediction as the breeder's equation when the association between traits and fitness does not differ between the genetic and phenotypic levels.) This equation predicts the change in mean breeding values; but mean phenotypes can also change for other reasons, such as phenotypic plasticity. These additional effects can be considered by expanding the Robertson-Price Identity into the "Price equation." The Price equation states that the change in mean phenotype is the sum of the change due to the additive genetic covariance between the trait and relative fitness (as above) and a "transmission bias" that includes all other factors influencing changes in mean phenotype from one generation to the next (Heywood 2005):

$$\Delta Z = COV(z, w) + E(w\Delta z).$$

In concrete terms, factors causing a transmission bias can include correlations with other traits, environmental effects (plasticity), mutation, and changes in age structure—and the transmission bias term can be expanded to explicitly separate such different effects (e.g., Heywood 2005, Coulson and Tuljapurkar 2008, Ellner et al. 2011). Although we are here discussing changes in mean phenotype, the Price equation is general and can be used to consider changes in almost anything. For instance, I later discuss its application to community structure (chapter 8) and ecosystem function (chapter 9).

So why don't we just jettison the breeder's equation and the Lande equation in favor of the Robertson-Price identity and the Price equation? The reasons are several. First, the additive genetic covariance between a trait and fitness is exceedingly difficult to measure accurately in natural populations—it generally requires multiple generations, good pedigrees, very good fitness metrics, and large sample sizes (Morrissey et al. 2012). The Price equation is therefore most useful in long-term studies where individuals and their offspring are captured, measured, genotyped, and monitored over many generations—a level of detail that is out of reach in the majority of studies. Second, although the Robertson-Price identity might do a better job of predicting evolution, it does not measure selection per se (i.e., effects of phenotypes on fitness) and so it can't by itself inform *why* changes are taking place. Third, many contributions to the transmission bias aren't known until after the change takes place, and so the approach is more useful for post hoc decomposition of the contributions to trait change (e.g., Coulson and Tuljapurkar 2008) than it is for a priori prediction of trait change. For all of these reasons, the Lande equation remains an appropriate starting point for eco-evolutionary studies—although the above alternatives also should be applied when the data permit.

Adaptation in nature

In the sections that follow, I use elements of the above-described framework to address key questions about adaptation. Several of these questions relate to patterns of phenotypic change without explicit consideration of whether or not the change is adaptive: (Q3) how fast is evolution, (Q4) how do evolutionary rates change with time scale, and (Q5) how constrained is evolution. Although most change probably is adaptive, some alternatives exist, with the most obvious one being genetic drift.[3] Other questions focus more specifically on the adaptive nature of phenotypic change. They do so by asking whether measurements of selection and genetic variation can predict (Q1) short-term or (Q2) long-term evolutionary change, or by (Q6) quantifying the degree of (mal)adaptation. Throughout the chapter, I will discuss phenotypic change as though it has a genetic basis. However, most of the cited studies were based on wild-collected individuals, where a substantial amount of the change might well be plastic. I set this ambiguity aside for now, and return to it later in the chapters on genetics (chapter 10) and phenotypic plasticity (chapter 11).

QUESTION 1: IS SHORT-TERM EVOLUTION PREDICTABLE?
Many artificial selection studies, although not all of them, have found that $R = h^2S$ or $\Delta Z = G\beta$ can reasonably predict phenotypic change under controlled conditions (Hill and Caballero 1992, Roff 2007, Hill and Kirkpatrick 2010)—but is prediction also good in the messy natural world? Postma et al. (2007) imposed artificial selection on an otherwise natural population of great tits. Over an 8 year period, one partially isolated segment of the population was exposed to negative selection on clutch size and another segment to positive selection on clutch size. The outcome was that "clutch size evolved in

---

[3] I think that drift will rarely be important for phenotypic traits of ecological interest. To paraphrase Bret Weinstein as quoted by Royte (2001, p. 300): Adaptation is a better explanation than God, but God is a better explanation than drift.

a predictable manner when under strong directional selection, but only after the effects of overlapping generations, immigration, and both fecundity and viability selection were taken into account" (Postma et al. 2007, p. 1830). Thus, the messy natural world does indeed complicate basic predictions (more about this later)—but the discrepancy is correctable with enough information about the study system.

For un-manipulated wild populations, a formal test of the predictive ability of $\Delta Z = G\beta$ was provided by Grant and Grant (1995) for medium ground finches experiencing two drought periods on the small Galápagos island of Daphne Major. The additive genetic variance-covariance matrix (G) was estimated for six beak and body dimensions by regressing offspring trait values on parental trait values. The vector of selection gradients (β) was estimated from a multiple linear regression using the six traits to explain interannual survival. These estimates were then combined to predict ΔZ for the six traits, which was then compared to actual changes by the next generation. A close association between predicted and observed changes was found, although the fit was better between drought periods than within them (fig. 3.5). In a different analysis, Grant and Grant (2002) found a strong association between predicted and observed changes in beak size (PC1) and beak shape (PC2) for two different Darwin's finch species on Daphne Major. Agreement between prediction and observation also has been found in several other studies of natural populations (Roff 2007). These findings suggest that $\Delta Z = G\beta$ provides a reliable way to predict adaptive responses to selection—even in natural populations.

Merilä et al. (2001b) came to a very different conclusion. They reviewed long-term studies of six bird and mammal populations where additive genetic variance and selection were both estimated, yielding a total of 14 trait-by-study-by-sex time series. Despite estimated directional selection on heritable traits in all cases, those traits never changed in the predicted direction. Some subsequent empirical studies have reported much the same outcome (e.g., Miehls et al. 2015). At first glance, these results suggest not only that $\Delta Z = G\beta$ fails to make good quantitative predictions, but that it even fails to make good qualitative predictions. The discrepancies between predicted and observed responses might be the result of several factors (Merilä et al. 2001b, Hadfield et al. 2011). First, G might be biased by estimation methods that fail to account for maternal effects or environmental characteristics that are shared by parents and offspring (see above). Second, selection estimates might be poor owing to statistical problems, inadequate fitness surrogates, or spatiotemporal variation (chapter 2). Third, selection might act on the environmental, rather than genetic, component of a trait (Price et al. 1988, Merilä et al. 2001b, see above). Fourth, the focal trait might be influenced by selection on correlated traits not included in the analysis (see above). Fifth, genetic responses might be masked by opposing environmental effects. For example, selection might favor larger body size but body size might not increase if competition for resources also increases (Cooke et al. 1990, Larsson et al. 1998, Hadfield et al. 2011). Sixth, agreement between prediction and observation can be poor when selection acts in a multivariate phenotypic direction that is not closely aligned with the multivariate axis of genetic variation (Roff 2007) (more about this below).

I suggest that $\Delta Z = G\beta$ will be reasonably predictive on short time scales as long as its parameters are estimated accurately and other effects, such as environmental or demographic change, are controlled or evaluated. In addition, post hoc decompositions of the contributors to trait change can reveal the factors that should be measured when attempting to make evolutionary predictions. For instance, Ozgul et al. (2009) used the

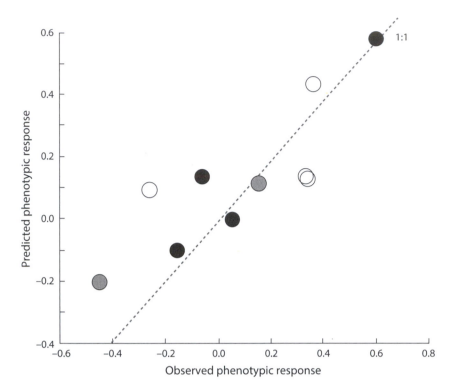

Fig 3.5. For Darwin's finches on Daphne Major, observed phenotypic responses to selection are very similar to those predicted from empirical estimates of **G** and β in the equation $\Delta \mathbf{Z} = \mathbf{G}\beta$. Observed responses are based on measurements of birds born before versus after episodes of selection. Predicted and observed values are both standardized by the phenotypic standard deviation. Data points include *G. scandens* beak size (open circles), *G. fortis* beak shape (filled gray circles), and *G. fortis* beak size (black circles). The authors used only statistically significant episodes of selection and also excluded estimates thought to be biased by introgression between species. Predictions were based on univariate estimates of selection differentials and heritabilities. This figure is redrawn from Grant and Grant (2002)

Coulson and Tuljapukar (2008) extension of the Price equation to argue that most of the change in body size for Soay sheep (*Ovis aries*) was caused by demography rather than selection. **Beyond *quantitative* prediction, it is clear that *qualitative* predictions of evolutionary change are often quite robust** as will be illustrated in the next chapter by reference to experimental introductions and studies of parallel evolution.

QUESTION 2: IS LONG-TERM EVOLUTION PREDICTABLE?

Can $R = h^2 S$ or $\Delta \mathbf{Z} = \mathbf{G}\beta$ also predict evolution on longer time scales, such as hundreds, thousands, or even millions of generations? This possibility is certainly implied in efforts to retrospectively estimate net strengths of selection necessary to cause observed changes given a **G** matrix estimated in contemporary populations (e.g., Schluter 1984). Also, It is directly tested in efforts to determine whether or not long-term evolutionary trajectories are biased by the structure of the **G** matrix (Schluter 1996a, Chenoweth et al. 2010). The specific prediction in this latter case is that evolution should be biased in the direction of the major multivariate axis of genetic variation—G_{max}. Both of these efforts rely on the assumption that selection and genetic (co)variances are reasonably constant through time. We already know this assumption is violated for selection (chapter 2) and I will now argue that it is also violated for G (see also Pigliucci 2006).

G might change through time owing to genetic drift, finite population size, mutation, or selection (Phillips et al. 2001, Agrawal et al. 2001, Steppan et al. 2002, Jones et al. 2003, 2004, 2007, Arnold et al. 2008). I will here focus on selection, which should be the most important mechanism of change (Jones et al. 2003, 2004, 2007, Arnold et al. 2008). Envision an adaptive landscape for two traits that has a single peak in the shape of an ellipse. Imagine that this landscape is colonized by a population with a broad distribution of phenotypes orthogonal to the ellipse of the adaptive landscape (fig. 3.6). Individuals with phenotypes that fall off of the high fitness ellipse will be eliminated from the population and the phenotypic distribution will therefore shrink along the maladaptive axis. At the same time, any recombination or mutation that generates new phenotypes further along the high fitness ellipse will persist. As a result of these joint effects, the phenotype distribution will evolve to line up with the adaptive landscape to an extent that depends on the steepness of the adaptive landscape (fig. 3.6). Of course, some effects can cause deviations from this adaptive expectation—but they are expected to be small if the adaptive landscape is reasonably steep (Jones et al. 2003, 2004, 2007). Overall, then, theory predicts instability of the G matrix until it becomes aligned with the adaptive landscape and it predicts stability of the G matrix thereafter only if the adaptive landscape remains stable (Jones et al. 2003, 2004, 2007, Arnold et al. 2008).

The key issue, of course, is how stable the G matrix is in real populations, which can be considered by reference to studies that have compared G between conspecific populations and closely related species. Such matrix comparisons can be implemented through several different approaches, one of which is the Flury Hierarchy (Phillips and Arnold 1999, Steppan et al. 2002, Arnold et al. 2008). This method tests whether matrices are equal (eigenvectors equal, eigenvalues equal), proportional (eigenvectors equal, eigenvalues not equal but proportional), have common principal components (eigenvectors equal, eigenvalues neither equal nor proportional), or are unrelated (eigenvectors not equal, eigenvalues neither equal nor proportional). (Eigenvectors correspond to the orientation of the matrix whereas eigenvalues correspond to its shape.) Arnold et al. (2008) summarized 20 studies that included 129 matrix comparisons based on the Flury Hierarchy (fig. 3.7). For conspecific populations, the most common outcome was that matrices were not equal or proportional but shared some principal components. For related species, the most common outcome was that matrices were equal or shared some principal components. The upshot of this review was that matrices are sometimes similar and sometimes not, and the degree of similarity varies widely. Even more to the point, G matrices can change dramatically on short time scales within populations, likely as a result of selection (Eroukhmanoff and Svensson 2011, Björklund et al. 2013).

G matrices evolve and selection changes through time. These realities together indicate that point-in-time estimates of selection and genetic (co)variance are unlikely to allow reasonable prediction of evolutionary change over more than just a few generations (see also Pigliucci 2006). An exception might occur when populations remain in stable environments for long periods of time. In such cases, populations should often be well adapted and so measured selection will be weak, which will correspond to a

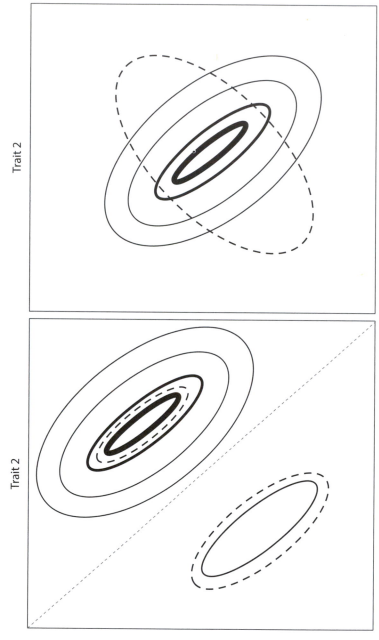

Fig 3.6. Expected evolution of the genetic variance/covariance (**G**) matrix for two traits on an adaptive landscape. The top panel illustrates a hypothetical starting situation, where orientation of the **G** matrix (dashed oval) in the colonizing population is very different from the orientation of the adaptive landscape in the new environment (solid contours, with thicker contours indicating higher fitness). The bottom panel illustrates how the **G** matrix is expected to evolve to line up with the adaptive landscape and is also expected to be narrower on a steeper adaptive landscape (above the diagonal) than on a shallower adaptive landscape (below the diagonal)

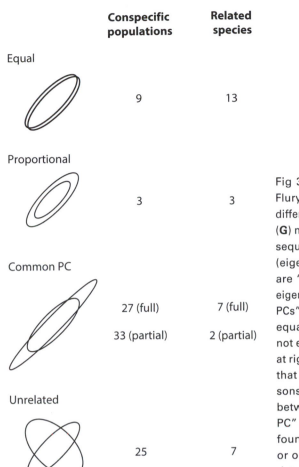

	Conspecific populations	Related species
Equal	9	13
Proportional	3	3
Common PC	27 (full)	7 (full)
	33 (partial)	2 (partial)
Unrelated	25	7

Fig 3.7. Summary of studies using the Flury Hierarchy to test for similarities and differences in genetic variance/covariance (**G**) matrices. The Flury Hierarchy tests, in sequence, whether matrices are "Equal" (eigenvectors equal, eigenvalues equal), are "Proportional" (eigenvectors equal, eigenvalues proportional), have "Common PCs" (eigenvectors equal, eigenvalues not equal), or are "Unrelated" (eigenvectors not equal, eigenvalues not equal). Shown at right are the number of published tests that fall into each category for comparisons between conspecific populations or between related species. The "Common PC" category is divided into studies that found equality for all eigenvectors (full) or only some eigenvectors (partial). The data are from Arnold et al. (2008)

prediction—and observation—of no change for long periods of time (Haller and Hendry 2014). Another exception can occur over moderate time frames when populations are introduced into new environments to which adaptation takes some time (see Question 4 below). I will return to this general question in chapter 5, where I ask whether phenotypic *divergence* among populations is biased by ancestral **G** matrices.

QUESTION 3: HOW FAST IS EVOLUTION?

Darwin (1859) seemingly argued that natural selection is very weak and evolution therefore very slow: "we see nothing of these slow changes in progress until the hand of time has marked the long lapse of ages" (p. 84) and "she can never take a leap, but must always advance by the shortest and slowest steps" (p. 194). Despite a few exceptions noted below, this view of evolution as a sedate force prevailed for the next hundred years or so. Before considering the sea-change that followed, it is worth noting that Darwin's "slow" evolution might be closer to our "rapid" evolution than is normally thought. For instance, Darwin (1859, p. 120–123) described the origin of 14 new species, as well as considerable variation within each, over less than 14,000 generations. Although we know that evolution works even more quickly than this, Darwin clearly realized that evolution could accomplish quite a bit on modest time scales.

One of the earliest and most famous examples of so-called "rapid" or "contemporary" evolution[4] was the dramatic change in frequency of melanic moths in the United Kingdom as a result of industrial pollution (Tutt 1896, Kettlewell 1973, Majerus 1998). The 1960s then saw the emergence of several more examples, all at the phenotypic level, including skeletal characters and body size in mice (*Mus musculus*) introduced to islands (Berry 1964), color and body size in house sparrows (*Passer domesticus*) introduced to North America (Johnston and Selander 1964), and pollution tolerance in several plant species on mine tailings (Jain and Bradshaw 1966). Numerous studies confirming a genetic basis for contemporary phenotypic change then emerged in the 1980s, including the color and life history of introduced guppies (Endler 1980, Reznick and Bryga 1987), the life history of introduced mosquitofish (*Gambusia affinis*) (Stearns 1983), and the body and beak size of Darwin's finches experiencing a drought (Boag 1983). Since then, many more examples have emerged of contemporary evolution in natural populations (reviews: Hendry and Kinnison 1999, 2001, Reznick and Ghalambor 2001, Stockwell et al. 2003, Hendry et al. 2008), including humans (e.g., Milot et al. 2011).

Comparative analyses of rates of phenotypic change require a common metric, and two such metrics are in wide use (Haldane 1949, Gingerich 1993, Hendry and Kinnison 1999). One—the "darwin" (Haldane 1949)—is the proportional change in the trait per million years, calculated as the natural logarithm of the mean trait value in one sample minus that in the other sample, divided by the length of time in millions of years. The other—the "haldane" (Gingerich 1993)—is the change in mean trait value in units of standard deviation per generation, calculated as the mean in one sample minus that in the other, divided by the pooled standard deviation and the number of generations. When these rates are estimated based on samples from two different times in the same population, they are considered "allochronic" (Hendry and Kinnison 1999). When they are calculated based on samples from different populations that had a common ancestor at a known time in the past, they are considered "synchronic" (Hendry and Kinnison 1999).

Large databases have now been compiled for rates of phenotypic change in natural populations (e.g., Hendry and Kinnison 1999, Kinnison and Hendry 2001, Hendry et al. 2008, Darimont et al. 2009, Westley 2011, Gotanda et al. 2015). Overall, the distribution of observed rates is roughly negative exponential, with most values being near zero and the remaining few tapering off toward higher values (fig. 3.8), as was the case for selection coefficients (fig. 2.4). In parallel with arguments advanced in the previous chapter, we might therefore conclude that **(1) phenotypic change in most natural populations is quite slow for most traits, and (2) phenotypic change in some populations can be quite fast for some traits.** As was the case for selection, however, it is hard to decide whether these rates should be considered "fast" or "slow." Given this ambiguity, comparative inferences are most informative. For example, rates of phenotypic change appear to be greater in human-disturbed situations (Hendry et al. 2008, fig. 3.9), particularly when humans act as predators (Darimont et al. 2009) but perhaps not when they cause species invasions (Westley 2011). Also, although I will not discuss it further, Kinnison

---

[4]A variety of terms might be used to describe evolution occurring over time spans of fewer than a few centuries, including "rapid," "contemporary," or "ongoing," as well as "evolution on ecological time scales." I will henceforth use "contemporary" for the reasons detailed elsewhere (Hendry and Kinnison 1999, Kinnison and Hendry 2001, Stockwell et al. 2003)

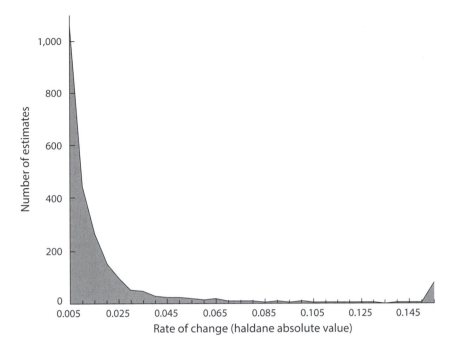

Fig 3.8. Frequency distribution of rates of phenotypic change in natural populations. Rates of change are the difference in the mean trait value divided by the standard deviation and the number of generations (i.e., the "haldane"). The data are from Hendry et al. (2008). Similar patterns are seen for various subsets of the data

and Hendry (2001) asked which types of traits (e.g., life history vs. morphology) evolved more quickly, in analogy to a similar questions regarding selection (previous chapter) and heritability (chapter 10).

QUESTION 4: HOW DO EVOLUTIONARY RATES CHANGE WITH TIME SCALE?

Do traits change in a constant and gradual fashion, as some authors attempted to caricature the "neo-Darwinian paradigm," or do they instead show long periods of stasis interspersed by brief periods of rapid change, as in the suggested alternative of "punctuated equilibrium" (Eldredge and Gould 1972, Gould and Eldredge 1977, Charlesworth et al. 1982)? A first important point is that Darwin did not hold the caricatured view (Charlesworth et al. 1982): for example, ". . . it is far more probable that each form remains for long periods unaltered, and then again undergoes modification" (Darwin 1866, p. 132). Nor did the founders of the "modern evolutionary synthesis": George Simpson, for example, described a "pattern of step-like evolution" (Simpson 1944, p. 194) involving relative stasis within "adaptive zones" punctuated by rare and relatively rapid transitions between zones.

If evolution is constant and gradual, rates of phenotypic change should remain relatively similar across different time scales: years to decades to centuries to millennia. In contrast, all analyses of the above-described databases instead find a strong negative correlation between rates of change and the length of the time interval (Gingerich 1983, 1993, 2001, Hendry and Kinnison 1999, Kinnison and Hendry 2001). However, analyses of this sort (rate vs. time) are bedeviled by a mathematical artifact that arises when time interval is present

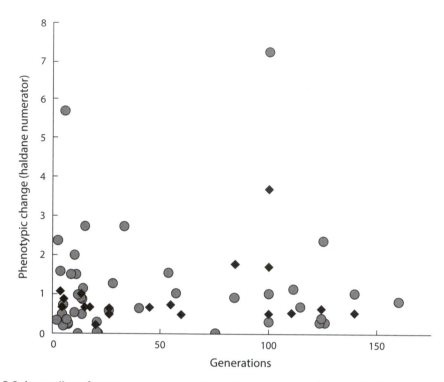

Fig 3.9. In studies of contemporary evolution, the amount of phenotypic change is (often) not correlated with the length of the time over which the change was measured, which also means that evolutionary rates are faster on shorter than on longer time scales. Each data point in the figure is the mean value (across traits and population comparisons) from a given study "system" (a unique combination of species and geographical area). The gray circles show studies that examined wild-caught individuals ("phenotypic"), whereas the black diamonds show studies that examined lab-reared individuals or used animal model approaches ("genetic"). The data are from Hendry et al. (2008)

on both the x-axis (time) and the y-axis (change over time) (Kinnison and Hendry 2001, Sheets and Mitchell 2001). One way around this problem of "spurious self-correlation" is to plot rate numerators (amount of change) versus denominators (time) (fig. 3.9). In this representation, evolution is slower over longer time intervals if the relationship between change and time levels off, perhaps reaching a plateau. Just such a pattern is present in most analyses (e.g., fig. 3.9), even to the point that simple linear regressions of the amount of change versus the length of time are generally nonsignificant over dozens to thousands of generations (Estes and Arnold 2007, Hendry et al. 2008, Uyeda et al. 2011, Westley 2011).

These whole-database analyses are not a very strong test because the data come from different species and populations from different places, and therefore do not represent true time series. A better test is to ask how the amount of phenotypic change varies with time interval for single populations sampled at multiple times. Kinnison and Hendry (2001) examined 16 such time series and found that the amount of change did increase with time in 15 of those studies but showed a pattern of diminishing returns, consistent with the idea that evolutionary rates on short time scales are not maintained on longer time scales. Two primary mechanisms could explain this result: temporal variation in

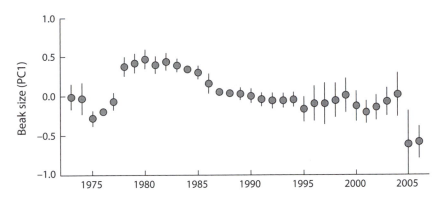

Fig 3.10. Evolution of beak size in *G. fortis* on Daphne Major over a 33-year period. Shown are the mean and 95% confidence interval for beak size of adult birds in each year. Beak size is the first principal component (PC1) of three beak dimensions: length, depth, and width. The figure is redrawn from Grant and Grant (2006)

selection and asymptotic evolution toward a new peak. (A third explanation—that genetic variation is depleted during adaptation—will be considered further in chapter 10.)

Temporal variation in selection, such as when an adaptive peak shifts around some long-term average, is certainly important in at least some instances. A clear example comes from the time series of beak size for medium ground finches on Daphne Major (Grant and Grant 2002, 2006): selection is sometimes positive and sometimes negative, with the net result being little cumulative directional change over the entire time period (fig. 3.10). Whether or not this variation in selection is a general phenomenon is still a matter of debate, as described in the previous chapter. Abrupt shifts in the phenotypic optimum, followed by initially rapid and then asymptotic evolution toward the new optimum (as shown theoretically in fig. 3.3), also appear to be important. One particularly clear example is the evolution of armor traits in stickleback exposed to different predation conditions. In particular, a well-armored population that colonizes a site with few predators evolves asymptotically toward a low-armored state in contemporary populations (Bell et al. 2004, Bell and Aguirre 2013, Lescak et al. 2015) and in fossil lineages (Bell et al. 2006, Hunt et al. 2008) (fig. 3.11). A similar pattern has been described for the evolution of increased body size in silvereyes (*Zosterops lateralis chlorocephalus*) following the formation of an island (Clegg et al. 2008). Of course, some exceptions are known: three species of invasive plant continue to evolve 200 years after introduction (Flores-Moreno et al. 2015). On much longer time scales, adaptive radiations are often characterized by rapid initial bursts followed by very little change, with an exemplar being *Anolis* lizard ecomorphs (Losos et al. 1994, Losos 2009).

Even though evolution thus seems to slow with increasing lengths of time, it must nevertheless sometimes accumulate if we are to bridge the sorts of gaps seen between mice and elephants. Uyeda et al. (2011) tackled this problem by combining databases of contemporary rates (Kinnison and Hendry 2001), paleontological rates (Gingerich 2001), and divergence among species on time-calibrated phylogenies. Two main observations emerged. First, and in confirmation of the above findings, the first two databases showed that divergence over decades to centuries is about the same as that over many thousands

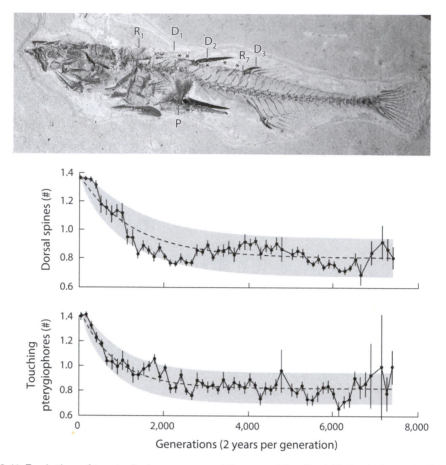

Fig 3.11. Evolution of two traits in an armored lineage of fossil stickleback that colonized a new lake environment. The top panel shows a representative fossil stickleback on which the traits were measured: D_1–D_3 indicate dorsal spines, R_1 and R_7 (R_2 – R_6 are not labeled) indicate pterygiophores (those with asterisks are "touching"—i.e., supporting the dorsal spines), and P is the pelvis. The bottom two panels show means and standard errors in each time period, and the x-axis gives the number of elapsed generations since the armored lineage first colonized the lake. The dotted lines show the best-fit adaptive models (Orstein-Uhlenbeck approach to a new optimum) and the gray areas their 95% probability envelopes. The data were originally from Bell et al. (2006) and the model fitting was done by Hunt et al. (2008), from which this figure is modified

of years—evolution is slower over longer time intervals. Estes and Arnold (2007) earlier came to the same conclusion and suggested conformation to a "displaced optimum" evolutionary model: the optimum phenotype is rather constant for a long time and then shifts abruptly to a new optimum (i.e., the second mechanistic explanation from the above paragraph). Hendry (2007) speculated that a model of fluctuating selection in the optimum (i.e., the first mechanistic explanation above) would yield a similar result and Uyeda et al.'s (2011) more recent analysis leaned in that direction. Adding the third database revealed that divergence over even longer time scales (starting at about a million years of elapsed time) can be much greater than that over short time scales, meaning that bursts of evolutionary change characterize divergence among species. Of all the models that Uyeda et al. (2011) fit to the three data sets combined, multiple shifts in the optimum provide the best fit to the data.

Evolution is typically characterized by long periods of relative stasis separated by shorter periods of rapid adaptation. As an extension, macroevolution appears to be microevolution not simply "writ large" but rather "writ in fits and starts" (Kinnison and Hendry 2001). Although this conclusion agrees with the punctuated equilibrium pattern originally espoused as an alternative to neo-Darwinism, just such a pattern has always been a part of the Darwinian perspective.

QUESTION 5: HOW CONSTRAINED IS ADAPTATION?[5]

Although evolution can accomplish remarkable change, and can sometimes do so quickly, it isn't omnipotent. Constraints on adaptation might occur for several reasons. (1) Transitional states between current phenotypes and better-adapted phenotypes might have low fitness: that is, fitness valleys might be present on the adaptive landscape (chapter 4). (2) Gene flow among populations might compromise their capacity for independent evolution (chapter 5). (3) Strong selection can have demographic costs that can cause extinction (chapter 7). (4) Genetic variation might be limited in the direction of selection. The first three possibilities are considered in the indicated chapters, whereas I here take up the last.

Genetic variation might limit adaptation in several ways. First, a particular trait under selection might have a low heritability. As will be seen in chapter 10, this constraint is unlikely to be severe in most cases given the near ubiquity of standing genetic variation in fitness-related traits. When standing genetic variation is limited, however, adaptation will have to wait for the production of new variants, and this can take some time. Second, traits can be genetically correlated in ways that limit their independent evolution.[6] These genetic correlations can arise and be maintained through (1) epistatic interactions among genes, (2) genes with pleiotropic effects, and (3) linkage disequilibrium between alleles at loci influencing different traits (Lynch and Walsh 1998). Some authors have argued that such correlations can substantially alter evolution in response to selection (Blows and Hoffmann 2005, Hansen and Houle 2008, Kirkpatrick 2009, Walsh and Blows 2009).

Meta-analyses reveal frequent, and sometimes strong, genetic correlations among traits (Roff 1996), suggesting the potential for substantial impacts on evolutionary trajectories. Whether or not these correlations impede evolution depends on how much genetic variation is present in the multivariate direction of selection (Hellmann and Pineda-Krch 2007, Hansen and Houle 2008, Kirkpatrick 2009, Walsh and Blows 2009, Agrawal and Stinchcombe 2009). Stated another way, we need to know how well the multivariate vector of selection lines up with the multivariate axis of genetic variation. Based on this logic, Teplitsky et al. (2014) estimated that the evolutionary response of 10 populations of birds in seven species was constrained by multivariate correlations among traits by an average of about 28%. Setting aside whether or not this is a noteworthy constraint,

---

[5]This section is adapted from that published in Hendry (2013).

[6]In the "evo-devo" literature, ties between trait correlations and evolutionary potential (often called evolvability) are considered in the context of "modularity," where correlations are strong between traits within a module but weak between traits in different modules (Wagner and Altenberg 1996, Hansen 2003).

we can ask is this result general? Agrawal and Stinchcombe (2009) surveyed studies that measured genetic or phenotypic (co)variances among traits, as well as selection acting on those traits. They then estimated the rate of adaptation (increase in mean fitness) in the presence of the measured correlations relative to their hypothetical absence. The upshot was that genetic correlations among traits were sometimes expected to influence the rate of adaptation, and this influence was as frequently positive (speeds adaptation) as it was negative (slows adaptation) (fig. 3.12). The reason trait correlations often aided evolution was that the axis of selection was often aligned with a major axis of genetic variation. Hansen and Houle (2008) proposed an alternative metric of constraint: a comparison of evolvability (potential rate of evolution) in directions where evolution happened relative to directions where it didn't happen: i.e., "conditional evolvability." They find the former to be often greater than the latter, which can imply a constraint. I find this metric less useful because what matters is how evolution proceeds in relation to the direction of selection and (separately) because genetic covariances can be shaped by selection (evolution might have occurred on an axis of low variation which then became an axis of high variation).

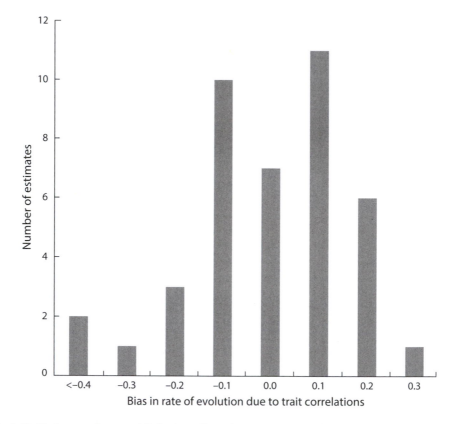

Fig 3.12. Estimates from published studies of the extent to which the rate of evolution will be biased by trait correlations. The x-axis values are the log of the ratio of (1) the rate accounting for trait correlations, to (2) the rate not accounting for trait correlations. Negative values correspond to situations where correlations decrease the rate of evolution, whereas positive values correspond to situations where correlations increase the rate of evolution. The data are from Agrawal and Stinchcombe (2009) and were provided by A. Agrawal

The above analyses consider trait correlations on a pair-wise basis, whereas we would ideally consider the n-dimensional multivariate space representing all traits (Bürger 1986, Blows and Hoffmann 2005, Hansen and Houle 2008, Kirkpatrick 2009, Walsh and Blows 2009, Brommer 2014). One approach to this problem is to measure the matrix of additive genetic (co)variances for traits, and then use the resulting **G** matrix to estimate the number of effectively independent trait dimensions that could respond to selection (eigenvectors or principal components of the matrix). Some studies adopting this approach have reported only a few effective trait dimensions, which argues that correlations could substantially constrain adaptive evolution (Kirkpatrick 2009, Walsh and Blows 2009). Other studies, however, suggest that the number of dimensions can be reasonably high (Mezey and Houle 2005). It seems to me that the number of dimensions must nearly always be high, certainly much higher than suggested by the current analyses of suites of very similar traits, such as cuticular hydrocarbons (Blows et al. 2004) or wing shape (Mezey and Houle 2005, McGuigan and Blows 2007). The reason is that overall adaptation to a given environment will inevitably involve a host of morphological, life history, physiological, and behavioral changes, which will not collapse down to only a few dimensions. Of course, this assertion remains to be confirmed through formal analysis.

Although genetic constraints are certainly present in some situations, and these constraints can slow the rate of adaptation, such constraints are not universal and might not even be common. Overall, I suggest that limitations to adaptation owing to patterns of genetic variation are probably not that overwhelming, at least not in the short term. The other constraints listed above are probably more pervasive and important, as will be discussed in other chapters of this book.

QUESTION 6: HOW WELL ADAPTED ARE POPULATIONS TO THEIR LOCAL ENVIRONMENTS?

Although the above questions evaluated the process of adaptation, and although I have several times asserted that most populations should be well adapted, I have yet to formally evaluate that assertion. At one extreme, most populations might be well adapted to their environments, with their phenotypes positioned close to a fitness peak (Simpson 1944, 1953, Cain 1964, Endler 1986, Schluter 2000a, Estes and Arnold 2007, Hendry in Hendry and Gonzalez 2008). On the other hand, adaptation might be so strongly constrained by other factors that most populations have phenotypes some distance from the nearest fitness peak (Gould and Lewontin 1979, Crespi 2000, Gonzalez in Hendry and Gonzalez 2008). The truth, as is so often the case, certainly lies somewhere between the two extremes, with different populations falling at different places along a (mal)adaptation continuum. The key is to find an objective way to determine how far populations are from local optima and what are the resulting fitness consequences (Crespi 2000, Hendry and Gonzalez 2008).

As outlined by Hendry and Gonzalez (2008), the degree of (mal)adaptation of a population can be quantified with respect to overall fitness (height along the y-axis of the adaptive landscape in fig. 3.1) or with respect to trait values (position along the x-axis on the adaptive landscape in fig. 3.1). (Of course, I here mean a nearby local fitness peak—not some overall global theoretical maximum.) In the first case, we might compare the average fitness of individuals in a population to the average fitness they would enjoy if optimally adapted to the local fitness peak. In the second case, we might compare the observed average trait value in a population to the average trait value that would correspond to the local fitness peak.

It is very hard to quantify (mal)adaptation in natural populations from the perspective of *fitness*. Problems including the logistical challenge of accurately quantifying fitness, temporal variation in environments, the confounding influence of density- and frequency-dependence, and the difficulty of implementing replication and controls. As a qualitative example, however, it is obvious that maladaptation must be the reason for range limits in many species (Bridle and Vines 2007). Similarly, maladaptation must be the ultimate cause of population declines in the face of disturbances such as global warming, pollution, invasive species, and harvesting (Bradshaw and McNeilly 1991, Both et al. 2006, Futuyma 2010). A corollary to these two points is that the continued persistence of most populations implies they are at least reasonably well adapted to their local environments: that is, they aren't so far from local adaptive peaks that their fitness is persistently below replacement. Moreover, several studies have shown that fitness improves rapidly in natural populations as they adapt to new environments (Kinnison et al. 2008, Gordon et al. 2009, chapter 7). Logic thus simultaneously dictates that most extant populations are currently well adapted (because they are extant) but that a time will come when this is not the case (because most populations that existed in the past have been extirpated).

Quantifying (mal)adaptation in natural populations from the perspective of *traits* is simpler and can proceed along several different paths. The three that I summarize here are phenotypic manipulations, variation among populations, and selection-based estimates of the adaptive landscape.

Phenotypic manipulations can be used to test how deviations from typical trait values might reduce fitness. As an example, Sinervo et al. (1992) evaluated adaptation in the size of lizard (*Uta stansburiana*) hatchlings by removing yolk from the eggs of some females (thus decreasing hatchling size) and surgically removed eggs from other females (thus increasing size of the remaining hatchlings). No meta-analyses have been performed of the fitness consequences of such phenotypic manipulations, and so I will have to rely on examples. On the one hand, phenotypic manipulations sometimes reveal that typical trait values yield higher fitness than manipulated trait values. Examples include the above lizard hatchling study, and also flowering time in some plants (O'Neil 1999) and reproductive life span in salmon (Hendry et al. 2004a). On the other hand, some phenotypic manipulations suggest higher fitness for atypical phenotypes. The most striking examples come from preferences of females for male trait values beyond the range seen in natural populations, including tail length in some birds (Andersson 1982) and fishes (Basolo 1990), gonopodium length in some fishes (Langerhans et al. 2005), and novel call elements in some frogs (Ryan et al. 1990). However, these results don't reveal much about overall adaptation because female choice is only one of many contributors to fitness. As a result, the apparent sexual benefit of exaggerated traits might well be counterbalanced by opposing natural selection, as seems to be the case owing to predation in at least some of the above examples (Rosenthal et al. 2001, Langerhans et al. 2005). Alternatively, exaggerated traits might well be adaptive but simply haven't yet arisen in the population—implying a genetic constraint on adaptation. In short, no definitive general insight emerges from phenotypic manipulations.

Variation among populations can be used to infer trait (mal)adaptation by first determining the trait values that should be optimal in a given environment, and then measuring the extent to which particular populations deviate from that expectation. One way to estimate optima is to consider specific populations not expected to be constrained short of adaptation. This approach is frequently applied in studies of the extent to which adaptation is constrained by gene flow (more in chapter 5). Optimal trait values in a given environment are here assumed to be those in the most divergent populations or in populations subject to the lowest gene flow. Using this approach, substantial maladaptation has been inferred for particular populations of mosquitofish (Stearns and Sage 1980), *Agelenopsis aperta* spiders (Riechert 1993), *Timema cristinae* stick insects (Bolnick and Nosil 2007), and stickleback (Hendry and Taylor 2004, Moore et al. 2007). A point of uncertainty in a number of these efforts is the need to assume the same optimum for all populations in a given environment, which is not necessarily true (see the discussion of nonparallel evolution chapter 4). Moreover, these studies do not involve a random selection of all possible populations—but rather focus specifically on those that already appear maladapted. Such studies cannot provide insight into the overall prevalence and strength of adaptation in nature—they are instead illustrative examples of special cases.

Selection-based estimates of (mal)adaptation use measurements of linear and quadratic selection to estimate the shape of the adaptive landscape in the vicinity of a population. This shape can be used to estimate the location of the nearest fitness peak if (1) selection estimates indicate negative curvature consistent with a nearby peak, and (2) one assumes a particular landscape shape, such as Gaussian (Estes and Arnold 2007). Applying this method to a trimmed version of the Kingsolver et al. (2001) selection database, Estes and Arnold (2007, p. 241) concluded that "phenotypic means are typically very close to the adaptive peak (46% are within 1 phenotypic standard deviation of the optimum, and 65% are within 2 standard deviations . . ."). Aside from the open question as to whether these particular milestones really should be interpreted as being "very close" to adaptive peaks, several limitations need to be kept in mind. First, not only must one assume a particular landscape shape, but the selection estimates need to be precise and accurate, which we have seen is hard to accomplish (chapter 2). In addition, the degree to which a particular trait deviates from its optimum (x-axis in fig. 3.1) does not necessarily reveal the fitness cost (y-axis in fig. 3.1) (Hendry and Gonzalez 2008). Indeed, we might expect substantial trait maladaptation in stable populations only when the trait is *not* closely related to fitness.

Populations and traits will differ dramatically in the degree to which they are locally (mal)adapted but, at present, we have no objective and reliable assessment of the frequency and magnitude of (mal)adaptation in nature. This question is nevertheless central to evolution and conservation and it therefore should be a major focus of future efforts (Crespi 2000, Hendry and Gonzalez 2008).

Conclusions, significance, and implications

In this chapter, I argued that (1) evolution can be reasonably predictable on the short term but not the long term, (2) evolution is usually slow but can sometimes be comparatively rapid, (3) evolution is faster on short than long time scales and is characterized by abrupt spurts separated by periods of relative stasis, (4) evolution is sometimes—but perhaps not often—constrained by genetic variation and trait correlations, and (5) most populations are reasonably well adapted for their environments—although the frequency and severity of (mal)adaptation in natural populations is not yet clear. All of these results conform nicely to Simpson's (1944, 1953) view of "adaptive zones" (fig. 3.13). That is, selection and evolution usually bounce subtly (but still rapidly on short time scales) back and forth

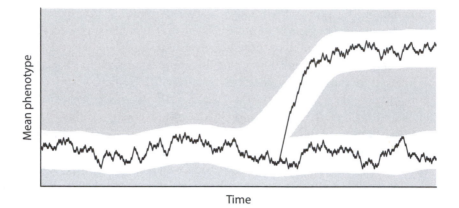

Fig 3.13. An illustration of Simpson's (1944, 1953) adaptive zones. Shaded areas show zones of low fitness whereas white areas show zones of high fitness (adaptive zones). The jagged line shows the mean phenotype of a species as it evolves through time. Environmental and other sources of variation cause phenotypes to routinely vary, but within the bounds of adaptive zones. Only occasionally does a route to a new adaptive zone open up, which then causes evolution that is faster and more consistent (directional) until a new adaptive zone is reached. Adaptation of a population to a new adaptive zone can lead to the formation of a new species. This speciation event is illustrated here by the split into two mean phenotypes, one of which adapts to the new zone and one of which remains in the original zone. This figure was produced by Ben Haller

within adaptive zones and populations are generally well adapted to those environments. An extension of this idea is "niche conservatism," wherein species retain their niches for long periods of time (Webb et al. 2002, Wiens and Graham 2005). Occasionally, however, environmental change causes abrupt shifts in trait optima that result in maladaptation and impose strong selection that—if constraints are not strong—allow rapid evolutionary shifts. When these shifts are extreme, populations can enter new adaptive zones, a likely route to speciation and adaptive radiation (chapter 6).

This view of evolution reinforces the previous chapter's implications for eco-evolutionary dynamics. First, the pragmatic way to study eco-evolutionary dynamics will be to focus on populations/periods where environmental change is greatest. In such cases, substantial phenotypic changes occur and thus could have profound influences for ecological processes at the population, community, and ecosystem levels. Second, these instances might be a relatively rare component of eco-evolutionary dynamics, which instead might be even more important in promoting eco-evolutionary stability.

Chapter 4

Adaptive Divergence

The previous chapter outlined the process of adaptation within populations. By extension, the adaptation of different populations to different environments[1] causes "adaptive divergence." This divergence involves the evolution of phenotypic traits that improve fitness for individuals in their home environments relative to individuals from other environments (Endler 1986, Schluter 2000a, Kawecki and Ebert 2004, Hendry and Gonzalez 2008). This adaptive divergence can then contribute to the evolution of reproductive isolation among populations, a process called "ecological speciation" (Schluter 2000a, Rundle and Nosil 2005, Nosil 2012). If adaptive divergence and reproductive isolation are substantial, the different populations are often considered to have attained the status of different species, and we have the beginnings of an "adaptive radiation."[2] The present chapter focuses on the first part of this process, adaptive divergence, whereas chapter 6 will focus on the second part, ecological speciation. Adaptive radiation as a whole, and in particular taxa, is the focus of excellent books by Lack (1947), Simpson (1953), Schluter (2000), Grant and Grant (2008), and Losos (2009).

The importance of adaptive divergence to eco-evolutionary dynamics is made clear by considering the causal effects of ecology on evolution and vice versa. In the first direction (eco-to-evo), many related species clearly differ from each other in having adapted to different environments, and so environmental differences have been a substantial part of evolutionary diversification. In the second direction (evo-to-eco), different species clearly have different ecological effects. Remembering that all species are the product of past evolution, adaptive divergence is a key foundation of the great diversity of species' ecological effects. In short, differences in selection among environments, and the resulting adaptive divergence of populations and species, are the root cause of the evolution of biological diversity and all of its attendant ecological effects.

---

[1] In this book, different "environments" are broadly intended to mean any combination of biotic or abiotic conditions, and the term therefore subsumes "resources" and "habitats." I sometimes use these more specific terms when they are better suited to a given discussion but "environments" will be the general term.

[2] Schluter (2000) asserted that "adaptive radiation" should be reserved for situations where diversification has been particularly rapid. Although knowing whether or not an adaptive radiation was rapid is certainly interesting, and while most adaptive radiation could well be rapid, I see no reason why slower radiations could not be also considered adaptive.

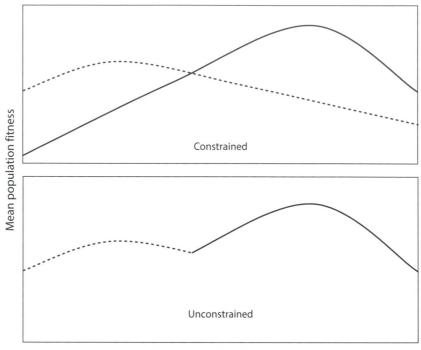

Fig 4.1. Alternative views of a phenotypic adaptive landscape with two optima. In the constrained landscape (top panel), individuals using one environment continue to use that environment even if their phenotypes would be better suited for the other environment. In the unconstrained landscape (bottom panel), individuals use the environment for which their phenotypes are best suited. In both panels, the curves give mean population fitness for a given mean population phenotype. The figure is modified from Hendry and Gonzalez (2008)

My first goal in the present chapter is to outline an established conceptual model for how adaptive divergence proceeds: this model builds on the preceding chapter's introduction of the adaptive landscape. My second goal is to discuss how adaptive divergence is empirically studied, focusing as always on natural populations. My final goal, as in all chapters, is to present several key questions relevant to adaptive divergence and to then provide some answers, which are sure to be debatable.

Adaptive landscapes with multiple peaks

A profitable way to conceptualize adaptive divergence is to return to the adaptive landscape introduced in the previous chapter. Instead of a single fitness peak, however, now consider multiple peaks that correspond to alternative combinations of mean phenotypes that yield high fitness. These different peaks can arise owing to alternative environments (i.e., alternative niches[3]) that are best exploited by different phenotypes. Alternatively, the

---

[3]Many uses of the term "niche" exist in the literature. I use will use the term in the sense usually intended by evolutionary biologists: a particular set of resources or environments that might be occupied by a single species (or population) having a particular phenotype. For more details, see Schluter (2000, p. 19).

different peaks can arise when the same environment can yield high fitness for multiple alternative phenotypes. In the former case, the adaptation of a population to one peak would not cause the other peaks to disappear. In the latter case, by contrast, adaptation to one peak could well mean disappearance of the other peak. With some exceptions, the discussion of alternative peaks in this book refers to the first cause: alternative niches.

I find it useful to conceptually discriminate between two ways of viewing these landscapes. A "constrained" view shows a different landscape (curve or surface) for each environment, resource, or location. Thus, in fig. 4.1, two environments are present and the relationship between trait values and fitness is illustrated separately for each environment. The reason this view is useful, and why I call it constrained, is that the fitness of a population with a given phenotype depends on the environment experienced (or the resource used) by that population. Thus, a population in the dashed-line environment with a too-large trait value for that environment might have higher fitness if it switched to the solid-line environment, but this might not be possible owing to spatial structure (perhaps only the dashed-line environment is available locally) or properties intrinsic to the organism (perhaps individuals are philopatric to a given location or imprint on a particular environment). An "unconstrained" view, which is more typical in the literature, represents the fitness of a given mean phenotype in the environment for which that phenotype would have highest fitness—that is, ignoring whether or not populations switch among environments in an optimal fashion. In addition to the above distinction, adaptive landscapes can be considered locally (e.g., all of the peaks available in a particular geographical area where the organism is found) or globally (e.g., all of the peaks available across geographical areas regardless of whether or not the organism is found there) (Hendry et al. 2012).

Adaptive landscapes can be considered also with respect to their topography and dimensionality (Hendry et al. 2012). By topography, I mean the number, position, gradient, and elevation of surface features such as peaks, ridges, moats, pits, and valleys. By dimensionality, I mean the number of traits on which selection is acting (Nosil et al. 2009b, McPherson et al. 2015). Consideration of these properties leads to many interesting questions, such as how many peaks are out there, how many have been colonized (occupied by a well-adapted population), and how do populations colonize new peaks?

How are valleys bridged?

Adaptive divergence and adaptive radiation depend on the ability of a single starting population to colonize and diverge onto multiple peaks, thus realizing divergent evolutionary trajectories. The potential for this diversification should be enhanced by "rugged"[4] adaptive landscapes composed of many distinct peaks separated by deep and wide fitness valleys (Schluter 2000a). The number of peaks is important because it dictates the maximum number of phenotypically divergent populations that might arise. The depth and width of valleys is important because these properties influence the potential for gene flow to reduce divergence (chapter 5) and for divergence to reduce gene flow (chapter 6). For instance, a deep fitness valley between peaks reduces gene

---

[4]"Rugged" is sometimes used simply to imply a large number of peaks—irrespective of the topography around them. However, different topographies will have different effects on evolution for a given number of peaks. I will therefore use "rugged" to mean not only many peaks but also deep valleys between them.

flow and enhances the opportunity for populations to fully diverge onto those different peaks and become reproductively isolated. The hitch is that deep and wide fitness valleys can make it more difficult for alternative peaks to be colonized in the first place. That is, a population adapted to one peak would have low fitness during any transition toward another peak. In essence, then, rugged adaptive landscapes both promote and constrain adaptive divergence. So a key question becomes: how can fitness valleys be crossed such that an ancestral population adapted to one peak is able to bud off a new population adapted to a different peak? Three solutions have been suggested: a reduction in the depth of existing fitness valleys, the bridging of fitness valleys even if they do not become less deep, and the creation of new valleys under an existing population. These mechanisms are mostly discussed in relation to a single population transitioning from one peak to another—and so could have been described in the previous chapter. However, they are also critical in the current context of adaptive divergence.

Several effects can decrease the depth of existing fitness valleys. One effect involves spatial variation in adaptive landscapes, where different populations of a given species colonize different places with different environments. In such cases, the original peak disappears for the colonizing population, leaving them in the domain of attraction of a new peak. Clear examples of this process come from instances where populations from one location are introduced into new locations (reviews: Hendry and Kinnison 1999, Reznick and Ghalambor 2001, Cox 2004, Strauss et al. 2006). Another effect invokes temporal variation in adaptive landscapes. For instance, a key resource could become scarce in a given location, leaving the population in the domain of attraction of a different peak. Once on this new peak, the population might persist even if the original resource eventually recovers. Or an intermediate resource may temporarily appear, thus erasing the valley long enough for a peak shift to occur. A third scenario invokes an increase in phenotypic variance (Kirkpatrick 1982, Whitlock 1995)—because greater variance causes adaptive landscapes to become smoother.

Several other effects can allow populations to bridge fitness valleys that have not become shallower. For instance, genetic drift might allow populations to leap by chance to a new fitness peak, although this is unlikely unless the population is very small and the fitness valleys narrow and shallow.[5] Another possible effect is macro-mutation, although such "hopeful monsters" (Goldschmidt 1940) encounter at least two major difficulties. First, most large-effect mutations will be deleterious—either intrinsically by causing genetic or developmental problems or extrinsically because the new phenotypes will only rarely hit a new peak. Second, new mutations will be initially rare and therefore susceptible to loss through drift and, in sexual organisms, a rarity of similar individuals with which to mate. A third effect that can cause large jumps in phenotype is hybridization or changes in ploidy, which can reduce some of the above problems in that more individuals can be affected and because genes present as standing genetic variation have a previous history of success in natural environments. Indeed, hybridization and polyploidy are now thought to have generated many forms adapted to new environments (Arnold 1997, Rieseberg et al. 2003, Seehausen 2004, Mallet 2007, Roy et al. 2015). Fourth, sexual selection can potentially bridge fitness valleys caused by natural selection (Bonduriansky 2011).

---

[5]The role of drift in adaptive radiation has received considerable attention in several contexts that I don't have space to detail, including Sewall Wright's shifting balance theory (Coyne et al. 1997, Wade and Goodnight 1998), variance-induced peak shifts (Whitlock 1995), and various forms of founder effect speciation (Templeton 2008).

Finally, new fitness valleys can sometimes emerge under existing populations, potentially driving their adaptive divergence. In particular, competition for shared resources can impose disruptive selection that favors extreme phenotypes within a single starting population (Dieckmann and Doebeli 1999, Rueffler et al. 2006). If positive assortative mating is present, the result can be multiple populations potentially en route to becoming separate species (more details are in chapter 6). Stated another way, a single adaptive peak can split through competition into two adaptive peaks, although this process is easiest when the resource distribution is bimodal to start with (Doebeli 1996, Thibert-Plante and Hendry 2011a). Beyond competition, changing environments can cause a population's current peak to disappear from the local landscape, leaving different parts of the population in the domains of attraction of different peaks, such as when warming climate leads to the extirpation of a phytophagous insect's preferred host plant making less preferred host plants the only remaining option. The result then could be adaptive divergence of different components of the original insect population to specialize on different host plants.

So many mechanisms exist for how populations might shift between adaptive peaks (for yet another see Price et al. 1993) that one wonders why it was such a concern in the first place. Of course, this doesn't mean that all possible peaks can be colonized: some simply will be too far away for evolution to achieve. Nevertheless, I am continually impressed by what appear to be populations on those lonely mountains, with phenotypes we might not have imagined possible were it not for their existence. Even familiar animals such as hippos, giraffes, and elephants are so distinct from other extant organisms that they would have been difficult to predict if they did not already exist, or if related forms were not known from the fossil record. Just how often fitness valleys are a major impediment is therefore not known.

How to infer adaptive divergence?

Several methods can be used to infer the presence and strength of adaptive divergence. These methods are variously applied to two types of inference: divergence in traits versus divergence in overall adaptation or fitness (Endler 1986, Hendry and Gonzalez 2008). In the first case, the goal is to evaluate whether specific trait differences among populations are the result of adaptation to different environments. In the second case, the goal is to evaluate how all integrated (and often unmeasured) aspects of phenotype improve *fitness* in local environments. On another level, some methods consider adaptive divergence between populations irrespective of whether or not a fitness valley is present between them, whereas other methods can inform the presence and depth of any such valleys. This distinction is important because diversification along an adaptive ridge (i.e., no fitness valleys between high-fitness phenotypes) will proceed quite differently from diversification among distinct adaptive peaks (i.e., fitness valleys present between high-fitness phenotypes). On ridges, strong adaptive correlations can be present between traits and environments—but transitions from one location on the ridge to another might not be opposed by selection (Schluter 2000a, Gavrilets 2004). By contrast, transitions between distinct peaks can be opposed by selection—as described above.

TRAIT-ENVIRONMENT CORRELATIONS

The classic (e.g., Cain and Sheppard 1950), and still common (Endler 1986, Schluter 2000a, MacColl 2011), method for inferring the adaptive significance of trait divergence is to test for associations between mean phenotypes and environmental conditions. It helps if these associations are predictable, such as when divergence matches clear a priori expectations: for example, increased parasitism should select for increased parasite resistance (but see Dargent et al. 2013). It also helps if they are repeatable, such as when independently derived populations converge on similar phenotypes in similar environments: that is, "parallel" or "convergent" evolution (Endler 1986, Schluter 2000a, Arendt and Reznick 2008, Losos 2011). As an example, body size increases with latitude among *Drosophila subobscura* populations in Europe, North America, and South America—the latter two being introduced ranges (Gilchrist et al. 2004). Repeatable trends such as these are almost certainly adaptive, even if the specific cause is not clearly established.

Trait–environment correlations are often based on samples collected from natural populations ("wild-collected"). In this case, inferences about adaptive divergence are weakened owing to uncertainty regarding the genetic versus plastic basis for observed trends. This uncertainty can complicate interpretation in several ways (details in chapter 11). First, phenotypic *differences* among natural populations can arise even when they are genetically *similar*, owing to different environmentally induced plastic effects. In a classic example, James (1983) showed that geographical variation in the morphology of nestling birds was due mainly to environmental, rather than genetic, effects. Second, and opposite to the first, phenotypic *similarity* among natural populations might reflect genetic *differences*, such as in the case of counter-gradient variation (Conover and Schultz 1995). For example, similar body sizes across latitudes in wild Atlantic silversides (*Menidia menidia*) result from higher intrinsic growth rates in the north (genetic effect) offsetting the shorter growing season (the opposing environmental effect) (Conover and Present 1990). The solution to disentangling such effects is common-garden or reciprocal transplant experiments, with the former being used in the above work on *Drosophila* (Gilchrist et al. 2004) and silversides (Conover and Present 1990), and the latter being used in the above work on nestling birds (James 1983). These types of experiments will be considered further below and in subsequent chapters.

METHOD 2: RECIPROCAL TRANSPLANT EXPERIMENTS

Following Clausen et al. (1940), the litmus test for overall adaptive divergence long has been whether or not fitness in a given environment is higher for "local" individuals from that environment than it is for "nonlocal" individuals originating from other environments (Schluter 2000a, Kawecki and Ebert 2004, Hereford 2009). This sort of comparison is best implemented through reciprocal transplant experiments, where the fitness of individuals from multiple environments is assessed in each of those environments. In some cases, these experiments can be performed in the laboratory by testing representatives from each population in treatments that mimic their natural environments. For instance, populations living in water of low or high acidity in the wild can be tested under high and low acidity in the laboratory (e.g., Räsänen et al. 2003, Derry and Arnott 2007). These laboratory experiments can reveal adaptation to a particular environmental factor, or sometimes multiple factors (e.g., acidity crossed with temperature), whereas field experiments assess overall adaptation to all of the environmental differences in nature.

An exemplar for reciprocal transplants in nature comes from studies of phenotypically divergent coastal and inland subspecies of the California annual plant *Gilia capitata* (Nagy 1997, Nagy and Rice 1997). The authors transplanted individuals of each subspecies between inland and coastal sites, finding that local types outperformed nonlocal types with respect to most fitness measures, including the probability of emergence, the probability of flowering, and the number of inflourescences. Overall adaptive divergence was thus inferred to be strong. In addition to reciprocal transplants, the authors used several other methods for inferring adaptive divergence, and so I will return to *G. capitata* several times in the sections that follow.

Several issues attend the implementation and interpretation of reciprocal transplant experiments. One is the source of experimental material. Many experiments use wild-collected individuals, in which case fitness differences after transplantation could reflect phenotypic plasticity, as opposed to genetically based adaptive divergence. An alternative is to use individuals raised for their entire lives in a common-garden environment. Although this approach controls for a number of environmentally induced effects, multiple generations in the common garden are sometimes needed (Bernardo 1996, Plaistow et al. 2006). In addition, interactions between the genetics of organisms and properties of the environment in which they are raised (genotype-by-environment interactions, GxE) can mean that phenotypic and fitness differences in the experiment depend on the particular common garden (chapter 11). As a result, the use of laboratory-reared individuals in field experiments might alter patterns of fitness from those that would have occurred in nature. Other issues include the aforementioned concerns (chapter 2) about the suitability of different fitness metrics (more inclusive surrogates are better) and spatiotemporal variation (more replication is better). An additional complication is coevolution,[6] where local individuals might not have an advantage over nonlocal individuals because biotic environments (competitors, predators, parasites, or pathogens) are themselves "adapted" to the local individuals being tested (Kaltz and Shykoff 1998, Hoeksema and Forde 2008). Methods to disentangle these effects have been suggested in the context of host-parasite coevolution (Nuismer and Gandon 2008).

METHOD 3: PHENOTYPIC MANIPULATIONS

Phenotypic manipulations increase the range of variation and thus increase the potential for detecting selection and inferring the extent to which existing phenotypes are well adapted. These manipulations are also informative in the context of adaptive divergence. In particular, the generation of intermediate phenotypes, and the assessment of their fitness, can inform the extent to which different populations are separated by a fitness valley—as opposed to a fitness ridge (Schluter 2000a). One way to generate intermediate phenotypes is to directly or digitally manipulate the trait values of individuals, as previously noted for egg size in lizards, flowering time in plants, tail length in birds, and gonopodium length in fishes (chapter 3). Another method is hybridization, because hybrids are often, although not always, phenotypically intermediate between parental forms.

Once intermediate phenotypes are generated, the key question is whether or not they have lower fitness than parental phenotypes, and whether any such fitness reduction is the result of adaptation, as opposed to intrinsic genetic problems. To inform this last question, a common approach is to assess the fitness of hybrids and parentals in benign

---

[6]By "co-evolution," I mean the evolution of one species influences evolution of another species and vice versa. For a detailed history, summary, and critique of this idea and its extensions, see Carmona et al. (2015).

laboratory conditions versus in nature. As an example, hybrids between benthic and limnetic forms of threespine stickleback show no problems in the laboratory but show reduced growth and survival in nature (review: Hendry et al. 2009a). These different patterns suggest that hybrids fall into fitness valleys in nature because they are poorly adapted for parental environments rather than because they have intrinsic genetic problems. However, ambiguity can remain because intrinsic genetic problems might be manifest only under stressful conditions (Coyne and Orr 2004). One way to address this last possibility is to test whether the fitness of backcrosses is intermediate between F1 hybrids and pure parental types (Rundle and Whitlock 2001, Rundle 2002).

Experiments of the above sort normally test hybrids and backcrosses in the parental environments, but those same hybrids and backcrosses would presumably have high fitness in intermediate environments. Thus, the experiments described above can't reveal the presence of a fitness valley unless they also confirm the absence of intermediate environments. Perhaps the best route to this determination is to examine the success of free-ranging hybrids and backcrosses in natural environments, as has been done for the benthic-limnetic stickleback system (Gow et al. 2007).

METHOD 4: DIVERGENT SELECTION

If trait differences among populations are the result of adaptive divergence, selection acting on those traits should be divergent. That is, selection should favor large body size where residents are large but should favor small body size where residents are small. This expectation means that the methods described in chapter 2 can be used to test whether or not phenotypic divergence is adaptive (e.g., Bolnick and Nosil 2007, Carlson et al. 2009, Weese et al. 2010). However, a fundamental difficulty in doing so is that selection is not expected to be strong in populations that are well adapted (Haller and Hendry 2014; chapters 2 and 3). A solution to this conundrum is to exploit natural, or create artificial, environment-phenotype mismatches—because selection should then be detectable. For example, divergent selection should be stronger when adaptation is constrained by gene flow, as has been demonstrated for *Timema* stick insects by Bolnick and Nosil (2007). Recent work, however, reveals that this expectation is not universal (Rolshausen et al. 2015a). Other natural phenomena that create mismatches and thus maintain selection are antagonistic coevolutionary dynamics, opposing sexual selection, functional or genetic constraints, high mutation rates, or recent environmental change. Artificial routes to generating environment-phenotype mismatches include phenotypic manipulations and experimental introductions. As an example of the latter, Thorpe et al. (2005) placed *Anolis oculatus* lizards from coastal xeric and rainforest populations into enclosures in xeric habitats, finding that selection was much stronger on the nonlocal rainforest-origin lizards than on the local xeric-origin lizards. Inferences are even stronger when the transplants are reciprocal. For instance, Nosil and Crespi (2006) transplanted *Timema* stick insects between different hosts, finding that selection on a given plant was toward the *Timema* phenotypes typically found on that plant.

A potential limitation of the transplant approach is that the range of phenotypes within a population might not be large enough for a robust estimate of selection. One solution is to mix individuals from multiple populations—as in Nosil and Crespi (2006)—although strong trait covariance across populations can make it difficult to isolate selection on specific traits. Another solution is to use intermediate phenotypes produced using the methods discussed earlier. Returning to *G. capitata*, Nagy (1997) crossed the coastal and

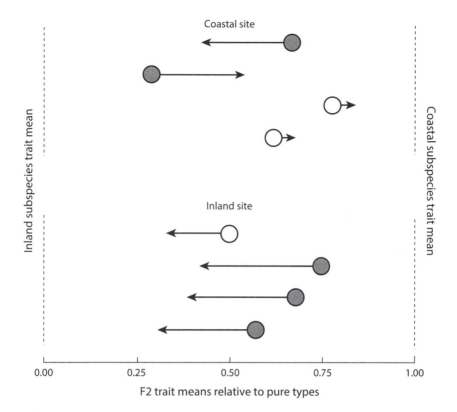

Fig 4.2. Selection on hybrids in two different ecological environments (coastal and inland) favors morphology characteristic of the locally adapted type. The circles show trait means (from the top in each panel: leaf diameter, inflorescence diameter, petal shape, and petal color) of F2 hybrids between the inland and coastal subspecies of *Gila capitata*. In each case, the mean is standardized to the difference between the two pure subspecies. The arrows show the direction and relative magnitude of selection gradients, with significant gradients indicated by gray shading. In all cases save the first row of the top panel, selection was in the direction of the locally adapted subspecies. The figure was redrawn from Nagy (1997)

inland subspecies to generate F2 hybrids, which showed enhanced phenotypic variation and a breakdown of trait associations.[7] These hybrids were then divided between the inland and coastal test sites and selection was measured on a number of phenotypic traits, with fitness assessed as the relative number of inflorescences per plant. As expected, selection in each environment generally favored hybrid individuals that were phenotypically most similar to the parental type normally found in that environment (fig. 4.2).

METHOD 5: EXPERIMENTAL EVOLUTION IN NATURE

Probably the ultimate confirmation of adaptation is to watch it take place de novo and in situ. The classic route to this determination is to introduce individuals from a common source population into different environments and then track their evolution.

---

[7]Another benefit of F2 hybrids (or naturally admixed populations) is that they can be used to examine divergent selection on the genes underlying phenotypic traits (e.g., Marchinko et al. 2014, Soria-Carrasco et al. 2014).

One prediction is that similar traits should evolve in similar environments, both among replicate introduced populations and in relation to natural populations. Another prediction is that fitness in a given environment should increase with the duration of evolution in that environment. This introduction approach is a staple of laboratory studies of experimental evolution (Bell 2008, Kawecki et al. 2012), and it also can be used to assess adaptive divergence in nature (Reznick and Ghalambor 2005).

One implementation of this approach takes individuals from one environment and introduces them into another environment. For example, guppies from high-predation environments that are introduced into low-predation environments generally evolve life history traits typical of natural guppy populations in low-predation environments (Reznick et al. 1997). In addition to this confirmation of an adaptive basis for divergence in *phenotypes*, these experiments can also assess *overall adaptation*. For example, survival rates in the new environment are higher for the introduced guppies evolving in that environment than for guppies from the ancestral environment (Gordon et al. 2009). Although planned and controlled experimental introductions are relatively rare (Reznick and Ghalambor 2005), uncontrolled introductions are common and can provide similar insights. The above described *D. subobscura* study provides one example because the North American and South American clines were both established after a single introduction of the flies to those continents (Gilchrist et al. 2004). Also, chinook salmon (*Oncorhynchus tshawytscha*) introduced to New Zealand (Quinn et al. 2001, Kinnison et al. 2001, 2008) and sockeye salmon (*Oncorhynchus nerka*) introduced to Lake Washington (Hendry et al. 2000b, Hendry 2001) evolved in ways expected from trait-environment associations in their native range.

Another implementation of this approach places individuals with intermediate phenotypes into different environments to see how their phenotypes evolve into the future. Returning again to *G. capitata*, Nagy (1997) examined naturally produced F3 individuals following the above-described F2 hybrid introduction experiment. Evolutionary change from the F2 to the F3 generations was, for all traits in both environments, in the direction expected to be adaptive. That is, the hybrid populations placed in the coastal (or inland) environment evolved in the direction of the typical coastal (or inland) phenotype (fig. 4.3). Based on all of its complementary approaches, the *G. capitata* study represents a particularly clear demonstration that phenotypes in different environments are the result of adaptive divergence.

METHOD 6: COMPARISONS TO NEUTRAL EXPECTATIONS

The above methods can be logistically or ethically difficult to implement, especially for large and long-lived organisms. A suggested alternative is to test whether genetic divergence in phenotypes deviates from that expected under neutrality. The most popular of these methods (for another, see Ovaskainen et al. 2011) is to test whether the proportion of additive genetic variance due to among-population differences (Q_{ST}) deviates from the equivalent estimate based on neutral genetic markers (F_{ST}). The rationale is that, when the traits are neutral, Q_{ST} and F_{ST} should diverge at similar rates and to similar extents, at least under a number of simplifying assumptions (Lande 1992, Spitze 1993, Merilä and Crnokrak 2001, McKay and Latta 2002, Leinonen et al. 2008, Whitlock 2008, Leinonen et al. 2013). By contrast, when the traits are under divergent (versus convergent) selection, Q_{ST} should be greater (versus lesser) than F_{ST}.

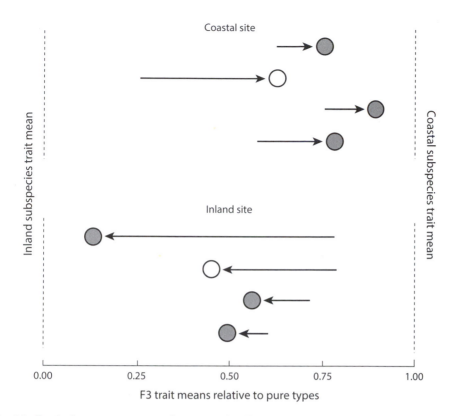

Fig 4.3. Evolutionary responses between the F2 and F3 generations of hybrids in two different ecological environments (coastal and inland) are toward the morphology characteristic of the locally adapted type. The circles show trait means (from the top in each panel: leaf diameter, inflorescence diameter, petal shape, and petal color) of F3 hybrids between the inland and coast subspecies of *Gila capitata*. In each case, the mean is standardized to the difference between the two pure subspecies. The arrow shows the shift in morphology from the F2 generation to the F3 generation, with significant shifts indicated by gray shading. The figure was redrawn from Nagy (1997)

Although Q_{ST} versus F_{ST} comparisons can certainly supplement the above-described methods, I do not find them very convincing for several reasons. First, accurate and meaningful estimates of Q_{ST} are difficult to obtain. Some studies use wild-collected individuals to estimate Q_{ST} (then called P_{ST})—but plasticity could here influence the outcome. Trait heritability can be added to the P_{ST} equation (Leinonen et al. 2006) but this term does not account for adaptive plasticity, only environmental noise. Other studies estimate Q_{ST} in common-garden experiments—but genotype-by-environment interactions make the relevance to nature uncertain. Second, the assumption that $Q_{ST} = F_{ST}$ under neutrality will not be correct in many instances (Hendry 2002, Edelaar et al. 2011). Third, when gene flow is low, F_{ST} can reach very high levels, making it impossible to infer that $Q_{ST} > F_{ST}$ even if divergent selection was the cause of trait divergence (Hendry 2002). For these reasons, and more (e.g., Ovaskainen et al. 2011), I prefer to compare Q_{ST} (or P_{ST}) values to themselves rather than to F_{ST}. For instance, the traits or populations with the greatest Q_{ST} values are likely under the strongest divergent selection (Kaeuffer et al.

2012). A nice application of this comparative approach is work on isopods in reed and stonewort habitats in lakes: pigmentation traits show the highest Q_{ST}s, those Q_{ST}s are much higher between than within habitats, and these patterns are replicated in two lakes (Eroukmannoff et al. 2009).

Adaptive divergence in nature

A century and a half of evolutionary studies have made clear that much of the variation among natural populations and species is adaptive: that is, organisms possess suites of traits that suit them at least reasonably well for their local environments and ways of life (Darwin 1859, Simpson 1944, Lack 1947, Simpson 1953b, Mayr 1963, Cain 1964, Endler 1986, Schluter 2000a, Losos 2009). The existence and importance of adaptive divergence thus is not in question. What remains uncertain, however, is just how precise adaptive divergence is, what drives it, and what constrains it (Gould and Lewontin 1979, Crespi 2000, Hendry and Gonzalez 2008). I will now summarize evidence relevant to key questions such as these by reference to data on traits and fitness in natural populations. These include (Q1) how common and strong is adaptive divergence, (Q2) how rugged are adaptive landscapes (i.e., how many peaks do they have), (Q3) how many peaks are "occupied" by populations/species, (Q4) to what extent do peaks multiply, such as when adaptive radiation of one group of organisms, promotes adaptive radiation of another group of organisms, (Q5) how predictable (i.e., "parallel" or "convergent") is evolution, (Q6) what is the role of sexual (as opposed to natural) selection, and (Q7) to what extent is adaptive divergence constrained by genetic (co)variance?

QUESTION 1: HOW COMMON AND STRONG IS ADAPTIVE DIVERGENCE?
Organisms persist across an astounding array of different environments, both within and between species. These different environments presumably impose very different selective pressures, which should lead to adaptive divergence in traits that improve fitness in local environments. But just how prevalent and strong is this adaptive divergence? Would organisms transferred between environments suffer a 1% fitness reduction, a 10% fitness reduction, or a 90% fitness reduction? The answer is important, and yet not obvious (Hendry and Gonzalez 2008), for several reasons. First, genetic drift might sometimes overwhelm adaptation, at least when population sizes are very small. Second, high gene flow or various other genetic or functional constraints might prevent populations from fully adapting to different environments. Third, environments that appear divergent to investigators might not, in reality, impose very strong divergent selection, particularly if the key traits determining fitness are under similar selection. Fourth, populations can sometimes persist in environments to which they are not well adapted owing to continued dispersal from other populations: that is, source-sink dynamics (Holt and Gomulkiewicz 1997). Fifth, spatiotemporal variation in selection (Siepielski et al. 2009, 2011), including Red Queen dynamics (Decaestecker et al. 2007, Hoeksema and Forde 2008), or frequent extinction-recolonization (Harrison and Hastings 1996, Kaltz and Shykoff 1998, Mopper et al. 2000, Hanski 2011) might mean that populations are not well adapted for their local environment at any particular time.

Objective insight into the prevalence and strength of adaptive divergence in nature can come from meta-analyses of fitness differences between local and nonlocal individuals in reciprocal transplant experiments. Schluter (2000a) reviewed 42 transplant studies that focused on populations showing divergence in phenotypic traits. Adaptive divergence was common given that local individuals had higher fitness (based on a variety of metrics) than did nonlocal individuals in 36 of 42 studies. Hereford (2009) followed with a much more comprehensive analysis (892 estimates of local adaptation from 74 studies) that did not require a priori phenotypic differences and that only used the most robust fitness surrogates (viability and fecundity). In this new analysis, 71% of the estimates were consistent with local adaptation, and the mean fitness advantage of local over nonlocal individuals was 45%. Similar results were obtained in an independent meta-analysis of plant reciprocal transplant experiments (Leimu and Fischer 2008).

These meta-analyses reveal that local adaptation is more common than not, but also that its frequency and strength is highly variable (fig. 4.4): that is, local individuals often have only a very small fitness advantage over nonlocal individuals or are, not infrequently, at an apparent fitness disadvantage. In a recent example, Rolshausen et al. (2015b) showed that guppies did not evince much, if any, local adaptation to oil polluted environments. Some of this variation will be owing to the factors listed in the first paragraph of this section—and some of these possibilities have been evaluated in the meta-analyses.

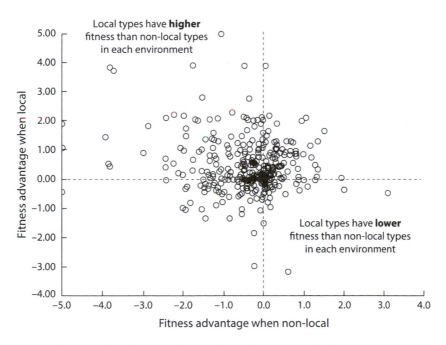

Fig 4.4. Compilation of estimates of local adaptation as revealed by reciprocal transplant experiments where individuals from each of two environments are tested in each of those two environments. Each data point is the relative fitness of a given type (relative to the other type) when that type is local (*y*-axis) versus non-local (*x*-axis). In the upper-left quadrant, local types have higher fitness than non-local types in both environments: that is, reciprocal local adaptation is present. In the lower-right quadrant, local types have lower fitness than non-local types in both environments: that is, reciprocal local maladaptation. The other two quadrants show cases where one type has higher fitness than the other type in both environments. The data are from Hereford (2009) and were provided by J. Hereford

Leimu and Fischer (2008) found that local adaptation was more common and stronger in larger (>1000 flowering individuals) than smaller (<1000 flowering individuals) populations, presumably because the latter are more susceptible to genetic drift, inbreeding, and low genetic variation. McPherson et al. (2015) found that local adaptation in reciprocal transplants was stronger when more traits were under divergent selection. In addition, the levels of local adaptation in the above broad meta-analyses are generally stronger than those in host-parasite systems (Hoeksema and Forde 2008), which implies a constraint due to temporal variation in selection—specifically antagonistic coevolution of hosts and parasites. Finally, gene flow (too much or too little) appears to be an important constraint in many systems, and will be considered in detail in chapter 5.

The above inferences need to be qualified through recognition of problems in the data sets and analyses. Some of these problems might cause a positive bias in favor of local adaptation. In particular, the populations chosen for study are usually not a random sample of all possible populations, but rather populations for which local adaptation was expected in the first place (e.g., large environmental or phenotypic differences). Other problems might cause a negative bias against local adaptation. For instance, many of the local adaptation estimates are based on questionable fitness metrics. Indeed, Hereford (2009) showed that evidence for adaptive divergence was stronger as fitness metrics became more inclusive, such as the combination of survival and fecundity. In addition, transplant studies are often conducted under relatively benign conditions (e.g., during the summer) and over only a short period of time (e.g., a few weeks). However, divergent selection might be stronger under more stressful conditions, which could occur in only part of the year (e.g., winter) or only in certain years, such as under drought conditions (Grant and Grant 2002, 2006).

Adaptive divergence is more common than not and sometimes can be very strong. However, considerable variation is present and adaptive divergence is probably almost always imperfect, being variously constrained by genetic drift, inbreeding, temporal variation, gene flow, antagonistic coevolution, and so on. At present, I suspect that the meta-data are biased against the detection of strong adaptive divergence, primarily because of poor fitness surrogates and the often short duration of reciprocal transplant studies.

QUESTION 2: HOW RUGGED ARE ADAPTIVE LANDSCAPES?

Adaptive divergence can occur along fitness ridges (no fitness valleys between populations) or among alternative fitness peaks (fitness valleys between populations)—as described above. Which situation typically characterizes adaptive divergence in nature—and just how many peaks (niches) are out there? The existence of alternative fitness peaks is obvious in a number of situations, with some classic examples being phytophagous insects on different host plants (Drès and Mallet 2002, Funk et al. 2002) and various instances of mimicry (Joron and Mallet 1998, Alexandrou et al. 2011, Pfennig et al. 2015). Alternative peaks are less obvious in other instances, such as birds adapting to different seed sizes (the seed distribution might be continuous) and plants adapting to different shade conditions (intermediate shade conditions might be present). Reciprocal transplants using pure parental types, as in the previous question, are of no direct help in making the distinction—because low fitness for nonlocal individuals would be expected with or without intervening valleys (Schluter 2000a). However, several other approaches –outlined earlier in this chapter—can reveal the presence or absence of fitness valleys between peaks.

A number of studies have generated phenotypically intermediate individuals, such as through artificial hybridization, and tested for their low fitness in nature (i.e., intermediates between pure types fall into a fitness valley). Nosil et al. (2005) reviewed 11 such studies, six of which reported lower fitness for intermediate forms (see also Lowry et al. 2008a). Of course, this tally also means that five studies did not find evidence of selection against intermediates—and, indeed, hybrids often can be superior to pure types in nature. A hybrid advantage is particularly likely to occur though "heterosis" when pure types suffer reduced fitness owing to inbreeding (e.g., Fenster and Galloway 2000, Ebert et al. 2002). In such cases, it is valuable to assess hybrid fitness across multiple generations, in which case heterosis often diminishes and hybrid problems emerge (e.g., Edmands 1999, Rhode and Cruzan 2005). Regardless, very few studies have employed the best approaches (described earlier) for establishing an ecological (as opposed to intrinsic genetic) basis for hybrid fitness differences. In addition, most studies test intermediate forms only in parental environments, whereas intermediate environments might be present in nature.

Other studies have taken advantage of naturally occurring intermediates, such as those produced by gene flow or hybridization, to test for fitness valleys between natural populations. (This method incorporates any intermediate environments as long as the individuals are free-ranging.) Within species, this approach has been used to demonstrate strong disruptive selection between beak size morphs in African *Pyrenestes* finches (Smith 1993) and in the medium ground finch (Hendry et al. 2009b). Among species, a number of studies have documented selection against natural hybrids (e.g., Gow et al. 2007), but it is often hard to confirm that such selection has an ecological basis. In one study to do so, hybrids between two Darwin's finch species (medium ground finch and cactus finch) were found to have low fitness in nature under stressful (low food) conditions but not under more benign (high food) conditions (Grant and Grant 1993, 1996). Overall, then, some studies have confirmed that naturally occurring intermediates have low fitness in nature because of a valley between fitness peaks. It is also true, however, that natural hybrids sometimes have equal or even higher fitness than do pure types (review: Arnold and Hodges 1995), again probably because of inbreeding or the existence of intermediate habitats.

Another approach is to formally estimate fitness landscapes so as to infer the presence of peaks and valleys. This approach is rarely attempted because it is exceedingly difficult—as noted earlier. However, *individual* fitness landscapes have been constructed in some situations, such as for beak traits (bill depth and grove width) of crossbills, *Loxia curvirostra* (Benkman 2003) (fig. 4.5). In this case, fitness landscapes were constructed from data on the feeding efficiency of crossbills with particular beak traits on cones of five different conifer tree species, in combination with mark-recapture data for wild crossbills that related feeding efficiency to survival. The estimated fitness landscape had five discrete peaks, each corresponding to beak characteristics best suited for a particular cone type. Another example of a fitness landscape with clear peaks and valleys is that estimated for Darwin's finches, as will be explained in more detail in the next question. Finally, the first approach (artificial production of intermediates that are then tested in nature) can be combined with the present approach (estimating fitness landscapes) to reveal valleys between peaks—as proved to be the case for body shape in an adaptive radiation of *Cyprinodon* pupfishes on San Salvador (Martin and Wainwright 2013).

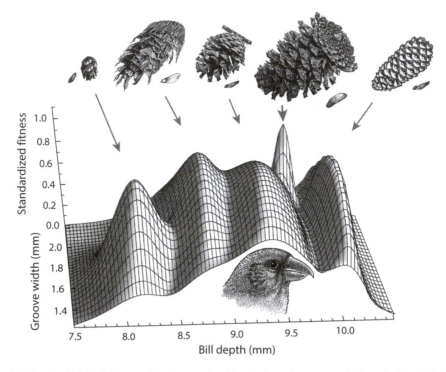

Fig 4.5. An individual fitness landscape for North American crossbills relating bill depth and bill groove width to estimated fitness; for details see Benkman (2003). The peaks correspond to the cones produced by different species of conifer (drawn to scale from left to right: western hemlock, Douglas fir, Rocky Mountain lodgepole pine, ponderosa pine, South Hills lodgepole pine). The figure is originally from Benkman (2003), the crossbill image was provided by C. Benkman, and both were modified and presented by Hendry et al. (2012)

Different populations and species occupy alternative fitness peaks separated by low fitness valleys—at least sometimes and perhaps often. Too few studies have applied the robust approaches that would enable well-supported general assertions about the ruggedness of adaptive landscapes. However, I propose that fitness valleys are probably common given the often discontinuous nature of resources/habitats. Moreover, theory suggests that fitness valleys can be generated by competition for shared resources even if the underlying resource distribution is continuous (Dieckmann and Doebeli 1999, Rueffler et al. 2006)—although it is easier to generate valleys if the resources are themselves multimodal (Doebeli 1996, Thibert-Plante and Hendry 2011a). However, it also remains likely that divergence at least sometimes proceeds along fitness ridges, which some theoreticians have argued should be much more common than classically considered fitness peaks (Gavrilets 2004).

QUESTION 3: HOW MANY OF THE FITNESS PEAKS ARE OCCUPIED?

If adaptive landscapes are rugged owing to multiple peaks separated by low-fitness valleys, a pertinent question is: how many of the peaks are occupied by different populations or species? This question starts from the classic Darwinian premise that a limited number of available peaks exist in nature, and that lineages radiate through time to occupy those diverse peaks. This progressive "filling" of peaks is—at least to some extent—inevitable given that classic adaptive radiations, such as *Anolis* lizards, Darwin's finches, Hawaiian honeycreepers, Hawaiian silverswords, and Lake Victoria cichlid fishes, each started from only one or a few species. Thus, one situation where some peaks would not be occupied is early in a radiation when not enough evolutionary time has elapsed (Seehausen 2006). However, it is also possible that some peaks remain unoccupied even when the radiation is essentially complete. These peaks might remain permanently unoccupied because the fitness valley (or moat) by which they are isolated can't be surpassed by a population adapted to one of the other peaks—that is, the problem of peak shifts discussed earlier in this chapter. This limitation is obviously true at a crude level—trees will never walk except in the movies and animals won't photosynthesize except by incorporating chloroplasts from plants (Rumpho et al. 2011). However, the answer is less obvious at a finer scale; that is, how many empty peaks exist that could, in principle, be colonized by a given lineage.

One way to infer the peaks that are occupied is to look for the matching of obvious potential niches with populations or species clearly specialized for those niches. For instance, some lineages of phytophagous insects have radiated onto many available plant species (e.g., Ferrari et al. 2006) or onto different parts of a given plant species (e.g., Joy and Crespi 2007). By extension, unoccupied plants or plant parts might represent unoccupied fitness peaks, although this inference is tenuous because they might not be true high-fitness peaks or, if so, they might be occupied by another lineage. Inferences are thus made easier when niches are replicated in multiple independent locations, which reveals how often the same suite of niches is occupied by independent radiations. For instance, limnetic and benthic forms of several fish species are found in many temperate lakes (reviews: Skúlason and Smith 1995, Schluter 1996b, Taylor 1999), similar *Anolis* lizard ecomorphs[8] occupy similar habitats on multiple islands in the Caribbean (Losos 2009), similar spider ecomorphs occupy similar habitats on multiple islands in Hawaii (Gillespie 2004) and in Galápagos (De Busschere et al. 2010), and so on. In these cases, it seems that many of the reasonably accessible niches have been filled through adaptive radiation. At the same time, however, such analyses always reveal at least some peaks that are seemingly unoccupied in some locations. As just one example, 9 of the 44 estimated possible island-by-ecomorph combinations in Hawaii are not occupied by a spider species (Gillespie 2004).

Another way to infer which peaks are or are not occupied by a given radiation is to formally estimate adaptive landscapes. As noted above, this estimation is very hard and rarely attempted. However, in the crossbill example, each of the five conifer cone peaks is clearly occupied by a well-adapted crossbill morph (Benkman 2003). As another example, Darwin's finches with appropriate beak sizes exist for nearly every fitness peak inferred from the available seed distribution on multiple Galápagos islands (Schluter and Grant 1984) (fig. 4.6). Again, however, not all of the inferred peaks were occupied, and the crossbill example might be biased because fitness was only evaluated on cone types they were already known to exploit.

A very different way to infer the existence—although not the identity—of unoccupied peaks is to assess whether new species can colonize a geographical location without

---

[8]"Ecomorph" was coined by Williams (1972, p. 72) to mean, in my rephrasing, species that independently evolve similar traits through adaptation to similar environments. Ecomorphs are the species-level equivalent of ecotypes.

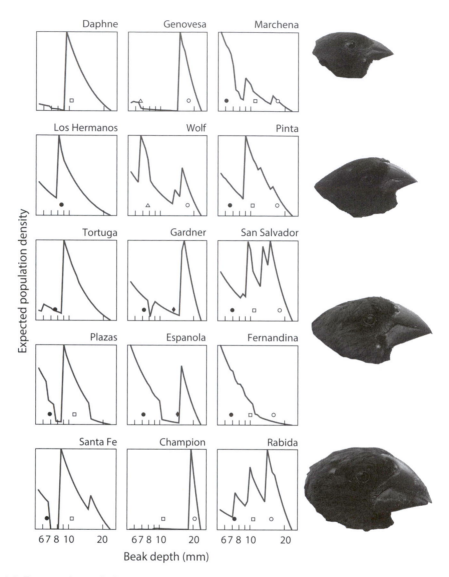

Fig 4.6. Expected population densities estimated for seed-eating finches of different beak sizes based on the distributions of seeds found on each of 15 Galápagos islands. The symbols near the x-axis of each figure correspond to the mean beak sizes (depths) of different finch species on each island: filled circles for *G. fuliginosa*, open boxes for *G. fortis*, open circles for *G. magnirostris*, open triangles for *G. difficilis*, and filled diamonds for *G. conirostris*. The images at right show the range of beak sizes within and among some of the species, with *G. fuliginosa* at top, small and large beak morphs of *G. fortis* in middle, and *G. magnirostris* at bottom. The figure is taken from Schluter and Grant (1984) and the finch photographs are by A. Hendry

causing the extinction of species already present. This question has been long considered in ecology in the contexts of "community saturation," "species packing," "limiting similarity," and "invasibility" (Lack 1947, Elton 1958, Hutchinson 1959, Macarthur and Levins 1967, Ricklefs 1987, Cornell and Lawton 1992, Tokeshi 1999, Shea and Chesson 2002). The basic idea is the classic one that two species with very similar niches can't coexist because competition would drive one of them extinct—and so colonization without corresponding extinction implies a previously unoccupied niche (Gause 1934, Hutchinson 1959, MacArthur and Levins 1967, Tilman et al. 1982).[9] After much debate, it is now well-established that numerous species can be added to communities without causing the extinction of species already present (Levine and D'Antonio 1999, Rosenzweig 2001, Davis 2003, Sax and Gaines 2008). (This effect is within a trophic level, whereas the introduction of predators or pathogens is often devastating for local species.) As one example, construction of the Panama Canal allowed 22–75% of the local freshwater fish species to cross the continental divide—without the extinction of a single species on either side. The net result was an increase in local species richness of 11–40% (Smith et al. 2004). As another example, introductions of vascular plants to oceanic islands have approximately doubled species richness on those islands (Sax et al. 2002). Of course, it remains possible that these systems are experiencing an "extinction debt" in which at least some of the native species ultimately will be driven to extinction (Gilbert and Levine 2013).

Some—perhaps many—available fitness peaks remain unoccupied in essentially all geographical locations. Indeed, this reality seems self-evident from the huge disparity in species numbers across time and space even when habitats are reasonably similar (Tokeshi 1999). For example, very different numbers of fish species can be present in independent, but physically similar and geographically proximate, locations (Seehausen 2006, Smith et al. 2010). These unoccupied peaks could be the result of high extinction rates, low dispersal rates, limited time for diversification, or difficulties in bridging fitness valleys. **Despite these arguments, many niches are repeatedly occupied in multiple independent locations—and so adaptive radiation must at least sometimes be constrained by available niche space**. For instance, the existing resource distribution and competition for those resources limits the number of seed eating species of Darwin's finches present on a given Galápagos island (Schluter and Grant 1984, Grant 1999, Hendry et al. 2009b), and sequential niche filling has been argued to limit the diversification of Himalayan songbirds (Price et al. 2014). The reconciliation between these seemingly contradictory assertions might be that adaptive radiations rapidly and (reasonably) predictably occupy easily accessible niches but have much more difficulty occupying less-accessible, yet nevertheless real, additional niches.

QUESTION 4: TO WHAT EXTENT DO PEAKS MULTIPLY?

The classic view of adaptive radiation, following from the previous question, is that a somewhat fixed number of peaks (niches) are available in a given location, and that species diversify until all of the accessible peaks are occupied (Darwin 1859, Simpson 1953b, Schluter 2000a, Phillimore and Price 2008, Losos 2009). Diversification then stops until extinction re-opens one of the original peaks. In this view, increasing diversity constrains further diversity. An alternative view is that ongoing diversification creates new fitness peaks that favor further diversification: that is, increasing diversity begets further diversity (Whittaker 1977). This effect is straightforward across trophic levels—because, for example, herbivores can't exist before plants, and parasites can't exist before hosts. To exemplify, the diversity of available host plants (apple, blueberry, snowberry, hawthorn) drove the diversification of phytophagous *Rhagoletis* flies, which

---

[9]An alternative body of work argues that coexistence of multiple species on the same resource is possible if the species are "ecologically equivalent"—because no single species then has an advantage in excluding the others (McPeek and Brown 2000, Hubbell 2001, 2006).

then drove the diversification of *Diachasma* wasps that are *Rhagoletis* parasitoids (Forbes et al. 2009).

What is less certain is whether "diversity begets diversity" (perhaps more correctly "diversity begets diversification") *within* a trophic level, an effect that might arise through several mechanisms. First, an increasing number of specialist species could impose competition-based selection on the remaining generalists to split into more specialists. Such effects of competition will be considered further in chapter 6 (ecological speciation) and chapter 8 (community structure). Second, an increasing number of species could increase environmental heterogeneity, and thus generate more niches (Harper 1977, pp. 746–747, Tokeshi 1999, pp. 62–65). Third, diversification within one trophic level could promote diversification at a higher trophic level (as above), which could then increase selection for further diversification at the lower trophic level (Whittaker 1977, Brown and Vincent 1992). For instance, diversification of prey might favor diversification of predators, which might then favor further diversification of prey. Fourth, positive interactions among species (e.g., facilitation) can promote coexistence, although such effects are most likely among lineages that differ in adaptive traits and do not compete severely (Valiente-Banuet et al. 2006, Vellend 2008).

One empirical test for whether diversity begets or constrains diversity employs a spatial comparison. If the beget view is true, then locations with more species should have higher speciation rates—even after controlling for external factors that influence speciation. In this spirit, Emerson and Kolm (2005) asked, for plants and arthropods on the Hawaiian and Canary islands, whether the number of native species on an island (a metric of diversity) was correlated with the number of endemic species on that island (a metric for speciation). After controlling for other variables, such as island age, size, elevation, and isolation, a positive association was seen between the number of native and endemic species (fig. 4.7), suggesting that diversity begets diversification. As might be expected, challenges to this analysis have been raised (e.g., Cadena et al. 2005, Kiflawi et al. 2007, Vellend 2008, Gruner et al. 2008) and defenses mounted (e.g., Emerson and Kolm 2007).

Another empirical test employs a temporal comparison. If the constraint view is true, speciation rates in a clade should start high (because ecological opportunity—*sensu* Yoder et al. 2010—is high) and should decrease as niches are filled through speciation. If the beget view is true, speciation rates should accelerate through time. The possibility of such "density-dependent speciation" has been tested a number of times by estimating the relative dates of branching points on phylogenies. Most such studies conclude that trait divergence and speciation rates decline as lineages radiate (Purvis et al. 2009, Harmon et al. 2010, Cooper and Purvis 2010), supporting the constraint view. However, a number of methodological limitations can bias conclusions in this direction (e.g., Pybus and Harvey 2000). Moreover, not all studies find declining speciation rates, and other methods have yielded even more variable results. As one example, Phillimore and Price (2008) found that only 13 of 23 clades of birds with more than 20 species show significant decreases in speciation rate with increasing clade age.

The debate about whether diversity constrains or begets diversity seems to me an instance of the "one model to rule them all" syndrome (X. Thibert-Plante, pers. comm.). **Instead, it seems obvious that diversity begets diversification through some mechanisms but constraints it through other mechanisms**. The key question then becomes when and where one set of mechanisms gains the upper hand and thereby drives increasing or decreasing rates of diversification. This refined question has yet to be seriously evaluated.

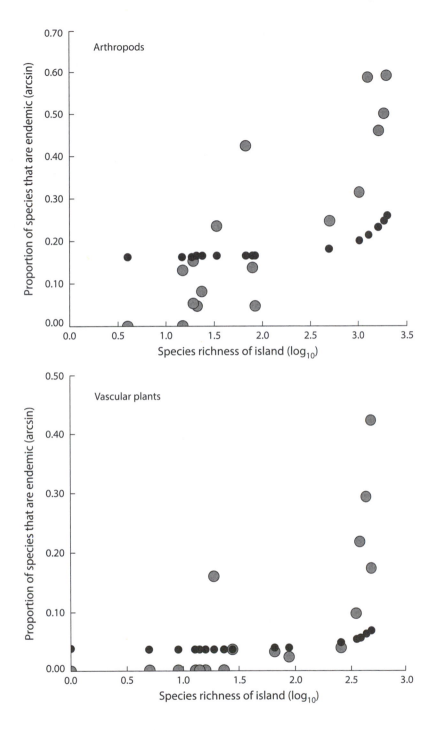

Fig 4.7. Evidence suggested to indicate that diversity begets diversity: the proportion of species that are endemic is higher on Hawaiian islands that have more species. The gray filled circles are the actual data and the black filled circles are expectations from a null model. The data are from Emerson and Kolm (2005) and were provided by B. Emerson

QUESTION 5: HOW PARALLEL (OR CONVERGENT) IS PHENOTYPIC EVOLUTION?

One of the strongest lines of evidence that phenotypic divergence is the result of adaptation to different environments is the demonstration that similar phenotypes evolve independently when different source populations colonize similar environments (Endler 1986, Schluter 2000a, Arendt and Reznick 2008, Elmer and Meyer 2011, Losos 2011, Rosenblum et al. 2014). That is, the same adaptive peaks (niches) are found in similar locations and organisms have independently adapted to them in similar ways. This pattern is variously called parallel or convergent evolution depending on whether it occurs from similar or different ancestors, or is based on similar or different genetic/developmental changes (Arendt and Reznick 2008, Losos 2011, Rosenblum et al. 2014). (I will use these terms in the first sense.) A particularly clear example is mimicry, where the phenotypes of either toxic (Mullerian) or nontoxic (Batesian) species converge on the phenotypes of sympatric toxic species (Joron and Mallet 1998, Alexandrou et al. 2011, Pfennig et al. 2015). Other examples include similar morphologies for *Anolis* lizard ecomorphs on different islands (Losos 2009, Losos and Ricklefs 2009, Mahler et al. 2013), similar life histories and morphologies for Poeciliid fishes experiencing similar predation regimes in different rivers (Reznick and Bryga 1996, Johnson 2002, Langerhans and DeWitt 2004), similar benthic-limnetic divergence for fishes in different lakes (Schluter and McPhail 1992, Robinson and Wilson 1994), similar traits in many plant species adapted to serpentine soils (Brady et al. 2005), and similar morphological and pigmentation differences between closely related cave and surface organisms (Jones et al. 1992). These examples are so common that one can get the impression that parallel/convergent evolution is almost ubiquitous.

But just how parallel/convergent are the phenotypes of different populations or species found in similar environments: that is, how repeatable or deterministic is the matching of phenotypes to environments? The typical approach to this inference is to survey multiple independent lineages that have each colonized multiple environment types, and to statistically partition the variance in traits among environment types (e.g., cave versus surface, serpentine versus nonserpentine), locations (e.g., different rivers or lakes or islands), and their interaction (Langerhans and DeWitt 2004). Although the main effect of environment is often significant in such analyses, thus implying parallelism/convergence, considerable phenotypic differences are often evident among lineages in a given environment type. An example comes from a study of three lizard species with ancestral populations on dark soil that each colonized white sand habitats within the last 2000–5000 years. Although some convergence was evident (lizards on white sands were always lighter-colored), the lineages differed considerably in the specifics of that divergence (Rosenblum and Harmon 2011). This outcome might be expected given that the three lineages were quite different to begin with, but substantial nonparallelism is also seen when ancestral populations are very closely related. Indeed, this is the case for each of the aforementioned examples of parallel/convergent evolution, including trophic traits in stickleback (Schluter and McPhail 1992, Kaeuffer et al. 2012), life history in guppies (Reznick and Bryga 1996, Fitzpatrick et al. 2014), and morphology and color in *Anolis* ecomorphs (Harmon et al. 2005, Langerhans et al. 2006). As another clear example, *Cepaea nemoralis* snails have adapted to open versus shaded habitats in many locations in Europe (Ożgo 2011). Although the frequency of dark-shelled individuals was higher in shaded habitats in 10 of 12 replicate comparisons (parallelism), the specifics of that divergence (% fused bands, % yellow) varied dramatically

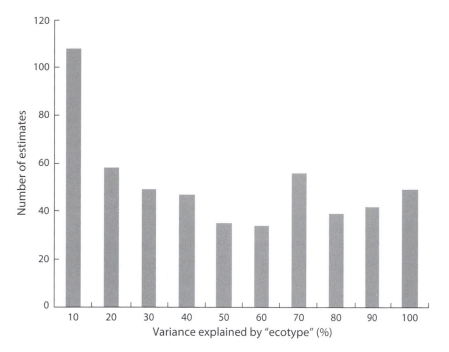

Fig 4.8. Compilation of the results from studies of parallel evolution in fishes. Data were collected by recording from the literature mean trait values for each population in each study, and then classifying those populations by habitat type, such as lake versus stream, benthic versus limnetic, high-predation versus low-predation, cave versus surface, and lava versus mud. The proportion of the variation among the population means that can be explained by habitat type was then calculated for each trait/species/study combination. The graph shows the frequency distribution of those estimates. The data were compiled by K. Oke, C. Leblond, and A. Hendry (unpublished)

(nonparallelism). Nonparallelism is also clear in different physiological mechanisms of adaptation to high elevation by humans in South America (Andes), Asia (Tibetan plateau), and Africa (Semien Mountains) (Beall et al. 2002, Beall 2007).

Beyond a diversity of specific examples, we might consider more generally and quantitatively just how parallel is parallel evolution?[10] To this end, we compiled data on the mean phenotypes of fish populations in divergent environments (i.e., "ecotypes": e.g., lake versus stream, benthic versus limnetic, high predation versus low predation), as often as possible from multiple independent locations. The compilation presented here includes 77 studies of 25 species, with multiple traits per study and multiple environment comparisons, altogether yielding 517 opportunities to quantify parallelism. For a simple assessment, we used ANOVA to calculate the variance among population means that could be attributed to environment type. This analysis revealed a huge range in the degree of parallelism (fig. 4.8). (Other metrics of parallelism yield similarly variable results.) One trend in the data was that a high degree of parallelism was often the case for life history traits associated

---

[10]I here use the term parallelism (rather than convergence) for convenience and because it is commonly used in some of the taxa I discuss—but changing the term would not change the points being made.

with high-predation versus low-predation environments—although even these traits can show striking deviations from parallelism (Fitzpatrick et al. 2014). Importantly, many of these studies were based on wild-caught fish and so plasticity might have influenced the degree of parallelism, as shown for lake-stream stickleback by Oke et al. (2015).

The inescapable conclusion is that parallel evolution is common but not necessarily that parallel. That is, populations in similar environments show some phenotypic similarity but they also show many differences. Studies of parallel evolution should therefore *quantify* the degree of parallelism and nonparallelism and explore the potential causes of each. The potential causes of nonparallelism might be grouped into four categories: ecological, genetic, sexual, and functional (Kaeuffer et al. 2012). Ecological explanations invoke variation in natural selection. For instance, the strength of divergent selection might differ among seemingly similar environmental contrasts and, indeed, trait divergence is greater when environments are more divergent (e.g., Berner et al. 2008, Landry and Bernatchez 2010, Kaeuffer et al. 2012). Genetic explanations include drift, different genetic backgrounds (chapter 11), and different levels of gene flow (chapter 5). Sexual explanations invoke variable sexual selection, which will be considered further in the next question. Functional explanations recognize that selection acts directly on performance (e.g., speed, endurance, force) (Arnold 1983, Walker 2007, Irschick et al. 2008) rather than morphology, and that similar performance can be achieved through alternative morphologies: that is, "many-to-one mapping" (e.g., Alfaro et al. 2005). In this last case, performance can vary in parallel even if morphological traits do not: that is, alternative phenotypic solutions are possible for the same ecological niche. Some of these and other influences on parallelism are discussed in Rosenblum et al. (2014).

QUESTION 6: WHAT IS THE ROLE OF SEXUAL SELECTION?

In this chapter, I have not yet explicitly distinguished natural from sexual selection, but most arguments implied the former owing to the focus on environmental differences. However, traits experiencing divergent natural selection might also experience sexual selection (Darwin 1871, Andersson 1994, Jennions et al. 2001), which can be as strong as, or even stronger than, natural selection (Kingsolver et al. 2001, Hoekstra et al. 2001, Siepielski et al. 2011). Sexual selection might substantially alter patterns of phenotypic divergence from those expected under natural selection alone (Bonduriansky 2011, Maan and Seehausen 2011). As one possibility, conserved sexual selection (similar even in different environments) could constrain trait divergence to be lower than expected under natural selection. As another possibility, spatial variation in sexual selection might cause divergence among populations that does not match spatial variation in natural selection. As a final possibility, correlated patterns of natural and sexual selection might enhance divergence beyond that expected from natural selection alone. So the key question might be: how much of the spatial variation in traits can be attributed to natural versus sexual selection?

Some nuances can be illustrated by attempts to explain why male guppies in populations exposed to strong predatory fishes (high predation) are generally less colorful than those in populations exposed to only weak predatory fishes (low predation). The classic explanation starts with the idea that female guppies generally favor more conspicuous males regardless of predation regime (Houde 1987). By contrast, natural selection is thought to be driven by predation regime: that is, strong (but not weak) predatory fishes select against conspicuous (usually more colorful) males. Thus, color variation among populations is classically thought to be driven by spatial variation in natural selection layered onto (mostly) conserved sexual selection. This explanation was supported by early surveys of natural populations (Endler 1978) and by experiments showing that guppies evolved high (versus low) color when introduced to low- (versus high-) predation environments (Endler 1980). However, more recent research has brought this simple interpretation into question. First, high- and low-predation guppy populations do not differ consistently in male color: that is, the former are sometimes *more* colorful than the latter (Weese et al. 2010, Millar and Hendry 2012). Second, other guppy introductions have not led to the expected evolution of color (Karim et al. 2007, Kemp et al. 2009). Third, selection estimates in nature have not found that high color is more strongly selected against in high-predation environments than in low-predation environments (Weese et al. 2010). Fourth, spatial variation in male color is not strongly correlated with spatial variation in natural selection (Weese et al. 2010), but seems to be with spatial variation in sexual selection (Houde and Endler 1990, Endler and Houde 1995). Thus, spatial variation in male color might be strongly influenced by spatial variation in sexual selection.

Several additional studies have examined spatial associations between phenotypic traits and sexual selection (Boughman 2001, Kwiatkowski and Sullivan 2002, Svensson et al. 2006)—and these studies have yielded variable results. Populations of the frog *Oophaga [Dendrobates] pumilio* in Panama's Bocas del Toro archipelago show spectacular spatial variation in color. These frogs are poisonous, and so the color patterns are aposematic, being under strong natural selection from predators. However, Maan and Cummings (2009) showed that although predators might cause the evolution of striking color, the particular color pattern at a given location might be driven by sexual selection. By contrast, Chenoweth et al. (2010) found little correspondence between spatial variation in sexual selection and spatial variation in cuticular hydrocarbons, which are sexually selected traits, among nine populations of *Drosophila serrata* in Australia. To explain this disconnect, the authors invoked constraints owing to different genetic architectures, which will be considered further in the next question. Additional work on this study system suggested that one constraint was a "lack of standing genetic variation that would allow an increase in sexual fitness while simultaneously maintaining high nonsexual fitness" (Hine et al. 2011, p. 3662).

A number of studies have thus argued that spatial variation in sexual selection can influence population divergence. Moreover, this result might be general in that the strength of preference divergence among populations and species is generally predictive of the divergence in display traits (Rodríguez et al. 2013). However, several caveats emerge. First, the Rodríguez et al. (2013) meta-analytical result might also apply to nondisplay traits. Second, most of the studies are correlational, and so cannot establish causation: that is, is sexual selection the cause or the consequence of trait divergence? Third, traits under divergent sexual selection might not be subject to strong natural selection (e.g., genitalia), in which case no conflict exists and the question disappears. Fourth, correlations between traits and selection coefficients are not, under many conditions, reliable indicators of whether selection drove divergence—because populations that have responded to selection should no longer experience strong selection (chapter 2, Haller and Hendry 2014).

Sexual selection sometimes, perhaps often, influences adaptive divergence among populations. What remains uncertain is the extent to which spatial variation in natural and sexual selection line up with each other, which force becomes more important in shaping adaptive divergence (Maan and Seehausen 2011), and whether sexual selection can promote the exploration of new phenotypic space (Bond)uriansky 2011).

QUESTION 7: IS DIVERGENCE CONSTRAINED BY GENETIC (CO)VARIANCE?

Adaptive divergence might be constrained, or at least altered, by patterns of genetic (co)variance within populations (Bürger 1986, Futuyma et al. 1995, Schluter 1996a, Hunt 2007, Hansen and Houle 2008). A specific prediction is that phenotypic variation *among* populations or species will be biased, relative to the direction favored by selection, toward the major axis of genetic variation in the ancestral population. In the first comprehensive consideration of this prediction for natural populations, Schluter (1996b) showed that multivariate phenotypic variation among freshwater populations of threespine stickleback was similar to the major axis of genetic (G_{max}) and phenotypic (P_{max}) variation estimated from laboratory crosses in one of those populations. This finding suggested that adaptive divergence was indeed biased by patterns of genetic (co)variance.

The reality turned out to be not quite so simple. One problem with Schluter's (1996b) analysis was that it examined only derived freshwater populations, when the ancestor (marine threespine stickleback)—and therefore the evolutionary trajectory toward each freshwater population—was quite different. Focusing on this distinction, Walker and Bell (2000) showed that divergence between marine and freshwater populations was almost orthogonal to that among freshwater populations. Another problem with the original analysis was that **G** and **P** matrices were not estimated for the ancestral marine population—but rather for a derived freshwater population. Two studies have since examined axes of variation within marine stickleback and related this to variation among freshwater populations. First, Berner et al. (2010) showed that divergence in gill raker traits among freshwater populations was not aligned with the ancestral marine P_{max} (fig. 4.9). The authors suggested instead that Schluter's (1996b) results arose because the evolution of variation is shaped by the same benthic-limnetic foraging axis present both within and among populations (see also Bolnick and Lau 2008, Berner et al. 2008). Further, Kimmel et al. (2012) found no evidence that **G** was constraining adaptive evolution of the shape of the opercle bone for marine stickleback repeatedly colonizing fresh water. Second, Leinonen et al. (2011) found that G_{max} for body shape in marine fish was a shallow-deep axis, which was also a major axis of divergence among stickleback populations. As above, however, it is hard to know if divergence was biased by G_{max} or whether selection similarly shapes genetic variation both within and among populations (Berner et al. 2008). Similar studies of other empirical systems have yielded mixed results regarding whether evolutionary divergence is (e.g., Hunt 2007, Chenoweth et al. 2010) or is not (e.g., McGuigan et al. 2005, O'Reilly-Wapstra et al. 2014) substantially associated with within-population **G** or **P**.

Beyond these diverse results, a number of inferential difficulties remain. First, **G** and **P** are strongly but not perfectly correlated (Roff 1996, Kruuk et al. 2008), and it isn't clear which matrix is the best to consider. On the one hand, **P** matrices could reflect environmental influences and therefore not reflect underlying genetic variation. On the other hand, **G** matrices are much harder to estimate with precision and are generally estimated in the laboratory, where G × E could cause substantial deviations from nature (Pigliucci 2006). Second, one must assume that current selection reflects historical selection and

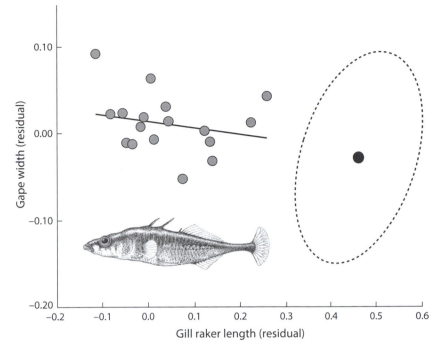

Fig 4.9. The major axis of divergence in foraging traits among derived lake populations of threespine stickleback (gray circles) is almost orthogonal to the major axis of trait covariance in the putative ancestral marine threespine stickleback. The foraging traits (gape width and gill raker length) were standardized to a common body size by calculating residuals. For the ancestral stickleback, mean trait values are represented by the black circle and the 50% confidence interval for covariance is represented by the dashed oval. The figure is redrawn from Berner et al. (2010). The image shows a generic freshwater stickleback and was drawn by D. Berner

also that derived, and therefore measured, **G** matrices reflect ancestral **G** matrices. Both of these assumptions are unlikely to hold in many instances, as described in chapters 2 and 3. Third, and as noted above, causality is hard to disentangle: that is, do **G** matrices influence adaptation or does adaptation influence **G** matrices—or both (Berner et al. 2008, 2010, Kolbe et al. 2011). One possible route around these last two uncertainties is to work with contemporary adaptation to new environments. Here it is easier to estimate the ancestral **G** matrix, the initial direction of selection, and the immediate trajectory of evolution. Along these lines, Eroukhmanoff and Svensson (2011) showed that the **G** matrix of isopods in an ancestral lake habitat (reed) did not have any apparent influence on adaptation of isopods to a new habitat (stonewort) over 40 generations. For an example on a slightly longer time scale, see Johansson et al. (2011).

On time frames of a dozen generations or more, strong selection will almost always overwhelm potential constraints imposed by the G matrix. First, such constraints are likely only when **G** matrices are ill conditioned (some trait combinations have far more genetic variance than others), and selection acts in a direction of low genetic variation (Chenoweth et al. 2010). Alternatively, patterns of genetic (co)variance could speed adaptation if they align with the direction of selection. For instance, the meta-analysis by

Agrawal and Stinchcombe (2009) that was mentioned in chapter 3 found no net tendency for **G** matrices to constrain the rate of adaptation. Second, the **G** matrix should evolve in response to selection, potentially removing any straightforward constraints (Arnold et al. 2008). Note that I am not saying that the pace and direction of generation-to-generation evolution are insensitive to **G**; rather that evolution over longer time periods will be much less so. And, of course, these arguments apply most directly to increases or decreases in trait values rather than the origin of novelties, where constraints are surely much more severe (Futuyma 2010).

Conclusions, significance, and implications

This chapter started by confirming the logical reality that the world is composed of a number of fitness peaks that are reasonably isolated from other such peaks by fitness valleys or moats. Adaptive divergence and adaptive radiation thus proceed as lineages diversify into multiple populations and species that occupy different peaks, which are often generated by discrete ecological niches (although different peaks sometimes can represent different phenotypic solutions to the same niche). Modifying this classic view, ongoing adaptive radiation can create new peaks, which can promote further radiation (diversity begets diversity). The various peaks on such rugged adaptive landscapes vary in their accessibility to a given lineage as a function of selection (wider and deeper valleys are harder to bridge) and genetic variation (which will be limited in some directions). Some of the peaks are reasonably easy to colonize, as revealed by the repeatability of replicate adaptive radiations, whereas other peaks are only rarely colonized, as revealed by unused resources or rare exceptional species. As a result, adaptive divergence and radiation show both parallel/convergent and nonparallel/nonconvergent components. This variation might be explained by idiosyncratic local selection (probably most important), sexual selection (perhaps also important), genetic constraints (particularly gene flow), and functional redundancy.

These conclusions are relevant to eco-evolutionary dynamics because much of biological diversity is the result of adaptive divergence: divergent ecological environments are the primary driver of evolutionary diversity. In addition, the different ecological effects of different populations and species are the result of adaptive divergence. The reason is that all traits that have ecological functions will obviously interact with the environment, and so also will be those traits most likely to be under ecologically based selection. Here, then, we have eco-evolutionary feedbacks at a very general level: ecological differences drive adaptive (and therefore functional) differences among populations, which are then likely to have different ecological effects—as will be discussed in more detail in chapters 7–9.

Chapter 5

Gene Flow

In the previous chapter, I outlined adaptive divergence as the primary determinant of population and species differences and therefore their divergent eco-evolutionary effects. It might have sometimes seemed I was arguing that adaptive divergence was ubiquitous and all powerful whereas, in reality, the success of adaptive divergence will depend critically on several factors. In the present chapter, I will focus on one of those factors: gene flow—the extent to which diverging groups are connected by genetic exchange. Gene flow can have a diversity of effects that will be discussed in detail below but the most obvious effect, by way of example, is its role in preventing the independent evolution of populations in different environments and thereby constraining their adaptive divergence (Lenormand 2002, Garant et al. 2007). Through this constraint, gene flow can also prevent speciation and therefore adaptive radiation (Felsenstein 1981). For these reasons, the effects of adaptive divergence on eco-evolutionary dynamics will be shaped and modified by gene flow.

I start with a consideration of what, precisely, gene flow is, including in relation to "dispersal" and "migration." I then discuss how gene flow is commonly measured and interpreted in nature. I go on to explain the diversity of effects, both positive and negative, that gene flow can have on adaptive divergence. I then review—through the usual "key" questions—empirical evidence for the effects of gene flow on adaptive divergence, speciation, demography, and species' ranges.

What are migration, dispersal, and gene flow?

These terms have been used in a variety of ways by a variety of authors (Endler 1977, Dingle 1996, Neigel 1997, Lenormand 2002, Garant et al. 2007). Instead of attempting to declare and defend some "correct" definitions, I will simply state definitions used consistently in the present book. Bear in mind that other publications sometimes use the same or different terms in similar or different ways.

"Migration" will be used for the back-and-forth, or cyclical, movement of individuals among locations. The term is thus intended to capture phenomena such as seasonal migrations between breeding and feeding locations and diel vertical migrations in lakes or oceans. "Dispersal" will be used for the more-or-less permanent displacement from one location to another of individuals, zygotes (e.g., seeds or eggs), or gametes (e.g., sperm or pollen). Dispersal thus differs from migration in the expectation of differences between parents and offspring in the location of reproduction—assuming the offspring survive to reproduce. Migration and dispersal can occur together, such as when individuals born in one location (e.g., salmon that hatch in a particular stream) migrate to another location (e.g., the ocean) for feeding and then disperse to a third location (e.g., a different stream) for breeding. In general, I will refer to individuals that disperse among locations as "dispersers." When dispersers move among established populations, I will sometimes (particularly in the context of ecological speciation—chapter 6) call them "migrants" or "immigrants"—the latter specifically referring to dispersing individuals that have arrived in a new population.

"Gene flow" will be used when reproduction following dispersal results in genetic exchange among populations.[1] Gene flow is not a genome-wide property; instead, it varies across the genome by chance and in accordance with variation in selection and recombination. For instance, loci influenced by divergent selection, whether directly or indirectly (the latter through physical linkage), should show low gene flow (Charlesworth et al. 1997, Barton 2000, Via and West 2008, Nosil et al. 2009a, Michel et al. 2010, Thibert-Plante and Hendry 2010). Conversely, loci influenced by common selection across multiple populations should show high gene flow for adaptive alleles (Slatkin 1976, Morjan and Rieseberg 2004, Mallet 2009). In addition, genome locations featuring reducing recombination, such as near centromeres or at the breakpoints of chromosomal inversions, might show reduced gene flow (Carneiro et al. 2009, Strasburg et al. 2009, Michel et al. 2010, Roesti et al. 2012, 2015, Renaut et al. 2014).

This chapter is about the general effects of dispersal and gene flow on adaptive divergence. For this reason, I will mostly ignore variation across the genome, and will instead consider a genome-wide average level of genetic exchange. This level is best reflected by neutral loci that are not linked to selected loci and are not in regions of restricted recombination. Conveniently, genetic markers with these properties are frequently targeted in population genetic studies attempting to quantify gene flow (Mills and Allendorf 1996, Waples and Gaggiotti 2006). Of course, these neutral, unlinked loci still could be influenced by overall selection against migrants and hybrids (chapter 6), leading to a generalized (or "genetic") barrier to gene flow (Barton and Bengtsson 1986, Gavrilets 2004). However, this generalized barrier is weak in most instances (Gavrilets and Vose 2005, Thibert-Plante and Hendry 2009, 2010) and, in any case, it reduces gene flow across the entire genome (variation within the genome becomes important only after recombination occurs in the gametes of hybrids). As a result, the above interpretation of "gene flow" as an average genome-wide property is appropriate for the present chapter.

Rates of dispersal and gene flow are unlikely to be the same. On the one hand, several effects can cause dispersal to be *higher* than gene flow. Among these effects is the generalized barrier to gene flow discussed above, where maladapted immigrants have lower fitness than adapted residents. In addition, several nongenetic effects can cause dispersers to have low fitness (Bensch et al. 1998, Marr et al. 2002, Pakanen et al. 2011, Thibert-Plante and Hendry 2011b). First, the dispersers might be generally inferior—perhaps they are

---

[1]Gene flow can also occur without physical dispersal, such as through a breakdown of reproductive barriers between sympatric populations. Also, the population genetic literature often uses the terms migration (between discrete populations) or dispersal (across continuous populations) in the sense I here use gene flow.

less competitive individuals that were unable to remain in the old environment. Second, the act of dispersal itself might be costly and could have carry-over effects after arrival in the new environment. Third, success in a new environment might first require familiarity with the local environment and its residents. Note that these three possibilities do not require ecological differences between the old and new environments, although such differences would likely magnify the effects.

On the other hand, several different effects can cause dispersal to be *lower* than gene flow, with the best studied possibility being inbreeding. In such cases, offspring production from inbred resident-resident matings can be lower than from outbred resident-immigrant matings, with the expected result being increased gene flow (Richards 2000, Whitlock et al. 2000, Ebert et al. 2002). Another effect increasing the success of dispersers can be negative frequency dependence, with examples including a "rare male" advantage during mating (Knoppien 1985, Hughes et al. 1999) and predator avoidance (Olendorf et al. 2006). Specifically, the traits of immigrants will often differ from those of typical residents and so stand out for females and deviate from predator "search images." Ecological differences could contribute to these effects because adaptive divergence often causes trait differences. Finally, divergence in mating traits but not in mating preferences can lead to asymmetric mating patterns, where resident individuals prefer to mate with immigrants over other residents (Ellers and Boggs 2003, Schwartz and Hendry 2006, Labonne and Hendry 2010).

Multifarious effects on adaptive divergence

Gene flow and dispersal can have a diversity of effects on adaptive divergence (reviews: Slatkin 1987, Garant et al. 2007, Kawecki 2008, Guillaume 2011, Weeks et al. 2011, Bourne et al. 2014) (fig. 5.1). The best-studied effect is that increasing gene flow decreases the independence of gene pools, which should prevent them from fully adapting to different environments (Haldane 1948, Slatkin 1973, Felsenstein 1976, Endler 1977, García-Ramos and Kirkpatrick 1997, Hendry et al. 2001, Lenormand 2002). Under this effect, increasing gene flow should (1) decrease phenotypic differences among populations from different environments, and (2) increase phenotypic differences among populations in similar environments. Beyond this classic negative effect, many other possibilities have been suggested, and a number of these effects can *promote* adaptation. First, gene flow can sometimes "rescue" populations from inbreeding depression, and thereby enhance fitness, population size, population persistence, and adaptive potential (Hedrick 1995, Tallmon et al. 2004, Willi et al. 2007, Fitzpatrick et al. 2016). Second, gene flow increases genetic variation within populations, and can thereby increase the potential for adaptation to new conditions (Swindell and Bouzat 2006, Perron et al. 2008, Bakker et al. 2010, Bell and Gonzalez 2011). Third, gene flow can increase the chance that beneficial alleles arising in one population will spread to other populations (Slatkin 1987, Morjan and Rieseberg 2004, Mallet 2009). Fourth, gene flow can decrease the negative effects of genetic drift in small populations (Alleaume-Benharira et al. 2006).

The potential positive effects of gene flow suggest that some is better than none, whereas the potential negative effects suggest that too much would be problematic, and so the expectation is for an intermediate, and probably low, "optimal" level of gene flow (Gomulkiewicz et al. 1999, Ronce and Kirkpatrick 2001, Alleaume-Benharira et al. 2006,

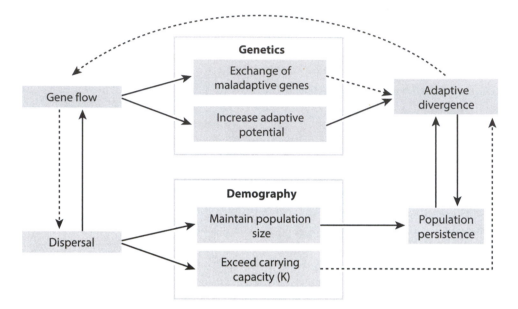

Fig 5.1. Some interactions between gene flow, dispersal, demography, and adaptive divergence (modified from Garant et al. 2007). Dashed lines indicate negative effects and solid
lines indicate positive effects. In brief, gene flow can increase the exchange of maladaptive genes between populations in different environments, which can decrease adaptive
divergence, but gene flow can also provide the genetic variants that promote adaptation,
which can increase adaptive divergence. Dispersal can cause demographic subsidies that
increase population size, which can enhance population persistence and thus adaptive
divergence. However, dispersal can also cause local carrying capacity to be exceeded,
which can decrease adaptive divergence. Increases (vs. decreases) in adaptive divergence
can then increase (vs. decrease) population persistence and decrease (vs. increase) gene
flow (i.e., ecological speciation). Finally, dispersal should usually increase gene flow but
might also reduce it by leading to selection for decreased dispersal

Garant et al. 2007, Lopez et al. 2009, Thibert-Plante and Hendry 2009). Several factors
should influence this optimum, and I here provide a few key examples. First, the optimum
should be higher when populations are smaller—because gene flow can counter some of
the problems experienced by such populations: inbreeding, limited evolutionary potential, and stochasticity. Second, the optimum should be lower when populations occupy
more divergent environments—because gene flow would be more likely to compromise
adaptive divergence. Third, the optimum should be higher in the case of antagonistic
coevolution, where species have to adapt to fitness peaks that are continually evolving:
that is, adaptation of one species favors counter-adaptations by the other species. In this
last case, the increased genetic variation brought by gene flow can enhance evolutionary
potential and rates of adaptation (Forde et al. 2004, Hoeksema and Forde 2008).

   Gene flow is strongly influenced by dispersal, and so dispersal can indirectly have all
of the effects ascribed above to gene flow. Complementing those indirect genetic effects,
dispersal can have several direct demographic effects, some of which can enhance adaptation. For instance, dispersal can maintain sink populations until they are able to adapt
(Holt and Gomulkiewicz 1997, Kawecki 2003, Holt et al. 2004). In addition, dispersal is

critical for maintaining the long-term persistence of metapopulations experiencing local extinctions (Gonzalez et al. 1998, Hanski and Ovaskainen 2000, Molofsky and Ferdy 2005). Other direct effects of dispersal can degrade adaptation. For instance, dispersal can increase the density of local populations beyond their carrying capacity, and thereby depress the average fitness of residents (Kawecki and Holt 2002). Finally, dispersal can interact with gene flow in a variety of interesting ways. As one example, dispersal among partially reproductively isolated populations can select for reduced maladaptive interbreeding (i.e., "reinforcement," chapter 6), which can thereby reduce gene flow.

The specific characteristics of dispersers can alter their effects on gene flow and adaptive divergence. For instance, dispersal can depend on the size, age, sex, dominance, or condition of individuals (Clobert et al. 2009), which can then influence success in the new environment (Bensch et al. 1998, Marr et al. 2002, Pakanen et al. 2011, Thibert-Plante and Hendry 2011b). Moreover, the dispersal of individuals can depend on the match between their phenotypes and the environment, a phenomenon called "matching habitat choice" by Edelaar et al. (2008). For example, Lin et al. (2008) showed that dispersers between beach-spawning and creek-spawning populations of sockeye salmon were those possessing phenotypes reasonably well suited for the new environment. Specifically, selection favors shallower bodies in creeks than on lake beaches, and the beach fish that dispersed into creeks had relatively shallow bodies whereas the creek fish that dispersed to beaches had relatively deep bodies (fig. 5.2). Analogous results were obtained in an experimental study with threespine stickleback in streams versus lakes (Bolnick et al. 2009) (fig. 5.2). In situations such as these, the putative negative effects of gene flow will diminish. Of course, examples also can be found where matching habitat choice does not seem important (Camacho et al. 2015) or is even maladaptive (Singer 2015).

Dispersal is not a static parameter but rather one that can evolve, including in contemporary time (reviews: Bowler and Benton 2005, Ronce 2007), and gene flow can evolve even if dispersal remains constant, such as through the evolution of reproductive barriers (chapter 6). Empirical examples for dispersal include the evolution of increased dispersal ability at expanding range margins (Hughes et al. 2003, Phillips et al. 2006) and the evolution of decreased (or increased) dispersal in fragmented landscapes (Schtickzelle et al. 2006, Cheptou et al. 2008). In addition, various models predict that increasing dispersal should evolve when selection for adaptive divergence is weak, the risk of inbreeding depression is high, environmental quality varies asynchronously among locations, the probability of local extinction is high, or competition among kin is high (reviews: Clobert et al. 2001, Hendry et al. 2004b, Bowler and Benton 2005, Ronce 2007). Any evolution of dispersal should then have corresponding effects on gene flow, and perhaps attendant effects on adaptive divergence. In short, interactions among dispersal, gene flow, and adaptive divergence are complex, variable, and interrelated.

How to quantify gene flow and its effects

Dispersal or gene flow can be quantified either between discrete populations or across continuous space. The discrete context is typical when organisms breed at distinct times or in distinct locations, such as separate watersheds, lakes, mountain tops, rock outcrops, or meadows. Estimates in this case often take the form of the number or proportion of individuals (or alleles or gametes) that move among populations. The continuous context

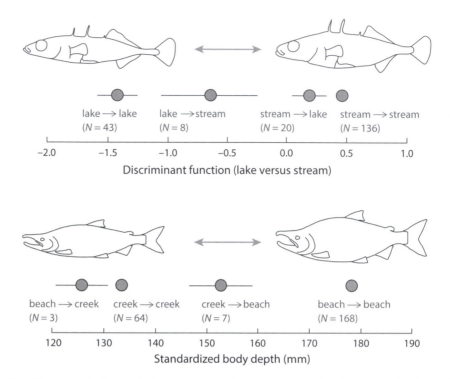

Fig 5.2. Dispersal between habitats can be phenotype-biased, with examples here from threespine stickleback (top panel) and male sockeye salmon (bottom panel). In both cases, the data points are means and standard errors. For stickleback, lake and stream fish tagged and released at the transition between the lake and stream generally returned quickly to their natal habitat (lake → lake; stream → stream) and the two forms are morphologically distinct (lake fish have shallower bodies). The individuals that did disperse to the alternative (nonnatal) habitat (lake → stream; stream → lake) tended to have morphologies biased in the direction of fish native to that alternative habitat. The images (provided by D. Berner) show generic lake and stream stickleback representing variation along the morphological function best discriminating lake stickleback from stream stickleback. The data are from Bolnick et al. (2009) and were provided by D. Bolnick. For salmon, beach and creek fish (as identified with genetic markers) generally return to their natal habitats to spawn (creek → creek; beach → beach) and the two forms are morphologically distinct (creek fish have shallower bodies). The individuals that did disperse to the alternative (nonnatal) habitat (beach → creek; creek → beach) tended to have morphologies biased in the direction of fish native to that alternative habitat. The images (by A. Hendry) show generic creek and beach salmon representing variation in body depth (standardized to a common body size). The data are from Lin et al. (2008) and were provided by J. Lin

is typical when organisms breed more or less evenly across space or time (Rousset 2000, Leblois et al. 2003), such as forest trees distributed across large areas or fish distributed along the length of a river system. Estimates in this case often take the form of probability distributions of dispersal by distance, summarized by parameters such as dispersal distance means, variances, or "kernels" (Rousset 2000, 2004).

Dispersal in plants is typically estimated by means of pollen traps (Bush and Rivera 1998), by monitoring the movement of pollinators (Van Rossum and Triest 2010), or by measuring seed dispersion (Stoyan and Wagner 2001). For animals, dispersal is typically estimated by marking large numbers of individuals and monitoring their movement across space (Koenig et al. 1996, Hanski 2000, Hendry et al. 2004b). Although marking can take place at any life stage, the most informative approach is to mark juveniles before dispersal and to then determine their breeding location as adults. An individual born in one location that breeds in a different location contributes to "natal dispersal," whereas an individual that breeds in multiple locations contributes to "breeding dispersal" (Greenwood 1980). Capturing and marking large numbers of juveniles can be logistically prohibitive, but an alternative is to use "natural tags" that simultaneously, and without intervention, mark all individuals in a given population. Examples include stable isotopes (Kennedy et al. 2000, Hobson 2005), temperature-induced banding patterns on bone (Quinn et al. 1999), and parasites (Lester 1990).

Gene flow estimates are sometimes based on dispersal estimates, an approach that Slatkin (1987) considered "direct." In reality, this approach is best considered "indirect" because, as outlined earlier, dispersal and gene flow are unlikely to be equivalent in nature. In addition, estimates of dispersal are normally biased because it is impossible or impractical to survey all locations to which an individual might disperse (Koenig et al. 1996, Hendry et al. 2004b). Gene flow is thus better estimated as the movement of alleles among gene pools, which can be obtained in two basic ways: (1) the real-time movement of alleles (i.e., "contemporary" estimates), or (2) the use of genetic differences to infer to long-term allele exchange (i.e., "historical" estimates).[2]

Contemporary estimates of gene flow involve genotyping a large number of individuals from multiple populations (or locations or times) and genetically assigning those individuals—or parts of their genomes—back to the sampled populations through various "assignment tests" (Manel et al. 2005, Broquet et al. 2009). An individual sampled from one population but genetically assigned to another population is thus a contemporary disperser. Gene flow estimates produced in this manner are sometimes very close to mark-recapture dispersal estimates (Berry et al. 2004), which is not surprising as both methods are here estimating the same parameter. As described above, however, dispersing individuals might not successfully reproduce, and so contemporary *gene flow* estimates can benefit from genetically assigning juveniles (or seedlings or eggs) to putative parental populations (Smouse and Sork 2004). Similarly, methods have been developed to genetically identify hybrids and backcrosses between populations or species (Anderson and Thompson 2002, e.g., Gow et al. 2006, 2007). Limitations of the contemporary method are several: (1) dispersal—and therefore gene flow—can vary dramatically through time (Hendry et al. 2004b), (2) statistical power can be low unless the populations are very genetically distinct (Waples and Gaggiotti 2006), (3) the estimates are not always accurate (Faubet et al. 2007), and (4) the long-term introgression of alleles is not measured.

Historical estimates of gene flow involve the application of a theoretical or statistical model of evolution to a measure of genetic differentiation. The classic case is Wright's (1931) island model: $F_{ST} = 1/(1+4N_em)$, where F_{ST} is the proportion of the total genetic variation attributable to differences among populations (see Holsinger and Weir 2009 for

---

[2]Slatkin (1987) considered historical estimates to be "indirect" because one is inferring a past process (gene flow) from a current pattern (genetic differences) based on a particular evolutionary model.

details), N_e is the effective population size,[3] and m is the proportion of alleles exchanged among populations. $N_e m$ is called the "effective number of migrants" and is the quantity that most directly influences neutral genetic differentiation (Wright 1931, Mills and Allendorf 1996), whereas m is the quantity that most directly influences adaptive genetic differentiation (e.g., Hendry et al. 2001, Lopez et al. 2008, Yeaman and Guillaume 2009, Débarre et al. 2015). In general, it is more difficult to estimate m than $N_e m$ because the former also requires an independent estimate of effective population size, which is often imprecise and can be biased (Wang 2005, Luikart et al. 2010).

Using Wright's model, any estimate of F_{ST} can be quickly and easily converted into an estimate of gene flow. However, a major problem with this simplistic approach is that it relies on a large number of assumptions not met in natural populations (Whitlock and McCauley 1999). Some of these assumptions have been relaxed in alternative models, such as the finite island model (Takahata 1983) and the isolation-by-distance model (Slatkin 1993, Rousset 2000). Even fewer assumptions attend more recent likelihood or Bayesian methods (Beerli and Felsenstein 1999, 2001, Hey and Nielsen 2004, Pinho and Hey 2010). Although these latter methods have proven very popular, they too are not infallible (Abdo et al. 2004, Slatkin 2005, Becquet and Przeworski 2009, Strasburg and Rieseberg 2010).

A critical limitation of many methods for estimating historical gene flow is the need to assume an equilibrium among selection, drift, and mutation. This limitation is important because reaching an equilibrium can take a very long time (Whitlock 1992), particularly for large populations (Waples 1998). Biases can result in several ways. First, founder effects can cause initial genetic differences that then take many generations to decay to equilibrium, especially if gene flow is low (Mopper et al. 2000, Labonne and Hendry 2010). Sampled short of equilibrium, genetic differences will be higher than expected, and so gene flow will be underestimated. Second, the accumulation of genetic divergence among new populations can be slow in the absence of founder effects, particularly when population sizes are large and gene flow is low (Manolio et al. 2009). Sampled short of equilibrium, genetic differences will be lower than expected, and so gene flow will be underestimated. Third, if gene flow suddenly increases (or decreases) among established populations, it can take many generations for genetic differences to decay (or increase) to the new equilibrium (Nei and Chakravarti 1977). In such cases, "historical" gene flow has changed and it will take some time for genetic differentiation to accurately reflect the new situation. Improvements in estimating historical gene flow under nonequilibrium conditions can be found in equations or simulations that take into account founding population sizes, time for divergence, and temporal variation in gene flow (e.g., Whitlock 1992, Chakraborty and Jin 1992, Pinho and Hey 2010).

In some situations, inferences are desired about additional aspects of dispersal or gene flow, such as the extent to which it is male-mediated versus female-mediated, or pollen-mediated versus seed-mediated. For dispersal, these inferences can be made by comparing the movements of males versus females or pollen versus seeds. For gene flow, some of the same inferences can be made by comparing genetic markers that are biparentally inherited (e.g., on autosomes) versus uniparentally inherited (mtDNA, cpDNA, Y chromosomes) (e.g., Ennos 1994, Ouborg et al. 1999, Goudet et al. 2002,

---

[3]Effective population size is the census population size of an idealized population having the same genetic properties as the population under study. A vast literature attends this concept (Wang 2005).

Lawson Handley and Perrin 2007). In other cases, inferences are desired as to the level of dispersal or gene flow in different directions, such as from population A into population B versus population B into population A. For dispersal, these directional estimates are straightforward—although they can still be difficult. For gene flow, these estimates can come from assignment tests (Manel et al. 2005) or some likelihood/Bayesian methods (Beerli and Felsenstein 1999, 2001, Hey and Nielsen 2004, Pinho and Hey 2010). In perturbed situations, directional gene flow can also be inferred from temporal shifts in the spatial distribution of alleles (e.g., Pringle et al. 2011).

As should now be clear, all methods for estimating dispersal and gene flow have a number of ambiguities, uncertainties, and caveats. The best practice is therefore to employ multiple methods, with congruence among estimates engendering increased confidence. In addition, inferences will be generally more robust in a relative sense (e.g., gene flow is higher here than there) than in an absolute sense (e.g., $Nm = 10$ or $m = 0.01$) (Bohonak 1999, Hendry et al. 2001). However, absolute estimates remain necessary for some inferences, such as the level of genetic independence among populations (Waples and Gaggiotti 2006) and—the topic taken up next—the influence of gene flow on adaptive divergence (Hendry et al. 2001, Moore et al. 2007, Yeaman and Guillaume 2009, Tack and Roslin 2010, Paul et al. 2011, Débarre et al. 2015; fig. 5.3). Also, as noted above and made clear in question 7, gene flow is not a single quantity for a given set of populations but rather varies across the genome in accordance with varying selection and recombination.

Beyond the estimation of gene flow, our main interest is in its consequences for adaptive divergence. The simplest approach to this inference is correlational: for example, one can examine multiple independent instances of divergence among environments (e.g., fish in lakes and adjacent streams in each of multiple independent watersheds) and test whether population pairs showing lower trait divergence (y-axis) are those showing higher gene flow (x-axis) (for examples see question 1 below). One limitation of this approach is that trait divergence might not accurately reflect adaptive divergence, such as when the studied traits are not closely linked to fitness, are highly plastic, or have diverged for reasons other than natural selection. In addition, variation in trait divergence might reflect variation in divergent selection, rather than gene flow. That is, population pairs showing greater trait divergence might occupy more divergent environments, in addition to exchanging fewer genes. Another limitation of the correlational approach is ambiguity regarding cause and effect: is gene flow constraining adaptive divergence (the question of interest here) or is adaptive divergence constraining gene flow (the question of interest in ecological speciation—chapter 6) (Räsänen and Hendry 2008)? One potential way to resolve this causal ambiguity is to plot trait divergence (y-axis) against the *potential* for gene flow (x-axis), such as geographical distance (corrected for environmental differences), physical barriers to dispersal, and relative population sizes (e.g., Langerhans et al. 2003, Bolnick and Nosil 2007). Bi-directional causality is here broken because (for example) barriers to dispersal can causally influence gene flow and therefore adaptive divergence, whereas adaptive divergence is less likely to causally influence barriers to dispersal.

More powerful approaches for inferring causality between gene flow and adaptive divergence employ experimental manipulations. For example, laboratory studies of antagonistic coevolution between hosts and pathogens have used experimental manipulations to demonstrate that gene flow strongly influences local adaptation (see question 4 below).

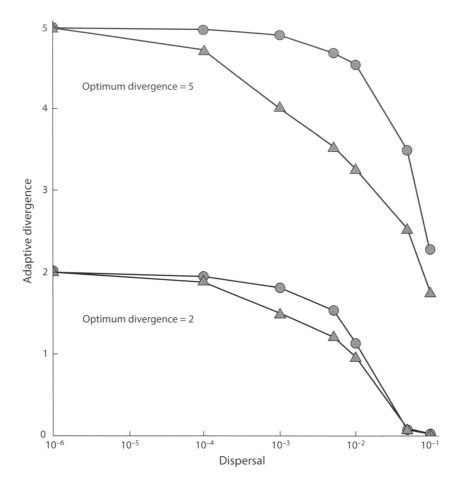

Fig 5.3. Theoretical relationships between dispersal and adaptive divergence. Dispersal is the proportion of individuals that move between two populations each generation, and adaptive divergence is the difference between the populations in mean phenotype. Results are shown for two differences (large = 5 and small = 2) in adaptive optima between the populations. In each case, the upper curve is the result for the general additive model of Hendry et al. (2001) and the lower curve is the simulation result of Yeaman and Guillaume (2009). In all cases, the strength of stabilizing selection around the optimum is $\omega^2 = 25$

And, in agriculture, gene flow is used to control the evolution of resistance to pesticides (see question 3 below). Experimental manipulations are much less common in "natural" populations—but some examples do exist. Riechert (1993) identified riparian populations of spiders that exhibited maladaptive behavior more typical of nearby arid populations. Drift fences were used to block the dispersal of arid population spiders into the riparian population and behaviors in the latter rapidly shifted toward the riparian expectation, implying that gene flow was causing the original maladaptation. Nosil (2009) recorded a similar outcome when a new road reduced gene flow between two formerly maladapted populations of *Timema* stick insects. More experiments of this sort are needed because they directly inform the extent to which naturally occurring levels of gene flow constrain adaptive divergence.

Effects of gene flow in nature

I will not here review levels of gene flow in natural populations. First, reviews on this topic have already been completed for many taxa (e.g., Bohonak 1999, Hendry et al. 2004b, Morjan and Rieseberg 2004, Weersing and Toonen 2009, Kisel and Barraclough 2010, Pinho and Hey 2010), and these reviews all indicate that gene flow varies dramatically among populations and species. This variation is often attributable to geographical distance, barriers to dispersal, or organismal properties, such as body size or mode of dispersal (e.g., dispersal of seeds by wind, water, or animals). Second, absolute (as opposed to relative) estimates of gene flow (e.g., m or $N_e m$) often will be unreliable for the reasons outlined above. Third, gene flow is most obviously relevant to eco-evolutionary dynamics when it has consequences for adaptation, which will therefore be my focus in the questions that follow. As noted earlier, gene flow can have a variety of influences that are difficult to disentangle (reviews: Slatkin 1987, Garant et al. 2007, Kawecki 2008), and very few empirical studies have attempted to do so. As a result, I will have to consider separately each of the potential effects: for example, I will first sequentially ask whether gene flow constrains (Q1) adaptive divergence, (Q2) speciation, (Q3) demographic success, and (Q4) species ranges. I will then sequentially ask whether gene flow promotes (Q5) adaptation (specifically in the context of antagonistic coevolution) and (Q6) genetic rescue. Finally, I will ask (Q7) how variable gene flow is across the genome, which is a question that has driven an important change in perspective in the way that gene flow is considered.

QUESTION 1: TO WHAT EXTENT DOES GENE FLOW CONSTRAIN ADAPTIVE DIVERGENCE?

Theoretical models have repeatedly demonstrated that gene flow should constrain adaptive divergence (see above), but theory doesn't tell us anything about what actually happens in nature. For that inference, we need studies of natural populations—from which come a number of specific examples of gene flow having an important constraining influence. The key question, however, is just how prevalent and strong are such constraints in typical populations on typical landscapes? Perhaps gene flow rarely matters: indeed, arguments have been made that selection is usually strong enough to overwhelm gene flow (Ehrlich and Raven 1969) or that the positive effects of gene flow typically outweigh the negative (Hoeksema and Forde 2008, Frankham et al. 2011). Alternatively, perhaps gene flow is often important: for example, the spatiotemporal cohesion of species and the rarity of speciation events both imply that genetic exchange generally limits population divergence (Futuyma 1987, 2010, Slatkin 1987, Morjan and Rieseberg 2004).

One way to assess the *general* effects of gene flow on adaptive divergence is to sample a large number of populations in different environments, and then relate divergence in adaptive traits to the level of dispersal or gene flow: that is, the above-described correlational approach. Studies of this sort suggest that gene flow constraints might be common (fig. 5.4). As one example, Nosil and Crespi (2004) found that morphological divergence between *Timema* stick insects on different host plants was negatively associated with the geographical potential for gene flow and with genetic estimates of gene flow. Bolnick and Nosil (2007) reached the same conclusion for *Timema* color patterns. Similarly, Hendry and Taylor (2004) found that divergence between lake and stream stickleback in foraging-related traits (gill rakers and body shape) declined sharply with increasing gene flow. Analogous associations have been reported for antipredator behavior in stream-dwelling

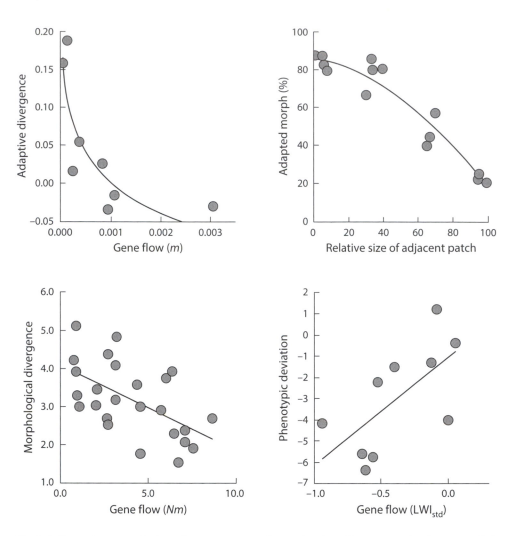

Fig 5.4. Negative associations between gene flow adaptive divergence. In the upper left panel, the relative difference in body depth between lake and stream stickleback is negatively correlated with the extent of gene flow (m) estimated using neutral markers. Each point is an independent lake-stream population pair and the curve is logarithmic. The data are from Hendry and Taylor (2004). In the upper right panel, the frequency of adapted *Timema* stick insect morphs in a host plant patch is negatively correlated with the extent of nearby alternative host plants, a likely indicator of gene flow. Each point is a patch and the curve is a second-order polynomial. The data are from Bolnick and Nosil (2007) and were provided by D. Bolnick. In the lower left panel, the multivariate morphological difference (Euclidean distance based on five morphological characters) between populations of the little greenbul (*Andropadus virens*) in different habitats (forest versus ecotone) is negatively correlated with the extent of gene flow (Nm) estimated using neutral markers. Each point is a pairwise comparison between populations in the different habitats. The data are from Smith et al. (1997) and were provided by T. Smith. In the lower right panel, ten *Mimulus cardinalis* populations show a positive association between their phenotypic deviation (optimal temperature for growth versus local temperature—a measure of maladaptation) and the estimated amount of gene flow from other populations (LWI_{std}). The data are from Paul et al. (2011), wherein further details are given, and were provided by J. Paul

salamanders (Storfer and Sih 1998), morphological traits in forest versus ecotone birds (Smith et al. 1997), and in many other situations (review: Räsänen and Hendry 2008). Pointing toward the generality of these system-specific findings, a meta-analysis of 75 studies found that trait local adaptation was higher among populations that likely exchanged fewer genes (Urban 2011). Although all of these studies were correlative, and thus subject to ambiguity about cause and effect, many were able to provide additional evidence pointing to the role of gene flow. It therefore seems likely that many natural populations are, to at least some extent, held short of adaptive trait optima as a result of gene flow from other populations.

Given that gene flow appears to be a common constraint, the next question becomes: how *strong* is that constraint? In other words, what is the magnitude by which traits deviate from adaptive optima as a result of gene flow? As described above, the ideal route to this inference is to eliminate gene flow and then monitor subsequent population divergence. In the few cases where this approach has been taken, an experimental reduction in gene flow quickly led to increased adaptive divergence (Riechert 1993, Nosil 2009). A somewhat less ideal—but mechanistically much easier—approach is to estimate the trait values best suited for a given environment and thereby also estimate the extent to which populations experiencing gene flow deviate from that optimum. If we take divergence at low gene flow to approximate optimal divergence, then published correlations (e.g., fig. 5.4) suggest strong constraints for many populations. In a quantitative assessment, Moore et al. (2007) examined a stream stickleback population experiencing high levels of gene flow from a lake population. Phenotypic divergence was here estimated to be only 14–20% of the divergence expected without gene flow. A constraint of this magnitude is certainly very large, but then the study population was specifically chosen because such a constraint was expected. By contrast, S. Fitzpatrick et al. (2015) argued that gene flow, while present, was not a substantial constraint on predation-associated adaptive divergence in guppies. Extending this quantitative approach, Paul et al. (2011) examined 10 populations of scarlet monkeyflowers (*Mimulus cardinalis*) along a latitudinal cline in western North America. For each population, the authors estimated the deviation of observed trait values from optimal trait values—and these deviations were often large. Moreover, the deviations were closely related to the level of gene flow from populations in different environments.

The adaptive divergence of traits is frequently constrained to some extent by gene flow between populations in different environments. The magnitude of this constraint is highly variable, and depends on the level of gene flow (greater when gene flow is higher), spatial variation in selection (greater [or perhaps lesser] when adjacent populations are in more divergent environments), and various other factors. **In at least some cases, the constraint is so high that trait divergence is far below that expected in the absence of gene flow.** Studies quantifying adaptive divergence in nature should more frequently and carefully consider the potential constraints imposed by gene flow. In addition, simply finding that traits are divergent among environments does not mean that gene flow is unimportant: that is, divergence could be even greater in the absence of gene flow.

QUESTION 2: TO WHAT EXTENT DOES GENE FLOW CONSTRAIN SPECIATION?

If gene flow constrains adaptive divergence (Question 1), and if adaptive divergence is a primary driver of speciation (chapter 6), then gene flow might also constrain speciation and therefore adaptive radiation. Constraints that gene flow places on speciation traditionally have been addressed in the context of sympatric versus parapatric versus allopatric speciation. The definitions of these terms have been highly variable (Mayr 1963, Bush 1994, Via 2001, Coyne and Orr 2004, Gavrilets 2004, Fitzpatrick et al. 2008, Butlin et al. 2008) and I will not here get into the debate. The key question is simply: to what extent does increased gene flow decrease progress toward speciation?

Gene flow is expected to be particularly high when divergence starts from a single initial population in a single physical location: that is, traditional (geographical) sympatry. Speciation has certainly occurred in this context, most obviously for phytophagous insects adapting to different host plants (Bush 1969, 1994, Via 2001, Berlocher and Feder 2002), plants adapting to different habitats on isolated islands (Papadopulos et al. 2011), and fishes adapting to different resources in lakes (Schluter 1996b, Jonsson and Jonsson 2001, Schliewen and Klee 2004, Barluenga et al. 2006, Siwertsson et al. 2010, Gordeeva et al. 2015). Despite such examples, it is clear that, of all speciation events, a relatively low percentage occur in strict geographical sympatry. For instance, meta-analyses have revealed that sister species do not show much spatial overlap (e.g., Barraclough and Vogler 2000) and that mobile organisms rarely speciate on islands (Losos and Schluter 2000, Coyne and Price 2000, Losos and Ricklefs 2009). Of course, exceptions do exist, including Tristan da Cunha *Neospiza* buntings (Ryan et al. 2007), *Oceanodromo castro* storm-petrels (Friesen et al. 2007), and several plant species on Lord Howe Island (Papadopulos et al. 2011). Generally, however, the few organisms that do speciation on islands tend to have low dispersal (and therefore low gene flow) relative to island size (Kisel and Barraclough 2010) (fig. 5.5). That is, speciation on islands is most likely for organisms that can experience spatial isolation even on a single island. Similarly, at least some of the explosive speciation of cichlids within large African lakes (Seehausen 2006) is influenced by spatial isolation within those lakes: gene flow is limited by distance and inhospitable areas, such as deep water or sandy habitats (Fryer and Iles 1972, van Oppen et al. 1997, Rico and Turner 2002). All of these patterns, and others, suggest that speciation is much more difficult when gene flow is very high than when it is somehow restricted.

What about speciation under low-to-modest levels of gene flow, such as between populations showing only partial spatial or temporal separation and exchanging some dispersers, a context often called parapatry. Theory suggests that speciation is much easier in such instances than in full sympatry (Endler 1977, Gavrilets et al. 2000, Doebeli and Dieckmann 2003, Thibert-Plante and Hendry 2009). However, formal demonstrations of parapatric speciation or "speciation-with-gene-flow" have been slow to accumulate (Coyne and Orr 2004), at least partly as a result of inferentially difficulties. One difficulty is that current species distributions are unlikely to mirror past distributions (Losos and Glor 2003), such that two species currently in parapatry might have diverged in allopatry (the same problem attends inferences of sympatric versus allopatric speciation). Also difficult is inferring gene flow that occurred during a speciation event in the past. Despite these problems, it is clear that many species pairs were exchanging genes at various stages of the speciation process, with examples including Darwin's finches (Grant et al. 2005, De León et al. 2010), lake-stream stickleback (Berner et al. 2009), *Timema* stick insects (Nosil 2007), Telmatherinid sailfin silversides (Herder et al. 2006), *Gyrinophilus* cave salamanders (Niemiller et al. 2008), and a variety of marine fishes (Puebla 2009). Moreover, many putative instances of allopatric speciation, such as among different islands, actually took place in the presence of at least some gene flow, with Darwin's finches again providing a

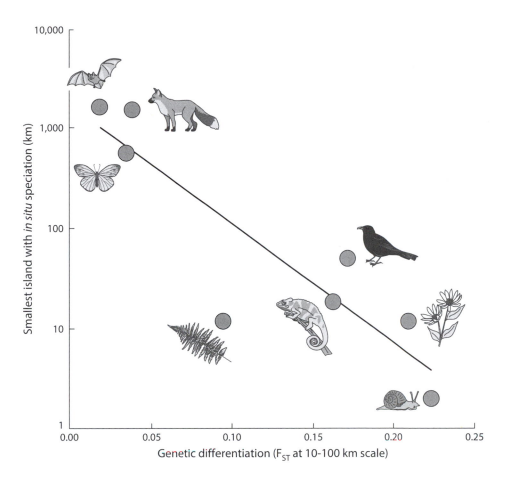

Fig 5.5. Taxonomic groups with lower gene flow (higher genetic differentiation) per unit distance are more likely to speciate on smaller islands. The *y*-axis shows the size of the smallest island or archipelago on which speciation has occurred for that group. The *x*-axis shows the typical level of genetic differentiation (F_{ST}) for each taxon between populations separated by 10–100 km. Bats, carnivores, and macrolepidoptera have high gene flow at this scale and typically do not speciate on small islands. Ferns, birds, lizards, angiosperms, and snails have low gene flow at this scale and are more likely to speciate on small islands. The data are from Kisel and Barraclough (2010) and were provided by Y. Kisel

clear example (Grant et al. 2005, Petren et al. 2005). Thus, even though incontrovertible evidence is spotty, speciation often takes place in the presence of some gene flow.

High gene flow constrains speciation in many instances, but low-to-modest gene flow is not a strict prohibition. That is, the effect of gene flow on speciation is not an all-or-nothing affair and is instead, as in adaptive divergence (question 1), a continuum of effects: that is, increasing gene flow makes speciation increasingly difficult. At the same time, gene flow can sometimes promote speciation. As one example, gene flow can strengthen reproductive barriers that first evolved under low gene flow, such as in character displacement and "reinforcement" on secondary contact (chapter 6). As other examples, gene flow can introduce variants that promote adaptation to new environments and thereby facilitate ecological speciation (e.g., Xie et al. 2007), and gene flow can lead to hybrid speciation (e.g., Jiggins et al. 2008). Further work will be necessary to determine the overall relationship between progress toward speciation and levels of gene flow at various times during that process.

QUESTION 3: DOES GENE FLOW GENERATE DEMOGRAPHIC COSTS?

Given that gene flow causes substantial constraints on *trait* divergence (question 1), how might this constraint matter for *overall adaptation*, such as the average absolute fitness of a population? Stated another way, to what extent can gene flow cause migration load[4] that leads to population declines—perhaps even to the point of extirpation? This question enters the realm of evolutionary influences on population dynamics, the topic that chapter 7 considers in detail. However, the present question fits well here because it flows logically from the questions above and into those below.

The basic prediction from theoretical models (Kirkpatrick and Barton 1997, Boulding and Hay 2001, Ronce and Kirkpatrick 2001, Tufto 2001, 2010, Bourne et al. 2014) is that maladaptation resulting from gene flow can depress individual fitness, mean population fitness, and population size. However, incorporating additional considerations into such models can reduce or eliminate these negative effects. First, gene flow (and dispersal) can have positive effects on adaptation, as discussed earlier. Second, deviations of some traits from adaptive optima might have minimal fitness consequences, most obviously when the traits are not under strong selection. Indeed, severe maladaptation seems likely only for traits that are *not* under strong selection. Third, trait maladaptation might not influence demography when population regulation is unrelated to the traits, as will be discussed further in chapter 7. As an illustrative example, imagine that the number of mature individuals exceeds the number of available breeding sites. Maladaptation might decrease the number of mature individuals, but—as long as they still exceed the number of breeding sites—the reproductive rate of the population might not decline.

Probably the best route to inferring potential demographic costs of gene flow is to experimentally manipulate dispersal and track any subsequent changes in population size. An interesting applied example is the use of "refuges" for transgenic Bt crops (Tabashnik et al. 2008, Carrière et al. 2010). The motivation in this context is that widespread use of Bt (*Bacillus thuringiensis*) crops imposes selection for resistance in pests, and the evolution of such resistance would decrease the efficacy of Bt crops. The solution has been to grow non-Bt crops (the "refuge") in close proximity to Bt crops, which then generates a cascade of effects. First, resistance to Bt is costly for the pest, and so selection in the refuge should reduce local Bt resistance (Gassmann et al. 2009). Second, the refuge-origin nonresistant pests should often disperse from the refuge and mate with pests from the Bt crops, thus generating "hybrids." Third, resistance to Bt is recessive and so the hybrids should be nonresistant and selected against on the Bt crops. As a result, much of the reproduction of Bt-resistant pests will be used up in hybrids that die owing to low resistance (Carrière and Tabashnik 2001, Tabashnik et al. 2005, 2008). Gene flow thus here delays the evolution of Bt resistance and sustains the ability of Bt crops to reduce pest population sizes, as has

---

[4]Various types of load have been described (Lande 1976, Barton and Partridge 2000), including standing genetic variation in fitness (standing load), new maladaptive variation owing to mutations (mutation load) or gene flow (migration load), natural selection within a generation (selection load), and deviations of the mean phenotype from the optimum (lag load).

been confirmed in small-scale experiments (Liu and Tabashnik 1997, Shelton et al. 2000, Tang et al. 2001). Experimental confirmation from the field is not extensive, but the fitting of empirical data to models suggests that refuges have been a primary factor delaying the evolution of resistance in pest populations (Tabashnik et al. 2008, 2009, Carrière et al. 2010), and a large-scale field study showed the expected effects of "refuges" from insecticide application (Carrière et al. 2010). As an alternative (or complement) to refuges, mass releases of nonresistant individuals could have similar effects: that is, gene flow from released non-resistant individuals should reduce the evolution of resistance in the wild population and therefore keep pest populations low (Alphey et al. 2007). Beyond these applied contexts, however, the few manipulations of gene flow in natural populations (Riechert 1993, Nosil 2009) have not assessed effects of altered adaptation on population size.

Given the rarity of experimental manipulations, we are left with correlative approaches that assess whether populations experiencing higher levels of gene flow have lower population densities. For example, Tack and Roslin (2010) showed that the abundance of phytophagous insect larvae on trees was negatively correlated with the fraction of those larvae that had immigrated from other trees. As always, such correlative analyses suffer from ambiguity as to cause and effect: are small populations the result or the cause of high gene flow? However, Farkas et al. (2013) were able to show that the negative association between maladaptation and gene flow in natural *Timema* stick insect populations was also attained in an experimental manipulation: an increase in maladaptive gene flow decreased population size. At a cruder level, we might note that all populations in which *traits* appear to be constrained by gene flow (i.e., those considered in question 1) are currently persisting in their supposedly maladaptive state. This observation suggests that gene flow does not have severe negative consequences for population dynamics, but it remains possible that other populations experiencing high gene flow did go extinct and the event simply wasn't recorded.

We currently have little knowledge of the extent to which high gene flow causes population declines or extirpation in natural populations. Although this paucity of information might partly reflect the difficulty of confirming such effects, it is also possible that the positive effects of gene flow typically balance or outweigh the negative effects. These and other possibilities will be considered in further detail in the questions that follow.

QUESTION 4: DOES GENE FLOW CONSTRAIN SPECIES RANGES?

As a special case of the above suggestion that maladaptive gene flow imposes demographic costs, Haldane (1956) and Mayr (1963) suggested that high gene flow from large, well-adapted populations at the center of a species range could swamp local adaptation at the range peripheries. Kirkpatrick and Barton (1997) confirmed that this swamping could depress peripheral population sizes and thereby prevent range expansion. Case and Taper (2000) then showed that this constraint could be exaggerated by interactions among species. This general idea that gene flow can cause range limits has proven very influential, although its basic prediction has vacillated from one model to another depending on the assumptions. For instance, the above models assumed a linear spatial gradient in the optimum trait value, no temporal variation in that spatial gradient, constant genetic variance, and no stochasticity. Barton (2001a) relaxed the assumption of constant genetic variance and showed that gene flow can be beneficial for peripheral populations by increasing the genetic variation that allows adaptation to new conditions. In this case, gene flow does not limit species' ranges. Polechová et al. (2009)

added temporal variation in the optimum and also found that gene flow did not limit species' ranges. Conversely, a simulation model by Bridle et al. (2010) added stochasticity by allowing the carrying capacity of local populations to be small. In this case, species ranges could again be limited by gene flow under a certain range of parameter space. In addition, Filin et al. (2008) showed that the outcome depended on the form of density regulation; and the models keep coming (e.g., Holt and Barfield 2011). Theory thus shows that gene flow might or might not constrain species ranges depending on the specific conditions. Thus, as is generally the case, the final answer must come from nature.

If gene flow constrains species' ranges, several criteria would be met: (1) individuals would disperse beyond the range limit (i.e., dispersal limitation is not the reason for range limits), (2) dispersers would be unable to establish new populations specifically because they are maladapted, and (3) maladaptation would be the result of gene flow (as opposed to, for example, limited genetic variation). The first criterion is certainly met in many instances given frequent observations of individuals beyond an established range, as well as the lack of physical barriers at many range limits. The second criterion can be evaluated by testing whether individuals at range margins are poorly adapted and would have difficulty surviving/reproducing beyond those margins. Supporting this idea, Savolainen et al. (2007) showed that, in at least some species, marginal populations are less well adapted than central populations. In addition, Sexton et al. (2009) reviewed 39 studies that transplanted individuals beyond range margins, and 28 of them reported fitness reductions relative to within the range. However, fitness reductions were not always present and, when present, were not always substantial—suggesting that at least some species should be able to expand their range (Samis and Eckert 2009, Sexton et al. 2009). The third criterion is the most difficult to evaluate and would, in principle, require an experimental manipulation of gene flow: cutting peripheral populations off from gene flow should trigger range expansion. Such experiments would be logistically challenging and seem not to have been attempted. (Introduced species are not especially informative here because they do not represent peripheral source populations incrementally expanding their range.) Getting part of the way to this goal, Sexton et al. (2011) used transplant experiments to show that gene flow from the center of a species' (*Mimulus laciniatus*) range would have negative fitness effects on populations at the warm limit of the range, but whether such effects were occurring naturally was not explored. In the absence of direct evidence, the tendency has been to argue for gene flow constraints when marginal populations appear maladapted, do not lack relevant genetic variation, and experience high gene flow from the main part of the range (Savolainen et al. 2007).

The processes generating range limits are difficult to study on large scales, such as across a continent, whereas the problem becomes much more tractable on small scales. As one example, Moore and Hendry (2009) examined the distribution of stickleback in a stream population experiencing very high gene flow from an upstream lake population. The authors found that maladaptation increased with increasing distance from the lake (see also Moore et al. 2007—as described above), and was closely associated with decreasing population density to a distribution limit about 2 km from the lake (fig. 5.6). Mark-recapture experiments showed that dispersal was present along the stream (dispersal limitation is not severe) and transplant experiments showed that local adaptation was present, but weak, near the distribution limit. These results were consistent with gene flow causing a constraint on the within-stream distribution of stickleback. In another small-scale study,

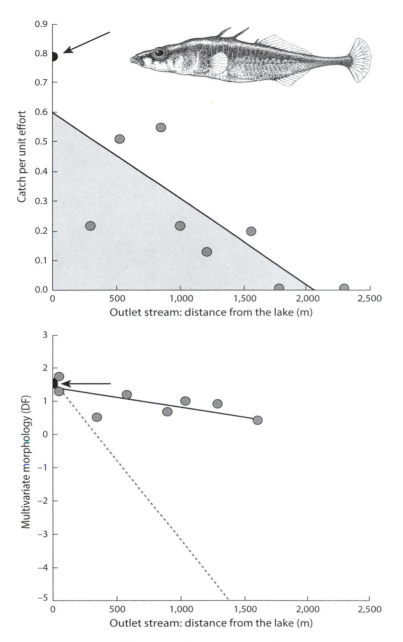

Fig 5.6. A putative example of high gene flow causing maladaptation that leads to small scale range limits in threespine stickleback. Black circles at the far left show, for lake fish, catch per unit effort (upper panel: number of fish per trap hour) and average multivariate morphology (lower panel: centroid from a discriminant function maximizing differences between lake and stream fish). Gray circles show the same for stream fish captured at different distances from the lake. The dashed line shows how the estimated optimal morphology for outlet fish varies with distance from the lake (for details see Moore and Hendry 2009). The graphs show that the deviation between optimal and observed trait values increases with increasing distance from the lake and that population density decreases correspondingly. The data are from Moore and Hendry (2009) and were provided by J.-S. Moore. The stickleback image was provided by D. Berner

Bridle et al. (2009) examined *Drosophila birchii* range limits along a steep altitudinal desiccation gradient. The authors found that the expected adaptive cline in desiccation resistance was absent, likely because gene flow swamped local adaptation. The implication is that high gene flow prevents adaptive clines along steep environmental gradients and therefore limits range expansion. However, a clear causal connection between gene flow and range limits has not been experimentally established in these (or other) small-scale studies.

Although species' range limits often might be the result of maladaptation (Hoffmann and Blows 1994, Bridle and Vines 2007, Kawecki 2008, Sexton et al. 2009, Moeller et al. 2011, Hargreaves et al. 2014, Lee-Yaw et al. 2016), gene flow seems unlikely to be the primary driver (except perhaps on some small scales). The reasons are dual. First, dispersal and gene flow have clear positive effects on range expansion (Barton 2001a, Thomas et al. 2001, Phillips et al. 2010). Second, widespread species tend to be patchily distributed, such that some patches should escape gene flow long enough to allow geographical expansion. Thus, although gene flow might in some cases constrain adaptation (Question 1) and decrease population size (Question 2), these effects seem unlikely to place long-term, large-scale limits on species' ranges (see also Paul et al. 2011).

QUESTION 5: DOES GENE FLOW ENHANCE ADAPTATION DURING ANTAGONISTIC COEVOLUTION?

I have thus far concentrated on the negative effects of gene flow that can result from the swamping of local adaptation, but I have also added frequent parenthetical allusions to positive effects of gene flow. I will now—and in the next question—focus on some of these latter effects. One situation where gene flow might be especially beneficial is antagonistic coevolution, such as host-parasite interactions. In these cases, ongoing evolutionary potential should be particularly important because the fitness peak to which a population is adapting is continually moving away. That is, adaptation by a parasite to a host promotes the evolution of resistance by the host, which then favors counter-adaptation by the parasite. The resulting need for ongoing evolutionary potential can be met through enhanced genetic variation, which might be brought about by gene flow (Gandon et al. 1996, Gandon and Michalakis 2002, Gandon and Nuismer 2009).

Experimental studies in the laboratory have provided a number of specific examples where gene flow increases local adaptation in host-parasite interactions. For example, Morgan et al. (2005) manipulated parasite (bacteriophage) and host (*Pseudomonas fluorescens*) dispersal rates among replicate populations, and found that parasite local adaptation (proportion of bacterial clones sensitive to the phage) was higher when dispersal occurred for the parasite but not the host. Another example is provided by Forde et al. (2004). But just how general is this benefit of gene flow? Hoeksema and Forde (2008) performed a meta-analysis of studies examining local adaptation of interacting species. They collated data from 24 cross-infection studies that evaluated the performance of parasites on local versus foreign hosts—dividing those studies into cases where gene flow was higher, lower, or similar for parasites relative to hosts. In general, parasites were better adapted to their local hosts when gene flow was higher for the parasites than for the hosts (fig. 5.7), supporting the idea that gene flow is an important source of genetic variation for ongoing adaptation.

During antagonistic coevolution, gene flow is a creative force that aids local adaptation. This effect has been clearly demonstrated for host-parasite systems and

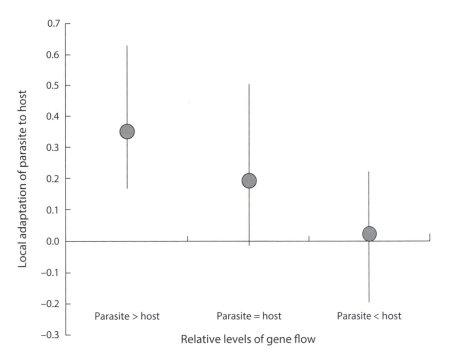

Fig 5.7. Results from a meta-analysis of reciprocal cross-infection studies showing that higher levels of gene flow in parasites relative to their hosts allows increased adaptation by the parasites. The data are means and 95% confidence intervals. Local adaptation is measured as the log response ratio of the performance of parasites on hosts with which they are sympatric relative to hosts with which they are allopatric. The data are from Hoeksema and Forde (2008) and were provided by J. Hoeksema

might also apply to other contexts, such as predators and prey. This positive effect of gene flow is most likely when gene flow is relatively low, whereas high gene flow likely starts to hamper local adaptation. In addition, the positive effects of gene flow should be strongest when population sizes are modest and adaptation to the other species is more important than is adaptation to abiotic factors (Nuismer 2006, Nuismer and Gandon 2008, Gandon and Nuismer 2009).

QUESTION 6: CAN GENE FLOW SAVE POPULATIONS FROM EXTINCTION?

I here consider two basic mechanisms by which gene flow can aid population persistence. These effects are in addition to the demographic benefits of *dispersal* in (1) maintaining sink populations that would otherwise go extinct, and (2) increasing population size and therefore the supply of new mutations (Gomulkiewicz et al. 1999, Holt et al. 2003, Alleaume-Benharira et al. 2006). First, small and isolated populations can suffer from inbreeding depression that results from the accumulation of recessive deleterious mutations or reductions in heterozygosity (Charlesworth and Charlesworth 1987, Husband and Schemske 1996, Keller and Waller 2002). Gene flow can reduce these effects by introducing novel genetic variation: so-called "genetic rescue" (Tallmon et al. 2004, Whiteley et al. 2015). Second, small and isolated populations can have low evolutionary potential (low V_A and h^2), which can limit adaptation to changing environments.

Gene flow can here supply new genetic variation that enhances evolutionary potential (Barton 2001a, Swindell and Bouzat 2006, Bell and Gonzalez 2011). Although these two mechanisms are conceptually different (inbreeding depression can reduce fitness even without changing environments), they will often be related in practice (e.g., inbreeding can reduce evolutionary potential).

To consider the effects of gene flow on inbreeding, three basic approaches have been used: (1) various levels of dispersal are implemented among experimentally inbred populations in the laboratory (e.g., Bijlsma et al. 2010), (2) naturally inbred populations are experimentally crossed to see if hybrids can enhance fitness (e.g., Bossuyt 2007, Willi et al. 2007), and (3) foreign individuals are introduced into inbred natural populations (e.g., Hedrick 1995, Westemeier et al. 1998, Ebert et al. 2002, Hogg et al. 2006, Fitzpatrick et al. 2016). In many of these instances, gene flow is found to have positive effects on fitness—and perhaps even on population size (Ebert et al. 2002, Tallmon et al. 2004, Pimm et al. 2006, Hedrick and Fredrickson 2010). In short, gene flow can facilitate genetic rescue when populations are very small, very isolated, show high inbreeding depression, and are not adapted to dramatically different environments (Richards 2000, Willi et al. 2007, Lopez et al. 2009, Bijlsma et al. 2010, Sexton et al. 2011, Whiteley et al. 2015)—although the last of these qualifiers is not universal (Fitzpatrick et al. 2016). Importantly, just a few migrant individuals per generation (i.e., very low gene flow) seems sufficient to eliminate, or at least greatly reduce, inbreeding depression (Hedrick 1995, Lopez et al. 2009).

To consider the effects of gene flow on adaptive potential, one approach has been to manipulate gene flow among experimental laboratory populations, which are then tested for adaptation to altered selection pressures. Although results are variable, a number of studies have shown that populations experiencing low levels of gene flow have a higher adaptive potential than populations not experiencing any gene flow (e.g., Swindell and Bouzat 2006, Perron et al. 2008, Bakker et al. 2010). Most of these studies did not, however, test whether the increased adaptive potential influenced population demographics and the probability of persistence. This latter possibility of "evolutionary rescue"[5] by gene flow was explicitly considered by Bell and Gonzalez (2011). Different laboratory populations of baker's yeast (*Saccharomyces cerevisiae*) were allowed to adapt to different levels of salt stress under different levels of dispersal. The evolved populations were then tested for their ability to adapt to even more stressful salt levels. Populations with a past history of dispersal were better able to adapt to the new conditions and were sometimes able to recover from the stress-induced initial population decline. In nature, correlative studies have shown that more isolated populations often (but not always) have lower levels of (neutral) genetic variation (Young et al. 1996, DiBattista 2008) but it isn't certain if the same is true for fitness-related traits (Kawecki 2008). Moreover, no empirical studies have formally examined evolutionary rescue by gene flow in nature (see Norberg et al. 2012 for a theoretical study).

Gene flow can increase fitness in populations suffering from inbreeding depression, and it can also increase the potential for evolutionary rescue in populations facing environmental change. These positive effects will not, however, be universal—because

---

[5]The effect of gene flow on inbred populations is usually called "genetic rescue," whereas the ability of adaptive evolution to save populations facing environmental change is usually called "evolutionary rescue." Evolutionary rescue in general will be considered in more detail in chapter 7.

"outbreeding depression" can occur if (1) populations are adapted to different conditions (i.e., the negative effects of gene flow on adaptation discussed earlier), or (2) the populations show intrinsic genetic incompatibilities (Tallmon et al. 2004, Edmands 2007, Frankham et al. 2011). The challenge, then, is to determine when increased gene flow will be beneficial, and just how much gene flow—and from what populations—will be optimal (Hedrick 1995, Tallmon et al. 2004, Edmands 2007, Weeks et al. 2011). In general, this balance is most likely to be achieved at very low gene flow levels, such as a few individuals per generation.

QUESTION 7: HOW VARIABLE IS GENE FLOW ACROSS THE GENOME?

The preceding questions have treated gene flow as a single genome-wide property. Earlier in this chapter, however, I noted that gene flow will vary across the genome depending on localized selection and recombination. The present question will consider how this variation is manifest in natural populations. The usual goal in such analyses is to infer regions of the genome that are under selection. For example, gene flow should be reduced in the vicinity of genes where different alleles are favored in different populations, but should be enhanced in the vicinity of genes where similar alleles are favored in different populations.

Many studies have surveyed large numbers of molecular markers in multiple populations and used variation in genetic divergence (e.g., F_{ST}) to infer gene flow across the genome. A review of 20 studies employing this "genome scan" approach revealed that 5–10% of loci are outliers that show especially high genetic divergence (Nosil et al. 2009a). These loci tend to be distributed widely across the genome, suggesting to some authors that genetic divergence conforms to a model of "genomic islands" (Wu 2001, Turner et al. 2005). In this model, most of the genome is not under divergent selection and so flows relatively freely among populations, whereas the few regions under strong divergent selection show noteworthy divergence. These results—and many more that have since followed—make clear that gene flow varies dramatically across the genome, but they are not necessarily indicative of the proportion of the genome that is under selection. One reason is that genome scans specifically target "outliers," and so won't easily reveal situations where selection is acting on many loci distributed widely across the genome (because selection can then alter the baseline). An early study highlighting this problem showed that genome scans detect many fewer regions under selection than do manipulative selection experiments and analyses of genotype-environmental associations (Michel et al. 2010). The implication is that much more of the genome is under selection than had previously been thought, promoting a model of "genomic continents." In addition, more powerful and precise sequencing methods are finding many more regions that appear to be under divergent selection than were revealed by earlier low-resolution methods (e.g., Lawniczak et al. 2010, Bergland et al. 2014, Soria-Carrasco et al. 2014). Some controversy attends the specific criterion, such as F_{ST} to identify such islands (Cruickshank and Hahn 2014) but, regardless, it is clear that some areas of the genome are much more differentiated than others.

What about the reverse effect: Do some regions of the genome show especially high gene flow? This pattern might arise if, for example, advantageous mutations arise in one population and spread to other populations through strong positive selection that is common across populations. This question is relevant to the long standing paradox of how species can evolve collectively when they are composed of multiple populations among which gene flow is restricted (Ehrlich and Raven 1969, Slatkin 1987, Morjan and Rieseberg 2004, Futuyma 2010). The proposed solution is that gene flow could be low across most of the genome but could be high for (and near) genes carrying alleles that are beneficial across the species range (Slatkin 1976, 1987, Morjan and Rieseberg 2004). This hypothesis is difficult to test because it is hard to distinguish regions where gene flow has been high owing to positive selection on new mutations from regions where positive selection has maintained similar ancestral alleles in the absence of high gene flow. Another difficulty is that such regions will be easiest to detect only when divergence is very high (and gene flow very low) across the rest of the genome.

Gene flow varies dramatically across the genome—and selection probably has a major role in this variation. However, it remains uncertain just how much of the genome is under divergent (or similar) selection, partly for the reasons described above and partly because genomic regions can have high F_{ST} for reasons unrelated (or at least not simply related) to divergent selection (Noor and Bennett 2009, Cruickshank and Hahn 2014, Lotterhos and Whitlock 2014, Roesti et al. 2014, 2015, Fraser et al. 2015). Related topics regarding the genetics and genomics of adaptation, speciation, and eco-evolutionary dynamics will be taken up in chapter 10.

Conclusions, significance, and implications

Gene flow influences eco-evolutionary dynamics by constraining or promoting adaptation and adaptive divergence. The classic perspective is that gene flow constrains adaptation by preventing the independent evolution of gene pools, which should impose "migration load" that depresses population size and also hinders speciation and (perhaps) range expansion. These negative effects of gene flow can have several consequences for ecological dynamics. First, instances of high gene flow can drag populations off fitness peaks, which can increase selection and create the dynamic conditions under which eco-evolutionary dynamics are most readily apparent (e.g., Farkas et al. 2013). Second, high gene flow can constrain dramatic changes in ecological effects by hampering adaptive radiation and preventing the evolution of novel forms that might have novel ecological effects. And yet it is also becoming clear that gene flow can have positive consequences for adaptation, particularly for small/isolated populations and when environmental change is severe or frequent (e.g., antagonistic coevolution). In these cases, gene flow can promote adaptation and will again enhance the ability to detect eco-evolutionary dynamics.

Given that gene flow can have both positive and negative effects, most theoretical analyses suggest that intermediate levels of gene flow should be "optimal" for local adaptation. This optimal level is expected to be very low, perhaps just a few migrants per generation. The positive versus negative effects of gene flow, and thus the optimal level of gene flow, will be modified by many factors, such as whether migrants are phenotypically biased relative to residents and whether they come from similar or different ecological environments. Finally, it is important to recognize that gene flow is not a single genome-wide property but is instead variable across the genome in response to varying selection and recombination. Much theoretical and empirical work remains to be conducted before we can determine optimal levels of gene flow under different conditions and the extent to which variable genetic divergence across the genome reflects variable gene flow and its responsiveness to selection and other factors.

Chapter 6

Ecological Speciation

The previous chapter outlined how gene flow can constrain adaptive divergence, which can thereby hinder the origin of new species. This constraint can be escaped through reproductive barriers that limit gene flow and thereby allow greater population divergence. "Extrinsic" reproductive barriers are properties of the environment; especially dispersal-limiting geographical barriers such as oceans, rivers, deserts, or mountain ranges (Dobzhansky 1940, Mayr 1963, Coyne and Orr 2004). In contrast, "intrinsic" reproductive barriers are evolved properties of organisms. A particularly important driver of intrinsic barriers is adaptive divergence that causes reductions in gene flow, a process now called ecological speciation (Schluter 2000a, Rundle and Nosil 2005, Nosil 2012).

Ecological speciation—beyond just adaptive divergence—can have several fundamental ecological consequences. First, reductions in gene flow that result from ecological divergence can then, in turn, drive greater ecological divergence (Räsänen and Hendry 2008). Second, and by extension, reproductive isolation allows a population adapting to a specific ecological situation to embark on an evolutionary trajectory that is no longer constrained by the collective evolution of multiple populations comprising the species as a whole (Futuyma 1987, 2010). For both of these reasons, ecologically driven reproductive isolation might be critical for evolutionary (and therefore ecological) divergence to proceed to substantial levels.

The present chapter will first describe modern conceptions of ecological speciation and how it can be inferred in natural populations. I will then address some key questions in the study of ecological speciation. As usual, coverage of the overall topic (ecological speciation) will be incomplete—focusing mostly on questions and issues particularly relevant to eco-evolutionary dynamics. For a more complete review of speciation in general, see Coyne and Orr (2004). For more detailed analyses of ecological speciation in particular, see Schluter (2000a) and Nosil (2012). In addition, ecological speciation in particular groups has been treated in depth in taxon-specific treatises, such as Grant and Grant (2008) for Darwin's finches, Price (2008) for birds in general, Tilmon (2008) for herbivorous insects, and Losos (2009) for *Anolis* lizards.

What is ecological speciation?

Ecological speciation occurs when the adaptation of populations to different environments drives the evolution of reproductive barriers among them (Schluter 2000a,

Coyne and Orr 2004, Rundle and Nosil 2005, Schluter 2009, Nosil 2012). To exemplify, the adaptation of threespine stickleback to different predation and foraging environments has contributed to the evolution of many reproductive barriers (McKinnon and Rundle 2002, Hendry et al. 2009a, details below). The same outcome has emerged through adaptation to different moisture conditions by *Mimulus guttatus* (Lowry et al. 2008b), to different elevations and pollinators by other *Mimulus* species (Ramsey et al. 2003), to different host plants by *Rhagoletis* flies (Bush 1969, Filchak et al. 2000) and *Timema* stick insects (Nosil 2007), to different "model" species by mimetic butterflies (Jiggins 2008) and hamlet fishes (Puebla et al. 2007), and to different diets by birds (Ryan et al. 2007, Grant and Grant 2008, Price 2008). Beyond these specific examples, ecological speciation is now assumed to be a major contributor to the diversity of life on earth (Mayr 1963, Schluter 2000a, Coyne and Orr 2004, Rundle and Nosil 2005, Funk et al. 2006, Price 2008, Losos 2009, Givnish 2010, Sobel et al. 2010).

In order to emphasize and highlight the unique features of ecological speciation, we must also outline—by way of contrast—some alternative speciation mechanisms.[1] First, the adaptation of different populations to *similar* environments can cause the fixation of different beneficial mutations that are incompatible on secondary contact, a process now called "mutation order" speciation (Porter and Johnson 2002, Schluter 2009, Nosil and Flaxman 2011). Second, genetic drift or founder effects can, independent of adaptation, similarly cause the fixation of different incompatible mutations (Nei et al. 1983, Gavrilets 2003). Third, chromosomal rearrangements or changes in ploidy can, under some circumstances, generate reproductive isolation without contributions from adaptation or drift (Otto and Whitton 2000, Rieseberg 2001, Faria and Navarro 2010). Fourth, sexual selection that contributes to reproductive isolation might diverge among populations in ecologically independent ways (Lande 1981, West-Eberhard 1983, Maan and Seehausen 2011). Among these potential speciation mechanisms, drift is probably the least important (Rice and Hostert 1993, Gavrilets 2003, Coyne and Orr 2004).

Whatever the speciation mechanism, a classic question is whether the process occurs in allopatry, parapatry, or sympatry. A variety of definitions for these terms have been suggested (Mayr 1947, 1963, Bush 1994, Via 2001, Coyne and Orr 2004, Gavrilets 2004, Fitzpatrick et al. 2008, Butlin et al. 2008, Mallet et al. 2009), with some relating to the degree of geographical overlap (e.g., Mayr 1947) and others to the amount of gene flow (e.g., Gavrilets 2003). Although the latter consideration is most important, I will sometimes need to invoke the geographical context. In particular, allopatric speciation occurs when populations are geographically separated to the extent that their dispersal ability does not allow frequent reproductive interactions. Sympatric speciation is the opposite extreme: populations are proximate enough that dispersal ability would allow frequent reproductive interactions, although dispersal *behavior* might restrict those interactions. Parapatric speciation is the broad range of intermediate possibilities, and is most commonly invoked when populations interact along a relatively narrow zone of

---

[1]Although "ecological speciation" *sensu stricto* requires divergent ecological environments, "ecology" considered more broadly can play a role in any speciation mechanism (Mayr 1947, Sobel et al. 2010, Maan and Seehausen 2011). For instance, divergent natural selection resulting from different genetic backgrounds independent of ecological differences is not ecological speciation. As another example, divergent sexual selection that results from ecological differences would be considered ecological speciation, whereas divergent sexual selection that arises solely for other reasons (e.g., Fisher's runaway process) would not.

contact. Of course, it is important to remember that current geographical context might not reflect the geographical context that prevailed (and perhaps varied) during the course of speciation (Losos and Glor 2003, Feder et al. 2005, Xie et al. 2007, Aguilée et al. 2013).

A final introductory point is that speciation can be considered as a process (speciation is occurring) or as an endpoint (speciation has occurred). I will emphasize the former perspective because it invokes process and because the relevant endpoint is often unclear. That is, although speciation is undoubtedly complete when a population has evolved total and irreversible reproductive barriers, many taxonomically recognized species still show hybridization and introgression with other species (Ellstrand et al. 1996, Mallet 2008). In short, the intraspecific to interspecific transition is variable and its delineation often arbitrary (Hendry et al. 2000a, Mallet 2008), as Darwin (1859, p. 486) took pains to make clear. It is therefore useful to think of speciation as "progress" along a continuum from panmixia to complete and irreversible reproductive isolation (Smith and Skúlason 1996, Drès and Mallet 2002, Hendry 2009, Nosil et al. 2009b, Peccoud et al. 2009, Berner et al. 2009, Hendry et al. 2009a, Siwertsson et al. 2010, Merrill et al. 2011, Renaut et al. 2011, Gordeeva et al. 2015). It is also sometimes useful to define stages along this continuum, while recognizing that the boundaries among the stages

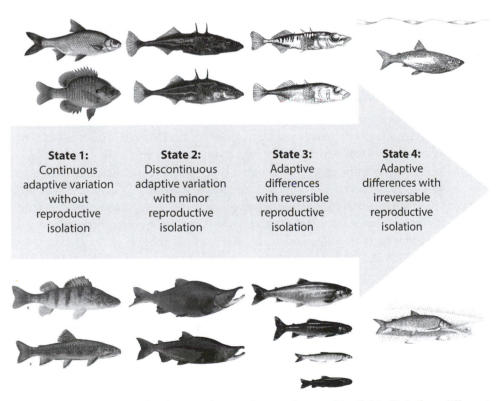

Fig 6.1. Ecological speciation is a continuum that can be roughly divided into four different states. Although populations are expected to proceed along this continuum, they might get stuck at any intermediate state and they might move in either direction. The images show species or ecotypes that represent different states along the continuum. The figure is from Hendry (2009), where additional details are provided

are not abrupt and that "progress" can occur in either direction (greater or lesser isolation). My attempt at heuristically applying such stages to postglacial fishes is shown in fig. 6.1.

How to infer ecological speciation[2]

When evaluating methods for studying ecological speciation, it is useful to consider the expected sequence of events (Schluter 2000a, Rundle and Nosil 2005, Räsänen and Hendry 2008, Nosil 2012). First, different groups of individuals, typically different populations, experience an ecological contrast, such as the occupancy (or use) of different environments, habitats, or resources. Second, this ecological contrast imposes divergent/disruptive selection among the groups. Third, this divergent/disruptive selection causes adaptive divergence. Fourth, this adaptive divergence causes reproductive isolation. Thus, the crucial extra requirement in inferring ecological speciation (beyond just adaptive divergence) is to confirm that reproductive isolation has evolved as a consequence of adaptive divergence. This requirement has engendered two specific study design elements.

One design element is a focus on the very early stages of speciation, such as among conspecific populations that show only partial reproductive isolation. This focus is intended to illuminate the mechanisms that actually drive speciation, as opposed to those that act after speciation is complete. As an example of why this focus is important, intrinsic genetic incompatibilities between two species might have arisen long after they stopped interbreeding for some other reason. These incompatibilities would certainly contribute to current reproductive isolation, but they would not be the initial cause of speciation. Focusing on these early stages does, however, cause several inferential limitations. One limitation is that partial reproductive isolation might never accumulate to the point of complete speciation, leaving uncertain whether the processes causing divergence within species are the same as those that permanently sunder them. Another limitation is that it can be difficult to infer how divergent selection was involved in older taxonomic splits. That is, species that formed long ago could well have arisen through ecological speciation—but the recognized definitive signatures of this process might have been obscured by subsequent evolution. Of course, this problem is not specific to ecological speciation: it is hard to conclusively infer any mechanism when considering older taxonomic splits. A potential solution to this second limitation is to compare, within a single taxonomic group, the nature of divergence and reproductive barriers within and among species (e.g., Langerhans et al. 2007, Nosil and Sandoval 2008, Peccoud et al. 2009, Hendry et al. 2009a, Merrill et al. 2011).

The second design element is motivated by the reality that intrinsic reproductive barriers can be assessed only for groups that actually interact. Most studies of ecological speciation therefore focus on populations that come into contact in nature (sympatric or parapatric) or in experiments. In the first case, any reproductive isolation must be intrinsic and can be studied in the natural context. In the second case, different groups can be brought into experimental contact in the laboratory or the wild. Benefits of the experimental approach (and thus limitations of studying unmanipulated populations) include increased replication, experimental controls, enhanced ability to isolate particular barriers (e.g., mate choice), potential to manipulate specific environmental factors, and suitability for allopatric populations (e.g., Funk 1998, Vines and Schluter 2006, Langerhans et al. 2007). Limitations of the experimental approach (and thus benefits of

---

[2]Parts of this section are modified from Hendry (2009).

studying unmanipulated populations) can include difficult logistics, questionable ethics (e.g., releasing individuals in nature), and the fact that artificial secondary contact, particularly in the laboratory, might not accurately mimic natural interactions.

Bearing in mind these two study design elements, I now consider the types of evidence typically provided for ecological speciation. Some types of evidence, although supportive and valuable, are not sufficiently robust to be definitive. For instance, ecological speciation is sometimes inferred simply when different populations or species show adaptive divergence— but uncertainty remains as to whether or not adaptive divergence was the *cause* of any reproductive barriers. Similarly, ecological speciation is sometimes inferred simply when different populations or species occupy/use different ecological environments—but it is also possible that the ecological divergence occurred after speciation (Rundell and Price 2009). Types of evidence that provide more definitive insight into ecological speciation fall into two main categories. One category explores integrated signatures of reproductive isolation, such as measures of genetic differentiation or estimates of gene flow. The other category assesses the extent to which reproductive barriers are associated with ecological differences, divergent selection, or adaptive divergence. I now consider these two approaches in detail.

INTEGRATED SIGNATURES OF ECOLOGICAL SPECIATION

Reproductive barriers should reduce gene flow among potentially interbreeding populations, an expectation that generates several testable predictions about patterns of genetic differentiation. I briefly describe these predictions, and then consider methodological issues relevant to their evaluation.

Gene flow should be lower than dispersal. If ecological speciation is occurring, adaptive divergence also must be present. If adaptive divergence is present, fitness should be higher for (local) individuals in their home environments than for (nonlocal) individuals arriving from other environments, as was explained in chapter 4. A consequence of this fitness difference is that the rate of dispersal by individuals, gametes, or zygotes should be higher than the resulting rate of gene flow because nonlocal immigrants[3] should perform poorly relative to local residents. This prediction can be tested by comparing the rate of dispersal as estimated by natural or artificial marks/tags to the rate of gene flow as estimated by neutral genetic markers (e.g., Hendry et al. 2000b). In essence, this test is an application of the traditional idea that reproductive barriers must be present if populations or species retain their genetic integrity despite potential interactions, most obviously in the case of full sympatry.

Gene flow should be lower when ecological differences are greater. Extending the above logic that ecological speciation requires adaptive divergence, we can further recognize that adaptive divergence requires an ecological contrast. Thus, a reasonable test for ecological speciation is whether genetic divergence is higher (and gene flow lower) among populations in different environments than among populations in similar environments (e.g., Crispo et al. 2006, Nosil et al. 2008, Lee and Mitchell-Olds 2011, Bradburd et al. 2013). As a specific example, Ogden and Thorpe (2002, see also Thorpe et al. 2010) showed that neutral genetic differences were greater among *Anolis roquet* lizards sampled from different habitats than among those sampled from similar habitats (fig. 6.2). This

---

[3]In this chapter, I will sometimes use the terms "migrants" and "immigrants" in the same sense as I used "dispersers" in chapter 5.

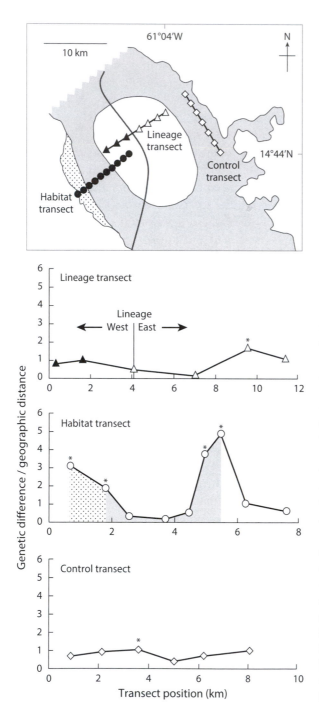

Fig 6.2. Neutral genetic differences between *Anolis roquet* populations on the island of Martinique (the southern and northern ends of the island are not shown) are greatest across habitat transitions. The top panel shows the three transects analyzed (sample locations are shown with symbols): a habitat transect that crosses three different habitat types, a lineage transect that crosses two different phylogenetic lineages, and a control transect within a single lineage and habitat type. The lower three panels show genetic differences scaled by geographical distance ($[F_{ST}/(1-F_{ST})]$ / distance in km $\times 10^{-2}$) between adjacent sampling sites along each transect. Symbols and shading are the same as on the map. Significant genetic differences are indicated with asterisks. The figure is redrawn from Ogden and Thorpe (2002) based on files provided by R. Thorpe

prediction has been termed "isolation by ecology" (Shafer and Wolf 2013) and "isolation by environment" (Wang and Bradburd 2014).

Gene flow should be lower when adaptive divergence is greater. If greater adaptive divergence leads to greater reproductive isolation (as expected under ecological speciation), populations showing greater adaptive divergence should show lower gene flow. A

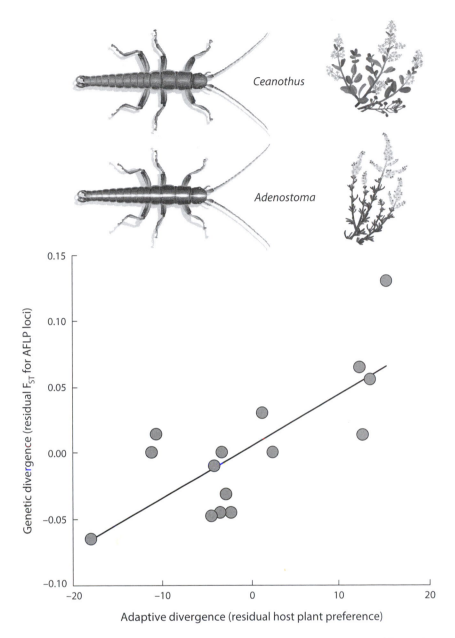

Fig 6.3. Illustration of "isolation-by-adaptation" in *Timema* stick insects. The points represent different pair-wise comparisons among six population pairs of stick insects on similar and different host plants. The x-axis indicates the degree of adaptive divergence as the degree to which stick insects from the different populations differ in the extent to which they prefer their local host plant type (host preference). The y-axis indicates genetic divergence (F_{ST}) for the 10% of AFLP loci showing the strongest association with adaptive divergence. Both axes are residuals from correlations with geographical distance. The images at top show the host plants (*Ceanothus* and *Adenostoma*) and the stick insect ecotypes adapted to them (drawings by Rosa Marín-Ribas). The data were taken from Nosil et al. (2008)

number of studies have therefore tested whether populations showing greater phenotypic differences in presumed adaptive traits also show greater genetic divergence, with early examples including Gíslason et al. (1999) and Bernatchez et al. (1999). More recent studies test this prediction through genome scans (see chapter 10) and have dubbed the expected pattern "isolation-by-adaptation" (Nosil et al. 2008, Orsini et al. 2013) (fig. 6.3). An important caveat in this approach is that divergence in specific phenotypic traits might not accurately reflect overall adaptive divergence (see chapters 4 and 5).

The above tests are all correlative, which leaves ambiguity regarding cause and effect: that is, adaptive divergence might be the cause or the consequence of low gene flow (Räsänen and Hendry 2008). Several supplementary analyses can reduce this uncertainty. For the first prediction (gene flow is lower than dispersal), some options include (1) demonstrating similar patterns in multiple independent replicates (parallel evolution) and adding controls (dispersal versus gene flow among populations in similar environments), (2) showing that dispersers have lower reproductive success than residents (e.g., Peterson et al. 2014), and (3) documenting specific reproductive barriers that are tied to adaptation (the approach described below). For the second prediction (gene flow is lower with greater ecological differences), parallel evolution and the study of reproductive barriers are again useful—and investigators can also statistically control for nonecological factors, such as geographical barriers, distance, or distinct lineages (Ogden and Thorpe 2002, Crispo et al. 2006, Thorpe et al. 2010, Lee and Mitchell-Olds 2011, Bradburd et al. 2013). For the third prediction (gene flow is lower with greater adaptive divergence), these supplementary analyses remain helpful and one can also seek to rule out the opposite casual pathway, with some suggestions provided in Räsänen and Hendry (2008).

These "integrated signatures of ecological speciation" have the benefit, over the reproductive barrier approach described below, of considering overall reproductive isolation in nature. This approach thereby circumvents concerns about whether or not particular reproductive barriers accurately reflect overall isolation, and whether or not experimental assays are relevant to natural interactions. But other concerns remain, which I here summarize only briefly (for more details see chapter 5). First, estimates of dispersal and gene flow are often imprecise and can be biased. Second, the success of dispersers among environments can be influenced by factors other than adaptation. Third, genetic differentiation will vary across the genome depending on a variety of factors, especially linkage to genes under selection (Charlesworth et al. 1997, Barton 2000, Nosil et al. 2009a, Feder and Nosil 2010). To get around this last complication, many studies focus on markers that are neutral and unlinked to any genes under selection, thereby hoping to quantify a "generalized barrier to gene flow" (Gavrilets 2004). Yet uncertainty often persists as to whether or not this neutral/ unlinked assumption is met; and, when met, whether such markers are informative with regard to ecological speciation (Gavrilets and Vose 2005, Thibert-Plante and Hendry 2010).

These and other complications make it tricky to infer ecological speciation from neutral genetic markers alone. It is therefore important to supplement such analyses with a consideration of specific reproductive barriers, the approach to which I now turn.

THE ECOLOGICAL BASIS FOR REPRODUCTIVE BARRIERS

Many types of reproductive barrier could evolve as a result of adaptive divergence (Schluter 2000a, Coyne and Orr 2004, Nosil et al. 2005, Rundle and Nosil 2005, Lowry et al. 2008a, Schemske 2010, Nosil 2012). In introducing some of these barriers, my goal is not to be comprehensive or definitive—but rather illustrative. After this listing of specific barriers, I will discuss general methodological considerations that arise when using reproductive barriers to infer ecological speciation.

Habitat preference: populations adapted to a particular habitat (or environment or resource) should evolve a preference for that habitat over alternative habitats (Rice and Hostert 1993, Fry 2003). The reason is that selection should often favor individuals with a genetic proclivity to remain in, or return to, a habitat to which they are well adapted. The resulting evolution of habitat preference should reduce cross-population encounters during reproduction, which should reduce gene flow. Assessments of habitat preference are particularly common in phytophagous insects and take two general forms: nonrandom associations between ecotypes and habitat types in nature (e.g., Via 1999) and laboratory tests for nonrandom choices among alternative host plants (e.g., Ferrari et al. 2006). Besides habitat preference, another way that individuals can remain in (or return to) an environment to which they are well adapted is philopatry or "natal homing" (Greenwood 1980, Johnson and Gaines 1990, Hendry et al. 2004b).

Selection against migrants: when adaptive divergence is present, individuals dispersing among environments (or habitats or resources) should be less well adapted than residents, and the dispersers should therefore suffer low fitness (Hereford 2009, chapter 4), which should reduce gene flow (Hendry 2004). Two likely routes to this "selection against migrants" are reduced survival ("immigrant inviability"; Nosil et al. 2005) and reduced fecundity ("immigrant infecundity"; Smith and Benkman 2007). Informing these possibilities, many studies have tested whether fitness or "performance" (e.g., foraging, growth, survival, or reproduction) in a given environment is higher for (local) individuals from that environment than for (nonlocal) individuals from other environments (chapter 4). In the context of ecological speciation, such tests have been performed in the laboratory (e.g., Schluter 1993), in field enclosures/plots (e.g., Schluter 1995, Lowry et al. 2008b, Moser et al. 2016), and in natural populations (e.g., Via et al. 2000, Peterson et al. 2014, Pfennig et al. 2015).

Temporal (allochronic) isolation: abiotic and biotic properties of the environment often have different phenologies in different locations, such as the timing of spring plankton blooms, bud break, flowering, wet/dry seasons, or freshets/droughts. Organisms whose successful reproduction depends on those properties will experience selection on reproductive timing that is divergent among environments. When reproductive timing is heritable, the initial outcome can be "isolation-by-time" (IBT: lower gene flow with greater time differences) and "adaptation-by-time" (ABT: adaptation to the conditions experienced at different times) (Hendry and Day 2005). IBT and ABT can then jointly reduce gene flow among organisms breeding at different times (Coyne and Orr 2004), leading to "allochronic speciation" (Alexander and Bigelow 1960). This barrier is often assumed to be important when sympatric/parapatric species that show low gene flow (1) have different reproductive timing (e.g., Thomas et al. 2003, Lowry et al. 2008b), (2) experience divergent selection on reproductive timing (e.g., Filchak et al. 2000, Hall and Willis 2006), or (3) have evolved divergent timing independently in multiple locations (e.g., Friesen et al. 2007, Yamamoto and Sota 2012). Interestingly, temporal isolation appears to evolve quite quickly (Quinn et al. 2000). (Of course, even very different habitats sometimes do not generate much allochronic isolation: Hanson et al. 2015.)

Pollinator (floral) isolation: in flowering plants, reproduction requires pollination and different plant traits can be differentially attractive to different pollinators, such as bees versus hummingbirds (Schemske and Bradshaw Jr. 1999, Ramsey et al. 2003), different wasps or moths (Weiblen 2002), and different bees (Schiestl and Schlüter 2009). Selection should favor adaptation to the locally most abundant or effective pollinators, which can thus lead to reproductive isolation among plant populations. This isolation can be the result of differences in floral structure (morphological isolation) or differences in pollinator behavior (ethological isolation) (Grant 1949, 1994, Schiestl and Schlüter 2009, Givnish 2010). Pollinator isolation is commonly inferred in the case of species-by-species matching between pollinators and plants, such as in figs and fig wasps (Weiblen 2002) and yuccas and yucca moths (Pellmyr 2003). It can also be inferred by measuring pollinator visitation or pollen transfer rates (Goulson and Jerrim 1997).

Mate choice: if the traits involved in adaptive divergence also influence mate choice, individuals adapted to a given environment might prefer to mate with individuals from the same environment: that is, assortative mate choice[4] (Nagel and Schluter 1998, Schluter 2000a). To consider this possibility, a number of studies have tested whether females (or males) are more willing to mate with males (or females) from similar environments than with those from different environments. These assessments have been conducted through controlled laboratory or field trials (e.g., Funk 1998, Rundle et al. 2000, Nosil et al. 2002, McKinnon et al. 2004, Schwartz et al. 2010, Merrill et al. 2011) or by examining mating patterns in nature (e.g., Cruz et al. 2004, Puebla et al. 2007, Huber et al. 2007). In addition to direct mate choice, ecological divergence in a mating trait (e.g., body size) can sometimes cause mechanical mating isolation as a byproduct (Richmond et al. 2011).

Extrinsic selection against hybrids: if hybrids are phenotypically divergent from parental types, as is often but not always the case (Rieseberg and Ellstrand 1993), they might be poorly adapted for either parental environment. A number of studies have therefore examined how well hybrids "perform" in relation to locally adapted parental types in the parental environments, bearing in mind that some hybrid crosses can have high fitness owing to heterosis (Edmands 1999, Barton 2001b). These studies have used either artificially generated hybrids in laboratory treatments or field enclosures/plots (Schluter 1993, 1995, Nagy 1997, Via et al. 2000, McBride and Singer 2010, Martin and Wainwright 2013, Arnegard et al. 2014, Moser et al. 2016), or have monitored the success of naturally produced hybrids in the wild (Grant and Grant 1996, Gow et al. 2007). Although the focus is usually on F1 hybrids, later generation hybrids or backcrosses should also perform poorly in the parental environments (Rundle and Whitlock 2001).

Other reproductive barriers: a number of other potential barriers, such as gametic incompatibility or intrinsic genetic incompatibility, could be involved in some cases of ecological speciation (Coyne and Orr 2004). However, these other barriers tend to be more difficult to assess in the context of divergent selection because (1) they are less clearly related to environmental differences, (2) they can take a long time to evolve, and (3) they act relatively late in the life cycle and thus are less likely to be the primary factors driving ecological speciation (Nosil et al. 2005, Lowry et al. 2008a, Schemske 2010, Nosil 2012). Finally, although geographical isolation is not often considered as an ecologically driven intrinsic barrier,

---

[4]I will use the general term "assortative mating" when mating is more common within than between population/species types: that is, positive assortative mating. Assortative mating can also refer to within-population mating based on phenotypic similarity (Kirkpatrick and Ravigné 2002, Jiang et al. 2013). I will use the specific term "assortative mate choice" when assortative mating is due to mate choice, as opposed to (for example) habitat choice or temporal isolation.

botanists have emphasized that adaptive divergence can be the *cause* of geographical isolation: that is, "ecogeographic isolation" (Ramsey et al. 2003, Sobel et al. 2010, Schemske 2010).

Regardless of the particular reproductive barrier under investigation, an important choice in experimental studies is whether to use individuals collected from the wild or those raised in a common-garden environment. This distinction is important because effects documented in the former are much more likely to involve contributions from plasticity, including prior experience, imprinting, developmental plasticity, and maternal effects. Hence, although patterns documented using wild-caught individuals might be most relevant to reproductive isolation in nature, the use of common-garden individuals is necessary to reveal any contributions of evolutionary divergence. As both contexts are interesting, more studies should quantify reproductive barriers using both types of individuals.

Given our specific interest in *ecological* speciation, it is important to establish that a given reproductive barrier has evolved because of adaptive divergence. A common route to this inference in studies of mate choice is to test for the parallel evolution of mating isolation (i.e., "parallel speciation"): specifically, mating should be assortative by environment type across multiple independent lineages (Funk 1998, Schluter 2000a). (Comparable experiments have only rarely been used for plants—see Ostevik et al. 2012). For example, benthic stickleback females should prefer benthic stickleback males over limnetic stickleback males regardless of the lake from which the specific males and females come (Rundle et al. 2000). Such a pattern would imply that adaptation to benthic versus limnetic environments *per se* has driven the repeated and deterministic evolution of mating isolation. As further examples, the same test has been applied to freshwater versus anadromous stickleback (McKinnon et al. 2004), *Timema* stick insects from different host plants (Nosil et al. 2002), guppies from high-predation versus low-predation environments (Schwartz et al. 2010), and freshwater isopods from reed versus stonewort habitats (Eroukhmanoff et al. 2011). A similar design could be applied to other reproductive barriers. For example, benthic stickleback should grow faster than limnetic stickleback in benthic enclosures regardless of the lake of origin and the lake of testing.

Testing for the parallel evolution of reproductive isolation can be very difficult or even impossible, and so other methods for inferring causality are more common. One is to establish whether specific reproductive barriers are caused by particular traits that have undergone adaptive divergence. Continuing with stickleback, body size is clearly under divergent selection between benthic and limnetic environments, and female stickleback often prefer to mate with male stickleback closest to their own size (Nagel and Schluter 1998, Boughman et al. 2005, for an example from lizards see Richmond et al. 2011). Likewise, beak size is under disruptive selection in Darwin's finches (Grant 1999, Hendry et al. 2009b) and beak size, along with its pleiotropic effects on song (Podos 2001), influences reproductive isolation (Ratcliffe and Grant 1983, Huber et al. 2007, Podos 2010). Similar approaches have been applied to color patterns in mimetic butterflies (Jiggins 2008) and hamlet fishes (Puebla et al. 2007, 2012).

Focusing more specifically on extrinsic selection against hybrids, an important question is whether or not reproductive isolation depends on the ecological context (Rice and Hostert 1993, Schluter 2000a, Coyne and Orr 2004). That is, hybrids should perform poorly because they are not well adapted for local conditions (relative to parental types), rather than owing to ecologically independent genetic incompatibilities. The simple and common approach to this determination is to show that hybrids have problems in nature but not under benign laboratory conditions; yet ambiguity can persist because some intrinsic genetic incompatibilities might be manifest only under stressful natural conditions (Coyne and Orr 2004). Several approaches to resolving such ambiguity have been suggested and implemented (Rundle and Whitlock 2001, Rundle 2002, Craig et al. 2007, Egan and Funk 2009, Arnegard et al. 2014).

An exclusive focus on reproductive barriers can leave uncertain the amount of overall progress toward ecological speciation—because overall reproductive isolation is a combination of all potential barriers (Nosil et al. 2005, Lowry et al. 2008a, Schemske 2010, Nosil 2012). Thus, the absence of a particular barrier does not rule out the possible existence of other barriers. Conversely, the presence of a particular partial barrier does not ensure substantial progress toward ecological speciation. First, other barriers might be absent. Second, introgression might be *increased* through an opposing reproductive "enhancer" (Hendry 2009). For example, colorful low-predation male guppies moving downstream over waterfalls into high-predation sites, where resident males are drab, might have lower survival (barrier: Weese et al. 2011) but higher mating success (enhancer: Schwartz and Hendry 2006, Labonne and Hendry 2010). Third, reproductive barriers are often assessed in artificial situations, such as the laboratory, and thus might not accurately reflect isolation in nature. For example, assortative mate choice in stickleback can be evident in the lab (McKinnon et al. 2004) but not in mesocosms or in nature (Jones et al. 2006, 2008). Fourth, reproductive barriers/enhancers are often asymmetric, constraining gene flow from one group into another while sometimes enhancing it in the reverse direction (Ellers and Boggs 2003, Forister 2004, Räsänen and Hendry 2014).

For all of the above reasons, the best studies examine multiple potential barriers in multiple populations in multiple ecological and geographical contexts (e.g., McKinnon and Rundle 2002, Nosil 2007, Jiggins 2008, Lowry et al. 2008b), and yet additional reproductive barriers and contexts could be considered in all natural systems. Given this necessary limitation, it is useful to focus first on barriers that would act early in a potential introgression event, such as habitat preference, selection against migrants, and temporal isolation. The reason is that later-acting barriers can only reduce gene flow that remains following early acting barriers—and meta-analyses have shown that early acting barriers do much of the job in ecological speciation (Coyne and Orr 2004, Nosil et al. 2005, Lowry et al. 2008a, Schemske 2010). In addition, it is good practice to combine "integrated measures of ecological speciation" with the reproductive barrier approach.

Ecological speciation in nature

Ecological differences have been thought to promote speciation ever since Darwin (1859), and this recognition was an important part of the modern synthesis (Dobzhansky 1940, Simpson 1944, 1953b, Lack 1947, Mayr 1947, 1963). However, it wasn't until the last 20 years or so that the field really took off with the formal exposition of the ecological theory of adaptive radiation and the role of ecological speciation therein (Schluter 2000a, Coyne and Orr 2004, Rundle and Nosil 2005, Nosil 2012). At present, evolutionary biologists don't question the fact that divergent selection can cause adaptive divergence which can contribute to the evolution of reproductive isolation. Yet many open questions remain that we can here consider, such as (Q1) how often is speciation ecological? (Q2) how often do ecological differences cause speciation? (Q3) how fast is ecological speciation? and (Q4) how "fragile" (prone to collapse) is it? Other topics address potential contributors to ecological speciation, including the importance of (Q5) competition within and among species, (Q6) "reinforcement" of mating preferences to avoid maladaptive between-type mating, (Q7) "magic" traits that are both under divergent selection and promote reproductive isolation, (Q8) the dimensionality (number of independent axes) of selection and adaptation, and (Q9) sexual selection that might promote or constrain mating isolation. As usual, I will address these questions by reference to meta-analyses whenever possible; otherwise I will provide key examples and speculate on their possible generality.

QUESTION 1: WHEN SPECIATION OCCURS HOW OFTEN IS IT ECOLOGICAL?

Ecological differences that cause divergent selection and thus drive adaptive divergence have clearly contributed to the evolution of reproductive isolation in a number of taxa (Lack 1947, Mayr 1963, Coyne and Orr 2004, Rundle and Nosil 2005, Price 2008, Grant and Grant 2009). However, the question remains: just how much of the current diversity of life has been the result of this particular speciation mechanism, as opposed to alternative mechanisms like uniform selection, polyploidy, or genetic drift? Note that we are here specifically interested in the role of *divergent* ecological environments (i.e., ecological speciation), while recognizing that "ecology" cast more broadly probably plays some role in most speciation mechanisms (Mayr 1947, Sobel et al. 2010). Although the current tide of opinion seems to favor ecological speciation as the predominant driver, some authors have argued that the question remains very much unresolved (Rundell and Price 2009, Svensson 2012).

One way to assess the general prevalence and importance of ecological speciation is to quantify associations across species/populations between ecological divergence and reproductive isolation. In so doing, one can obtain an estimate of effect size: How much of the variation in reproductive isolation can be explained by ecological divergence? An early demonstration comes from the work of Bolnick et al. (2006) on centrarchid fishes (fig. 6.4). The authors demonstrated that, after controlling for divergence time (using genetic differences), hybrid inviability was positively correlated among species with body size differences (fig. 6.4). The interpretation was that greater body size divergence reflects greater ecological divergence, which has generated greater reproductive incompatibilities. Funk et al. (2006) extended this approach by relating multiple reproductive barriers (prezygotic, postzygotic, total) to ecological divergence (habitat, diet, body size) among species within each of eight taxonomic groups, including fishes, birds, insects, and plants. Nearly all of the associations were positive, implying a role for ecological differences in the evolution of reproductive isolation. In this case, however, the correlations were very weak (mean $r = 0.12$). Along the same lines, a meta-analysis by Shafer and Wolf (2013) showed that ecological differences explained about five percent of the neutral genetic differences among conspecific populations (see also Orsini et al. 2013, Sexton et al. 2014, Wang and Bradburd 2014).

Meta-analyses thus suggest that ecological differences often contribute to reproductive isolation but, given how weak are the correlations, those differences might not be the primary driver. Such conclusions have to be tempered by serious caveats—some specific to the analysis methods and others more general. First, if ecological divergence drives reproductive isolation, it should reduce gene flow and thereby increase genetic differences.

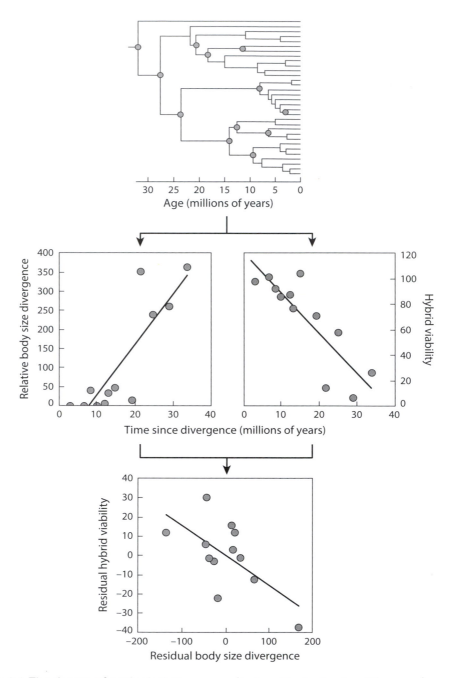

Fig 6.4. The degree of ecological divergence (indexed by body size differences) among centrarchid fishes is negatively correlated with the degree of hybrid viability. The top panel shows a phylogeny of centrarchids. The middle two panels show how increasing time of divergence corresponds to increasing ecological divergence and decreasing hybrid viability. The bottom panel shows how, after correcting for divergence time using residuals from the middle two panels, centrarchids with greater ecological divergence show lower hybrid viability. The data points are weighted averages for body size differences and hybrid viability estimates for phylogenetically independent nodes, indicated with circles in the top panel. The data are from Bolnick et al. (2006) and were provided by D. Bolnick

Hence, the use of genetic differences to estimate and control for "time since divergence" potentially conflates time and reproductive isolation: in short, the analysis might partly "control" for precisely the phenomenon it seeks to test. Second, most measures of ecological divergence (e.g., body size) are very crude and almost certainly underestimate total ecological divergence. Third, it is often uncertain whether ecological differences were the cause of reproductive isolation or rather arose after the fact (Rundell and Price 2009). Fourth, ecological differences will interact with other speciation mechanisms, such as sexual selection or polyploidy, and so the observed correlations do not necessarily reflect divergent selection *per se*. Note that these caveats are just specific illustrations of the usual problem that correlation does not necessarily reveal causation.

We still have very little understanding of how important ecological speciation has been, in general, to the evolution of biological diversity, whether alone or in combination with other speciation mechanisms. It is certainly clear that a number of speciation events were driven by divergent selection and a number weren't: the trick is to figure out which is which, how often, how much, and why or why not.

QUESTION 2: WHEN ECOLOGICAL DIFFERENCES ARE PRESENT, HOW OFTEN DO THEY CAUSE ECOLOGICAL SPECIATION?

The previous question asked how much of the reproductive isolation in nature has evolved owing to ecological differences. That is, when progress has been made toward speciation, what was the primary driver? The present question is the converse: How often do ecological differences in nature cause reproductive isolation? Stated another way, when divergent selection, the proposed driver of ecological speciation, is present, how effective has it been in causing substantial reproductive isolation. At a crude level, the quick answer must be "not very effective"—because most species are composed of many populations (Hughes et al. 1997) in different environments that nevertheless maintain a reasonably cohesive identity (Futuyma 1987, Morjan and Rieseberg 2004). Although we might consider these populations as having made some progress toward ecological speciation or to be "species in waiting," it is nevertheless clear that few of them will ever progress so far as to attain evolutionary independence. To be more definitive, however, we need quantitative analyses. No meta-analyses exist but several studies have addressed the problem by examining multiple population pairs with species, and I will here provide two examples.

Berner et al. (2009) used eight independent parapatric pairs of lake and stream stickleback to consider how ecological differences were associated with adaptive divergence and reproductive isolation. For lake fish versus stream fish at varying distances from each lake, ecological differences were estimated based on diet (limnetic versus benthic), adaptive divergence was estimated based on trophic morphology, and reproductive isolation was estimated based on neutral genetic markers. The eight pairs revealed a diversity of outcomes (fig. 6.5). For two pairs (Misty and Morton), reproductive isolation was very weak and lake/stream fish formed a single genetic cluster.[5] For the six other pairs, two genetic clusters were inferred, implying at least some progress toward ecological speciation. One cluster

---

[5]The contrast here is between Misty Lake and Misty *Outlet*, which other analyses have confirmed are nearly genetically homogeneous (Hendry et al. 2002, Moore et al. 2007). By contrast, the Misty Lake versus Misty *Inlet* comparison shows very high divergence, as has been detailed in other articles (Hendry et al. 2002, 2011, Berner et al. 2011) and in some other places in this book.

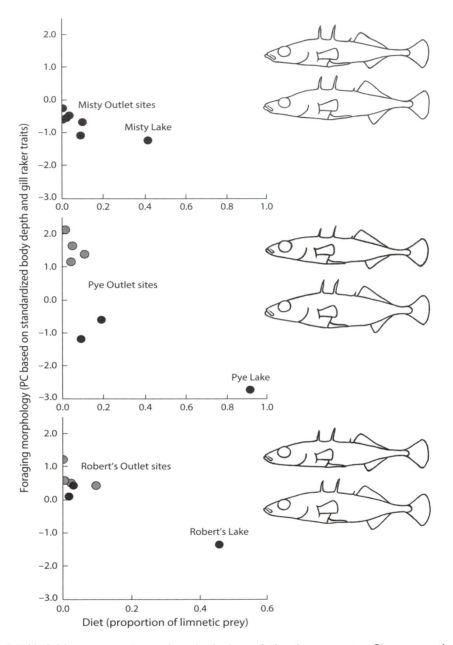

Fig 6.5. Variable progress toward ecological speciation in parapatry. Shown are three representative watersheds from among the eight analyzed by Berner et al. (2009). Each point (one lake and six outlet stream sites per watershed) shows the mean for 20 fish, with shading (black or grey) indicating the genetic cluster to which it belongs within that watershed. The top panel shows a watershed (Misty) where divergent selection based on diet is present (x-axis) but divergence in trophic traits (y-axis) is minor and all lake and outlet sites form a single genetic cluster. The middle and bottom panels show watersheds (Pye and Robert's) where divergent selection based on diet is strong, divergence in trophic traits is high, and two genetic clusters are present. Images to the right of each panel show the average body shape for lake stickleback (top) and for outlet stickleback farthest from the lake (bottom)—all scaled to the same body length (drawn by D. Berner). The data are from Berner et al. (2009) and were provided by D. Berner

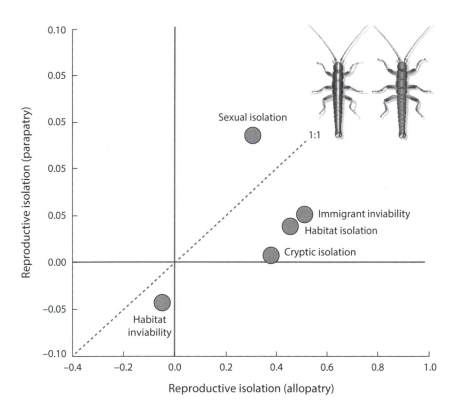

Fig 6.6. Reproductive barriers between host races of *Timema* stick insects. Each of five different barriers (the different points) was quantified for multiple population pairs and then averaged for allopatric population pairs (patches of the two host plants are isolated from each other) and parapatric population pairs (patches of the host plants are in contact). A value of zero indicates no barrier between host races, a value of unity would indicate a complete barrier between host races, and negative values indicate an enhancement of gene flow between (relative to within) host races. The fact that barriers tend to be stronger in allopatry than in parapatry (most points are below the 1:1 line) shows that gene flow constrains the evolution of reproductive isolation. Sexual isolation is the exception, probably because gene flow promotes the evolution of assortative mating, such as through reinforcement. The data were taken from Nosil (2007) and the images (*Adenostoma* ecotype on left, *Ceanothus* ecotype on right) were drawn by Rosa Marín-Ribas

always included the lake and at least one nearby stream sample, whereas the other cluster included more distant stream samples. However, the magnitude of divergence between the clusters, and its association with the lake-stream habitat transition, varied considerably (see also Roesti et al. 2012). For instance, large morphological/genetic differences were found in the Robert's and Pye pairs, but the transition was less abrupt and farther from the lake in the latter (the transition was even less abrupt and further away in other pairs). This variable progress toward ecological speciation could be attributed to two main factors (Berner et al. 2009): the enhancing effect of strong and abrupt environmental transitions and the constraining effect of high dispersal.

Nosil (2007) summarized the influence of dispersal on reproductive barriers between population pairs of *Timema* stick insects where the two host plants were either geographically isolated (allopatric) or adjacent (parapatric). Greater dispersal (i.e., parapatry) was found to constrain the evolution of several reproductive barriers (habitat preference, immigrant inviability, and "cryptic" isolation), whereas it promoted the evolution of sexual isolation (fig. 6.6). Negative effects of dispersal on the first three barriers presumably reflect the constraining role of gene flow (chapter 5), whereas a positive effect for the last barrier might reflect selection to avoid maladaptive mating (i.e., "reinforcement"—see question 6). As noted earlier, these effects among populations within species might not reflect the processes that permanently sunder species. It is therefore useful to combine among-population and among-species analyses. Continuing with stick insects, Nosil and Sandoval (2008) compared reproductive barriers between host races within two *Timema* species to the barriers between those species. The key result was that the transition from ecotypes to species was associated with a novel ecological axis: that is, only crypsis differentiated the host races whereas crypsis *and* physiology differentiated the species. These results were used to argue that substantial progress toward ecological speciation in *Timema* requires selection along multiple ecological axes (see question 8).

Although the data are still sparse, they clearly reveal that **progress toward ecological speciation is highly variable even when ecological differences are present**. Ecological differences thus sometimes do and sometimes do not cause the evolution of substantial reproductive isolation. Research emphasis should now be directed toward describing how specific factors influence progress toward speciation, as was illustrated above for dispersal (Nosil 2007), ecological transitions (Berner et al. 2009), and dimensionality (Nosil and Sandoval 2008, Nosil et al. 2009b, McPherson et al. 2015). Two additional factors, from among many, that are likely to be important are ecological opportunity (number and size of available niches—Yoder et al. 2010) and historical contingency (e.g., postglacial colonization history)—as illustrated for European whitefish, *Coregonus lavaretus*, by Siwertsson et al. (2010) and Arctic char, *Salvelinus alpinus*, by Gordeeva et al. (2015).

QUESTION 3: HOW FAST IS ECOLOGICAL SPECIATION?

Speciation by most mechanisms is thought to be very slow, typically requiring millions of years (Coyne and Orr 2004, Price 2008). Ecological speciation, however, might be relatively fast. This expectation derives from the intersection of two basic facts (Hendry et al. 2007). First, adaptive divergence can occur quite rapidly (Hendry and Kinnison 1999, Reznick and Ghalambor 2001), as described in previous chapters. Second, adaptive divergence drives the evolution of reproductive isolation, as described in the present chapter. Thus, progress toward ecological speciation might occur on the same short time scales over which contemporary evolution occurs.

Several theoretical models have addressed the rate of ecological speciation and suggested that substantial progress can occur after only dozens to hundreds of generations (Dieckmann and Doebeli 1999, Kondrashov and Kondrashov 1999, Hendry 2004, Thibert-Plante and Hendry 2009, Labonne and Hendry 2010, Rettlebach et al. 2013). For example, Thibert-Plante and Hendry (2009) used individual-based simulations of a population adapted to one environment that colonizes a new environment to show that "by 100 generations, essentially zero hybrids were formed." This result might seem in conflict with statements from other modeling studies that ecological speciation requires on the order of 10,000 generations (Gavrilets and Vose 2007, Gavrilets et al. 2007). However, as Thibert-Plante and Hendry (2009) point out, "this apparent difference of opinion is illusory because the different studies examined different spatial contexts (parapatry versus sympatry) and different degrees of reproductive isolation (partial versus nearly complete)." The reality, then, is that theory typically demonstrates substantial progress toward ecological speciation on very short time scales as long as divergent selection is strong and mate choice (or some other route to assortative mating) can evolve. Confirming these expectations, studies of divergent selection in the laboratory sometimes (although not always) find that partial reproductive isolation can evolve after dozens to hundreds of generations (Rice and Hostert 1993, Higgie et al. 2000, Dettman et al. 2007).

For natural populations, substantial progress toward ecological speciation is known to have occurred on the order of a few thousand generations, as exemplified by postglacial fishes diverging into benthic versus limnetic environments (Schluter 1996b, Taylor 1999, Hendry 2009, Siwertsson et al. 2010, Gordeeva et al. 2015). This sort of time scale was speculated by Darwin (1859, p. 120-123): "After fourteen thousand generations, six new species, marked by the letters n^{14} to z^{14}, are supposed to have been produced." Yet even here, however, divergence could have occurred much more rapidly, perhaps soon after colonization. To consider this possibility, we need to examine progress toward ecological speciation in younger population pairs. Although work on these shorter time scales was rare until recently, we can now point to a number of instances in which partial reproductive isolation evolved within only dozens to hundreds of generations (Hendry et al. 2007). Examples include fishes adapting to new environments (Hendry et al. 2000b, Pearse et al. 2009, Pavey et al. 2010, Elmer et al. 2010, Furin et al. 2012, Phillis et al. 2016), birds evolving new migration routes (Bearhop et al. 2005, Rolshausen et al. 2009), plants adapting to mine tailings (Davies and Snaydon 1976) and fertilizer treatments (Silvertown et al. 2005), native insects colonizing introduced plants (Filchak et al. 2000, Sheldon and Jones 2001), and a number of other cases (Byrne and Nichols 1999, Eroukhmanoff et al. 2011, Montesinos et al. 2012). Many more such examples are likely to accumulate in the near future.

Noteworthy progress toward ecological speciation can take place on the same time scales as adaptive divergence, perhaps only dozens to hundreds of generations. However, rapid and substantial progress is not inevitable, nor perhaps even common (see the above questions). Indeed, many laboratory studies show no evolution of reproductive isolation despite divergent selection over multiple generations (e.g., Rundle 2003, Kwan and Rundle 2010). Moreover, the evolution of complete and irreversible reproductive isolation as a result of divergent selection probably takes considerably longer—leading us to the next question.

QUESTION 4: HOW FRAGILE/REVERSIBLE IS ECOLOGICAL SPECIATION?

Adaptive divergence can lead to the evolution of reproductive barriers whose effectiveness depends on the current environment, a phenomenon variously called "environment-dependent" (Rice and Hostert 1993), "ecologically dependent" (Rundle and Whitlock 2001), or "extrinsic" (Coyne and Orr 2004). Two contexts in which this phenomenon might be manifest are (1) dependency on current *differences* among the parental environments, or (2) dependency on current conditions that are *shared* across environments. One example of the first context is when hybrids are ecologically different from parental forms and therefore have lower fitness in both parental environments (Schluter 2000a, Rundle and Whitlock 2001, Rundle 2002). Another example is when migrants or hybrids have low mating success because they are adapted to divergent parental signaling environments (Boughman 2001, Seehausen et al. 2008b). An example of the second context is when mate choice depends on general environmental conditions, such as fish visual communication in clear water (Seehausen et al. 1997, van der Sluijs et al. 2011). These dependences on current ecological conditions can mean that reproductive isolation is fragile and can break down when environments change, potentially causing "reverse speciation" (Seehausen et al. 2008a, Gilman and Behm 2011). (This effect is separate from the known role of range changes, such as through introductions, on promoting hybridization and thereby causing speciation reversal.) For instance, the emergence of an intermediate environment, or a decrease in the distinctiveness of parental environments, could increase hybrid fitness. Or mate choice might become less precise if the environment for signal transmission is degraded (Seehausen et al. 1997). No meta-analyses have considered the fragility of ecological speciation, and so I will simply here outline three clear examples: for another, see Vonlanthen et al. (2012).

Darwin's finches are reproductively isolated through assortative mate choice and ecologically dependent selection against hybrids (Ratcliffe and Grant 1983, Grant 1999, Huber et al. 2007, Grant and Grant 2008, Hendry et al. 2009b, Podos 2010). The latter barrier depends on a scarcity of foods suitable for hybrids and is potentially exacerbated by competition with the parental species. If, however, environments become more benign through an increase in food availability, selection against hybrids might decrease and introgression therefore increase. This effect has been documented in two instances. First, La Niña conditions that dramatically increased the availability of seeds eliminated the survival disadvantage formerly suffered by hybrids between the medium ground finch and the cactus finch on the island of Daphne Major (Grant and Grant 1993, 1996). The outcome was ongoing morphological (Grant and Grant 2002) and genetic (Grant et al. 2004) convergence of the species. Second, increasing human population density at Academy Bay on Santa Cruz Island altered the available foods and precipitated the fusion of two formerly distinctive beak size morphs within the medium ground finch (Hendry et al. 2006, De León et al. 2011) (fig. 6.7).

The rare benthic-limnetic sympatric species pairs of threespine stickleback are reproductively isolated through assortative mate choice and ecologically dependent selection against hybrids (Rundle et al. 2000, Rundle 2002, Gow et al. 2007, Hendry et al. 2009a, Arnegard et al. 2014). In one of the lakes (Enos), the two formerly distinctive species have merged into a single hybrid swarm (Taylor et al. 2006), reflecting a breakdown of ecologically dependent reproductive barriers. In particular, studies have documented reduced selection against hybrids and reduced assortative mate choice (Gow et al. 2006, Behm et al. 2010, Gilman and Behm 2011, Lackey and Boughman 2013). A likely culprit for the critical ecological change is an invasive signal crayfish (*Pacifasticus leniusculus*) that has altered water clarity and changed stickleback reproductive behavior (Velema et al. 2012). Interestingly, recent work has shown how this particular species collapse might alter community and ecosystem properties (Rudman and Schluter 2016), thus providing links to the concepts discussed in chapters 8 and 9.

A number of haplochromine cichlid fishes in Lake Victoria are reproductive isolated through assortative mating (Seehausen et al. 1997, van der Sluijs et al. 2008). In particular, strong isolation by mate choice is achieved through dramatic differences in male color pattern and corresponding female preferences. This reproductive barrier depends on clear water providing a broad spectrum of wavelengths necessary to see a broad spectrum

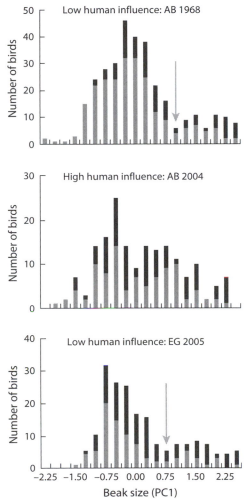

Fig 6.7. Illustration of how human influences can reverse the process of speciation. The image at top shows variation in *G. fortis* beak size (small versus large morphs) on Santa Cruz Island (photo by A. Hendry). The middle two panels show the frequency distribution of beak sizes (PC1 based on beak length, depth, and width) in this species at Academy Bay before (1968) and after (2004) a large increase in human population density. The bottom panel shows the frequency distribution of beak sizes in 2005 at another site (El Garrapatero) that does not experience strong human influences. The black portions of bars are mature males and the gray portions are all other adults. The vertical arrows highlight the partial discontinuity between beak size modes when human influences are low. The same discontinuity is absent when human influences are high. The data are from Hendry et al. (2006)

of color. However, eutrophication over the past 50+ years has reduced water clarity and the spectrum of light available at each depth. The resulting change in the efficacy of mate choice appears to have caused the fusion of formerly distinct species into a hybrid swarm (Seehausen et al. 1997, 2008b, Konijnendijk et al. 2011).

Ecological speciation is fragile, at least in its early stages, primarily because reproductive isolation is often ecologically dependent, and therefore sensitive to changing conditions. Thus, although divergent ecological environments can sometimes cause rapid progress toward ecological speciation (the previous question), the resulting reproductive isolation can remain dependent on current environmental conditions. Escape from this ecological-dependency through the evolution of absolute and irreversible reproductive isolation seems to take considerably longer, and might often depend on nonecological mechanisms (Coyne and Orr 2004, Price 2008). Even so, the nonecological "completion" of speciation could find its beginnings in ecological divergence. First, ecological differences might reduce gene flow to the point that complete and permanent barriers can evolve. Second, ecological speciation begins whenever populations colonize/use different environments, thereby increasing the number of chances for the (likely) rare events that permanently sunder evolutionary lineages. Third, divergent selection might directly cause the evolution of permanent ecologically independent reproductive barriers, such as genetic incompatibilities (Agrawal et al. 2011).

QUESTION 5: HOW IMPORTANT IS COMPETITION TO ECOLOGICAL SPECIATION IN SYMPATRY?

Sympatric speciation is most likely when strong disruptive selection is coupled to a mechanism for assortative mating (Kirkpatrick and Ravigné 2002, Bolnick and Fitzpatrick 2007); but what drives the disruptive selection? One possibility is the existence of discrete, or at least reasonably discrete (i.e., multimodal), habitats or resources, such as different host plants on which phytophagous insects can specialize (Drès and Mallet 2002, Funk et al. 2002). The other potential source of disruptive selection arises when the most common phenotypes experience the strongest competition and thereby have the lowest fitness (Rueffler et al. 2006). Under some conditions, this competition-based disruptive selection can lead to speciation even in the absence of initially multimodal resource distributions (Dieckmann and Doebeli 1999, Doebeli and Dieckmann 2003, Bolnick 2004, Rueffler et al. 2006, Doebeli et al. 2007, Rettelbach et al. 2013). The question I consider here is which of these two potential sources of disruptive selection (multimodal resources or competition) is most likely to drive sympatric speciation?

Theoretical models tend to conclude that sympatric speciation is easier in the case of multimodal resources than competition on unimodal resources (Doebeli 1996, Thibert-Plante and Hendry 2011b, Débarre 2012); but how might we test the relative effects of these two sources of disruptive selection in nature? One approach has been to test natural populations for quadratic selection (chapter 2) and its dependency on competition. In stickleback, for instance, Bolnick and Lau (2008) used field surveys to show that some lake populations experience disruptive selection (although seemingly only some of the time—Bolnick and Araújo 2011), and Bolnick (2004) used enclosure experiments to show that this selection was stronger when density, and therefore competition, was higher. However, this approach (see also Martin and Pfennig 2009) has several limitations. First, disruptive selection is expected to be strong only during the speciation event itself, not afterward (Rueffler et al. 2006); and so the absence of disruptive selection within populations of established species does not negate its potential importance during species formation. Second, the presence of disruptive selection within panmictic populations (e.g., Bolnick and Lau 2008) suggests that speciation does *not* proceed by that mechanism—because it would be expected to eliminate panmixia (Bolnick 2011). A route around these limitations is to study population/species in the very early stages of divergence. For instance, the above-mentioned sympatric beak size morphs of the medium ground finch are at least partly maintained by disruptive selection (Hendry et al. 2009b), although the importance of competition has not been tested. More directly, Martin and Wainwright (2013) used F2 hybrids between *Cyprinodon* pupfish species to show that disruptive selection in experimental enclosures was strongest when densities were highest.

Despite the above hints, we can currently point to only a few cases from natural populations where competition on unimodal resources was clearly the primary cause of ecological speciation in sympatry. This paucity of empirical evidence might reflect the relative unimportance of this mechanism or simply the difficulty of conclusively demonstrating its role. By contrast, one context in which the importance of competition is indisputable is secondary contact between partially divergent groups. In this case, individuals of each group with phenotypes closest to the other group will experience elevated competition and hence lower fitness (Schluter 2003, Pfennig et al. 2007). The expected result is competition-mediated divergent/disruptive selection that causes ecological character displacement, a phenomenon arguably common in nature (Brown Jr. and Wilson 1956, Schluter 2000b, Pfennig and Pfennig 2009, 2012, Stuart and Losos 2013, Beans 2014).[6] I will later reconsider competition and character displacement in the context of community structure (chapter 8). For the present chapter, the key point is that competition between partially divergent groups can generate selection that promotes ecological speciation (Schluter 2000a, Rundle and Nosil 2005, Pfennig and Pfennig 2009, 2012).

Although many sympatric speciation models have studied competition on unimodal resources, **speciation probably rarely proceeds solely in this context—at least in comparison to adaptation on multimodal resources. In the presence of multimodality, however, competition might well speed divergence**. One reason is that competition in a population established on one resource reduces the fitness of individuals on that resource and therefore favors colonization of other, underutilized resources (Bolnick 2001, Winkelmann et la. 2014). **In addition, competition can enhance divergence on secondary contact by causing ecological character displacement.**

QUESTION 6: HOW IMPORTANT IS REINFORCEMENT?

The above question considered the effects of ecological interactions (competition) on speciation, whereas the present question considers the effects of *reproductive* interactions. In general, interbreeding among populations is expected to place a constraint on speciation (Felsenstein 1981, Coyne and Orr 2004, Gavrilets 2004); but an exception occurs in the case of direct or indirect selection for reduced interbreeding.

---

[6]If the two species are very similar in their resource use, secondary contact could favor *convergence,* at least in theory (Abrams 1987, Fox and Vasseur 2008, terHorst et al. 2010b).

Direct selection could occur if costs of the mating act itself are higher for interspecific than for intraspecific matings, with possible examples including mechanical damage or increased risk of predation (Servedio 2001, Albert and Schluter 2004). Indirect selection could occur if hybrids have low fitness (Dobzhansky 1940, Noor 1995). The result in either case could be an evolutionary increase in assortative mating when individuals from different populations interact: that is, "reinforcement" in the broad sense (Servedio 2001, Servedio and Noor 2003, Ortiz-Barrientos et al. 2009). Reinforcement is classically considered to be driven by ecologically independent reproductive incompatibilities, but theoretical models have shown it can be just as important in the context of ecological speciation (Thibert-Plante and Hendry 2009, Aguilée et al. 2013, Rettelbach et al. 2013).

Many studies have documented situations where closely related species in parapatry/sympatry—relative to allopatry—show enhanced divergence in reproductive characters (i.e., reproductive character displacement) and/or enhanced reproductive barriers (i.e., potential reinforcement) (Servedio and Noor 2003, Coyne and Orr 2004, Ortiz-Barrientos et al. 2009, Yukilevich 2012, Turelli et al. 2013). These patterns suggest that reproductive interactions among populations/species can strengthen reproductive isolation, but most studies focus on the role of intrinsic genetic incompatibilities whereas we are here more interested in the role of ecological differences. In this latter context, some putative examples of reinforcement exist, such as greater mating isolation among sympatric/parapatric than among allopatric populations of threespine stickleback (Rundle et al. 2000), *Timema* stick insects (Nosil et al. 2003) (fig. 6.6), and Trinidadian guppies (Schwartz et al. 2010). In addition, ecological and nonecological barriers can interact, such as when intrinsic incompatibilities drive the evolution of ecologically based premating barriers. An example is flower color divergence that generates nonrandom pollinator movement that reduces hybridization between two species of *Phylox* that manifest intrinsic incompatibilities (Hopkins and Rausher 2012). Yet counter-examples are also known: reinforcement failed to reduce gene flow following experimental secondary contact between two mosquito species (Urbanelli et al. 2014) and after human disturbances increased interactions between two herring species (Hasselman et al. 2014). Overall, we have few clear examples where ecological speciation has been enhanced by reinforcement, although this paucity may only be for want of looking; and we also don't know how often ecological differences *fail* to cause reinforcement.

Although reproductive interactions are known to enhance ecological speciation in at least some instances, the general importance of this effect is not known. It is also uncertain to what extent reproductive character displacement is an indirect consequence of ecological character displacement (see above; Pfennig and Pfennig 2012) versus a direct consequence of selection to reduce interbreeding. In addition, debate persists as to whether or not reinforcement can complete speciation, given that its evolution reduces maladaptive interbreeding and thereby decreases selection for further reinforcement (Bank et al. 2012, Rettelbach et al. 2013). Finally, it has been suggested (Nosil et al. 2003, Coyne and Orr 2004, Nosil 2013) that the evolution of reinforcement is maximal at intermediate levels of reproductive interactions (e.g., intermediate dispersal) because infrequent interactions will not strongly select for reduced interactions (as noted above) whereas too frequent interactions can lead to homogenizing gene flow (chapter 5).

QUESTION 7: HOW IMPORTANT ARE MAGIC TRAITS?

Ecological speciation, particularly in the presence of ongoing gene flow, is expected to proceed most readily when adaptive divergence and reproductive isolation are based on the same (or closely linked) genes, such as in "magic trait" models[7] (Fry 2003, Gavrilets 2004, Thibert-Plante and Gavrilets 2013). In a broad sense, any trait involved in local adaptation could be an "automatic" magic trait (Servedio et al. 2011) because it should, by definition, cause selection against migrants and (often) hybrids (Coyne and Orr 2004, Hendry 2004, Nosil et al. 2005, Thibert-Plante and Hendry 2009, Schemske 2010). Automatic magic traits are thus likely to be common and important contributors to ecological speciation, at least in its early stages. However, substantial progress toward ecological speciation additionally requires the evolution of assortative mating (Kirkpatrick and Ravigné 2002, Gavrilets 2004, Bolnick and Fitzpatrick 2007). In this case, the focus is on "classic magic traits," which experience divergent/disruptive selection and also promote positive assortative mating (Servedio et al. 2011).

A number of strong candidates have emerged for classic magic traits (Servedio et al. 2011), and I here provide just a few examples. First, adaptation to different environments should favor habitat preferences for those environments (as described above), which should increase assortative mating (Rice and Salt 1990, Fry 2003). Second, divergence in ecological traits (e.g., bird beaks) can lead to pleiotropic divergence in mating signals (e.g., songs) that promote assortative mate choice, an outcome that is particularly likely when offspring imprint on parental signals—as is common in birds (Price 1998, 2008). Third, color traits are often involved in adaptation to different environments and also in mate choice, with examples including mimetic butterflies (Jiggins 2008) and hamlet fishes (Puebla et al. 2007). Fourth, adaptation of floral traits to different pollinators can generate pollinator isolation as a byproduct (Schemske and Bradshaw Jr. 1999, Schiestl and Schlüter 2009). Fifth, MHC genotypes are under divergent selection among environments with different parasite faunas and those genotypes influence mate choice (Eizaguirre et al. 2009). These and other cases are almost certain to represent classic magic traits—and many more examples are likely to be forthcoming.

Although magic traits could be common, the critical question is: How important are they for ecological speciation (Haller et al. 2012)? Stated another way, just how dependent is ecological speciation on magic traits? A first key point is that magic traits are not necessary for ecological speciation given that many theoretical models have shown reproductive isolation can evolve even in their absence (e.g., Kondrashov and Kondrashov 1999, Fry 2003, Gavrilets 2004, Thibert-Plante and Hendry 2009). In such nonmagic cases, speciation should proceed more readily when reproductive isolation involves fixation of the same allele (e.g., "mate with my own type"), as opposed to different alleles (e.g., "mate with blue males" versus "mate with red males"), in different populations (Felsenstein 1981, Kirkpatrick and Ravigné 2002). A specific one-allele mechanism occurs when the expression of a trait in males (e.g., plumage color) is conditional on strong performance in local environments, such that females (in all populations) preferring males with high

---

[7]The definition of a magic *trait* is actually the definition of a magic *gene*: a gene under divergent/disruptive selection that also influences reproductive isolation (Gavrilets 2004, Servedio et al. 2011). In many cases, it is instead more useful to talk about magic phenotypes (Shaw and Mullen 2011).

values of that indicator trait select locally adapted males and thereby discriminate against maladapted immigrants (van Doorn et al. 2009, Thibert-Plante and Gavrilets 2013). Yet a number of models also have shown that speciation can occur even in two-allele models, depending on the circumstances (e.g., Fry 2003).

Given that many options remain open in theory, we can really only address the importance of classic magic traits by examining their contribution in nature. This examination is exceedingly difficult because, in principal, it would require quantifying the contribution of divergent selection acting through a specific gene to total reproductive isolation that evolved during (as opposed to after) a speciation event.[8] However, a good place to start is to identify traits that are under divergent selection and that contribute to reproductive barriers among conspecific populations (representing the early stages of speciation). The hard part comes in then estimating the relative contribution of that trait to total reproductive isolation. This estimation can start by quantifying the contribution of each barrier to overall reproductive isolation (Ramsey et al. 2003, Nosil 2007, Sobel and Chen 2014) and then estimating the contribution of the trait to each barrier. This last estimation could start through multifactorial analyses of multiple traits that might influence the barrier and could finish with experimental manipulations, such as song playbacks in birds and insects, tail length manipulations in birds and fish, and color manipulations in many taxa. In addition, if we are to determine how crucial the trait is for speciation, we would like to know the distribution of its effects across a range of potential speciation events that differed in outcome. Finally, multiple magic traits might contribute to a single speciation event, and so it would be beneficial to quantify their combined contributions to overall "magic speciation." Accomplishing these tasks for traits, or accomplishing the same for specific genes, is an exceedingly tall order that has not been attempted for any organism.

Automatic magic traits are surely very important in ecological speciation; by contrast, we currently lack information on the general importance of classic magic traits. A formal determination of this importance will not be easy, as noted above. However, a useful starting point would be to enumerate, for many putative instances of ecological speciation, the situations where at least one trait is clearly under strong divergent/disruptive selection and makes a strong contribution to reproductive isolation (Servedio et al. 2011). Encouragingly, theoretical analyses have indicated that classic magic traits are expected to evolve when ecological differentiation is present (Thibert-Plante and Gavrilets 2013).

QUESTION 8: WHAT IS THE ROLE OF DIMENSIONALITY?

It has been suggested that increasing the number of trait "dimensions" (e.g., color, morphology, physiology, behavior) on which divergent/disruptive selection acts might enhance the evolution of reproductive isolation (Rice and Hostert 1993, Nosil and Sandoval 2008, Nosil et al. 2009b; Svardal et al. 2014). This outcome might arise through several different mechanisms. First, multiple dimensions might simply increase the total strength of selection promoting phenotypic divergence, which should increase

---

[8]We could distinguish between "trivial magic traits" (or "squib traits"—Ben Haller, pers. comm.) that have small effects on speciation from "important magic traits" that have large effects on speciation; and we might call nonmagic traits "muggle traits" (Eva Kisdi, pers. comm.). More generally, Nosil and Schluter (2011) discuss effect size calculations for "speciation genes," although they do not specifically emphasize magic traits.

the strength and number of reproductive barriers. Second, multiple dimensions should cause divergence in multiple traits, each of which could be a potential target for mate choice and thereby contribute to assortative mating. Third, multiple dimensions could increase the number of places in the genome where divergence occurs, making more likely the evolution of assortative mating through linkage (i.e., genetic hitchhiking). Fourth, multiple dimensions should increase the phenotypic variance in hybrids, which could reduce their fitness through "transgressive incompatibilities" (Chevin et al. 2014).

Few studies have specifically considered whether multifarious selection promotes ecological speciation, but work on *Timema* stick insects provides an illustrative example. As mentioned earlier, host races within species diverge mainly in crypsis and show only minor reproductive isolation, whereas different *Timema* species diverge in both crypsis and physiology and show much greater reproductive isolation (Nosil and Sandoval 2008, Nosil et al. 2009b). Beyond stick insects, many populations that show progress toward ecological speciation are clearly adapted to environments that differ in a multitude of ways (Nosil et al. 2009b, McBride and Singer 2010, Singer and McBride 2010, Rosenblum and Harmon 2011). In threespine stickleback, for example, environments differ in diet, predation, parasites, light availability, temperature, osmolarity, density, and a host of other factors. These diverse selective pressures presumably target many different traits—and, indeed, many traits show divergence between these young species (McKinnon and Rundle 2002, Hendry et al. 2009b).

Yet the jury is still out as to the importance of dimensionality. First, causality is hard to establish in all of the above cases because the evolution of reproductive isolation along one dimension should allow trait divergence along additional dimensions: that is, isolation might be driving dimensionality rather than the reverse. Second, it isn't clear in many situations which axes of ecological/trait divergence are the main drivers of reproductive isolation. Perhaps divergent selection acting on only foraging traits was sufficient to do the job in stickleback, even though other traits diverged coincidentally. Third, increasing the strength and dimensionality of selection might increase selection load so as to make extinction more likely (chapter 7). Fourth, some situations where trait divergence is highly multidimensional show only minor progress toward ecological speciation. High-predation versus low-predation guppies, for example, differ in trophic morphology, color, many life history traits, swimming performance, physiology, many aspects of behavior, and seemingly countless other traits (Endler 1995, Magurran 2005). Whereas these populations do show selection against migrants (Weese et al. 2011), they also show only weak mating isolation (Schwartz et al. 2010) and they maintain high gene flow in nature (Crispo et al. 2006). Sixth, some of the clearest examples of ecological speciation are associated with magic traits, such as those described above, where only one or a very few dimensions are clearly under divergent selection: for example, butterfly color, hamlet fish color, floral traits, and Darwin's finch beaks. So perhaps ecological speciation proceeds more easily if one collapses all of the selection onto one or a few trait dimensions.

The importance of dimensionality to ecological speciation, and the specific mechanism by which it might work, is entirely unknown. For natural populations, more studies should—like the above *Timema* example—quantify the number of traits contributing to reproductive isolation at various stages along the speciation continuum, although establishing causality will remain tricky. In addition, it will be very difficult to determine which mechanism is more important: increasing the number of dimensions *per se* (multifarious selection hypothesis) or increasing the total strength of selection (total selection hypothesis) (Nosil et al. 2009b). This determination is likely to be most easily accomplished in laboratory systems where the strength and dimensionality of divergent selection can be independently manipulated and progress toward speciation measured.

QUESTION 9: DOES SEXUAL SELECTION HELP OR HINDER ECOLOGICAL SPECIATION?

Sexual selection could theoretically enhance or constrain progress toward ecological speciation (Panhuis et al. 2001, Ritchie 2007, Maan and Seehausen 2011). One enhancing effect could occur when sexual selection and natural selection are correlated, such that sexual selection speeds or enhances adaptive trait divergence (see question 6 in chapter 2) and thereby promotes ecologically based reproductive barriers (as above), at least when gene flow is quite low (Servedio 2015). Another enhancing effect could occur if sexual selection within populations provides the basis for assortative mating among populations.[9] A clear example is speciation by sensory drive (Boughman 2002), which can occur when signaling environments differ among populations. In such cases, divergent sexual selection drives divergence in signaling traits that thus coincidentally promotes discrimination against males moving among environments (Boughman 2001, Seehausen et al. 2008b). Another example occurs when sexual selection for locally adapted traits generates assortative mating among populations adapted to different environments, with a putative example being female preferences for MHC genotypes that are best suited for local parasites (Eizaguirre et al. 2009).

On the constraining side, rare phenotypes can be important. For instance, theory has shown that sexual selection can reduce the success of rare male phenotypes and thereby hinder progress toward sympatric speciation (Kirkpatrick and Nuismer 2004, Otto et al. 2008). However, this constraint can be overcome by other advantages to rare phenotypes, such as a reduction in competition (see above) or predation (Olendorf et al. 2006, Hoso et al. 2010). In other cases, females actually prefer rare male phenotypes (Partridge 1988, Hughes et al. 1999, 2013), which can enhance gene flow among populations showing phenotypic divergence. This last effect can also provide a counterpoint to the above-suggested enhancing role of MHC, whereby female preferences for rare MHC genotypes (Penn and Potts 1999) could increase gene flow among populations adapted to different parasite environments.

An additional route by which sexual selection might constrain ecological speciation is when male traits diverge as a consequence of natural selection but female preferences are conserved. For instance, particular traits might indicate generalized good genes or sexy sons benefits (Prokop et al. 2012). In such cases, among-population trait divergence could increase among-population mating in certain directions (Pfennig 1998, Ellers and

---

[9]Sexual selection can have many components (Andersson 1994, Clutton-Brock 2009) but I here—for convenience and consistency—focus mainly on female choice among males. Sexual selection is not equivalent to assortative mating: the former implies differences in mating success among individuals with different traits, whereas the latter implies an association during mating between female traits and male traits (Maan and Seehausen 2011). For example, females and males might pair up on the basis of similarity in their traits (strong assortative mating) even if all males and females obtain mates (no sexual selection). Mate choice can drive both assortative mating and sexual selection, and these two outcomes might or might not be coupled.

Boggs 2003, Schwartz and Hendry 2006). As an illustrative example, divergent predation environments drive color divergence of male guppies (Endler 1980), whereas female guppies generally favor high male color (Endler and Houde 1995). Thus, females in high-predation environments, where males are usually drab, might favor male immigrants from upstream low-predation environments, where males are usually colorful, which would enhance gene flow (Schwartz and Hendry 2006, Labonne and Hendry 2010). Interestingly, selection to avoid mating with such males can partly reduce this constraint (Schwartz et al. 2010). As a recent example, Comeault et al. (2015) showed that a melanic morph of *Timema* stick insects was favored by females (and survived better) on both host plants, thus providing a genetic "bridge" between ecotypes that hampered speciation. Related to these ideas, theoretical analyses have shown that because mate preferences might often introgress among divergent populations more quickly than alleles underlying the traits experiencing divergent natural selection, sexual selection can counteract adaptive divergence and ecological speciation (Servedio and Bürger 2014).

I have thus far presented ideas, anecdotes, and examples because comprehensive meta-analyses are lacking for the role of sexual selection in ecological speciation in general and assortative mating in particular. However, qualitative analyses provide hints as to some contexts where such meta-analyses might be profitable. One context would be to test whether sexual selection and assortative mating are based on the same traits, such that the former would be predictive of the latter. Ptacek (2000) reported that such associations are present in many organisms but, at the same time, they are not universal. Another context would be to test for correlated divergence in male traits and female preferences, which should promote assortative mating. Schwartz and Hendry (2006) found such associations in only about half of the studies they reviewed: instead male traits often diverged among populations, whereas female preferences did not. However, a more recent and comprehensive meta-analysis of the same sort of data found more frequent (but not universally) positive associations between preference and display divergence (Rodríguez et al. 2013).

Sexual selection will sometimes help and sometimes hinder progress toward ecological speciation. Although quantitative data are still limited, they currently trend toward the idea that sexual selection more frequently has positive than negative effects. At the simplest level, we can point to the frequency of mate choice for conspecifics over heterospecifics, although this is not universal (Mayr 1963, Coyne and Orr 1997, Sasa et al. 1998, Mendelson 2003, Mendelson and Shaw 2005, Price 2008, Ord and Stamps 2009). However, the relevance of this result depends on the relationship between sexual selection and assortative mate choice, which is not at all certain (Servedio 2015). In addition, a number of studies have shown that speciation rates can be elevated in taxonomic groups where sexual selection is prevalent (Barraclough et al. 1995, Hodges and Arnold 1995, Panhuis et al. 2001, Ritchie 2007). Similarly, Seddon et al. (2013) analyzed 84 recent speciation events in 23 passerine bird families and found that sexual dichromatism (a proxy for the strength of sexual selection within species) was positively correlated with divergence in male plumage traits (but not female plumage traits or other male traits). Cause and effect associations remain uncertain in these analyses but evidence seems to be accumulating that sexual selection might not, at the least, pose much of a constraint on speciation. Although we can't be certain these analyses involve cases of ecological speciation, divergent ecological situations could enhance such positive effects (Price 1998, Funk et al. 2006).

Conclusions, significance, and implications

Ecological speciation occurs when ecological differences drive divergent selection that causes adaptive divergence, which then leads to the evolution of reproductive barriers. The particular significance of this process for eco-evolutionary dynamics is twofold. First, it provides the mechanism whereby ecological differences drive the formation of new species, and hence the evolution of biological diversity. Ecological speciation thus provides a route whereby ecology can have its maximal effect on evolution (eco-to-evo)— because large evolutionary changes are really only possible when different populations escape the constraint imposed by gene flow. Second, and following from the first, the large evolutionary changes allowed through the formation of new species provide the basis for highly divergent ecological consequences (evo-to-eco). Importantly, ecological speciation thus also provides a likely route to eco-evolutionary feedbacks, wherein ecological differences promote large evolutionary changes that are then likely to have large ecological consequences. In short, ecological speciation provides an important potential transition from modest (among population) to dramatic (among species) eco-evolutionary effects.

It is clear that ecological speciation has been important in the adaptive radiation of many taxonomic groups (remembering that it can play out in allopatry, parapatry, or sympatry), and the early stages of this process can unfold very quickly. Yet it is hard to ascertain the general importance of this process as opposed to other speciation mechanisms. Moreover, ecological differences often fail to generate substantial progress toward speciation, and any such progress can long remain reversible. At present, then, we lack a comprehensive understanding of the prevalence and importance of ecological speciation in driving the diversity of life. We also have an only modest understanding of the factors that promote ecological speciation: (1) competition is likely important but perhaps less so than multimodal resource distributions, (2) reproductive interactions that promote reinforcement of mating preferences can be important but links to *ecological* speciation have only rarely been explored, (3) magic traits are strongly associated with some instances of ecological speciation but certainly not all of them, (4) increasing niche dimensionality might promote ecological speciation but perhaps only because it increases the total strength of divergent selection, and (5) sexual selection can promote or constrain ecological speciation but inferring the relative importance of these opposing effects awaits formal meta-analysis. In short, much work remains to be done in this exciting dimension of eco-evolutionary dynamics.

Chapter 7

Population Dynamics

The previous chapters all focused on the eco-to-evo side of eco-evolutionary dynamics: that is, how ecological changes/differences cause evolutionary changes/differences. The next three chapters will shift to focus instead on the reverse direction of causality (evo-to-eco), starting with population dynamics in the present chapter and moving to community structure and ecosystem function in the following chapters. The basic idea behind the present chapter is that the evolution of a population will influence the survival and reproduction of its members, which will then alter demographic parameters. As a straightforward example, a poorly adapted population should show reduced survival/reproduction, which could lead to population declines and extirpation. However, adaptation of the population should increase survival and reproductive success, which could increase population size.

Population dynamics is one area where links between evolution and ecology have long been considered (Chitty 1960, Pimentel 1968, Antonovics 1976, Roughgarden 1979). For example, many theoretical models have explicitly considered interactions between population size and natural selection (e.g., Levin 1972, León and Charlesworth 1978, Lively 2012, Ferriere and Legendre 2013), and a number of empirical studies have tested the predictions of these theories (e.g., Agashe 2009, Turcotte et al. 2011a, Travis et al. 2013). Work in this area is sometimes called "evolutionary demography" (e.g., Pelletier et al. 2007). This overall body of work has shown quite clearly that evolutionary change and population dynamics reciprocally influence each other. Although this literature is certainly relevant to the present chapter, I will not—for several reasons—review it in any detail. First, very little of the work has been done in nature (Strauss et al. 2008), which is the focus of this book. Second, much of the previous work concentrates on frequencies of alternative genotypes or phenotypes, whereas my main focus is on continuously varying phenotypes. Third, most of the previous work does not fit easily into the general framework for predicting changes in composite ecological variables, which is the focus of much contemporary work and which forms a key theme of this book. As a result, my outline of interactions between evolution and population dynamics will look different from classic (e.g., Roff 1992, Stearns 1992) and some contemporary (e.g., Morris and Lundberg 2011, Smallegange and Coulson 2013, Vindenes and Langangen 2015, Childs et al. 2016) treatments.

I start by considering the conditions under which evolution might or might not influence population dynamics—because strong effects are not inevitable. I then consider various approaches for inferring and measuring effects of evolution on population dynamics, particularly in natural populations. I then turn to some key questions related to the effects of adaptation and maladaptation on the fitness of individuals, the growth of populations, and the persistence and spread of species. In most cases, I will equate phenotypic effects on population dynamics with evolutionary effects, when they might often include a substantial plastic component. Although I have already discussed this issue (chapter 1), it seems useful to briefly remention it here in light of our shift in causality from eco-to-evo (chapters 2–6) to evo-to-eco (chapters 7–9). In short, nongenetic changes are important to include within a broad consideration of eco-evolutionary dynamics because (1) it can be difficult to confirm that phenotypic change has a genetic basis (even when it does), (2) nongenetic phenotypic effects on population dynamics can be just as important as the genetic effects, (3) phenotypic plasticity will often be the product of past evolution, and (4) plasticity is likely to influence future evolution. These topics are considered at greater length in chapter 11.

When might evolution influence population dynamics?

Several conceptual models exist for considering the conditions under which evolution influences population dynamics (León and Charlesworth 1978, Hairston Jr. et al. 2005, Coulson et al. 2006, Saccheri and Hanski 2006, Kinnison and Hairston Jr. 2007, Childs et al. 2016). The following presentation will center on the particularly simple and well-known logistic population growth model:

$$\frac{dN}{dt} = r_{max} N \left(1 - \frac{N}{K} \right),$$

where dN/dt is the rate of change in population size (population growth rate), r_{max} is the maximum per capita population growth rate (maximum intrinsic growth rate), N is current population size, and K is the maximum population size the current environment could support (carrying capacity). One way to think of r_{max} is as the maximum growth rate possible when individuals are fully adapted to the current environment and density dependence is absent. However, it is here more useful to think of r_{max} as the maximum growth rate possible in the absence of density dependence given *the current state of (mal) adaptation in the current environment* (see also Kinnison and Hairston Jr. 2007). This second interpretation allows maximum population growth rate to evolve as organisms adapt to their local environments.

In this simple framework, population dynamics depend on three parameters. First, current population size (N) will influence current population growth rate for a given r_{max} and K. Specifically, very small populations will grow at their maximum possible rate (r_{max}) and growth rate will slow as population size increases toward carrying capacity (K). Of course, these statements are simplifications given that small populations can show negative density dependence whereby decreasing population size decreases per capita fitness (Allee effects). In addition, population sizes will be variable even at the theoretical carrying capacity. Second, an increase (or decrease) in r_{max} will cause an increase (or decrease) in population growth rate, especially when population sizes are small. Third, an increase (or decrease) in K will cause an increase (or decrease) in maximum population size and therefore also a higher (or lower) growth rate for a given population size.[1]

Both r_{max} and K can be influenced by environmental features, by organismal properties, and by their interaction (i.e., the level of adaptation). These factors can influence r_{max} by altering birth or death rates in ways that do not depend on population size. An example might be severe temperature conditions that increase mortality by exceeding the critical thermal maxima/minima for an increasing subset of the population (Pörtner and Knust 2007). Evolution could here modify r_{max} through selection for increased temperature tolerance (Huey and Kingsolver 1993, Skelly et al. 2007, Barrett et al. 2011, Leal and Gunderson 2012). The same factors can influence K by altering the number of individuals that can be successful in a particular life stage. An obvious example is the addition of nest boxes that allow more cavity-nesting birds to breed (Newton 1994). Evolution could here modify K (Lively 2012) by favoring the evolution of traits/tendencies that increase the successful use of alternative nesting habitats.

The degree to which evolution influences r_{max} or K will vary among different scenarios. Figure 7.1 illustrates some of the possibilities for a hypothetical organism with two life stages, conceptualized as juveniles (life stage 1) and adults (life stage 2). Scenario A illustrates a situation where mortality in each stage is determined by density-independent factors acting during that stage (i.e., density-independent probability of survival)—because population size is well below carrying capacity in each stage. In this scenario, an evolutionary increase in r_{max} (for simplicity; an increase in birth rate) would increase population size for both life stages. By contrast, an evolutionary increase in K at either stage would not increase population size because the population is not density limited. Scenario B illustrates the opposite situation, where density dependence is present at both life stages: the number of individuals reaching each stage is greater than the number of individuals that can make it through that stage. In this scenario, an evolutionary increase in r_{max} would not influence population size at the end of either life stage. By contrast, an evolutionary increase in carrying capacity at the first life stage (K_1) would influence the number of individuals making it through that stage (N_1) but not the number of individuals making it through the next stage (N_2), whereas an increase in carrying capacity at the second life stage (K_2) would increase the number of individuals making it through that stage (N_2), as well as the number of propagules/offspring, but not the number of individuals making it through the first life stage of the next generation.

Scenarios C and D illustrate density dependence at only one life stage or the other. In Scenario C, survival through stage 1 is density independent but survival through stage 2 is density dependent, and so (1) the evolution of increased r_{max} (birth rate) would increase N_1 but not N_2, (2) an evolutionary increase in K_1 would have no influence, and (3) an evolutionary increase in K_2 would increase both N_2 and N_1 (in the next generation). An example of this scenario could be the cavity-nesting birds noted above, where the evolution of traits allowing successful use of alternative nesting habitats should increase the number of breeders, the number of offspring produced, and the numbers of juveniles and adults. In Scenario D, survival through stage 1 is density dependent but survival through stage 2 is density independent, and so (1) the evolution of higher r_{max} (birth rate) would

---

[1] In this r-K formulation of the logistic model, K can still depend on r. As reviewed by Mallet (2012), this conflation is not present in the original formulation: $dN/dt = r_{max} - \alpha N$, where α is the "density-dependent crowding effect." The general concepts presented in this section do not depend on the specific formulation.

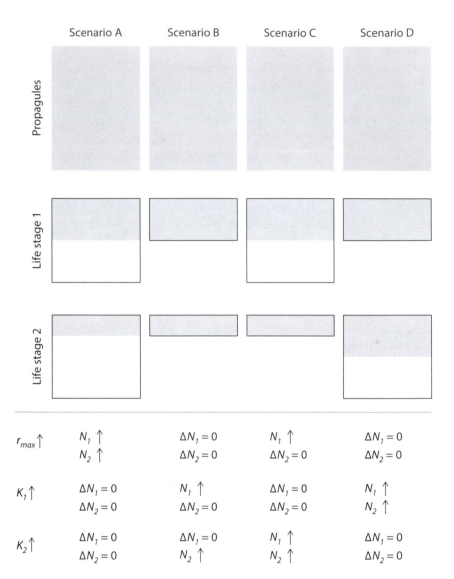

Fig 7.1. Representation of the four scenarios described in the text. The top row represents the number of propagules produced by a population in a given year. The second row represents the number of individuals that survive through the first life history stage (gray area) in relation to the carry capacity in that stage (area encompassed by black lines). Similarly, the third row represents the number of individuals that survive through the second life history stage in relation to the carrying capacity in that stage. The symbols below represent, for a given increase in population growth rate (r_{max}) or carrying capacity in a given stage (K_1 or K_2), the expected change in population size at each life stage (N_1 or N_2)

have no influence, (2) the evolution of higher K_1 would increase both N_1 and N_2, and (3) the evolution of higher K_2 would have no effect. An example of this scenario could be organisms where density is limited by juvenile territories rather than adult breeding sites (e.g., Atlantic salmon, *Salmo salar*; Grant and Kramer 1990). In short, the evolution of r_{max} or K might or might not alter population sizes at different life stages depending on when density regulation occurs (León and Charlesworth 1978).

The above scenarios are not intended to represent any particular realities, and many reasons exist why the specific outcomes might not be as described (see below for some of the reasons). Instead, the scenarios are intended to illustrate as simply as possible the basic points that (1) effects of evolution on population dynamics can occur through multiple routes (e.g., r_{max} or K), but (2) such effects are not inevitable and can be specific to a particular life stage.

Increasing complexities

The preceding section outlined several simple possibilities for how evolution might influence population dynamics, as well as how those possibilities can be organized under a particular conceptual framework. At the same time, however, the simple outline ignored many complexities likely present in the natural world. These complexities can modify all of the above expectations in varied and interesting ways, several of which I will here describe. These particular complexities were chosen to be instructive and illustrative rather than comprehensive.

1. The above outline considered how an arbitrary evolutionary increase in r_{max}, K_1, or K_2 (each considered individually) might influence population size under density-dependent or density-independent conditions. The reality, however, will be less arbitrary because the traits influencing r_{max} and K are expected to have evolved under those conditions in the first place. For instance, populations facing high density-independent mortality often evolve traits that increase r_{max}, including early reproduction, high reproductive effort, and small/numerous offspring (Pianka 1970, Roff 1992, Stearns 1992, Reznick et al. 2002). By contrast, populations facing density-dependent mortality often evolve traits that increase competitive ability, including late reproduction, low reproductive effort, and few/large offspring (Pianka 1970, Roff 1992, Stearns 1992, Reznick et al. 2002). In short, density-dependent versus density-independent mortality should impose divergent selection on life history traits, which should shape the evolution of r_{max} and K accordingly (Hairston et al. 1970, Roff 1992, Stearns 1992). This evolution should have population dynamic consequences within established populations and also in populations colonizing new environments.

   Some of these realities can be illustrated by reference to guppy populations adapted to high-predation versus low-predation environments (Bronikowski et al. 2002, Reznick et al. 2002, 2004). High-predation environments are generally not resource limited but instead have a suite of dangerous predators that cause high guppy mortality (Reznick et al. 1996a, 2001, Weese et al. 2010). As expected from the above ideas, these high-predation populations evolve classic "r-type" life histories: early reproduction, high reproductive effort, and many/small offspring (Reznick and Bryga 1996, Reznick et al. 1996b, 1997). By contrast, low-predation environments are often resource limited but have only a few predators, and these predators do not cause high guppy mortality (Reznick et al. 1996a, 2001, Weese et al. 2010). As expected, these low-predation populations evolve classic "K-type" life histories: late reproduction, low reproductive effort, and few/large offspring (Reznick and Bryga 1996, Reznick et al. 1996b, 1997). For established populations, the consequence should be higher r_{max} in high-predation environments and higher

K in low-predation environments. For colonizing populations, the consequences should play out in establishment rates and subsequent evolutionary trajectories. Indeed, low-predation guppies introduced into high-predation sites generally fail to establish viable populations (Reznick et al. 2004), suggesting that the high r_{max} in established high-predation guppies helps maintain population size in their mostly density-independent environments. By contrast, high-predation guppies introduced into low-predation sites always establish themselves and rapidly increase in abundance. Then, when density dependence comes into play, they evolve traits better suited for density-dependent situations (Bronikowski et al. 2002, Reznick et al. 2002, 2004, Bassar et al. 2014), which should increase carrying capacity.

2. I posited above that changes in r_{max} were relatively unlikely to influence population size under strong density dependence, but this result will not always occur. First, the evolution of traits influencing r_{max} should simultaneously influence *K*, as discussed immediately above, which should then influence population size even under density dependence. Second, high r_{max} can cause an overshoot of *K*, which can then lead to a sharp decrease in per capita reproductive success and, hence, a decrease in population size at subsequent stages. A classic example of this effect is the "Ricker Curve" of stock-recruitment, where increasing numbers of breeders eventually causes a *decrease* in the number of recruits (fig. 7.2). For Pacific salmon, the taxon that inspired this curve (Ricker 1954), a likely reason is that too-numerous breeders can dig up and destroy each other's nests, which dramatically reduces egg survival (Fukushima et al. 1998, Essington et al. 2000, Hendry et al. 2004a). Several other mechanisms can also lead to negative density dependence at high densities (Walters and Juanes 1993).

3. The above two complexities show how population density and life history evolution can influence each other, a situation that can lead to eco-evolutionary *feedbacks* (Lankau and Strauss 2011, Bassar et al. 2013, Sanchez and Gore 2013, Duckworth and Aguillon 2015). For example, the size of a population in relation to carrying capacity will influence selection for *r*-type versus *K*-type life histories. The evolution of these life histories should then modify population density, which could influence further selection on those life histories, and so on. These feedbacks can be positive (reinforcing) or negative (opposing). Positive feedbacks should lead to accelerated and exaggerated responses for both life history traits (evolution) and population dynamics (ecology), whereas negative feedbacks should lead to slower and more subtle responses that stabilize dynamics. An example of negative feedback comes from genetically based morphs of female side-blotched lizards: orange-throated females have few/large offspring (*K*-type), whereas yellow-throated females have many/small offspring (*r*-type) (Sinervo et al. 2000). In years of low density, *r*-type females are favored, and their many offspring cause an increase in population growth that overshoots carrying capacity. These new conditions favor *K*-type females, and their few offspring cause a decrease in population growth that undershoots carrying capacity. The result is a cycle in population size driven by compensating changes in the frequency of the two morphs (Sinervo et al. 2000). As another example, maladaptation in the crypsis of *Timema* stick insects causes increased predation, which decreases population size, which increases selection against the noncryptic color pattern (Farkas et al. 2014).

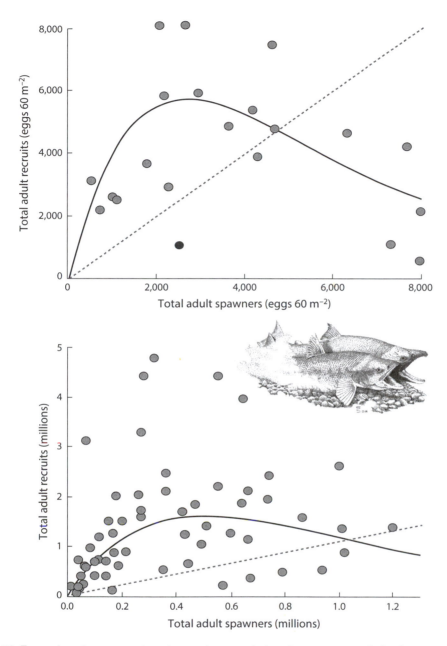

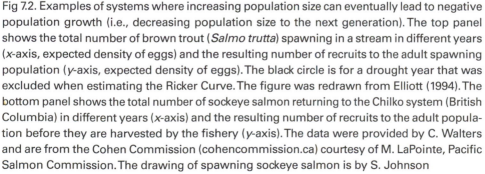

Fig 7.2. Examples of systems where increasing population size can eventually lead to negative population growth (i.e., decreasing population size to the next generation). The top panel shows the total number of brown trout (*Salmo trutta*) spawning in a stream in different years (*x*-axis, expected density of eggs) and the resulting number of recruits to the adult spawning population (*y*-axis, expected density of eggs). The black circle is for a drought year that was excluded when estimating the Ricker Curve. The figure was redrawn from Elliott (1994). The bottom panel shows the total number of sockeye salmon returning to the Chilko system (British Columbia) in different years (*x*-axis) and the resulting number of recruits to the adult population before they are harvested by the fishery (*y*-axis). The data were provided by C. Walters and are from the Cohen Commission (cohencommission.ca) courtesy of M. LaPointe, Pacific Salmon Commission. The drawing of spawning sockeye salmon is by S. Johnson

4. It is usually assumed that increasing mortality rates will decrease population size, yet this assumption won't always hold. For instance, high mortality can release individuals from competition, which can dramatically increase their growth rate and thus reproductive output, which can then increase population density and even biomass ("overcompensation"). Thus, increased mortality during a particular life stage can *increase* population size in other life stages and sometimes even in the first life stage by the next generation. This expectation has been outlined in theoretical models (De Roos et al. 2007, Persson and De Roos 2013) and confirmed in some experiments (Nicholson 1957, Schröder et al. 2009) and observational/modeling studies (Ohlberger et al. 2011). Overcompensation then can have a variety of nontrivial consequences for interacting species and alternative stable states (Persson et al. 2007, Persson and De Roos 2013). The effects just described are nonselective in the sense that they can occur even if different phenotypes/genotypes do not differentially influence mortality. However, it also has been argued that, at least for marine fishes, most of the early life mortality is selective (for example, with respect to body size) and that this selection can influence recruitment within that generation (Johnson et al. 2014). As a counter-example, the "load" imposed by increasing selection when populations are maladapted sometimes can be compensated by a reduction in competition, thus having no net effect on population size (Reed et al. 2013).

Some ways in which evolution might influence population dynamics have been here illustrated through the lens of density-dependent versus density-independent population regulation (see also Kinnison and Hairston Jr. 2007). An alternative, but related, distinction (Saccheri and Hanski 2006) is that between soft selection (frequency- and density-dependent) and hard selection (frequency- and density-independent) (*sensu* Wallace 1975). In a metapopulation context, hard versus soft selection can be considered as global versus local population regulation (Saccheri and Hanski 2006). These concepts have become rather muddled in the literature and selection can be density-dependent without being frequency-dependent and vice versa (Wallace 1975, Heino et al. 1998). It therefore seemed most intuitive to use the simple logistic equation framework, but the basic ideas do not depend on any specific model (although various details and nuances certainly might).

How to detect evolutionary effects on population dynamics

Three general approaches have been used for studying evolutionary or, more generally, phenotypic influences on population dynamics in nature: experimental manipulations, empirical modeling, and observational studies. (Other ways to organize the various approaches have been suggested elsewhere—see Strauss 2014). Although these approaches are sometimes combined within a particular study system, I will present them separately. Another approach—not discussed in detail here—is to identify particular signals in population dynamic data that are indicative of evolutionary contributions (Hiltunen et al. 2014).

The experimental approach involves manipulating phenotypic/genotypic distributions and monitoring the resulting population dynamics. For instance, a researcher can take two groups (different genotypes/ecotypes/populations/species) and compare their population dynamics under controlled conditions, such as in the laboratory or seminatural mesocosms/enclosures (Travis et al. 2013). Any differences in dynamics under these common-garden conditions would indicate that evolutionary divergence has shaped the organismal traits that influence population dynamics. In some cases, these experiments can track the effects of real-time evolution on population dynamics by implementing "evolution" (genetic variation present) and "no evolution" (genetic variation absent) treatments (Yoshida et al. 2003, Turcotte et al. 2013). More recently, Kasada et al. (2014) examined how predator-prey dynamics are influenced by different types of genetic variation, specifically the nature of the trade-off between competitive ability and predator defense.

Of course, population dynamics under controlled conditions are unlikely to replicate those that occur in nature, which can be better studied through reciprocal transplants that ask how groups evolving under different conditions exhibit population dynamics under those conditions. These experiments can be similar in form to those used for testing local adaptation (chapter 4)—but additional response variables are needed. That is, reciprocal transplant experiments normally focus on traits and individual fitness, whereas they here need to also monitor population dynamics. For example, Siepielski et al. (2016) used a reciprocal transplant experiment coupled with a density manipulation in damselflies to determine how local adaptation influenced the strength of intraspecific competition. They found evidence that increasing maladaptation resulted in a reduction in the strength of negative density dependence on per capita growth rates. Such results show how eco-evolutionary dynamics can affect the demographic forces regulating populations. Additionally, such experiments can use natural groups, such as ecotypes, that diverged relatively recently, which allows insight into how quickly evolution can change population dynamics (Kinnison et al. 2008, Gordon et al. 2009). Transplant experiments manipulate the degree to which a group is adapted to the local environment. A similar effect can be accomplished by manipulating maladaptive gene flow or environmental conditions, after which population dynamics can be monitored. Experiments of this latter sort have been performed in the laboratory (Bell and Gonzalez 2009, 2011, Turcotte et al. 2011a, 2013, Cameron et al. 2013) and, even better but less frequently, in nature (Farkas et al. 2013).

The empirical modeling approach measures factors that should influence population dynamics (traits, heritabilities, environmental change, selection, plasticity, etc.), and then uses those measurements to parameterize models predicting population dynamics. These empirically based eco-evolutionary models are common in laboratory settings (e.g., Yoshida et al. 2003, Fischer et al. 2014, Ellner 2013, Kasada et al. 2014), but we are here more interested in their application to natural settings (van Tienderen 2000, Knight et al. 2008, Baskett et al. 2009, Zheng et al. 2009, Hanski et al. 2011, Reed et al. 2011, Farkas et al. 2013, Vedder et al. 2013). Even here, of course, the resulting model predictions do not actually *demonstrate* evolutionary effects on population dynamics, but rather provide evidence that such effects are possible. Of greatest utility, such models help to reveal the conditions under which evolutionary change is most and least likely to have substantial effects on populations. If model predictions and empirical observations are closely matched, confidence increases that the processes shaping model dynamics, including evolution, are also shaping natural dynamics. If predictions and observations are not closely matched, the model is missing key components that must be explored through further research. A particular value of the empirical modeling approach is its utility in examining how future environmental change might influence future population dynamics, with or without evolutionary change. In short, these models allow for manipulative experiments to be performed *in silico*, and the development of flexible eco-evolutionary modeling platforms (e.g., Bocedi et al. 2014) will make the approach accessible to a broader range of biologists. The obvious limitation of this approach is that experiments are not actually performed, and so whether or not the expectations will hold in nature remains uncertain.

The observational approach monitors the dynamics of natural populations and statistically relates year-to-year ("real time") changes in population dynamic parameters to year-specific phenotypic trait values or allele frequencies. These statistical models can also include other factors, such as climate, that influence population dynamics. As a specific example, Hanski and Saccheri (2006) showed that interannual changes in the size of local butterfly populations were influenced by local frequencies of an allele that influences migration, with additional dependencies on patch size and connectivity within the larger metapopulation. As another example, several studies of individually marked mammals have used the Price equation (see chapter 3) to quantify the relative contribution of phenotypic traits (e.g., birth size), demographic variables (e.g., age structure), and climate variables (e.g., rainfall) in a given year to interannual population growth (Coulson and Tuljapurkar 2008, Ozgul et al. 2009, 2010). Similar partitioning approaches can be implemented in other statistical frameworks (Hairston Jr. et al. 2005, Coulson et al. 2006, Pelletier et al. 2007, Ezard et al. 2009, Ellner et al. 2011), as will be discussed later in the chapter.

The above approaches relate changes/differences in phenotypes or genotypes to changes/differences in population dynamics; yet a *lack* of changes/differences in population dynamics also could reflect ongoing evolution. The reason is that populations must constantly evolve if they are to maintain their success when faced with the degrading effects of environmental change, gene flow, and mutation (Burt 1995, Hadfield et al. 2011). That is, without ongoing adaptation, populations would become increasingly maladapted and would not be able to maintain their abundance; hence, population dynamic *stability* can be the product of (and is probably an indicator of) the underlying importance of contemporary evolution. A classic example is antagonistic coevolution (e.g., parasites and hosts), where host evolution represents environmental degradation for the parasite, which must adapt in response, which then has reciprocal effects on the host (Van Valen 1973, Lively and Dybdahl 2000). These "Red Queen" dynamics mean that adaptation has to be continual if fitness, and therefore population size, is to remain relatively constant. This stabilizing effect of evolution is likely substantial: Burt (1995) used data on additive genetic variance for fitness, as well as fitness declines owing to gene flow and mutation, to estimate that selection increases fitness by about 1–10% per generation. Importantly, this evolution can be "cryptic"—that is, not visible as phenotypic changes (Merilä et al. 2001a, Veen et al. 2001; Kinnison et al. 2015). For example, intraspecific competition means that continual evolution might be necessary just for individual body size to remain relatively constant through time (Cooke et al. 1990, Hadfield et al. 2011). In short, adaptation might be having strong effects on population dynamics even if phenotypes and population sizes are reasonably stable (see also Yoshida et al. 2007, Weese et al. 2011).

The main goal of this chapter is to consider evolutionary/phenotypic effects on population-level parameters, such as the size, growth, stability, or persistence of populations. Stated another way, we want to consider *the evolution of population dynamics*, an endeavor the previous paragraphs have made clear is difficult to implement for natural populations. A simpler proxy might be within-generation measures of individual fitness (although they too can be hard to accurately and reliably estimate—chapter 2), because an increase in individual fitness would be expected to increase population size. This assumption is quite reasonable if the mean absolute fitness of individuals has been assayed in the natural environment over an entire generation, but it is otherwise more tenuous. Unfortunately, most such studies focus on relative fitness (fitness of individuals in relation to each other) for only part of the life cycle and often outside the natural environment. Such fitness estimates might not predict population dynamics if density regulation occurs for some other reason (see above). In addition, adaptive responses to individual-level selection can *decrease* mean population fitness (Hairston et al. 1970, Schlaepfer et al. 2002, Denison et al. 2003). In agriculture, for instance, competition among plants often selects for high investment in somatic growth (root, stem, and leaf), whereas productivity at the field level is maximized at intermediate root and leaf sizes (Schieving and Poorter 1999, Zhang et al. 1999). In animals, a similar effect occurs when territoriality leads to selection for increased competitive ability, which can increase territory sizes and thus reduce population densities. Taking these arguments to the extreme, several authors have suggested that adaptive responses to individual-level selection can lead to extinction (Matsuda and Abrams 1994, Gyllenberg and Parvinen 2001, Webb 2003).[2] In summary, individual fitness can yield insights into population dynamics (some examples are given below), but links between the two levels are not always straightforward.

Eco-evolutionary population dynamics in nature

The effects of evolutionary/phenotypic change on population dynamics are perhaps easiest to consider in the context of environmental change. I will therefore first ask to what extent maladaptation resulting from environmental change decreases individual fitness and thus causes population declines, range contractions, and extirpations (Q1). I will then consider how adaptive evolution responds by increasing individual fitness (Q2), which might sustain or recover population sizes (Q3), which could make the difference between persistence and extirpation (i.e., "evolutionary rescue"—Q4) and could promote range expansion (Q5). Finally, I ask how intraspecific diversity (variation) influences population dynamics, such as through "portfolio effects" (Q6). Ample evidence for all of the above effects is present in agricultural and medical contexts: that is, the dynamics of weeds, pests, and pathogens reflect their (mal)adaptation in the face of control efforts (Heap 1997, Palumbi 2001, Bergstrom and Feldgarden 2008). However, as rationalized in chapter 1, I will here focus on more "natural" contexts.

---

[2]These so-called "Darwinian extinction" and "evolutionary suicide" effects are separate from the idea that adaptation to a current environment can reduce the genetic variation that enables adaptation to future environments, which can then cause range contractions, extirpations, and extinctions. A recent branding of this latter idea is "evolutionary traps" (Schlaepfer et al. 2002).

QUESTION 1: TO WHAT EXTENT DOES MALADAPTATION
CAUSE POPULATION DECLINES?

Environmental change should render populations maladapted for the new conditions, which should depress individual fitness and, under some conditions, cause population declines, range contractions, and extirpation/extinction (Bürger and Lynch 1995, Gomulkiewicz and Holt 1995, Boulding and Hay 2001, Björklund et al. 2009, Gonzalez et al. 2013). These expectations are clearly borne out when considering long time scales. For instance, changing environments have altered the distribution of species in accord with their environmental tolerances: that is, as environments disappear so too do the organisms adapted to them. As a concrete example, the distribution of temperate organisms contracts toward high latitudes/altitudes during warm periods, whereas the distribution of tropical organisms contracts toward low latitudes/altitudes during cold periods (Hewitt 2000, Davis and Shaw 2001). How do these sorts of dynamics play out on contemporary time scales?

Many populations and species have recently undergone range contractions and/or numerical declines (Ceballos and Ehrlich 2002, Stuart et al. 2004, Pimm et al. 2006, Lenoir et al. 2008, Lavergne et al. 2013), and both of these effects must ultimately stem from maladaptation. A particularly straightforward context is pollution: species that cannot tolerate a pollutant generally decline and often disappear when that pollutant is added to the environment (Bradshaw 1984, Posthuma and Van Straalen 1993, Tammi et al. 2003). Similar effects are also clear in the context of habitat loss: species that require a particular habitat often disappear when that habitat is lost (Fahrig 1997, Brooks et al. 2002). This last effect is the basis of modeling exercises that predict changes in species distributions/abundances as a function of projected climate change (e.g., Thomas et al. 2004, Thuiller et al. 2005, Skelly et al. 2007, Deutsch et al. 2008). In many other contexts, however, inferences are less straightforward. Indeed, relatively few studies have explicitly demonstrated that environmental change has rendered current trait values maladaptive, which has then caused population declines. However, some compelling examples exist in the climate change literature and I will now review three of them.

Both and Visser (2006) studied pied flycatcher (*Ficudela hypoleuca*) populations that breed in Europe following annual migration from their wintering grounds in Africa. Successful rearing of the chicks is heavily dependent on the availability of caterpillars, and so flycatchers generally time their breeding such that chick rearing occurs at the time of peak caterpillar abundance. (Recent work suggests that temperatures experienced by nestlings might be more important—Visser et al. 2015.) This peak is heavily dependent on the timing of budburst, which has recently advanced owing to climate warming—and the amount of this advance differs among flycatcher populations. Both and Visser (2006) showed that locations where caterpillar peak abundances are now the earliest have flycatcher populations showing the greatest declines in abundance (fig. 7.3). The interpretation is that adult birds in the declining populations are now arriving too late to breed at the appropriate time, so why don't they simply arrive earlier? The reason is that arrival time on the breeding grounds is linked to departure time from overwintering sites, and the latter is determined by genetically based endogenous rhythms rather than plastic responses to proximate climate variables (Both and Visser 2001). As a result, climate change has made the formerly adaptive departure dates/cues maladaptive for reproduction under changing conditions, which has then led to population declines. A similar scenario has been reported for other European birds (Møller et al. 2008) and for caribou (*Rangifer tarandus*) that are declining as they become mismatched to the advancing timing of plant growth in western Greenland (Post and Forchhammer 2008).

Pörtner and Knust (2007) studied eelpout (*Zoarces viviparus*), a temperate marine fish, at the southern limit of their range in the Wadden Sea. The abundance of these eelpout varied among years in relation to summer water temperatures: abundances were lowest when temperatures were highest (fig. 7.3). The authors attributed this association to the fact that increasing temperature increases metabolic oxygen demand while also decreasing oxygen supply until, at some threshold temperature, the growth and survival of eelpout becomes compromised. These physiological problems should be greatest for the largest fish and, indeed, large fish were rare and selected against at the highest temperatures. It thus seems that climate warming has increased temperatures beyond those to which eelpouts are adapted, which is then contributing to population declines.

In the mid-19th century, Henry David Thoreau kept detailed records of the occurrences and flowering dates of plants in the vicinity of Concord, Massachusetts, USA. Other naturalists performed similar surveys in subsequent years, right up to the present day. The resulting time series shows that a number of species dramatically declined in abundance and some were extirpated (Willis et al. 2008). These species turn out to be those that cannot easily use plasticity to adjust their flowering time to suit local temperature conditions—they instead show mostly genetically based reproductive timing (Willis et al. 2008). The implication is that local climate change (2.4°C over the past 100 years at Concord), which favors earlier flowering (Fitter and Fitter 2002, Parmesan and Yohe 2003, Menzel et al. 2006), has made the original genetically based flowering dates maladaptive to the point of population declines and extirpation. (The more plastic species advanced their flowering by an average of 7 days over 150 years, although it hasn't been confirmed that plasticity was the specific reason for these advances.) Remarkably similar results have been observed for European birds—species that are advancing their timing are not declining whereas those not advancing their timing are declining (Møller et al. 2008).

The extent to which the above examples reflect a general phenomenon remains to be established, but additional examples of maladaptation causing population declines continue to emerge (e.g., Lane et al. 2012). Combining the existing examples, including from the pollution and habitat contexts, with the long-term analogies noted earlier, it seems safe to assert that **altered environmental conditions frequently cause maladaptation that contributes to population declines and extirpation/extinction.** Of course, negative effects won't always occur because some environmental changes benefit some organisms. For example, environmental change that causes the extinction of one species can benefit a competitor or prey species that is not as susceptible (e.g., Dulvy et al. 2000, Rolshausen et al. 2015b). In addition, species held at a range limit owing to an inability to adapt beyond that limit can, with a change in conditions, increase in abundance and perhaps expand their range (Thomas and Lennon 1999, Thomas et al. 2001, Parmesan and Yohe 2003). My focus in the present question, however, was on instances where maladaptation causes population declines—because this situation leads naturally to the following questions about how ongoing adaptation might improve fitness and arrest population declines.

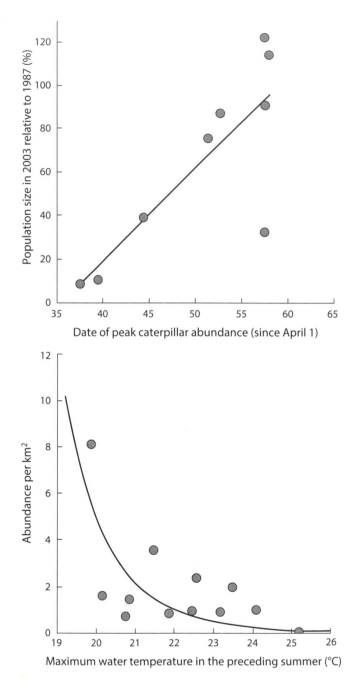

Fig 7.3. Two putative examples where increasing maladaptation leads to decreasing population size. The top panel shows that pied flycatcher populations at sites with early caterpillar emergence experienced dramatic population declines between 1987 and 2003. The data are from Both et al. (2006) and were provided by C. Both. The bottom panel shows that eelpout in the Wadden Sea, which is the warmest part of their distribution, show lower abundance in years following warmer summers. The figure was redrafted from Pörtner and Knust (2007)

QUESTION 2: TO WHAT EXTENT, AND HOW RAPIDLY, DOES ADAPTATION INCREASE INDIVIDUAL FITNESS?

Environmental change that causes maladaptation won't equally impact all individuals in a population. The reason is that different individuals have different trait values that render them more or less well adapted, and this variation should lead to selection for trait values better suited for the new conditions. If those traits are genetically based, they should evolve and individual fitness should increase as a consequence. Part of this expectation now has a firm empirical footing: populations experiencing environmental change often show noteworthy phenotypic change within years to decades (chapter 3, reviews: Hendry and Kinnison 1999, Reznick and Ghalambor 2001, Parmesan and Yohe 2003, Stockwell et al. 2003, Hendry et al. 2008). Many of these contemporary changes are adaptive and genetically based: for example, exposure to predators can lead to the evolution of antipredator behaviors (Magurran et al. 1995, O'Steen et al. 2002), heavy metal pollution can lead to the evolution of increased heavy metal tolerance (Antonovics and Bradshaw 1970, Levinton et al. 2003), earlier droughts can lead to the evolution of earlier flowering (Franks et al. 2007), loss of pollinators can lead to the evolution of selfing (Bobdyl Roels and Kelly 2011), and so on. However, the contribution of adaptive *trait* changes to improvements in *fitness* is less well understood. One issue is that multiple traits contribute to fitness, and so even small changes in each trait could add up across many traits to yield large changes in fitness. Another issue is that many traits might be only weakly correlated with fitness, in which case large changes in some traits might be necessary for even small fitness improvements. An understanding of these nuances requires data not just on trait changes but also on fitness changes: that is, the evolution of fitness. Although the definition and measurement of "fitness" is tricky (chapter 2, here) I will be inclusive to major fitness components, such as survival and reproductive success

Laboratory studies are often able to directly measure the evolution of fitness during adaptation (Burt 1995, Bell 2008, Kassen 2014). The classic approach is to take individuals from a "source" population that evolved in an ancestral environment and introduce them into new environments, thus generating a series of replicated "derived" populations. The derived populations are then allowed to adapt to the new environment, whereas representatives from the original source population are maintained in the ancestral environment. In ideal cases, representatives of the true source population can be held in a suspended no-evolution state, such as by freezing microbes (Bell 2008) or archiving resting eggs or seeds (Cousyn et al. 2001, Angeler 2007, Franks et al. 2007). After some period of adaptation to the new environment, individuals from the source and derived populations are compared to each other to assess their fitness in the ancestral and new environments. Although such studies are obviously much more difficult to conduct in nature, a few examples do exist.

For a first example, we can return to introductions of guppies from one predation environment (high or low) into the other. When the introduced populations persist, they typically show the contemporary evolution of many traits, including color (Endler 1980), life history (Reznick et al. 1997), and behaviors (Magurran et al. 1995), and this evolution occurs within just a few years (guppies have approximately 2–3 generations per year in the wild). What are the consequences of this trait evolution for individual fitness?

O'Steen et al. (2002) provided initial insight into this question by taking guppies from ancestral (source) high-predation populations and derived (introduced) low-predation populations and exposing them to predators typically found in high-predation sites. The experiment showed that survival was higher for ancestral high-predation guppies than for derived low-predation guppies. These results held for wild-caught and laboratory-reared guppies, confirming a genetic basis for survival differences that evolved over an estimated 26–36 guppy generations (O'Steen et al. 2002). This specific comparison does not show that evolution increased survival in the new environment—but rather that it decreased survival in the ancestral environment. It nevertheless illustrates how adaptation can alter survival probabilities on short time scales. The logical next step is to assess the evolution of fitness in nature.

Gordon et al. (2009) attempted this next step by exploring how the evolution of guppies following their experimental introduction into new predation environments changed their survival probabilities in nature. In 1997, 200 guppies were introduced from a high-predation site in the Yarra River into a low-predation site in the previously guppy-free Damier River, after which they self-colonized a high-predation site immediately downstream. Seven years after this initial introduction, guppies were captured from the ancestral (Yarra) and derived (Damier) populations, individually marked, and released into the Damier sites. Recaptures several weeks later revealed that foreign (ancestral) juveniles had only 2/3 the survival probability of local (derived) fish, and a similar trend was evident for males. Adaptation to new environments thus improved a major fitness component after less than 10 years (30 generations). One limitation of this study was that it used wild-caught fish, and so a genetic basis for survival differences could not be confirmed, although it is certainly expected based on prior evidence that trait divergence in guppies has a strong genetic basis (see earlier citations).

Work on a different study system (Kinnison et al. 2008) provided further insights into how contemporary evolution can improve major fitness components in nature—and here the investigators (1) confirmed a genetic basis, (2) replicated their fitness assays, and (3) employed inclusive fitness surrogates. Between 1901 and 1907, chinook salmon from California were introduced into New Zealand, where they quickly colonized several breeding sites. Approximately 90 years (26 generations) later, common-garden studies revealed adaptive trait divergence between two populations with different migration distances (Quinn et al. 2001, Kinnison et al. 2001, 2003). Kinnison et al. (2008) tested the fitness consequences of this trait divergence by releasing common-garden fish from the two populations from a location with a long migration distance and from another location with a short migration distance. When undertaking the long migration, but not the short migration, reproductive output (survival multiplied by average fecundity) was more than twice as high for the population adapting to the longer migration as the population adapting to the shorter migration (fig. 7.4). Evolution here more than doubled fitness over approximately 26 generations, and this fitness change was much larger than the measured trait changes (Kinnison et al. 2008).

More studies are needed before concrete generalizations can be drawn about the rate at which individual fitness increases during adaptation in nature. Nevertheless, **the existing work suggests that fitness improvements of up to 5% per generation are**

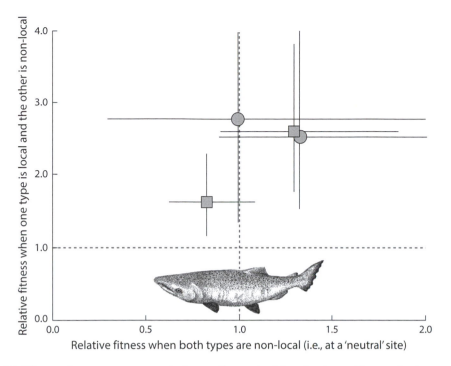

Fig 7.4. Effects of contemporary evolution on fitness in chinook salmon. The *y*-axis shows the ratio of fitness in Glenariffe Stream for fish from the population (Glenariffe) adapting to that stream (i.e., the local population) to the fitness of the population (Hakataramea) not adapting to that stream. The *x*-axis shows the same ratio (fitness of Glenariffe relative to Hakataramea salmon) at a "neutral" site (Sliverstream) to which neither are adapting. The fact that all estimates are above unity along the *y*-axis but not along the *x*-axis indicates that local adaptation over 26 generations has improved fitness. Squares show survival rates in two years and circles show total egg production (survival multiplied by fecundity) in the same years. The data (error bars are 95% confidence intervals, truncated in some cases) are from Kinnison et al. (2008) and were provided by M. Kinnison. The chinook salmon drawing is by S. Johnson

achievable following abrupt environmental change (Kinnison et al. 2008, Gordon et al. 2009). This estimate is within the 1–10% range suggested by Burt (1995), and even higher rates (for lifetime reproductive success) have been reported for human populations (Milot et al. 2011). Several nuances remain unexplored. First, fitness evolution will not be a linear function of time elapsed, and so a rate averaged over multiple generations might not reflect how rapidly fitness changed immediately after a disturbance. Second, changes in individual fitness might not translate directly into changes in population growth, as discussed early in this chapter (see also Reed et al. 2013). Third, studies of fitness evolution in nature must inevitably work with populations that persisted through the disturbance. This reality means that, in all such cases, the studied population persisted even when it wasn't well adapted, such as immediately after the disturbance or introduction. That is, the introduced individuals were already well-enough adapted to allow positive population growth or (alternatively or additionally) the small initial population sizes increased fitness despite maladaptation by reducing density dependence. In short, the examples considered in this section do not provide direct insight into the potential for evolutionary rescue,[3] a topic that will be considered further in question 4.

QUESTION 3: TO WHAT EXTENT DOES ADAPTATION INFLUENCE POPULATION GROWTH?

Chapter 3 summarized how fitness-related traits often evolve following environmental change, and the above question showed how this trait change can have important consequences for individual fitness. Yet the consequences of this trait change for population dynamics are not always transparent. For instance, individual traits seem to change less rapidly than do population sizes (DeLong et al. 2016)—but this could mean either that population dynamics are not driven by trait change, that unmeasured traits are more important, or that even modest trait changes can have large population dynamic consequences. Thus, it is now time to take up this topic more directly, with the present question asking how trait change alters population growth and the next question asking how it influences population persistence (i.e., evolutionary rescue). In this discussion, I will have to rely on specific case studies as concrete data from nature are still extremely rare (Gomulkiewicz and Shaw 2013, Vander Wal et al. 2013, Carlson et al. 2014).

Pelletier et al. (2007) analyzed the contribution of phenotypic variation to population growth for Soay sheep on the small island of St. Kilda, UK. Specifically, the authors estimated how much of the variation in the contribution of individuals to population growth could be explained by the traits of those individuals. The authors found that "body weight accounted for 4.7% of population growth, hind leg length 3.19%, and birth weight 1.69%." After accounting for trait heritability, they estimated that "Additive genetic variation for body weight contributed 0.88% of population growth, with values of 1.43% and 0.19% for hind leg length and birth weight, respectively." Of additional interest, these contributions varied dramatically among years (fig. 7.5), being highest in years when mortality was high—as would be expected (Pelletier et al. 2007). Although the estimated contributions were often statistically significant, the effect sizes were quite small (e.g., <1% for the additive genetic contribution of body weight). However, only a few traits were examined whereas it seems inevitable that population growth will influence many life history, morphological, behavioral, and physiological traits. In addition, the estimates represent per generation contributions—and so cumulative effects might be quite large over reasonable time frames.

The above concerns about the interpretation of effect size echo questions raised in chapters 2 and 3 about how best to judge the importance of a particular effect. The key is to have some standard of comparison. One possibility is to partition the different contributions to population growth into those that can be considered primarily "ecological" (i.e., changes in the environment) versus those that can be considered primarily "evolutionary" (i.e., changes in phenotype). Following Hairston Jr. et al. (2005), we can start by assuming that the rate of change in population growth rate (X) over time (t) is

---

[3] Recall that all introductions of high-predation guppies to low-predation sites are apparently successful, whereas all introductions in the opposite direction fail (Reznick et al. 2004). In the former case, evolution is apparently not necessary for population persistence. In the latter case, evolution is apparently insufficient. Between these two situations might be a range of conditions under which evolution really does make the difference in establishment success in a new (or dramatically altered) environment (Jones and Gomulkiewicz 2012).

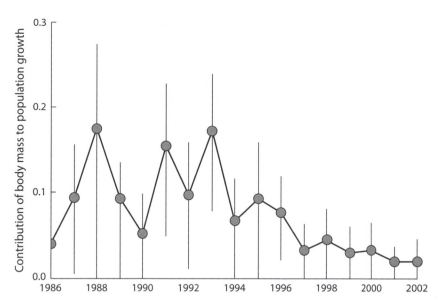

Fig 7.5. Contributions of the body mass of individual Soay sheep to overall population growth in each year. "Contributions" are the r^2 (means and 95% confidence intervals) between individual body weights and how population growth would change if that individual were not a part of the population. The data are from Pelletier et al. (2007) and were provided by F. Pelletier

a function of changes in an environmental variable (K) and a phenotypic variable (Z): that is, $X(t)=X[K(t), Z(t)]$. In continuous time, differentiation by the chain rule yields[4]:

$$\frac{dX}{dt} = \frac{\partial X}{\partial K}\frac{dK}{dt} + \frac{\partial X}{\partial z}\frac{dZ}{dt}$$

Here, the rate of change in population growth rate is the sum of an ecological effect (first term) and an evolutionary effect (second term). The ecological effect is the rate of change in population growth rate (X) associated with a change in the ecological variable (K; e.g., rainfall in a given year) multiplied by the rate of change in that ecological variable. The evolutionary effect is the rate of change in population growth rate (X) associated with a change in the phenotypic variable (Z; e.g., mean birth weight) multiplied by the rate of change in that phenotypic variable. (This basic approach will reappear in chapters 8 and 9.)

The two terms in the above equation can be estimated for actual populations by means of general linear models relating interannual changes in population growth rate to interannual changes in ecological and evolutionary (phenotypic) variables. Hairston Jr. et al. (2005) provided two examples. In one, they used data for Darwin's finches to estimate that changes in population growth were twice as strongly influenced by evolution (changes in beak size) as by ecology (changes in rainfall). In the other example, they used data for the freshwater copepod *Onychodiaptomus sanguineus* to estimate that changes in mean population fitness (per capita production of diapausing eggs) were one-quarter as

---

[4]This continuous time solution does not have an interaction term between k and z—because this cross-product term is assumed to be so small that it can be safely ignored (Hairston Jr. et al. 2005). The discrete time solution would retain the interaction term but I use the continuous version for ease of presentation.

strongly influenced by evolution (date at which females start producing diapausing eggs) as by ecology (predation intensity). In a subsequent application of the same partitioning method, Ezard et al. (2009) used long-term data sets for five ungulate populations to estimate how population growth was influenced by evolution (measures of body size) versus ecology (climate variables). Within each population, effects of ecology and evolution were roughly equivalent, but both effects varied dramatically among populations (fig. 7.6). One reason for this variation might be that it is easiest to statistically explain variation in population growth when that variation is highest, which was the case for Soay sheep.

The above partitioning approach has several limitations, although some of these can be mitigated. First, existing applications have considered only a single phenotypic trait and a single ecological variable, when in reality many such variables will influence population growth. Fortunately, multiple traits and ecological variables can be easily added to a statistical model, although increasingly long time series will be required. Second, it isn't

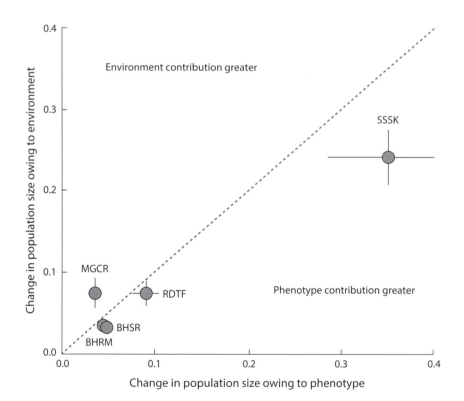

Fig 7.6. The relative contributions of phenotypic traits and environmental variables to interannual variation in population growth for five ungulate populations. One way to interpret these contributions is as coefficients from a linear model where values 0 and 1 are assigned to years t and $t + 1$ (equation 8 in Hairston Jr. et al. 2005). The populations are Soay sheep on St. Kilda (SSSK), bighorn sheep at Ram Mountain (BHRM), roe deer in Trois-Fontaines (RDTF), mountain goats at Caw Ridge (MGCR), and bighorn sheep at Sheep River (BHSR). The phenotypic traits (one per population) were various measures of body size and the environmental variables (one per population) were either sward height, vegetation index, rainfall, or the Pacific Decadal Oscillation. The data (means and 95% confidence intervals, truncated for SSSK) are from Ezard et al. (2009) and were provided by T. Ezard

always clear what to do with an interaction between ecology and evolution—should it be considered "evolution," "ecology," or both (Ellner et al. 2011)? Related to this problem, the equation does not allow for feedback between ecology and evolution. Fourth, current applications assume that phenotypic change is entirely genetically based, whereas population growth could also be influenced by plasticity (Barrett and Hendry 2012, Kovach-Orr and Fussmann 2013, Fischer et al. 2014) or by changes in (for example) age structure. This assumption has been relaxed through the development of partitioning methods based on the Price equation (Coulson and Tuljapurkar 2008, Ozgul et al. 2009, 2010, Collins and Gardner 2009, Ellner et al. 2011, Childs et al. 2016), which was introduced in chapter 2 and will be revisited in chapters 8 and 9. The few applications of these newer approaches to natural populations reveal a wide range of variation in the estimated contributions of phenotypes to population growth: for example, <6% in Ozgul et al.'s (2009, 2010) analyses of Soay sheep and the yellow-bellied marmot (*Marmota flaviventris*) versus 62% in Ellner et al.'s (2011) analysis of great tit data from Garant et al. (2004). Interestingly, the estimated contribution of evolution can *increase* after accounting for plasticity—because the effects of genetic change and plasticity can oppose each other (Ellner et al. 2011).

The above studies all examined the influence of continuous polygenic traits on population growth in nature. A different set of studies have done something similar for alternative morphs/alleles. For example, Turcotte et al. (2011b) used experimental populations in nature to show how evolutionary shifts in clonal frequencies of aphids (*Myzus persicae*) influenced population growth rate. (Laboratory studies of this sort will discussed in the context of predator-prey interactions in chapter 8.) Other examples come from observational studies of unmanipulated natural populations. One is the above-described work on side-blotched lizards (Sinervo et al. 2000). Another is Hanski et al.'s (2006) study of Glanville fritillary butterflies (*Melitaea cinxia*) distributed across approximately 4000 discrete habitat patches in Finland. In their words: "the allelic composition of the glycolytic enzyme phosphoglucose isomerase (*Pgi*) has a significant effect on the growth of local populations, consistent with previously reported effects of allelic variation on flight metabolic performance and fecundity in the Glanville fritillary and *Colias* butterflies. The strength and the sign of the molecular effect on population growth are sensitive to the ecological context (the area and spatial connectivity of the habitat patches), which affects genotype-specific gene flow and the influence of migration on the dynamics of local populations." To clarify this last point, the effects of variation in *Pgi* were greatest in large isolated habitat patches (fig. 7.7).

A persistent limitation of observational approaches is the need for long-term data on population sizes, traits/genotypes, and ecological variables. Although many such data sets exist (Clutton-Brock and Sheldon 2010), they still represent only a small fraction of the situations for which we would like to study evolutionary influences on population dynamics. When long-term data are lacking, several alternatives come to mind. One is a space-for-time substitution: that is, monitoring and correlating traits, population sizes, and environmental variables across populations at a given time rather than across time for a given population. Among several limitations of this approach (Fukami and Wardle 2005) is an even greater difficulty in inferring causation: for example, is population size the cause or the consequence of trait differences? The other alternative, one that is more conducive to inferring causation, is to experimentally manipulate phenotypes/genotypes in nature and then monitor any resulting changes in population size. For instance, gene flow can be manipulated to cause trait changes (Weese et al. 2011, Farkas et al. 2013) or

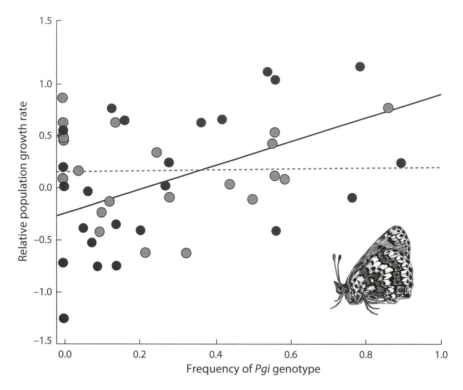

Fig 7.7. Growth rates of isolated local populations of the Glanville fritillary butterfly are correlated with the frequency of *Pgi* genotypes in small habitat patches (<0.3 ha, black symbols) but not in large patches (≥0.3 ha, gray symbols). Relative growth rate is the residual from the relationship between local population growth rate (proportional change in population size between generations) and the regional trend in population sizes. Frequency of *Pgi* genotype is the combined frequency of *ff* and *fd* genotypes, as described in Hanski and Saccheri (2006). The figure was redrawn from Hanski and Saccheri (2006) and the butterfly image was provided by I. Hanski

populations can be swapped among divergent environments. The expectation in both manipulations would be a decrease in population size relative to that typical of well-adapted populations in the same environment. The swapping idea is essentially a reciprocal transplant design, with the key addition of monitoring population dynamics, as opposed to just traits (chapter 4) or individual fitness components (question 2). Such experiments would be logistically challenging, and would have to be balanced against ethical concerns, but they would also be extremely informative regarding how (mal)adaptation influences population dynamics in nature. Of course, uncontrolled and nonreciprocal transplants have been performed many times through species introductions, the frequent failure of which points to the population dynamic consequences of adaptation, with more details discussed in question 5 below.

Phenotypic/genetic changes can have a substantial influence on the dynamics of natural populations. Yet such effects will not always be strong because (1) they can be swamped or obscured by environmental changes (see also Luo and Koelle 2013), (2) the most relevant traits and environmental variables might not have been measured, or (3) evolution might not be fast enough. At present, we have only examples, rather than detailed comparative analyses of when, and how much, contemporary evolution matters for population dynamics.

QUESTION 4: DOES ADAPTATION ALLOW EVOLUTIONARY RESCUE?

It is common sense that most populations must be reasonably well adapted for their local environments—otherwise they would not persist (chapter 3). Rapid environmental change, however, can cause maladaptation that reduces population size (question 1). This maladaptation should impose selection on phenotypes and thus promote adaptive evolution that improves individual fitness (question 2) and perhaps increases population size (question 3). Extending these ideas, a critical question becomes: to what extent can adaptive evolution arrest population declines that would lead to extinction and instead stabilize or even recover population size: that is, "evolutionary rescue" (Gomulkiewicz and Holt 1995, Bell and Gonzalez 2009, 2011, Gonzalez et al. 2013, Martin et al. 2013, Schiffers et al. 2013, Carlson et al. 2014, Botero et al. 2015).[5] Answering this question requires moving beyond a consideration of population growth (question 3) to a consideration of the probability of persistence. Notably, evolutionary rescue can be "friend" (when we hope it will save a species we like) or "enemy" (when we hope it won't save a species, such as a pathogen, we don't like) (Hendry et al. 2011, Carlson et al. 2014,

A number of theoretical models have considered evolutionary rescue, and all of them find that rescue sometimes does and sometimes does not occur. These models generally agree on several factors that increase the probability of evolutionary rescue: (1) larger initial population sizes, (2) less dramatic environmental change (lower initial maladaptation), (3) weaker stabilizing selection around the "optimum" trait value, (4) higher additive genetic variance in the direction of selection, and (5) shorter generation times (e.g., Pease et al. 1989, Bürger and Lynch 1995, Gomulkiewicz and Holt 1995, Boulding and Hay 2001, Orr and Unckless 2008, Björklund et al. 2009, Gomulkiewicz and Houle 2009, Norberg et al. 2012, Vincenzi 2014). These results are likely general because they hold across very different models: different types of environmental change, quantitative-genetic versus single-locus architecture, density dependence or not, and analytical versus simulation. Additional influences have been explored in specific models, and here is a partial listing. (1) Rescue in a continuously changing environment is more likely when noise in the environmental trend is positively autocorrelated (Björklund et al. 2009). (2) Rescue should be aided by adaptive plasticity, unless the costs of plasticity are too high (Chevin et al. 2010, Chevin and Lande 2010, 2011, Kovach-Orr et al. 2013). (3) Rescue can be hampered by depensatory density dependence or aided by compensatory density dependence (Holt et al. 2004, Björklund et al. 2009). (4) Rescue is more likely if negative interspecific interactions (e.g., predator-prey) most heavily impact individuals that are least well adapted (Jones 2008). (5) Rescue is less likely in the presence of competing species that are preadapted to the new conditions (de Mazancourt et al. 2008, see also Norberg et al. 2012). (6) Rescue is more likely if it involves quantitative changes

---

[5]Evolutionary rescue of one species can then stabilize entire communities: that is, community rescue (Kovach-Orr and Fussmann 2013). Also, note that evolutionary rescue is different from "genetic rescue" and "demographic rescue" as explained by Carlson et al. (2014). Remember also the earlier point that adaptive evolution can lead to extinction (Gyllenberg and Parvinen 2001, Webb 2003).

in an existing "mode of responses," such as the slope or elevation of a reaction norm, than when it requires qualitative shifts among modes, such as plasticity to bet-hedging (Botero et al. 2015). (7) Gene flow and dispersal might or might not aid evolutionary rescue depending on the context (Boulding and Hay 2001, Norberg et al. 2012, Bourne et al. 2014), as was described further in chapter 5.

Studies of evolutionary rescue in real organisms would ideally (1) monitor replicate populations some of which do and some of which do not then persist during environmental change, (2) determine whether or not these differential outcomes were influenced by adaptive evolution, and (3) explore potential determinants of the outcome (e.g., population size, magnitude of the disturbance, genetic variance, and gene flow). Studies of this type have started to appear for populations maintained in the laboratory, a context where replication, control, and manipulation is relatively straightforward (Dey et al. 2008, Bell and Gonzalez 2009, 2011, Cameron et al. 2013, Linsey et al. 2013). As an exemplar, Bell and Gonzalez (2009) exposed thousands of replicate yeast populations to high salt concentrations. Initial maladaptation caused large population declines, but contemporary evolution then sometimes allowed population recovery (fig. 7.8). The probability of this rescue depended on initial population size (fig. 7.8), with larger populations being less susceptible to stochastic extinction and having a greater potential for beneficial mutations (Bell and Gonzalez 2009). In follow-up experiments, Bell and Gonzalez (2011) showed that past adaptation to moderately stressful salt levels increased evolutionary rescue following exposure to very stressful salt levels. The authors also showed that, in the absence of this preadaptation, gene flow among replicates evolving under different salt levels aided evolutionary rescue following exposure to very stressful salt levels.

Similar experiments have not been performed in nature. However, evolutionary rescue must be extremely common, given the countless situations in which environments have changed dramatically and yet populations have persisted. For instance, evolution must have contributed to population persistence in numerous cases of industrial pollution (Bradshaw 1984, Medina et al. 2007), insecticide/herbicide application (Heap 1997, Whalon et al. 2008), biocontrol (Fenner 1983, Dwyer et al. 1990), rapid climate change, and biological invasions. In the context of climate change, a meta-analysis showed that European birds not showing numerical declines tend to be from taxonomic groups showing the fastest past rates of niche evolution (Lavergne et al. 2013). In the context of biological invasions, native species might need to evolve (Phillips and Shine 2006) but so too might the invaders. As an example, the Polynesian field cricket (*Teleogryllus oceanicus*) recently colonized Hawaii, where it came into contact with the parasitoid fly *Ormia ochracea*. The fly is attracted to the cricket's courtship song and its lethal effects caused, between 1992 and 2001 on Kauai, a dramatic cricket population decline to near extinction. Since 2001, cricket population sizes have increased dramatically coincident with evolutionary changes in wing structure of the crickets that eliminated their ability to produce courtship calls and thus reduced the rate of parasitism (Zuk et al. 2006, Tinghitella 2008). Although all of these examples might well—and indeed probably do—reflect evolutionary rescue, directly causal links between maladaptation and population declines and then between adaptation and population recovery are hard to confirm (Gomulkiewicz and Shaw 2013, Vander Wal et al. 2013, Carlson et al. 2014). As just one complication, decreasing population size owing to maladaptation will decrease competition and thus increase population fitness, potentially arresting population declines. Of course, Allee effects

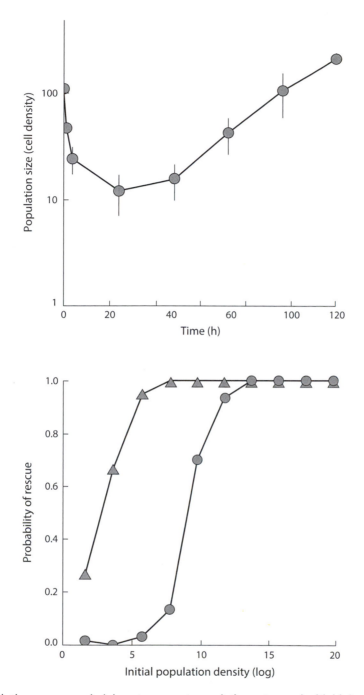

Fig 7.8. Evolutionary rescue in laboratory yeast populations stressed with high salt. The top panel shows the characteristic signature of evolutionary rescue: a rapid decline following exposure to the stressor followed by a recovery of population size as evolution allows adaptive responses. The data are means and 95% confidence intervals across replicates. The lower panel shows the probability across 60 replicate laboratory populations that evolution rescues populations from extinction depending on initial population density (dilution level). In this lower panel, triangles show trials without salt, where stochastic extinction can occur at low density, and circles show the trials with salt added. The data are from Bell and Gonzalez (2009) and were provided by A. Gonzalez

could lead to the opposite outcome, whereby declines in population size will reduce per capita fitness, thereby causing further declines in population size, potentially spiraling to extinction (Holt et al. 2004). Indeed, Allee effects represent an interesting context for future research on eco-evolutionary dynamics because contemporary evolution might be particularly critical near population sizes below which Allee effects begin.

Additionally, studies of natural populations can assess the *potential* for evolutionary rescue by (1) estimating or postulating values for parameters thought to be of importance (rate of environmental change, selection, additive genetic variance, demography), and then (2) using those parameters in analytical models or simulations that predict the probability of population persistence (Knight et al. 2008, Willi and Hoffmann 2009, Reed et al. 2011, Vedder et al. 2013). As an exemplar, Knight et al. (2008) studied populations of the herbaceous perennial understory plant *Trillium grandiflorum* that were declining owing to herbivory by white-tailed deer (*Odocoileus virginianus*). Herbivory causes selection against early flowering, and so the evolution of this trait might provide some resilience in the face of deer herbivory. Using individual-based simulations and extensions to the method of van Tienderen (2000), the authors found that evolutionary rescue would be possible except when populations were small or had low trait heritabilities. As another example, qualitatively similar results were obtained when Willi and Hoffmann (2009) analyzed the potential adaptation of *D. birchii* to climate warming. These sorts of predictions also can be made more generally. For instance, some studies have argued that rates of environmental change are often too rapid for local (or even global) persistence given typical trait heritabilities observed in nature (Bradshaw and McNeilly 1991, Jump and Penuelas 2005).

Many populations in altered environments would disappear without sufficiently rapid and effective adaptation. This evolutionary rescue is most likely when environmental change is small, initial population size is large, and appropriate genetic variation is high. These conclusions are based on theoretical models and laboratory studies, whereas we really have no idea when evolutionary rescue will or will not take place in nature. One limitation is that investigators typically focus on the supposedly "characteristic signature" of evolutionary rescue: a U-shaped population trajectory reflecting rapid declines after the disturbance followed by recovery back to the original population size (Gomulkiewicz and Holt 1995). Beyond the obvious problem that this signature applies only to abrupt (as opposed to gradual) environmental change, evolutionary rescue is likely occurring even without population declines: indeed, the absence of a population decline is probably just as indicative. The reason is that, as described earlier, fitness is expected to decline as a result of gene flow, mutation, and environmental change, necessitating constant evolution just to maintain population size. Thus, population stability is likely a good signature of evolution constantly rescuing populations from extinction (see also Vander Wal et al. 2013). This idea that *stability reflects eco-evolutionary dynamics* is a recurring theme in this book.

QUESTION 5: DOES ADAPTATION AID RANGE EXPANSION?

Many species show geographical variation in fitness-related traits (or genotypes), suggesting that adaptive divergence occurs across species ranges (Ashton et al. 2000, Gilchrist et al. 2004, McKellar and Hendry 2009, Colautti et al. 2009). Perhaps this divergence was instrumental in range expansion. Stated another way, perhaps colonization of, and persistence in, certain areas of the range was not possible before adaptation enhanced the fitness and thus growth rate of local populations. Certainly, adaptive variation along spatial gradients is a key part of theoretical models of the evolution of range limits (chapter 5). Moreover, some range limits clearly exist because species are maladapted beyond those limits and thus unable to persist: that is, they have negative population growth (Sexton et al. 2009, Eckhart et al. 2011). For example, Griffith and Watson (2006) showed that the weedy annual cocklebur (*Xanthium strumarium*) would be able to persist beyond its current range only through the evolution of a trait (repro-ductive timing) that would improve adaptation to novel conditions. Results of this sort appear common: 75% of 111 experiments transplanting plants beyond their latitudinal or elevational range limits revealed low fitness, suggesting that range limits reflect niche limits (Hargreaves et al. 2014, see also Lee-Yaw et al. 2016).

The above results indicate that adaptation would be necessary for future range expansion, but they don't confirm that past range expansion was aided by adaptive divergence. Instead, organisms might be able to persist across their range without any divergence, although divergence could certainly improve the fitness of residents relative to immigrants. That is, organisms from each part of the range might be able to persist in any other part of the range in the absence of better adapted local conspecifics (i.e., soft selection). Thus, adaptive clines in traits (or genotypes) and higher fitness for local than foreign individuals are necessary, but not sufficient, conditions for inferring that local adaptation has aided range expansion. That more difficult inference also requires demonstrating that populations from one part of a range could not persist in other parts of the range in the absence of locally adapted conspecifics. Many (but not all) reciprocal transplant experiments do indeed show that local individuals have higher fitness than foreign individuals (chapter 4)—but, as has been repeatedly emphasized in the present chapter, this local adaptation will not necessarily have straightforward consequences for population dynamics. Thus, reciprocal transplant studies also need to track the long-term success of transplanted populations in the absence of competition from locally adapted residents, as would be the situation attending range expansion.

In a sense, precisely this sort of "experiment" has been performed many times through the accidental or intentional introduction of species to places where they are not native. The fate of these populations can yield insights into the role of adaptation in the growth and expansion of species ranges in new environments. In such cases, most introduced species never become established in the new location and most that do become established never become so widespread and abundant as to be considered "invasive" (Williamson and Fitter 1996). In addition, the subset of species that do become invasive often ex-perience a long lag time between introduction and expansion (Crooks 2005). These patterns might be caused by several different effects, with initial (mal)adaptation being one of them. In essence, most introduced species might be so maladapted for the new conditions that they either go extinct or just barely hold on. Permanent establishment and subsequent expansion might then require adaptation that improves local fitness and thereby increases population size.

Several suggestions have been advanced for the sort of adaptation that allows introduced species to expand their ranges. One possibility is the "evolution of increased competitive ability" following release from enemies found in the native range (Blossey and Nötzold 1995). The argument here is that selection should favor reduced investment into defense, thereby allowing increased investment into competitive ability. A first meta-analysis of this hypothesis revealed that 8 of 19 studies recorded greater competitive ability in invasive than native plant species (Bossdorf et al. 2005). A more recent meta-analysis found less support: defense traits generally did not decrease, and competitive ability generally did not increase, for invasive species (Felker-Quinn et al. 2013). These ambiguous/negative results suggest that the hypothesis is not a strong (at best) explanation for the evolution of invasiveness, doubly so given methodological limitations of the studies (Bossdorf et al. 2005, Colautti et al. 2009). Another mechanism for the evolution of invasiveness could be adaptation to the new environmental conditions, such as an improved ability by introduced insects to use native plant species. Certainly, adaptation has been documented in many invasive species (Cox 2004, Hendry et al. 2008, Westley 2011, Felker-Quinn et al. 2013, Colautti et al. 2015, Flores-Moreno et al. 2015), although it isn't clear to what extent it has influenced the invasion spread.

Once an introduced species starts to spread, ongoing evolution might be important in the rate and extent of that spread. Certainly, many invasive species now show apparently adaptive spatial variation in traits (or genotypes) across their new ranges (Colautti et al. 2009, 2015) but, as explained above, it is hard to establish how this adaptive variation contributed to range expansion. Nevertheless, several additional lines of evidence support the idea. First, ongoing adaptation along spatial gradients can, in theory, increase the rate of spread of invasive species (García-Ramos and Rodríguez 2002, Perkins 2012). As an empirical example, it has been argued that the adaptive evolution of increased dispersal ability (e.g., longer legs) increased the rate of spread of cane toads (*Chaunus* [*Bufo*] *marinus*) in Australia (Phillips et al. 2006). Second, evolution might allow the expansion of a species beyond the range that would be possible without such evolution. This situation also has been argued for cane toads, given that their range already exceeds that predicted from environmental niches in their native range (Urban et al. 2007). That is, the toads have evolved an ability to tolerate environmental conditions they do not occupy in their native range. Finally, the spatial location where invasive species stop spreading seems to be at least sometimes be related to adaptive limits (Bridle and Vines 2007).

Adaptation probably often plays a role in range expansion, but the magnitude of this effect is hard to conclusively establish. For instance, the invasive knapweed *Centaurea maculosa* has clearly evolved a number of adaptations in its introduced range, including greater growth rates, stronger competitive ability, and greater defense against herbivores (Ridenour et al. 2008). However, it is difficult to ascertain to what extent these particular evolutionary changes were instrumental in shaping the original invasiveness and subsequent range expansion. Further insights are likely to come from more detailed studies of introduced species (as described above), responses to climate change (Atkins and Travis 2010), and both (Moran and Alexander 2014).

QUESTION 6: DOES INTRASPECIFIC DIVERSITY INFLUENCE POPULATION DYNAMICS?

Phenotypic/genetic variation can shape evolutionary responses (chapter 3 and above) and can thereby influence population dynamics. For instance, a greater amount of genetic variation in adaptive traits should allow faster adaptation to environmental change, which should then facilitate evolutionary rescue. Separate from this cross-generational evolutionary effect of intraspecific diversity is the potential for within-generation effects.

As a hypothetical example, a species with multiple ecotypes/phenotypes/genotypes might be able to exploit more environments and should therefore have access to more total resources, which might allow a greater population size. For other examples and discussions of the effects of intraspecific diversity, see Luck et al. (2003), Hughes et al. (2008), Bolnick et al. (2011), and Vindenes and Langangen (2015). These within-generation effects of intraspecific variation will be a common theme in the next two chapters about community and ecosystem level effects. It therefore makes sense to here briefly consider their potential influence on population dynamics.

One approach to this question involves establishing experimental plots or meso-cosms with groups of individuals that differ in their amount of phenotypic or genotypic diversity: for example, some plots with individuals of a single clone (monocultures) versus other plots with the same number of individuals spread across multiple clones (polycultures). Although this experimental approach will be discussed to a much greater extent in the next chapter, I can here provide an exemplar in the form of experiments with plots of seagrass (*Zostera marina*) differing in genotypic diversity, specifically the number of distinct clones. The plots were monitored for population-level responses to grazing by Brant geese (*Branta bernicla*), to grazing by three invertebrates, and to shoot clipping by the experimenters (Hughes and Stachowicz 2004, Hughes et al. 2010, Hughes and Stachowicz 2011). These studies found that more genetically diverse plots (1) were more resistant to disturbance by geese, (2) maintained higher biomass in the presence of one invertebrate but not the others, and (3) were more resilient to clipping under high disturbance but not low disturbance. In separate experiments with the same seagrass, increasing genotypic diversity within plots increased biomass production and stem density during a period of exceptionally high temperatures (Reusch et al. 2005). Similar effects have been seen in terrestrial plants (Drummond and Vellend 2012). For instance, more genetically diverse plots of *Arabidopsis thaliana* have higher productivity and survival in the presence and absence of herbivores (Kotowska et al. 2010). In addition, meta-analysis across taxonomic groups has shown that greater genotypic/phenotypic diversity of a founding population increases the probability of successful establishment (Forsman 2014). These examples serve to illustrate that intraspecific genetic diversity does influence population-level variables in some cases but not others.

In those instances where greater genotypic diversity influenced population dynamics in positive ways (e.g., higher resistance, resilience, survival, or productivity), we might ask about the specific mechanism. I will discuss these mechanisms in more detail in the next chapter and here select only a few for illustration. First, one important benefit of having more genotypes is the increased likelihood of including the "best" genotype. This is an additive "sampling effect" of diversity, rather than an effect of diversity per se. Additional nonadditive mechanisms can also be involved (Hughes and Stachowicz 2004, Crawford and Whitney 2010, Kotowska et al. 2010), such as different genotypes showing "complementarity." For example, different genotypes could use different resources, leading to reduced competition in polyculture that then allows greater productivity relative to monocultures. An example is the soil arthropod *Orchesella cincta*, where more diverse groups achieved a higher population size and biomass, and these effects were greatest for genotype mixtures with greater phenotypic dissimilarity (Ellers et al. 2011). Or the different genotypes might be less susceptible to different diseases or herbivores, mak-ing polycultures less susceptible than monocultures to an outbreak (Tooker and Frank

2012). In a remarkable experiment demonstrating this last effect, monocultures and polycultures of rice (*Oryzae sativa*) were planted over 3342 ha of fields in Shiping County, Hunan Province, China (Zhu et al. 2000). The prevalence of an important fungal blast disease of rice was much lower in the genetically diverse fields and "mixed populations produced more total grain per hectare than their corresponding monocultures in all cases" (Zhu et al. 2000).

Although the above studies were often performed in natural or seminatural conditions, they all used artificially generated mixtures of genotypes. The benefit of this approach is that causality is relatively easy to infer because diversity is an experimental treatment—but the limitation is uncertainty as to how such effects would play out in real populations (see also Tack et al. 2012). However, correlative evidence from real populations also suggests effects of intraspecific diversity on population dynamics. I will here emphasize work on the effects of diversity among local subpopulations within a metapopulation (Luck et al. 2003, Jetz et al. 2009, Schindler et al. 2010, González-Suárez and Revilla 2013), as opposed to diversity among individuals within local populations (as in the previous paragraph). One example comes from sockeye salmon, where productivity (escapement and fishery yields) are vital for conservation, management, and harvesting (Hilborn et al. 2003, Schindler et al. 2010). Each year, a major commercial fishery targets adult sockeye salmon returning from the open ocean to spawn in fresh water in Bristol Bay, Alaska. This metapopulation consists of nine major rivers, each containing dozens to hundreds of spawning populations adapted to different local environments. The number of fish returning to each local population and to the different major rivers varies asynchronously across years, likely as a result of spatiotemporal variation in environmental conditions. As a result, the size of the overall metapopulation (all fish returning to Bristol Bay) was about half as variable as would be expected if the population-specific interannual fluctuations were instead synchronized (Schindler et al. 2010). Diversity among local populations within a metapopulation thus has a dramatic positive influence on reducing variation in overall population density. As another example, meta-analyses have shown that mammal species with greater interpopulation variation have lower risk of extinction (González-Suárez and Revilla 2013).

Greater phenotypic/genetic diversity within and among populations can lead to higher productivity, greater resistance/resilience to disturbance, and lower temporal variation. These results mesh well with arguments for the importance of intraspecific diversity in altering a wide range of evolutionary and ecological processes (e.g., Hughes et al. 1997, Luck et al. 2003, Hughes et al. 2008, Bolnick et al. 2011, Vindenes and Langangen 2015). However, it is critical to recognize that the positive effects will be the product of *adaptive* variation, such as differences among individuals that allow the population to better exploit the full range of available resources (Rueffler et al. 2006, Bolnick et al. 2011) and differences among populations that reflect adaptation to different local conditions. *Maladaptive* variation within populations, such as might be produced by developmental or mutational noise (Burt 1995, Hansen et al. 2006), is instead likely to have negative effects, such as a reduction in population mean fitness owing to genetic load (see question 3 in chapter 5). Similarly, maladaptive differences among populations would be expected to depress local population sizes and therefore the entire metapopulation.

Conclusions, significance, and implications

This chapter was the first to explicitly focus on the evo-to-eco side of eco-evolutionary dynamics; that is, the effect of evolutionary change on ecological processes. The specific focus was on population dynamic variables such as population size (abundance or density), growth rate, or structure (e.g., age- or size-structure). Theoretical considerations suggest that the effects of evolutionary change on population dynamics will be highly variable, depending on factors such as when and how density dependence acts and the nature of feedbacks between evolutionary and ecological change. Of additional interest, these dynamics often might be cryptic, such that population dynamic *stability* might reflect underlying eco-evolutionary *dynamics* (Kinnison et al. 2015).

Eco-evolutionary dynamics and feedbacks at the population level have been elegantly demonstrated in a number of laboratory experiments but the evidence from nature is more circumstantial and indirect. Here, several studies have demonstrated that maladaptation causes population declines and, indeed, common-sense dictates that extinction events are ultimately the result of maladaptation. Such maladaptation usually will be the result of environmental change (biotic or abiotic), which is an ever-present challenge faced by all populations. This realization indicates that ongoing evolution must be a major reason why many populations maintain their abundance through time—if they weren't evolving they would be declining. Several studies have shown the *potential* for evolution to influence population dynamics by demonstrating the contemporary evolution of individual fitness in nature or by showing that year-to-year changes in population size or growth are correlated with phenotypic/genotypic characteristics. However, studies formally demonstrating evolutionary rescue in natural populations remain elusive, likely due to the difficulty of conclusively inferring this phenomenon. In addition to rescue, ongoing evolution appears important for range expansion although, again, conclusive demonstrations are difficult. Finally, intraspecific diversity within and among populations is clearly important for a wide range of population-level properties, such as colonization, resistance to disturbance/disease/herbivory, overall productivity, variation in productivity, and extinction risk. The next question is how such influences scale up from the population level to the community and ecosystem levels.

Chapter 8

Community Structure

We have has thus far considered the evolution of organisms (chapters 2–8) and the effects of that evolution on their population dynamics (chapter 9). My goal in the present chapter, and the one that follows, is to consider how organismal evolution can have effects beyond the organisms themselves. As a simple example, the evolution of one species, such as a predator, could demographically influence other species, such as their prey. On long time scales, such eco-evolutionary interactions are clearly important, even if they are incompletely understood (Odling-Smee et al. 2003, Johnson and Stinchcombe 2007, Erwin 2008, Vellend 2010). For example, herbivores that specialize on a particular plant species cannot exist before the plant first evolves. The topic at hand, however, is how such dynamics play out on contemporary time scales.

Evolutionary effects on communities are classically considered in the context of species interactions, such as between pairs of competing species, between specific predators and prey, or between specific hosts and parasites. This species-by-species approach has been very informative but is limited when considering the complicated effects of multiple interacting and potentially (co)evolving species. As an example, Tétard-Jones et al. (2007) showed that different barley (*Hordeum vulgare*) genotypes influence the performance (growth rate) of aphids (*Sitobion avenae*) and, simultaneously, different aphid genotypes influence the performance of barley. Moreover, both of these effects interact with the presence/absence of rhizobacteria (*Pseudomonas aeruginosa* 7NSK2), thus generating a GxGxE interaction. Building on this system, Zytynska et al. (2010) showed that the performance (body size as a proxy for fitness) of parasitoid wasps (*Aphidius rhopalosiphi*) was influenced by the interaction among barley genotype, aphid genotype, and rhizobacteria presence/absence. Many other species likely influence this system, and every other community, and it is clearly impossible to consider all (or even most) such species-by-species interactions.

An alternative to the above "species-by-species" approach is a "focal-species composite-response" approach. The idea here is to evaluate how the evolution of a focal species influences a composite community variable, such as species richness, species diversity, food web length, invasibility, or stability. This approach has been couched (Whitham et al. 2003, 2006, Wimp et al. 2005) in the context of organisms having "extended phenotypes" (Dawkins 1989, Bailey 2012). For example, the arthropod community on a given genotype of tree is an extended phenotype of that tree, just as a nest is to a bird and a dam to a beaver. Another way of expressing the same idea is to extend the concept of "indirect genetic effects"—the effects of an individual's genes on the phenotype or fitness of conspecifics (Moore et al. 1997, Wolf et al. 1998)—into the realm of "interspecific indirect genetic effects"—the effects of an individual's genes on the phenotype or fitness of heterospecifics or on composite variables (Shuster et al. 2006, Genung et al. 2011, Bailey et al. 2014).

In the present chapter, I will mainly invoke species-by-species approaches in two classic contexts: interactions among competitors and between predators and prey, the latter including host-parasite and plant-herbivore interactions. In most other contexts, I will emphasize the "focal-species composite-response" approach. The following chapter (chapter 9) will extend this latter approach from the level of communities to that of ecosystems. Given that composite-response variables at these two levels tend to grade into each other, distinctions betweem them—and hence between the two chapters—can be somewhat arbitrary. Indeed, the chapters could have been reasonably combined into a single long chapter. I have instead divided the discussion into two chapters because community-level variables tend to focus on ecological responses at the level of organisms whereas ecosystem-level variables tend to focus on ecological responses beyond organisms.

When and how might contemporary evolution influence community structure?

Evolutionary influences on community-level composite responses could occur through two general routes. First, evolution could alter population dynamics (chapter 7), which could then influence community structure. That is, the abundance of a particular species can influence community structure (e.g., the number of predators can influence prey community structure), and so evolution that changes the abundance of that focal species will indirectly initiate cascading effects on the community. Second, evolution that alters the phenotype of a particular species could directly influence per capita effects of that species on the community. For instance, the evolution of trophic traits in a predator can change its influence on prey community structure (Palkovacs and Post 2009, Moya-Laraño 2011). These indirect and direct routes from evolution to community structure partly echo the distinction in ecology between density-mediated and trait-mediated effects (Abrams 1995, Werner and Peacor 2003), except that here density changes are themselves trait mediated (although perhaps through different traits than those having the direct effects). I will now outline three conceptual frameworks for considering these effects and then discuss the expectations that derive from their joint consideration.

Extending the Lande equation: Johnson et al. (2009) presented a quantitative genetic framework to conceptualize and quantify direct effects as discussed above. This framework is an extension of the Lande equation (chapter 4), where changes in the mean values of traits (ΔZ) can be predicted from directional selection (β) and the genetic (co)variance matrix (G): that is, $\Delta Z = G\beta$. Johnson et al. (2009) suggested that, if one also knows how trait changes influence a community variable, evolution-induced changes in that variable (ΔC_1) can be predicted as:

$$\Delta C_1 = \begin{bmatrix} \alpha_{11}, & \alpha_{12}, & \cdots & \alpha_{1i} \end{bmatrix} \begin{bmatrix} G_{11} & G_{12} & \cdots & G_{1j} \\ G_{21} & G_{22} & \cdots & G_{2j} \\ \vdots & \vdots & & \vdots \\ G_{i1} & G_{i2} & \cdots & G_{ij} \end{bmatrix} \begin{bmatrix} \beta_1 \\ \beta_2 \\ \vdots \\ \beta_i \end{bmatrix}$$

where the Gs make up the (co)variance matrix for the traits and the βs are a vector of selection gradients on those traits (chapter 4). The new term is an α vector that represents the change in a community variable as a function of changes in the trait values. This modification of the Lande equation is the same as that developed for predicting the indirect genetic effects of trait change (Wolf et al. 1998) and thus "interspecific indirect genetic effects" (Shuster et al. 2006). The αs are estimated as partial regression coefficients from a multiple regression model relating changes in the community variable to changes in the traits, and the equation can be extended to multiple community variables (Johnson et al. 2009). To illustrate their framework, Johnson et al. (2009) provided a worked example based on empirical estimates of the terms on the right-hand side of the equality, although data were not available to test the resulting predictions for ΔC_1. This framework retains the limitations explained earlier in the context of using $\Delta Z = G\beta$ to predict trait change (chapter 4). An additional limitation is that the framework does not currently incorporate changes in population size. That is, it considers only the direct effects of trait change on community variables, while ignoring the possible indirect evolutionary effects acting through changes in population dynamics.

Extending the Hairston Jr. et al. method: a framework that does consider population dynamics is the Hairston Jr. et al. (2005) method introduced in chapter 7. In the present context, we can start by assuming that the change in some community variable (ΔC) might be influenced by changes in the mean value of a phenotypic trait (ΔZ) and the population size (ΔN) of a focal species. That is, $C(t) = C[N(t), Z(t)]$ and, by the chain rule in continuous time, we have

$$\frac{dC}{dt} = \frac{\partial C}{\partial N}\frac{dN}{dt} + \frac{\partial C}{\partial Z}\frac{dZ}{dt}$$

Here, the rate of change in a community variable (e.g., the number of arthropods) is the summation of a population size effect (first term) and a phenotypic trait effect (second term). (As noted in chapter 7, the discrete time form of the equation would have an interaction term.) The next step is to add the possibility that phenotypic traits can influence population size. For instance, we can postulate that $C(t) = C[N(Z(t),K(t)),Z(t)]$, for which the continuous time solution is

$$\frac{dC}{dt} = \frac{\partial C}{\partial N}\frac{\partial N}{\partial K}\frac{dK}{dt} + \frac{\partial C}{\partial N}\frac{\partial N}{\partial Z}\frac{dZ}{dt} + \frac{\partial C}{\partial Z}\frac{dZ}{dt}$$

In this case, the rate of change in a community variable is the summation of a direct phenotypic effect (last term) and two population size effects. The first of these latter effects occurs when a community variable changes as a function of changes in population size that result from changes in a phenotype-independent environmental variable (K, e.g., rainfall). The second of these latter effects occurs when a community variable changes as a function of changes in population size that result from changes in mean phenotype (i.e., the indirect effect discussed above). This partitioning approach can be expanded to any desired extent, such as by adding an environmental variable that directly influences the community variable. Some of these potential expansions will be considered to a greater extent in the context of ecosystem function (chapter 9). In a manner analogous to that described by Hairston Jr. et al. (2005) and Ellner et al. (2011), these equations can be applied to empirical data to statistically partition effects among the various terms. This framework has the same limitations as explained earlier in the context of population dynamics (chapter 7), such as (1) it generally focuses on only a single species, trait, and ecological variable; (2) it usually ignores nongenetic contributions to phenotypic change; and (3) it retains uncertainty regarding the interaction terms that would be present in discrete time. Additional developments are reducing these limitations (e.g., Ellner et al. 2011).

Extending the Price equation: The Price equation was introduced in chapter 3 as a means of partitioning changes in mean phenotype into the effects of selection (additive genetic covariance between the trait and relative fitness) and a "transmission bias" (including plasticity). The Price equation was also discussed in chapter 7 as a way of partitioning changes in population parameters among, again, selection and various forms of transmission bias. A similar sort of partitioning can be performed to explain changes in a community variable due to (for example) changes in the abundance of a focal organism, selection-driven changes in their phenotypic traits, and any number of other factors (Collins and Gardner 2009, Ellner et al. 2011, Genung et al. 2011). Ellner et al. (2011) provided one example of how this partitioning can be accomplished, explicitly linking its implementation to the above Hairston Jr. et al. (2005) approach.

As an aside, the above frameworks all consider the effects of changes in mean trait values that result from linear (directional) selection. Yet it also seems likely that changes in trait variance, whether resulting from linear or nonlinear selection, will influence community structure. A predictive framework has yet to be advanced for considering such nonlinear effects, although they do make an appearance in the context of how changes in genetic diversity can influence community structure (Question 4).

EXPECTATIONS

Before outlining some expectations, we need to acknowledge two important points. First, the net effect of evolution on a community variable will be shaped by multiple causal pathways (e.g., direct and indirect) that can reinforce or oppose each other. For instance, a given phenotypic change might generate a direct effect that increases some community variable (e.g., more arthropods) and an indirect effect acting through population dynamics that decreases the same variable (fewer arthropods). The total net effect of evolution thus would be less than the effects occurring along each individual pathway—to the point that the community variable might not change at all. Bassar et al. (2012) provided an example of such opposing effects, albeit independent of the population dynamic context emphasized here. Second, the evo-to-eco effect of evolution on communities can generate an eco-to-evo feedback that alters the dynamics of both (Post and Palkovacs 2009, Pregitzer et al. 2010, Genung et al. 2011, Becks et al. 2012, Strauss 2014). For instance, the evolution of larger beaks in granivorous birds could deplete larger seeds from the environment, which could lead to selection for smaller beaks (if the remaining seeds are smaller) or larger beaks (if the remaining seeds are larger). More generally, such positive (versus negative) feedbacks between ecology and evolution should reinforce (versus oppose) both evolutionary and ecological change. These ideas were introduced in chapter 7 and are considered in more detail in chapter 9.

The above concepts can be used to generate predictions for when evolutionary (or, more generally, phenotypic) change is most likely to influence community structure on contemporary time scales. Specifically, such influences might be strongest when . . .

1. . . . phenotypic change is greatest. When the trait change is genetically based, such effects will be greatest when selection is strongest (such as when environmental change is occurring) and when genetic variation is highest in the direction of selection (Hairston Jr. et al. 2005, Johnson et al. 2009, Jones et al. 2009, Becks et al. 2010). When the trait change is plastic, such effects will be greatest when environmental change is the most dramatic and when plasticity is the most responsive.

2. . . . phenotypic trait variation has strong direct effects on community structure (Wymore et al. 2011): that is, the αs of Johnson et al. (2009) are large. This result seems most likely for traits that have important ecological functions (e.g., trophic traits or "functional traits") in organisms that have large effects as individuals, such as "ecosystem engineers" (Jones et al. 1994), "niche constructors" (Odling-Smee et al. 2003), "foundation species" (Ellison et al. 2005), "keystone species" (Mills et al. 1993), or abundant migratory species (Bauer and Hoye 2014). Erwin (2008) has argued that the effects of evolution in such species have persisted over millions of years and have shaped macro-evolution and biological diversity.

3. . . . the focal organisms are very abundant. In this case, phenotypic change might have large effects when summed across all individuals even if the effect of each individual organism is small. This outcome seems most likely in the case of micro-organisms and abundant weeds, pests, and parasites.

4. . . . evolution influences population dynamics. In particular, the abundances of certain species (e.g., trees, elephants, locusts, salmon) clearly influence the structure of the communities in which they reside, and so phenotypic change that influences their population size should influence community structure.

5. . . . direct and indirect effects of evolution do not oppose each other. In this case, the net effect of evolution is less likely to be diminished by opposing effects that act through different causal pathways.

6. . . . feedbacks between ecology and evolution are present and positive. As explained above and in the next chapter, such positive feedbacks should enhance eco-evolutionary dynamics.

7. . . . external drivers, such as those represented by K in the Hairston Jr. et al. (2005) approach, are not overpowering. Otherwise, any contribution from evolutionary change can be swamped by nonevolutionary ecological drivers, with climate-driven variation in rainfall being an obvious candidate.

The above expectations are based on demonstrable evolutionary and ecological *change*. However, as discussed in chapter 7 and elsewhere in the book, a *lack of change* in traits, population dynamics, and community structure might—and likely does—reflect underlying "cryptic" eco-evolutionary dynamics. This situation is not encapsulated in the above expectations, and instead requires an alternative inferential framework (Hiltunen et al. 2014, Strauss 2014, Kinnison et al. 2015).

Common approaches

I now outline empirical approaches commonly used in the "focal-species composite-response" program for studying eco-evolutionary dynamics at the community (and ecosystem) levels.[1] Many of these approaches are also applicable to the alternative species-by-species program. The first two approaches involve measuring in nature (or manipulating in experimental plots/arenas/mesocosms), the diversity or identity of genotypes of a focal species and then monitoring composite-response variables. The third approach involves the use of common-garden experiments to compare the community-level effects of phenotypically/genetically divergent populations. The fourth approach relates temporal changes within populations to coincident changes in community-level variables. Additional approaches, such as community phylogenetics and experimental evolution in the laboratory, are a bit further from the focus of this book but will be mentioned where appropriate.

1. Genetic diversity within populations

This approach takes its inspiration from the "biodiversity-ecosystem function" (BEF) studies conducted at the among-species level (Loreau et al. 2001, Hooper et al. 2005) and extends them to the within-species level (e.g., Wimp et al. 2004, Reusch et al. 2005, Hughes et al. 2008, Kotowska et al. 2010, Drummond and Vellend 2012, Zuppinger-Dingley et al. 2014). This approach was partly motivated by the observation that large amounts of the phenotypic variation within plant communities is due to variation among individuals within species (Albert et al. 2010, 2012, Siefert et al. 2015). Hence, the intraspecific BEF studies examine how the amount of genetic diversity within a focal species influences local community and ecosystem variables (for details see Hughes et al. 2008). One way to implement this approach is to quantify, across natural populations of a focal species, associations between levels of genetic variation and attendant community variables. For instance, meta-analyses have shown that levels of genetic diversity within particular species are typically correlated with species diversity (Vellend et al. 2014). Limitations of this observational approach include difficulties establishing cause and effect, and the complications resulting from nontarget variables (e.g., variation in productivity) that cause spurious correlations or obscure real associations. An alternative is to design experiments in which genetic diversity is manipulated among plots/arenas/mesocosms and community responses are then monitored. Such experiments are most often implemented by varying the number of clones of a target species (Booth and Grime 2003, Reusch et al. 2005, Crutsinger et al. 2006, Johnson et al. 2006, Drummond and Vellend 2012).

Effects of genetic diversity on community structure could be the result of additive or nonadditive mechanisms (Hughes et al. 2008). Additive mechanisms arise when "the ecological response of individual genotypes measured in monoculture, and knowledge of the initial relative abundance of each genotype in a population, are jointly sufficient to predict the same ecological response for a genetically diverse population" (Hughes et al. 2008). Nonadditive mechanisms cause deviations from such predictions. One additive mechanism is the sampling effect, wherein a more diverse mixture of genotypes is more likely to include particularly influential genotypes. If, for example, a particular plant genotype harbors a great diversity of

---

[1]Other authors have suggested alternative ways to categorize empirical methods for inferring evolutionary influences on community and ecosystem variables (e.g., Genung et al. 2011).

arthropods, then mixtures with more plant genotypes are more likely to include this exceptional genotype and, hence, to have more arthropods. Another additive mechanism occurs when the effects of different genotypes are cumulative. For example, single-clone plots will not harbor more arthropod species than the maximum number of species found on a given clone. Two-clone plots, however, can have all of the arthropod species that are shared between the two clones as well as any species unique to each clone. One nonadditive mechanism is niche partitioning, wherein different genotypes use resources in different ways that complement each other. In this case, competition might be reduced in more genetically diverse populations. Another nonadditive mechanism is facilitation, where the presence of one species (e.g., a nitrogen fixer) improves the success of other species. A particularly important nonadditive mechanism, at least as far as this book is concerned, is that greater genetic diversity increases evolutionary potential (chapter 3), which can shape the evolution of species interactions and, thus, community structure (Vellend 2006, 2010).

Several caveats and limitations attend the intraspecific BEF approach. First, most studies focus on genetic diversity in neutral markers, whereas genetic diversity in functional traits is what matters. The assumption has been that the former is a reasonable proxy for the latter, but the actual association is often very weak (Reed and Frankham 2001, McKay and Latta 2002). Second, the effects of genotypes (and hence their diversity) will depend on the environment (GxE), the other genotypes present (GxG) and interactions between the environment and the genotypes (GxGxE) (Tétard-Jones et al. 2007, Zytynska et al. 2010, Genung et al. 2012a, 2012b). As a result, different studies of the same species could yield very different outcomes. Third, carefully controlled manipulations of clonal diversity confirm that ecological consequences exist, but not how strong they are in natural contexts. For instance, clonal diversity might be important in isolation, whereas its contribution might be trivial when compared to other drivers (more about this later). Fourth, effects of genetic diversity on community structure suggest that contemporary evolution could have ecological effects, but they do not confirm such effects—because the populations are artificial and evolution usually does not take place during the experiment. Such studies therefore cannot assess how important evolutionary *dynamics* are for ecological *dynamics*. Fortunately, evolution is increasingly considered during such experiments (e.g., Agrawal et al. 2012, 2013, Zuppinger-Dingley et al. 2014)

2. *Genetic identity within populations*

Ecology has a long tradition of using experiments to compare the ecological effects of different species, such as different predators (e.g., Schmitz and Suttle 2001) or different dominant plants (e.g., Emery and Gross 2006). A logical extension was to compare the ecological effects of different genotypes within species. At the simplest level, different genotypes of a focal species could influence the abundance of another species, such as a competitor, predator, or prey. Increasing the level of complexity, interactions between genotypes of a focal species and multiple other species might be important, as described earlier for the barley-aphid-rhizobia-wasp system (Tétard-Jones et al. 2007, Zytynska et al. 2010). Many such studies have now been conducted and are often grouped under the umbrella of "community genetics" (Fritz and Price 1988, Whitham et al. 2006, Tack et al. 2010).

The standard approach in community genetics is to place different genotypes in similar environments and then measure associated changes in a variety of ecological variables: Matthews et al. (2011b) dubbed such experiments "common gardening." As one example, Shuster et al. (2006) surveyed the leaf-modifying arthropods found on individual trees in a common-garden plot of North American cottonwood (*Populus*) trees. Owing to an experimental design that planted multiple clonal individuals per genotype (explained in more detail below), the authors could determine how much of the variation in arthropod communities was attributable to genetic variation among trees. As another example, Iason et al. (2005) showed that genetic differences among individual Scot's pine trees (*Pinus sylvestris*) in the chemical diversity of their secondary metabolite monoterpenes influenced the species richness of the ground flora. And, perhaps bringing us back to ecological speciation (chapter 6), *Enchenopa binotata* treehoppers reared on different host plant genotypes express different mating signals, which could promote evolutionary diversification (Rebar and Rodríguez 2014). Most community genetics studies focus on plants owing to the easy production of clonal genotypes and the ease of monitoring individuals in nature. Studies of animals instead tend to focus on comparing different ecotypes in enclosures or mesocosms, as will be explained in the next section. (In reality, studies that examine the different effects of individual animals, such as different foraging preferences or rates, could be recast as community genetics studies.)

The effects of genetic identity and genetic diversity often can be considered in the same studies. In BEF experiments, for example, differences among monocultures quantify the effects of genetic identity whereas differences between monocultures and polycultures quantify the effects of genetic diversity. As such, studies examining the effects of genetic identity suffer from many of the same caveats and limitations that attend studies of genetic diversity (see above)—although the former do often focus more explicitly on functional differences, such as cottonwood genotypes with different levels of condensed tannin (Whitham et al. 2006) or coyote bush (*Baccharis pilularis*) genotypes that encode prostrate versus erect morphs (Crutsinger et al. 2013, 2014). Among the limitations that remain, I need to again draw special attention to the fact that nearly all such studies do not use recently diverged genotypes or populations that can evolve during the course of the experiment: as a result they can only suggest, rather than demonstrate, how contemporary evolution might matter for ecological dynamics.

3. *Population comparisons*

At the close of both preceding sections, I noted how studies of genetic diversity and identity examine the effects of standing variation but typically do not consider the ecological effects of contemporary evolution per se. To get more directly at this later topic, a common approach is to compare the ecological effects of populations that diverged from a common ancestor at a known time in the past. At an observational level, one (e.g., Palkovacs and Post 2009) might compare ecological variables (e.g., zooplankton communities) among environments (e.g., different lakes) that harbor different ecotypes (e.g., freshwater or anadromous) of a focal fish species (e.g., alewife—*Alosa pseudoharengus*)—ideally including multiple independent evolutionary origins of the ecotypes. A limitation of such field surveys is that cause and effect are murky: (1) the different phenotypes might have caused the ecological differences (evo-to-eco), (2) the ecological differences might have caused the different phenotypes (eco-to-evo), (3) both pathways could be at play, or (4) phenotypes and ecological conditions might be associated for noncausal reasons (e.g., both are correlated with an environmental variable). It is also possible that the phenotypic effects are offset by environmental effects, leading to only small net ecological differences in nature despite substantial phenotypic effects. This outcome would be, in essence, counter-gradient variation (chapter 4) at the level of extended phenotypes.

To isolate particular causal effects (here evo-to-eco), controlled experiments are the logical next step. The typical approach here is to place different ecotypes into replicate common-garden environments and then monitor the resulting ecological changes. These experiments are also "common gardening" akin to experimental approaches for examining effects of genetic identity (above): in this case, however, one uses different evolved populations, rather than different individual genotypes (Matthews et al. 2011b). The most frequent context for such experiments is the placement of different conspecific fish populations into controlled mesocosms, followed by the monitoring of ecological responses (Palkovacs and Post 2009, Palkovacs et al. 2009, Harmon et al. 2009, Bassar et al. 2010, 2012, Lundsgaard-Hansen et al. 2014, Matthews et al. 2016). Of particular interest from the perspective of contemporary evolution are studies examining the effects of recently diverged populations, thus allowing an assessment of the time scale on which ecological effects evolve. Laboratory experiments can introduce populations into new conditions and directly test how the resulting evolution influences ecological variables (e.g., Lawrence et al. 2012, Ellner 2013), but our interest here is on natural populations. In this latter context, one could examine the effects of populations recently introduced into new environments, in comparison to their ancestral population from the original environment.

The experimental population-comparison approach has several limitations. First, it isn't known whether or not the phenotypic effects are genetically based unless plastic differences among populations are first minimized through common-garden rearing. To date, however, nearly all such studies simply use wild-caught individuals whose phenotypes represent an unknown combination of genetic and plastic effects (an exception is Lundsgaard-Hansen et al. 2014). However, due to GxE, raising different populations in a common garden is not a panacea because population differences likely depend on the specific common environment. Second, the specific phenotypic difference that causes a particular ecological difference is often unknown—because different populations typically show a multitude of phenotypic differences. In such cases, investigators tend to rely on statistical associations between phenotypes and ecological variables, as well as common sense regarding the functions of particular traits (e.g., Lundsgaard-Hansen et al. 2014). Third, studies of recently diverged populations reveal only the *maximum* time frame over which an ecological effect arose (i.e., time since the populations separated), whereas the effects might well have arisen much earlier. Finally, mesocosms are not natural, no matter how much we strive to make them so, and hence they can't reveal effects that actually occur in nature. A partial solution to this disconnect is to compare, for the same populations, ecological effects observed in mesocosms to ecological differences documented in nature, thus bringing together experimental and observational studies (Palkovacs and Post 2009).

4. *Real-time changes*

The first and second approaches described above emphasize standing variation within populations, whereas the third approach focuses on previously evolved differences among populations. The final approach to discuss is tracking the real time evolution of populations and its ecological consequences. In the laboratory, one implementation has been to manipulate evolutionary potential, such as by varying the number and nature of clones or the amount of gene flow, and asking whether the resulting evolutionary differences alter ecological dynamics. As examples, populations of algae and their rotifer predators show very different dynamics depending on whether the algae can evolve (multiple clones) or not (single clone) (Yoshida et al. 2003, fig. 8.1), on the nature of genetic variation in adaptive traits (Kasada et al. 2014), and on interactions between plasticity and evolution (Fischer et al. 2014). These experiments are often coupled with mathematical models that mimic the system and can thereby help to inform specific mechanisms (Tuda and Iwasa 1998, Fussmann et al. 2003, Yoshida et al. 2003, Jones et al. 2009, Becks et al. 2010, Ellner 2013, Fischer et al. 2014, Kasada et al. 2014).

Assessing the ecological consequences of real-time evolution is much harder in natural populations and yet several methods can be useful. First, organisms with dormant stages, such as resting eggs in *Daphnia* or seeds in some plants, can be used to directly compare ancestors and descendants—the so-called resurrection ecology program (Kerfoot et al. 1999). Such studies have documented evolutionary changes in phenotypic traits and in responses to factors such as predators, parasites, toxins, or abiotic conditions (e.g., Cousyn et al. 2001, Franks et al. 2007, Decaestecker et al. 2007, Sultan et al. 2013, Stoks et al. 2016). The next step in the present context would be to use these ancestor-descendent pairs to assess effects on community-level composite-response variables, such as in common-gardening experiments. Second, a number of studies have related real-time changes in phenotypes/genotypes to population dynamics (e.g., Sinervo et al. 2000, Hanski and Saccheri 2006, Pelletier et al. 2008, Ozgul et al. 2010) as described in detail in chapter 7. The next step would be to also monitor consequences for community variables, although cause and effect will be hard to determine without experimental manipulations. Third, some studies have used experimental manipulations of gene flow to alter evolutionary trajectories in nature (e.g., Riechert 1993, Nosil 2009). The same studies could just as well track community consequences. For instance, Farkas et al. (2013) used gene flow to experimentally manipulate the maladaptation of stick insect populations, finding within-generation effects on arthropod communities. The next step would be to track the effects of subsequent evolution across generations. Fourth, one can—with plants at least—create populations in nature with different evolutionary potentials or starting conditions, after which evolution and its community consequences can be monitored (Agrawal et al. 2012, 2013).

Some limitations attending particular implementations of this real-time approach are the same as those encountered for the earlier approaches, including the unrealistic nature of mesocosms, uncertainty regarding genetic versus plastic changes, ambiguity about cause and effect in the absence of manipulations, and

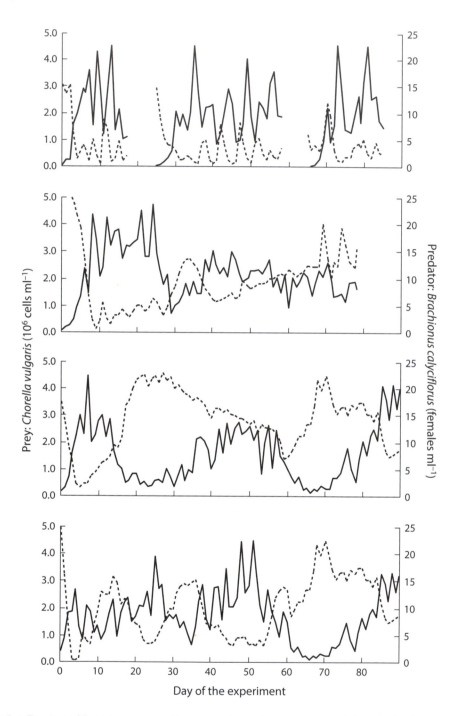

Fig 8.1. Predator (*Brachionus calyciflorus*) and prey (*Chorella vulgaris*) cycles in chemostats with either a single clone of prey (three separate experiments shown in the top panel) or multiple clones of prey (separate experiments in each of the three lower panels). The cycles are much longer and the predator-prey oscillations are out of phase in the experiments with multiple clones. The data presented here correspond to the first three experiments of each type (single clone or multiple clone) shown in figure 2 of Yoshida et al. (2003) and were provided by T. Yoshida

complications arising from GxE. Other limitations are new. In particular, the changes that take place over a short time frame might be too small to have measureable ecological effects—even if those effects would accumulate to be much larger over even modestly longer time frames. In addition, the effort required to manipulate and monitor dynamics through time in nature can be prohibitive. Finally, the previously discussed possibility that evolution promotes stability will be missed in approaches that focus explicitly on explaining ecological *changes*.

Eco-evolutionary community structure in nature

I will first briefly consider two classic questions in ecology that have long incorporated evolutionary considerations: the role of evolution in shaping interactions between predators and prey (Q1) and competitors (Q2). Both of these topics emphasize antagonistic interactions but mutualistic interactions will also be shaped by eco-evolutionary dynamics (e.g., Andrande-Dominguez et al. 2014, Buser et al. 2014). I will then (Q3) consider how contemporary evolution interacts with dispersal to alter community structure (i.e., "evolving metacommunities"), again with an emphasis on predators and competitors. These topics are usually investigated with the species-by-species approach, the results of which I will therefore emphasize. I will next turn to questions typically investigated through the focal-species composite-response approach. In particular, I will ask how genetic diversity/identity influences community structure (Q4), how rapidly these effects arise (Q5), how strong they are (Q6), the extent to which they are "heritable" (Q7), and whether they are driven by direct or indirect effects (Q8). As research emphasis on these questions is rather recent, I will have to rely more on case studies than on meta-analyses.

QUESTION 1: HOW DOES EVOLUTION INFLUENCE PREDATOR-PREY DYNAMICS?

Predators and prey, here inclusive of parasites-hosts and herbivores-plants, clearly impose selection on one another and the evolution of each is expected to enhance its persistence. That is, selection imposed by predators should favor the evolution of prey traits that reduce predation, which should enhance prey persistence. Similarly, selection imposed by prey should favor the evolution of predator traits that maintain predation success. Many theoretical models have considered these dynamics (reviews: Abrams 2000, Fussmann et al. 2007), and the outcomes have been diverse. A common result is that evolution should have stabilizing effects on predator-prey dynamics (e.g., Hochberg and Holt 1995, Loeuille 2010), although some conditions cause just the opposite (Matsuda and Abrams 1994, Abrams 2000). Overall, "it appears that there are likely to be few universally applicable generalizations regarding how adaptive evolution alters predator-prey dynamics" (Fussmann et al. 2007).

A series of laboratory experimental studies have considered how evolution influences predator-prey dynamics (reviews: Fussmann et al. 2007 and Ellner 2013). In an early example, Pimentel and coworkers (Pimentel et al. 1963, Pimentel 1968) compared the traits and population responses of houseflies (*Musca domestica*) to parasitoid wasps (*Nasonia vitripennis*) when the houseflies were allowed to evolve or not. The authors found that wasp populations were smaller and the system dynamics were more stable when evolution was allowed. A similar outcome has been observed for bean weevils (*Callosobruchus* spp.)

and their parasitoid wasps (*Heterospilus prosopidis*) (Tuda and Iwasa 1998). Destabilizing effects have been documented in other systems, including a three species (two prey and one predator) system where the evolution of one prey strain in response to predation fundamentally alters the outcome of interactions between that predator and another prey strain (Friman et al. 2014). Other laboratory studies considering how evolution shapes predator-prey dynamics include Yoshida et al. (2003, fig. 8.1), Jones et al. (2009), Becks et al. (2010), terHorst et al. (2010a), Kasada et al. (2014), Fischer et al. (2014), and Turcotte et al. (2015). Also, in an interesting twist, Hiltunen et al. (2014) showed that many laboratory consumer-resource studies that did not explicitly consider evolution nonetheless show cycles consistent with prey evolution.

Beyond the laboratory, what sort of evidence exists from nature that evolution influences predator-prey dynamics? On long enough time scales, such dependence is obvious given ample evidence that many predators are specialized on particular prey and that many prey have evolved traits that reduce susceptibility to particular predators. Plant-herbivore interactions provide a particularly obvious demonstration (Futuyma and Moreno 1988, Jaenike 1990). Moreover, these evolutionary interactions are often reciprocal, as embodied in the evidence for predator-prey coevolution (Vermeij 1987, Thompson 1994). Yet the key question here is: to what extent do such effects play out on contemporary time scales? In this context, substantial evidence exists for prey traits evolving in response to recently introduced predators. For example, changes in predation pressure have led to adaptive evolution on short time scales in guppies (Endler 1980, Reznick et al. 1997, O'Steen et al. 2002, Gordon et al. 2009), *Daphnia* (Cousyn et al. 2001, Fisk et al. 2007, Walsh and Post 2011), and various plants (Zangerl and Berenbaum 2005, Turley et al. 2013, Agrawal et al. 2012, 2013). Many more examples are summarized in Strauss et al. (2006). Also, when humans are the predators, especially dramatic phenotypic changes are evident—usually toward smaller size and earlier maturity (Darimont et al. 2009, Sharpe and Hendry 2009). On the flip side, some evidence exists that predator traits evolve in response to introduced prey. As one example, native soapberry bugs (*Leptocoris tagalicus*) in Australia have evolved traits that improve their feeding success on introduced plants (Carroll et al. 2005). Indeed, many native phytophagous insect species have evolved host races that specialize on introduced plants (Berlocher and Feder 2002, Drès and Mallet 2002).

The above paragraph makes clear that *traits* often evolve owing to changes in predation. In many cases, this trait evolution alters resistance to the predators. For example, invasive cane toads in Australia are highly poisonous for many native predators, including snakes, and the abundance of these predators has declined dramatically since the toad invaded (Phillips et al. 2003). At the same time, some snakes are evolving increased resistance to the toad toxin and the toads are evolving reduced toxicity (Phillips and Shine 2004, 2005, 2006). The effect of toads on depressing snake populations therefore should decline into the future. Similar expectations emerge from considerations of host-parasite dynamics, where costly parasitism leads to the contemporary evolution of host defense, and sometimes parasite offense (Lively and Dybdahl 2000, Decaestecker et al. 2007, Duffy and Forde 2009). Inverting the problem, a reduction in parasitism should cause the evolution of decreased resistance (as it is often costly), which has been observed in—for example—African village weaverbirds (*Ploceus cucullatus*) colonizing areas without brood-parasitic diederik cuckoos (*Chrysococcyx caprius*) (Lahti 2005), and in reed warblers (*Acrocephalus scirpaceus*) in areas where common cuckoos (*Cuculus canorus*) are declining (Thorogood and Davies 2013). However, intuitive results such as these (removal of an enemy leads to an evolutionary reduction in resistance to that enemy) are not always obtained. Dargent et al. (2013) showed that experimental reductions in *Gyrodactylus* parasitism on guppies in nature instead led to the evolution of *increased* resistance. The examples given here document the evolution of resistance, usually assayed in the laboratory, but they do not document the consequences for predator-prey dynamics in nature. One study to have taken this additional step is Agrawal et al. (2012, 2013), where the experimental relaxation of insect herbivory led to the evolution of resistance that altered herbivore abundance. In the context of parasitism, Duffy and Sivars-Becker (2007) argued that the rapid evolution by natural *Daphnia* populations of increased resistance to *Metschnikowia bicuspidata* parasites terminated disease epidemics in lakes.

A good place to look further for evolution influencing predator-prey dynamics might be biocontrol efforts, where a predator species (again including herbivores and pathogens) is introduced in hopes of controlling some prey species, often an invasive pest or weed (Louda et al. 2003, Hufbauer and Roderick 2005). First, we might ask whether biocontrol agents have, following their release into the new environment, evolved to use, and thereby influence the dynamics of, nontarget species. Second, we might ask whether biocontrol agents evolve to become more or less effective with time, as might be expected if resistance (or tolerance)[2] evolves in the target species or if virulence evolves in the biocontrol agent. For the first question, van Klinken and Edwards (2002) considered 352 introduced biocontrol agents of weeds and concluded that none of them appears to have evolved a propensity to use new hosts. Other authors have agreed in the sense that no concrete cases are evident, but it is also clear that the definitive tests are rarely performed (Holt and Hochberg 1997, Louda et al. 2003, Hufbauer and Roderick 2005). Moreover, a major consideration in selecting biocontrol agents in the first place is that they are unlikely to switch to nontarget species (Louda et al. 2003). With regard to the second question, some nice candidate examples exist. One is that of European rabbits (*Oryctolagus cuniculus*) introduced to Australia and the virus (myxoma) introduced to control them. As the story goes, introduction of the virus caused a rapid population crash of rabbits, but the rabbits then evolved to be more resistant and the virus evolved to be less virulent, with the outcome being reduced rabbit mortality (Fenner 1983, Dwyer et al. 1990). Several other examples have been advanced (Burdon et al. 1981) but the data are very sparse at present (Hufbauer and Roderick 2005).

Contemporary evolution influences predator-prey dynamics, and this evolution often has stabilizing effects. This expectation makes intuitive sense given that whichever party is suffering most from a given interaction should be the party under the strongest selection—and therefore the most likely to evolve dramatic fitness improvements. Many more studies from nature are needed before we can say just how strong, common, and rapid are such dynamics.

QUESTION 2: HOW DOES EVOLUTION INFLUENCE THE COEXISTENCE OF COMPETITORS?

The idea that competition shapes communities extends at least as far back as Darwin (1859, p. 67), summarized nicely in his wedge metaphor: ". . . Nature may be compared

---

[2]Resistance is the "ability to limit parasite burden" and tolerance is the "ability to limit the disease severity induced by a given parasite burden" (Simms and Triplett 1994, Råberg et al. 2007)

to a yielding surface, with ten thousand sharp wedges packed close together and driven inwards by incessant blows ..." The idea has since seen many manifestations (e.g., limiting similarity, species packing, community saturation, community invasibility), with a particularly influential and controversial version taking the form of "competitive exclusion": more than one species cannot occupy the same niche at the same location (Lack 1947, Elton 1958, Hutchinson 1959, Macarthur and Levins 1967, Strong et al. 1979, Connell 1980, Ricklefs 1987, Cornell and Lawton 1992, Tokeshi 1999, Shea and Chesson 2002). Another manifestation is whether predicting community structure requires knowledge of species differences in resource use or whether it can be explained by simple neutral models (Tilman 1982, 1994, 2004, Hubbell 2001, Silvertown 2004, Leibold and McPeek 2006, Adler et al. 2007, Rosindell et al. 2011). Notwithstanding these debates, ample evidence exists that species are not equal in their resource use and that these inequalities influence their distributions. Following from this empirical reality, I will here consider how evolution might influence the coexistence of competitors.

Evolution might enhance the coexistence of competitors in at least three ways (see also Lankau 2011). First, competitive interactions might drive evolutionary *divergence* in resource use: that is, selection can favor ecological character displacement that reduces competition (Brown Jr. and Wilson 1956, Hutchinson 1959, Cody and Diamond 1975, Schluter 2000b, Pfennig and Pfennig 2012, Beans 2014). Second, competitive interactions might drive evolutionary *convergence* in resource use and resource use efficiency, particularly when the key resources are nonsubstitutable (Abrams 1987, Fox and Vasseur 2008, Vasseur and Fox 2011).[3] Third, negative frequency dependent selection might maintain coexistence if "selection causes one species to be a superior interspecific competitor when it is rare and an inferior interspecific competitor when it is abundant" (Vasseur et al. 2011). Many theoretical analyses of such effects have been conducted and, as with predator-prey dynamics (above), the results are legion (review: Fussmann et al. 2007). What of the empirical outcomes?

One approach to assessing the role of competition in species coexistence is "community phylogenetics," wherein observed phylogenetic relationships among species in a community are compared to hypothetical patterns based on simulated random assembly from a regional species pool (Webb et al. 2002, Emerson and Gillespie 2008). Typical analyses distinguish three phylogenetic patterns: (1) randomness, (2) clustering (species are more closely related than expected at random), and (3) overdispersion (species are less closely related than expected at random). Clustering is often taken as evidence of "environmental filtering" wherein certain clades are best adapted to local conditions. Overdispersion is often taken as evidence that competition shapes communities because coexistence should be more difficult for closely related species that have similar niches. This assumption indirectly invokes "niche conservatism," which in the present context implies that closely related species have similar traits owing to reduced opportunity for divergence (Webb et al. 2002, Wiens and Graham 2005). In an early exemplar, Cavender-Bares et al. (2004, 2006) showed that Florida oak tree communities were phylogenetically clustered at large taxonomic and spatial scales but were overdispersed

---

[3]Substitutable resources (e.g., different prey species) occur when either resource can supply the needs of the consumer. Consumers can switch between such resources without major cost. Nonsubstitutable resources (e.g., nitrogen and phosphorous) are required to meet the needs of a consumer, and therefore cannot be forsaken.

at small scales. Combined with a number of subsequent studies, this pattern appears to be a somewhat general result (Haloin and Strauss 2008, Emerson and Gillespie 2008, Vamosi et al. 2009). The implication is that environmental filtering is more important to the assembly of regional communities, whereas competition is more important to the assembly of local communities; yet it remains hard to confidently infer process from pattern (Vamosi et al. 2009). As just one example of the ambiguity, clustering also can arise when very similar competitors have a hard time excluding one another.

Community phylogenetics, as noted earlier, invokes the idea that closely related species are more likely to compete. Although some empirical studies have supported this assumption, the pattern isn't very strong and it is highly context dependent (Clark 2010, Jiang et al. 2010, Burns and Strauss 2011, Violle et al. 2011). Relying less on this assumption, other studies consider communities from the perspective of resource use *traits*, which might more directly indicate competition. For instance, competition might be inferred to have shaped community assembly if resource use traits are overdispersed within guilds[4] (Dayan and Simberloff 2005). Supporting this inference, a number of studies have shown that resource use traits tend to be overdispersed within communities. Although such patterns can be hard to distinguish from random community assembly (Strong et al. 1979, Gotelli and Graves 1996), newer methods have confirmed overdispersion of traits in a number of instances (Davies et al. 2012). Of course, this trait-based approach has its own limitations, including the need to assume that the focal traits are the real determinants of competitive interactions.

The above approaches usually do not address the extent to which in situ evolution has shaped communities. Instead, community assembly could occur through environmental filtering or competitive exclusion that results from species traits that evolved previously in other locations/contexts. To get more directly at the role of in situ evolution shaping competitive communities, a classic approach has been to ask whether potentially competing species show greater divergence in resource use traits in sympatry than in allopatry: that is, ecological character displacement (Brown Jr. and Wilson 1956, Hutchinson 1959, Cody and Diamond 1975, Schluter 2000b). Just such a pattern has been observed in many natural systems, suggesting that competition drives in situ evolution within many communities (Robinson and Wilson 1994, Schluter 2000b, Dayan and Simberloff 2005, Pfennig and Pfennig 2012, Beans 2014). An intriguing example comes from independent lineages of catfishes (Corydoradinae) in South America that, in sympatry, converge in color (they are Müllerian mimics) but nearly always maintain ecological differences (Alexandrou et al. 2011). Of course, it remains hard to infer process from pattern because mechanisms other than competition can also contribute to apparent character displacement (Connell 1980, Dayan and Simberloff 2005, Stuart and Losos 2013). Moreover, evidence for character displacement does not equal evidence that the displacement was necessary for coexistence.

The above methods are usually applied to long-established communities, and hence cannot inform the importance of contemporary evolution. For this additional inference we must turn to dynamically changing communities. For instance, several laboratory experiments have shown that evolution enhances the coexistence of competitors. As an

---

[4]A guild is a "group of species that exploit the same class of environmental resources in a similar way", such as foliage gleaning birds (Root 1967, Simberloff and Dayan 1991).

early example supporting the importance of negative frequency dependence, Pimentel et al. (1965) showed that a numerically subordinate species (blowfly: *Phaenicia sericata*) evolved increased competitive ability in the face of a numerically dominant species (housefly). In addition, several laboratory studies have shown the importance of evolving character displacement. As just one example, Tyerman et al. (2008) showed that *E. coli* adapted to use different carbon resources showed phenotypic divergence when in competition but phenotypic convergence when removed from competition.

Studies considering the contemporary evolution of competition are less commonly implemented in nature, yet several have provided supportive evidence. For instance, a highly invasive plant (*Alliaria petiolata*) evolved reduced allelopathic effects on three competing native plants over the course of 50 years (Lankau et al. 2009) and at least one native species (*Pilea pumila*) evolved increased resistance to the invader (Lankau et al. 2011). Both of these evolutionary effects should increase the potential for coexistence (Lankau 2011). Evolution in this system likely stems from frequency dependence, whereas other studies invoke character displacement. For instance, Diamond et al. (1989) showed that two honeycreeper species (*Myzomela pammelaena* and *Myzomela sclateri*) underwent some, albeit minor, character displacement in body size within 300 years of joint colonization of an island. On an even shorter time scale, colonization of Daphne Major by the large ground finch shifted drought-induced evolution of the medium ground finch from increasing beak size without the competitor to decreasing beak sizes with the competitor (Grant and Grant 2006). Similar contemporary evolution of character displacement was reported for *Anolis* lizards by Stuart et al. (2014).

Evolution influences the coexistence of competing species and thereby strongly shapes community assembly. Losos (2009) argues that *Anolis* lizards provide a particularly compelling example. **Overall, it seems likely that the in situ evolution of competitors will most often enhance their coexistence** (Loeuille 2010, Lankau 2011, Vasseur and Fox 2011). However, the generality of this result is not yet clear and the only mechanism yet studied in considerable detail has been character displacement.

QUESTION 3: HOW DOES DISPERSAL INTERACT WITH EVOLUTION TO INFLUENCE COMMUNITY STRUCTURE?

In chapter 5, I discussed the many ways in which dispersal and gene flow can shape evolutionary dynamics within species. Here I would like to formally consider how interactions between dispersal and evolution can influence community structure, primarily through the lens of predator-prey and competitor interactions.

For predator-prey interactions, I previously alluded to several examples that I will here reiterate together. In chapter 5, I discussed how differential rates of gene flow for parasites versus hosts can influence the success of each party in the interaction. In particular, a number of laboratory studies have demonstrated that the player with the higher level of gene flow tends to gain the upper hand, presumably because gene flow in metapopulations increases the evolutionary potential that is so critical in antagonistic coevolutionary scenarios (Forde et al. 2004, Hoeksema and Forde 2008). In chapter 7, I discussed how an experimental manipulation of gene flow in stick insects (Farkas et al. 2013) caused maladaptation by reducing crypsis, and how the resulting increase in predation by birds reduced stick insect population sizes. Earlier in the present chapter, I noted that this same experiment demonstrated effects on other arthropods—and the reason was that the predatory birds attracted by mal-adapted stick insects also ate other arthropods (Farkas et al. 2013). These and other ways in which maladaptation, probably most often caused by gene flow, can shape community structure have been summarized in the context of island biogeography theory by Farkas et al. (2015).

For competitive interactions, I will extend the above question about the role of competition in community assembly to consider effects of the order in which species arrive. In particular, a number of ecological studies have emphasized the importance of priority effects, wherein early arriving species gain a numerical advantage that allows them to exclude later arriving competitors (Wilbur 1997). These priority effects seem to be strongest for species that are closely related, suggesting that competition is indeed the reason, supporting Darwin's so-called "naturalization hypothesis" (Jiang et al. 2010, Peay et al. 2012). This invocation of priority effects does not require in situ evolution, but such evolution can be important. A priority effect playing out on long time scales is that early-arriving species can sometimes radiate to occupy open niches that cannot then be colonized by later arriving ecological equivalents—with work on Hawaiian spiders (Gillespie 2004) and *Anolis* lizards (Losos 2009) being likely exemplars. A priority effect playing out on contemporary time scales is that in situ adaptive evolution following colonization of a new environment can enhance the advantage of early arriving species over later arriving species, leading to "community monopolization" (De Meester et al. 2002, Urban and De Meester 2009, De Meester et al. 2016). This last effect can extend to different genotypes of the same species, as has been clearly shown for *Daphnia* adapting to water temperatures (Van Doorslaer et al. 2009). That is, a period of adaptation by a given set of genotypes to local conditions improves their success in competition with later arriving (and thus not locally adapted) sets of genotypes.

These ideas have been woven together (Urban and Skelly 2006, Urban et al. 2008) into a theoretical and conceptual framework of "evolving metacommunities": "a set of local communities linked by the dispersal of multiple potentially interacting species in which genetically determined trait variation within species modifies the outcome of interspecific interactions" (Urban et al. 2008). Empirical work explicitly incorporating this framework is as yet rare. However, Howeth et al. (2013) showed how dispersal of zooplankton and phytoplankton among lakes provided a migrant pool that was then sorted by the direct and indirect effects of predation in the local environment. Importantly, this sorting effect was strongly modified by intraspecific variation in the predator, in this case whether the alewife were anadromous or resident (more details about this system appear below). More generally, Urban (2011) used a meta-analysis to show that as much variation in species interaction traits ("... any characteristic that directly modifies a species interaction including traits such as competitive ability, predator defense, prey preference and host resistance ..."—Urban 2011, p. 724) was due to regional-level selection (implying the importance of gene flow among local populations) as was due to local-level selection. Further work formally implementing this framework would be profitable.

Dispersal and gene flow clearly interact with adaptation to shape species interactions and therefore community structure. Some clear examples include the effects of gene flow in promoting or constraining adaptation and the influence of contemporary evolution on priority effects during community assembly.

QUESTION 4: DOES INTRASPECIFIC DIVERSITY/IDENTITY INFLUENCE COMMUNITY STRUCTURE?

A perennial topic in ecology and environmental science is the extent to which species diversity influences community and ecosystem properties, such as resistance/resilience, stability, productivity, and various "ecosystem services" (Loreau et al. 2001, Hooper et al. 2005). This topic is usually explored by creating experimental plots with different numbers of species and measuring the resulting differences in community/ecosystem variables: so-called "biodiversity-ecosystem function" (BEF) experiments. Effects observed in BEF studies (for example, greater species diversity often leads to higher productivity) are the result of genetic differences among species in functional traits. These interspecific differences must have had their origins in intraspecific differences. Indeed, differences among individuals within species can substantially exceed differences among species means (Clark 2010) and the consideration of intraspecific diversity can strongly alter interpretations about functional diversity in a community (Albert et al. 2010, 2012, Clark et al. 2011, Siefert et al. 2015). In accord with these findings, many BEF studies have now moved beyond manipulations of species diversity into manipulations of genotypic diversity within species (Haloin and Strauss 2008, Hughes et al. 2008).

Intraspecific BEF studies now have been performed for a number of ecosystems (although most are terrestrial), species (although most are plants), and community variables (although most are the diversity of other species). Here are some examples (for others, see Hughes et al. 2008). Assemblages of plant species with greater genetic diversity per species tended to retain higher species diversity through time (Booth and Grime 2003). More genetically diverse plots of the seagrass *Z. marina* had higher plant diversity and supported a greater abundance of invertebrate fauna during a period of heat stress (Reusch et al. 2005). More diverse plots of common evening primrose (*Oenothera biennis*) had more arthropods (Johnson et al. 2006). More genetically diverse stands of tall goldenrod (*Solidago altissima*) were more resistant to colonization by other plant species (Crutsinger et al. 2008b). With a step toward mechanism, community-level effects of genotypic diversity have been shown to involve nonadditive effects in a number of instances (Crutsinger et al. 2006, Johnson et al. 2006, Kotowska et al. 2010), although not all of them (Genung et al. 2012b, 2013).

The above studies examining effects of diversity (numbers of genotypes) typically also demonstrate effects of genetic identity. In addition, many community genetics studies have shown that different plant genotypes have different effects on arthropod communities and on plant competitors (reviews: Hughes et al. 2008, Bailey et al. 2009, Tack et al. 2012). Another context for assessing effects of genetic identity is the use of mesocosms that compare effects of different fish ecotypes, which I will now exemplify by reference to Trinidadian guppies. Guppy populations living with or without dangerous predators (high predation versus low predation) differ in a wide array of phenotypic traits (Endler 1995, Magurran 2005; chapter 4). Several studies have now considered how these ecotypes differ in their ecological effects (Palkovacs et al. 2009, Bassar et al. 2010, 2012). In each case, experimental channels beside a stream in Trinidad were seeded with invertebrates, and water was allowed to flow from the stream into the channels. Guppies of the two ecotypes were introduced into different channels (with control treatments in other channels), left alone for about a month, and then a number of ecological variables were measured in each channel. A common finding was that channels with high-predation guppies tended to have lower total invertebrate biomass than did channels with low-predation guppies—because the former ecotype feeds more often on benthic invertebrates as opposed to diatoms and detritus (Palkovacs et al. 2009, Bassar et al. 2010, 2012). Thus, evolutionary divergence of guppy ecotypes had an effect on community structure, at least when assessed in controlled channels. I will return to these studies in the context of the strength of effects (question 6) and corresponding changes in ecosystem function (chapter 9).

A caveat attending most mesocosm experiments with fish, including guppies (Palkovacs et al. 2009, Bassar et al. 2010, 2012), threespine stickleback (Harmon et al. 2009, Ingram et al. 2012, Des Roches et al. 2013, Rudman and Schluter 2016), and alewife (Palkovacs and Post 2009, Weis and Post 2013), is their use of fish captured from the wild, rather than reared in a common-garden environment. As a result, an *evolutionary* basis for the above effects remains to be formally demonstrated. It is known that phenotypic divergence in these systems is often (but not always) genetic, but this knowledge doesn't confirm that genetic differences were what caused the observed community effects. Instead, it could be that plastic phenotypic differences or behavioral differences associated with prior experience are the primary drivers of at least some community effects. In an exception, a mesocosm study of European whitefish was able to separate genetic and plastic effects by raising limnetic and benthic ecotypes under common-garden conditions, with different individuals of the benthic ecotype being raised on either limnetic or benthic diets (Lundsgaard-Hansen et al. 2014). Both genetic differences (among ecotypes raised on the same diet) and plastic effects (the same ecotype raised on different diets) influenced zooplankton communities, which highlights the need for other fish studies to conduct comparable experiments. Matthews et al. (2016) have reported similar outcomes for lake and stream ecotypes of threespine stickleback. Beyond fish, studies of plants can more easily separate genetic from plastic effects. As just one example, Barbour et al. (2009a, 2009b) demonstrated differences in herbivore and fungal communities associated with *Eucalyptus globulus* trees from eight different populations in Tasmania all planted in a common-garden environment.

Intraspecific genetic diversity/identity influences a wide variety of community-level variables. However, documenting the existence of an effect does not immediately reveal how important that effect is: perhaps it is so weak as to be unworthy of much attention in a larger context. In addition, some studies have failed to find effects of genetic diversity on some community-level variables (Crutsinger et al. 2008a, Johnson et al. 2008, Tack et al. 2012, Crutsinger et al. 2013). Thus, I will next ask how strong are the community-level effects of genetic diversity/identity in relation to other factors potentially influencing community structure.

QUESTION 5: HOW STRONG ARE THE COMMUNITY EFFECTS OF INTRASPECIFIC DIVERSITY/IDENTITY?

Throughout this book, I have repeatedly wrestled with assessing the strength/importance of a particular effect. Is selection strong or weak (chapter 2)? Is adaptation slow or fast (chapter 3)? Is adaptive divergence large or small (chapter 4)? To what extent does gene flow influence adaptation in nature (chapter 5)? Just how important is ecological divergence to speciation (chapter 6)? Are the contributions of evolution to population dynamics big or small (chapter 7)? In each case, external frames of reference were helpful in deciding just what magnitude of a given effect should be considered important

(Hairston Jr. et al. 2005, Hersch-Green et al. 2011, Matthews et al. 2014). For the present context (community-level evo-to-eco influences), useful frames of reference can include interspecific effects, the presence/absence of a focal species, variation in the density of a focal species, and spatial variation. In each case, we might ask two questions: (1) How important is the variation within populations, which also reflects the raw material for future eco-evolutionary dynamics; and (2) How important are the differences among populations, which also reflect the likely origins of interspecific effects?

The interspecific frame of reference has been applied in several studies. Focusing on genetic *diversity*, Crutsinger et al. (2006) used experimental plots to show that genotypic diversity (1, 3, 6, or 12 genotypes) in tall goldenrod influenced arthropod species richness (fig. 8.2) about twice as strongly as did plant species diversity (estimated in a different experiment). By contrast, Crutsinger et al. (2008a) showed in the same system that effects of intraspecific genetic diversity were minimal for litter-based arthropod communities. Similarly, Hargrave et al. (2011) found that the feeding rate of experimental *Daphnia* assemblages was influenced much more strongly by species diversity than by clonal diversity within species. Focusing on genetic *identity*, Shuster et al. (2006) planted, in

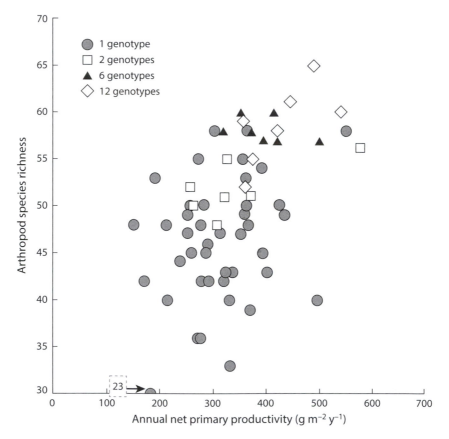

Fig 8.2. Experimental plots with more genotypes of *Solidago altissima* have more arthropods and higher aboveground net primary productivity. The data point indicated with the arrow has been moved (for clarity) from 23 arthropods to 30 arthropods. The data are from Crutsinger et al. (2006) and were provided by G. Crutsinger

a common garden, multiple individuals of each of multiple genotypes of each of four cross types: *Populus fremontii* (Fremont cottonwood), *Populus angustifolia* (narrowleaf cottonwood), F1 hybrids (*P. fremontii* x *P. angustifolia*), and backcross hybrids (BC; F1 X *P. angustifolia*). Ten years later, the authors surveyed leaf-modifying arthropods in the lower canopy of 76 individual trees. The variation among genotypes within a cross type explained more of the variation (57%) in arthropod communities than did variation among the cross types (19%). Focusing on *population (ecotype) differences*, Palkovacs et al. (2009) used mesocosms to compare effects of variation within and between two fish species: guppies from populations evolving with or without major predators, and killifish (*Rivulus hartii*) from populations evolving with or without guppies (fig. 8.3). For invertebrate biomass, the intraspecific effect of killifish was greatest (4.7 times the inter-specific effect), followed by the intraspecific effect of guppies (1.8 times the interspecific effect). Importantly, these comparisons don't necessarily show that interspecific effects are small; they instead show that intraspecific effects can be even larger. Post et al. (unpubl. data) subjected these and other similar studies to a meta-analysis and reported that the average effects of intraspecific variation (different genotypes or different ecotypes) and interspecific variation (different species) were roughly equivalent for plants, whereas the former were greater for animals. In addition, community structure in BEF studies was more strongly influenced by species richness than by genotype richness.

The species presence/absence and density frames of reference also appear in guppy and stickleback studies. Bassar et al. (2010, 2012) implemented a guppy ecotype treatment (high versus low predation), a guppy presence/absence treatment, and a guppy density treatment (twofold difference in density). For aquatic invertebrates (total biomass and biomass of chironomids), the effect of guppy presence/absence was greatest, followed by guppy phenotype (about 2/3 as large), and density (about 1/3 to 1/2 as large). Of additional interest was a strong effect of the guppy phenotype-by-density interaction: invertebrate biomass was the greatest for low-predation guppies at low density. This study has also been re-analyzed by Ellner et al. (2011), who obtained similar outcomes. Contrasting results were found for stickleback by Des Roches et al. (2013), where community effects of stickleback presence/absence and density were generally larger than those of stickleback ecotype (generalist, benthic, limnetic). Using different stickleback populations, Ingram et al. (2012) found that species presence/absence was more important than ecotype for some variables (benthic invertebrate biomass) but of roughly equal importance for others (zooplankton biomass). More recently, Rudman et al. (2015) found that the effects of cottonwood genotypes and stickleback ecotypes on community variables were similar to the effects of stickleback presence/absence. As a final example, the effects of experimental adaptation in *D. magna* on zooplankton communities were of similar magnitude to the effects of fish or macrophyte presence/absence (Pantel et al. 2015).

The spatial frame of reference typically involves testing the effects of genetic identity or diversity in common gardens in multiple locations, and then partitioning the total vari-ance in ecological parameters into that due to differences among genotypes (or diversity treatments) and that due to spatial location. For example, Tack et al. (2010) planted oak trees (*Quercus robur*) in multiple common gardens at the landscape scale (5 km^2) and at the regional scale (10,000 km^2). Analogous to Shuster et al. (2006), multiple cuttings per tree allowed the partitioning of variation in insect communities among the effects of different plant source populations, individual plant genotypes within source populations,

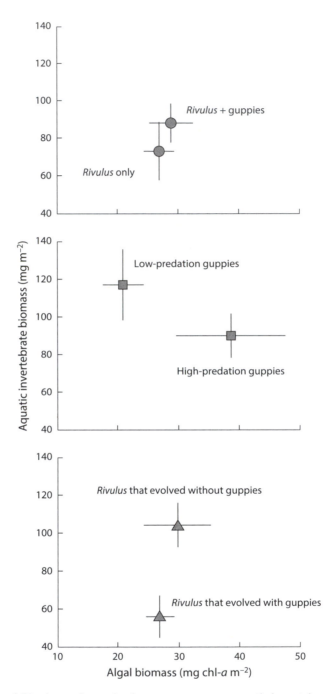

Fig 8.3. Effects of *Rivulus* and guppies in mesocosms on aquatic invertebrate biomass and algal biomass. The top panel shows the relatively small difference between mesocosms that only have *Rivulus* versus mesocosms that have both *Rivulus* and guppies. The middle panel shows the large difference between mesocosms that have low-predation guppies versus mesocosms that have high-predation guppies. The lower panel shows the large difference between mesocosms that have *Rivulus* from populations that coexist with guppies versus mesocosms that have *Rivulus* from populations that do not coexist with guppies. The data (means and standard errors) are from Palkovacs et al. (2009) and were provided by E. Palkovacs

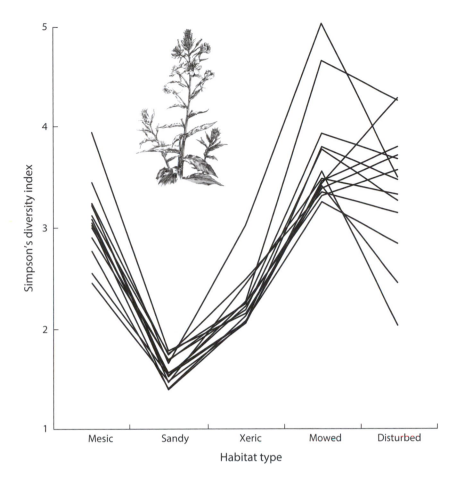

Fig 8.4. Diversity of arthropods on 14 different genotypes (different clonal families—each a different line in the graph) of evening primrose planted in each of five different habitats. Shown are best unbiased linear predictors (analogous to means) across plants of each genotype in each habitat. The data are from Johnson and Agrawal (2005) and were provided by M. Johnson. The evening primrose drawing is by D. Johnson

and different spatial planting locations. In all cases, source population and genotype explained much less of the variation than did spatial location. Similarly, when different genotypes of primrose were planted in different habitats, arthropod species richness was more strongly influenced by habitat than by primrose genotype (Johnson and Agrawal 2005) (fig. 8.4). Other studies have found varying results, with spatial effects sometimes more and sometimes less strong than genotypic effects (Tack et al. 2012). Overall, the relative importance of genotype increases for genotypes selected from more distant locations and the relative importance of spatial location increases with more distant testing locations (fig. 8.5). An important observation in these experiments is large genotype-by-habitat interactions, where the genotype having the greatest arthropod diversity in one habitat might be the one having the lowest arthropod diversity in another habitat (fig. 8.4).

Although specific outcomes are context-dependent, **the community effects of intraspecific variation in a focal species are perhaps one-half to twofold as important as are other ecological effects, such as interspecific variation, species presence/absence, and density**

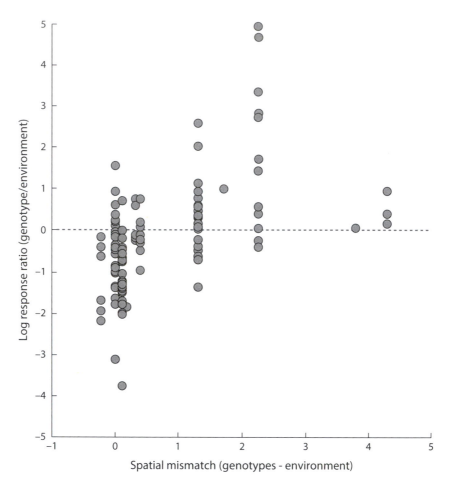

Fig 8.5. Relative importance of plant genotype versus spatial variation in shaping community variables, such as the abundance, diversity, or richness of arthropod species. Shown is the log response ratio for 15 studies of the variation attributed to plant genotype relative to spatial location (*y*-axis) plotted against the spatial mismatch between the scales of genotype collection and experimental planting (*x*-axis). Specifically, spatial mismatch is the logarithm of the maximum distance in meters between source locations of the plant genotypes minus the logarithm of the maximum distance in meters between the different experimental blocks or common gardens in the experiment. The dashed horizontal line shows the ratio at which genotype and spatial effects are equal. The data are from Tack et al. (2012) and were provided by A. Tack

of the focal species. By contrast, intraspecific variation appears much less important than spatial or habitat variation. As caveats, intraspecific effects can be overestimated in certain experimental designs (Tack et al. 2012) and testing conditions (Johnson et al. 2008), and many more studies are needed before broad generalizations are possible. As an aside, consideration of these different frames of reference should bear in mind that evolution will be important in many of them. That is, interspecific effects are also evolutionary, species presence/absence can be determined by adaptation, and population density will often be influenced by evolution. These last two effects were considered in more detail in chapter 7.

QUESTION 6: ON WHAT TIME SCALES DO EVOLUTIONARY CHANGES HAVE COMMUNITY EFFECTS?

Having already established that noteworthy *phenotypic* changes can occur on short time scales (chapter 3), we can now ask whether these changes have community-level effects. Stated another way, how fast are eco-evolutionary dynamics at the community level? Laboratory studies suggest that such effects can emerge very quickly (e.g., Bohannan and Lenski 2000, Yoshida et al. 2003, reviews: Fussmann et al. 2007 and Ellner 2013), but what of more natural contexts? Note that we are here interested in contemporary evolution, and so I do not present community genetics or BEF studies that do not also consider the effects of in situ evolution.

The community effects of recent population divergence have been considered for two ecotypes of alewife: one that completes its life cycle entirely in fresh-water lakes (resident) and the other that transitions from fresh-water lakes to the ocean and back again (anadromous). The two ecotypes differ in traits that influence foraging efficiency on zooplankton: anadromous alewife have a larger gape, fewer/longer gill rakers, and larger spaces between gill rakers (Palkovacs and Post 2008). As would be expected from these trait differences, anadromous alewife selectively prey on large zooplankton and, accordingly, lakes with anadromous alewife have smaller zooplankton than do lakes with resident alewife (Post et al. 2008). The rapidity with which such effects arise can be assessed by considering formerly anadromous populations that were landlocked from the ocean by dam construction in colonial New England (Palkovacs et al. 2008). Palkovacs and Post (2008) showed that phenotypes in these landlocked populations were similar (300 generations after dam construction) to phenotypes in the much older naturally resident populations. Palkovacs and Post (2009) then used mesocosm experiments (large plastic bags in lakes) to show that landlocked alewife no longer reduce zooplankton size to the extent typical of their recent anadromous ancestors. The differences in zooplankton community structure between alewife ecotypes in mesocosms were similar to zooplankton differences in natural lakes harboring the different ecotypes (fig. 8.6), and phytoplankton community structure also changed (Weis and Post 2013). Ecotype divergence leading to community and ecosystem effects (in mesocosms) have arisen on even shorter time scales for stickleback recently colonizing lake versus stream environments in Switzerland (Matthews et al. 2016) and for *D. magna* that evolved in experimental treatments with or without fish and with or without macrophytes (Pantel et al. 2015). These results suggest that community-level ecological differences can arise from phenotypic changes occurring within at least a few hundred generations.

Several studies of plants also inform the pace at which community-level consequences can arise, and these studies can, often more directly than fish mesocosm studies, confirm a genetic basis for the effects. Lankau and colleagues conducted a series of experiments comparing plant communities from locations that had experienced the aggressive allelopathic invasive plant *Alliaria petiolata* for different lengths of time (up to 63 years). As previously described (question 2), the studies showed substantial evolution of competitive ability in the invader and some native plants. Lankau and Nodurft (2013) then showed that evolution by one of the natives (*Pilea pumila*) more than doubled the species richness of its attendant arbuscular mycorrhizal fungi. Similarly, invasive elk (*Cervus canadensis nelsoni*) have caused the evolution of flowering time in *Solidago velutina*, which has had large cascading effects on arthropod communities (Smith et al. 2015). In experimental

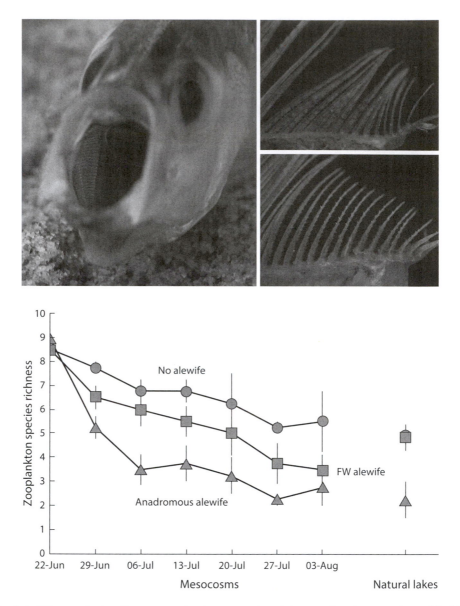

Fig 8.6. Effects of alewife on zooplankton species richness. The images (provided by E. Palkovacs) show an alewife (left) and the gill rakers of a freshwater alewife (top) and of an anadromous alewife (bottom). In the lower panel, the three lines show results through time in experimental mesocosms without alewife (circles), with freshwater landlocked alewife (squares), and with anadromous alewife (triangles). Also shown at the far right are average values from June to August for natural lakes without alewife, with freshwater alewife, and with anadromous alewife. The data (means and standard errors) are from Palkovacs and Post (2009) and were provided by E. Palkovacs

studies on a shorter time scale, Agrawal et al. (2012, 2013) planted plots with similar combinations of genotypes, tracked their evolution (clonal frequencies, life span, flowering time, herbivore defense, competitive ability) for three generations, and found that the final genotypic composition was highly predictive of the fourfold variation in the abundance of an herbivorous moth (*Mompha brevivittella*). However, we shouldn't expect such effects to always emerge so quickly. For instance, comparisons between heavily rabbit-grazed areas versus rabbit exclosures showed dramatic variation among plant species in the strength of evolutionary responses (Turley et al. 2013, Didiano et al. 2014), suggesting the likelihood of similarly dramatic variation in community consequences.

Substantial divergence in the community-level ecological effects of a focal species can occur on very short time frames. In these cases, the ecological change can some-times be clearly linked to phenotypic traits (e.g., foraging traits in fish and defense traits in plants) that underwent contemporary divergence among populations, replicates, or treatments. However, such effects are unlikely to be universal and will instead by highly context-dependent.

QUESTION 7: ARE ECO-EVOLUTIONARY EFFECTS DIRECT OR INDIRECT?

The consideration of direct versus indirect effects has loomed large in ecology. The basic idea is that a given species can directly influence another species, which can then (through density or trait mediated effects) indirectly influence other species (Abrams 1995, Werner and Peacor 2003). For example, fish predators can directly influence zoo-plankton communities and can thus indirectly influence phytoplankton communities through a "trophic cascade" (Paine 1980). Direct versus indirect effects of this sort will certainly be relevant to eco-evolutionary dynamics, where the evolution of species A can directly influence species B and thus indirectly influence species C. I will return to this classic context for direct versus indirect effects in chapter 9, whereas I here wish to focus on a different sort of direct-indirect dichotomy—the one based on population size as described at the outset of this chapter. As a direct effect, evolution can alter species interaction traits, which can then influence community structure independent of any changes in population dynamics that might occur in the focal species. As an indirect effect, evolution can alter the population dynamics of a focal species (chapter 7), which can then influence community structure independent of changes in species interaction traits.

Direct effects have been experimentally demonstrated by holding the density/biomass of a focal species constant and varying the distribution/frequency of phenotypes or genotypes. For instance, this approach is standard for the fish mesocosm experiments described above, although Bassar et al. (2010, 2012) also manipulated density. Direct effects are also the usual emphasis of community genetics and BEF experiments with plants. For example, arthropod communities are measured on *individual* plants of dif-ferent genotypes (Johnson and Agrawal 2005, Whitham et al. 2006, Johnson 2008, Tack et al. 2010) and among plots with different numbers of genotypes but similar numbers of individuals (Booth and Grime 2003, Crutsinger et al. 2006, Johnson et al. 2006). Thus, most of the previously described results focus on direct effects of genotypes or ecotypes, which could be strong or weak in relation to indirect effects acting through population dynamics.

Indirect effects are more difficult to demonstrate because they first require document-ing evolutionary influences on population dynamics (chapter 7), and then showing how those changes alter community variables. From this perspective, several experimental studies have documented the effects of genetic identity/diversity on the shoot density of plants, which can then cascade to effects on plant or insect communities. For example, Crutsinger et al. (2008b) showed that plots of goldenrod with greater genetic diversity had greater stem densities, which then reduced the potential for colonization by invasive plants. Similarly, Hughes and Stachowicz (2004) showed that more genetically diverse plots of *Z. marina* retained more shoots during a period of grazing by geese, which then influenced the diversity of epifaunal organisms. In addition, studies have yielded insight into the relative importance of direct and indirect effects. Crutsinger et al. (2008b) found that direct effects of plant genotypic diversity on arthropods were minimal after accounting for indirect effects acting through stem density. By contrast, Kotowska et al. (2010) reported that the higher insect biomass in polycultures of *A. thaliana* was not simply because those polycultures were more productive. More studies of this sort are needed.

The above studies are somewhat artificial in that they didn't examine "real" populations, and so the observed dynamics might not be representative of nature. However, I am not aware of any studies of natural populations that have formally quantified the relative importance of direct and indirect (through population dynamics) evolutionary effects on community structure—although some plausible narratives can be constructed. For instance, the Australian rabbits discussed earlier rapidly became epidemic and had drastic effects on local plant communities (Lange and Graham 1983). The introduced myxoma virus then rapidly reduced rabbit abundance (Fenner and Ratcliffe 1965, Fenner 1983), which again altered plant communities. Soon thereafter, however, evolution of the rabbits and the virus (Fenner and Ratcliffe 1965, Fenner 1983, Best and Kerr 2000) arrested the decline of rabbits and yet again altered plant communities. Although direct and indirect effects were not explicitly quantified, it does seem likely that the evolving trait (resistance) of the rabbits did not directly influence plant communities. Instead, the evolution of resistance influenced rabbit abundance, which then influenced plant communities. Other putative examples include situations where adaptation to pollution allows the persistence of species that have important ecological effects (Medina et al. 2007). In this evolutionary rescue context, Fussmann and Gonzalez (2013) analyzed models showing how the adaptive evolution of a focal species could have strong community-level consequences, including "community rescue."

Many studies have now documented how genetic variation/evolution can have direct and indirect effects on community-level variables, whereas the relative importance of each pathway has proven difficult to establish. In addition, careful studies have yet to be performed in nature. **I suggest that direct effects are more likely for organisms that have large individual effects (e.g., foundation species, keystone species, ecosystem engineers) and indirect effects are more likely for organisms that are very abundant (e.g., microbes, pests, weeds).**

Conclusions, significance, and implications

The present chapter was the first to move beyond focal organisms to consider the broader effects of evolution. Such influences have been long considered in the context of predator-prey interactions and interactions among competitors. Although results are highly variable, a frequently observed outcome is that evolution stabilizes these interactions, primarily because the player suffering the most from an interaction will be the player under strongest selection to improve its performance in that interaction. Predator-prey interactions, especially host-parasite interactions, are an arena where the importance of contemporary evolution is reasonably well established—that is, ongoing evolution clearly shapes the evolution of species interaction traits and the joint population dynamics of interacting species. Additionally, all of these eco-evolutionary interactions are strongly altered by the dynamics of dispersal and gene flow. For instance, the species that experiences the highest gene flow often has an advantage in antagonistic interactions, and the species that arrives first at a given location often has an advantage over later-arriving competitors.

More recently, studies have emphasized how the evolution of a particular focal species has important effects on composite community-level variables. These studies have documented effects of (1) genotypic diversity/identity (usually in plants) on aspects of community structure (usually the number and diversity of arthropods or other plants), and (2) population differences (usually in fish) on aspects of community structure (usually zooplankton and benthic invertebrate communities). These evolutionary effects can be large in relation to some other ecological drivers, such as the presence/absence of a species, species replacements, or changes in population density. However, evolutionary effects can appear small in relation to some other ecological drivers, such as spatial/habitat variation. In some instances, community-level effects have been shown to evolve quite rapidly, such as over only a few to dozens of generations, whereas in other instances little change is observed over similar time frames. Finally, most studies focus on direct effects of phenotypes on communities, whereas the greatest effects might often occur through evolutionary changes that alter the population structure of a focal species.

It might seem appropriate to here revisit the seven expectations presented at the start of the chapter for how eco-evolutionary effects at the community level are most likely to play out on contemporary time scales. However, the same expectations, plus a few more, hold in the ecosystem context, and so I will defer a discussion of the combined (community and ecosystem) expectations until the end of the next chapter.

Chapter 9

Ecosystem Function

The previous chapter started our consideration of the "focal-species composite-response" approach, wherein I asked how evolution in a focal species might influence composite variables at the community level, such as species richness or diversity. The present chapter extends this approach to composite responses at the ecosystem level, such as nitrogen fixation, phosphorous mineralization, carbon cycling, decomposition rate, primary productivity, food chain efficiency, energy or biomass flux, and so on. Although parsing ecological response variables into categories such as populations (chapter 7), communities (chapter 8), and ecosystems (present chapter) is not always straightforward, some useful distinctions do exist. Specifically, variables at the population level involve only the focal species, variables at the community level involve the properties of other species (often considered in aggregate), and variables at the ecosystem level move beyond species to consider fluxes, flows, and biomasses of nutrients and energy.

Evolutionary influences on ecosystem function are clear on long time scales (Erwin 2008). As a few examples, the evolution of photosynthesis had a huge impact through oxygen production, the evolution of terrestriality dramatically altered nutrient dynamics on land, and the evolution of nitrogen fixation reshaped the nitrogen cycle. The question at hand, however, is how such effects play out on contemporary time scales. Given that macroevolution is simply microevolution in fits and starts (chapter 3), it seems likely that at least some large eco-evolutionary effects, perhaps even those noted above, emerged rapidly in the place/time they evolved. Such micro-to-macro transitions are rare and unpredictable events, and so studies considering how *contemporary* evolution influences ecosystem function will necessarily deal with much smaller effects. Yet even these smaller eco-evolutionary effects might well have noticeable—perhaps even important—influences on ecosystem function. Although I will use the term "ecosystem function" throughout, the concepts discussed relate closely to the popular discourse on "ecosystem services" (Constanza et al. 1997, Díaz and Cabido 2001, Díaz et al. 2013). Indeed, given that all past, present, and future ecosystem services provided by organisms are the product of evolution, a more appropriate term would be *EVOsystem services* (Faith et al. 2010, Hendry et al. 2010).

Some of the eco-evolutionary concepts discussed here, and in the previous chapter, have been discussed in other guises. In particular, evolutionary effects on communities and ecosystems have been invoked and discussed in the context of "extended phenotypes" (Dawkins

1989, Bailey 2012), "interspecific indirect genetic effects" (Shuster et al. 2006, Bailey et al. 2014), "niche construction" (Odling-Smee et al. 2003, 2013, Matthews et al. 2014), "biological stoichiometry" (Elser 2006, Matthews et al. 2011b, Jeyasingh et al. 2014), "community genetics" (Whitham et al. 2006, Tack et al. 2012), "genes-to-ecosystems" (Schweitzer et al. 2008, Bailey et al. 2009, Crutsinger et al. 2014b), community and ecosystem evolution (Loreau 2010), and "functional traits" (Díaz and Cabido 2001, McGill et al. 2006). The umbrella of "eco-evolutionary dynamics" can encompass the dynamic aspects of all of these ideas.

When and how might contemporary evolution influence ecosystem function?

By extension from the previous chapter on community structure, evolutionary influences on ecosystem function could occur through indirect or direct routes. Indirectly, evolution could alter the population dynamics of a focal species (chapter 7) or the structure of its community (chapter 8), with either of these effects then modifying ecosystem function.[1] As a hypothetical example, adaptive evolution that increases the abundance of a nitrogen fixer can indirectly influence nitrogen cycles. Alternatively, or additionally, evolution can alter the phenotypes of a focal species in ways that directly influence ecosystem function, independent of any changes in population dynamics or community structure. As an established example, changes in the frequencies of tree genotypes with different tannin levels will alter litter decomposition (Whitham et al. 2006). In the following paragraphs, I first outline how the above effects can be considered in five different conceptual frameworks, the first three of which are simple extensions to those previously discussed (chapter 8). I then revisit the expectations outlined for community structure in the current context of ecosystem function.

Extending the Lande equation: the Johnson et al. (2009) framework for extending the Lande equation to consider community consequences also can be used to consider ecosystem consequences. In short, just as ΔC_1 can be a community variable, it also can be an ecosystem variable, with the αs representing the change in that variable as a function of change in trait values, which are themselves a function of $\mathbf{G\beta}$. As described previously (chapter 8), these parameters can be estimated for natural populations and used to predict changes in ecosystem variables (Johnson et al. 2009). As before, a particularly relevant limitation of this framework is that it does not currently consider indirect effects acting through, for example, population dynamics.

Extending the Hairston Jr. et al. method: to consider both direct and indirect effects, we can turn to further extensions of the Hairston Jr. et al. (2005) approach (see also Ellner et al. 2011). For community dynamics (chapter 8), I provided the following example based on the expectation that $C(t) = C[N(Z(t),K(t)), Z(t)]$:

$$\frac{dC}{dt} = \frac{\partial C}{\partial N}\frac{\partial N}{\partial K}\frac{dK}{dt} + \frac{\partial C}{\partial N}\frac{\partial N}{\partial Z}\frac{dZ}{dt} + \frac{\partial C}{\partial Z}\frac{dZ}{dt}$$

---

[1] I will emphasize pathways that move from populations to communities to ecosystems. However, it is also possible that effects can propagate in the reverse direction. For example, phenotypes might influence population dynamics, which might influence ecosystem function, which might then influence community structure.

As in the Johnson et al. (2009) approach, the *C* in this equation could represent an ecosystem variable, perhaps now indexed as *E*. In this case, the equation would already account for indirect effects acting through population size. However, it would not consider a route whereby ecosystem function was influenced by changes in community structure, which themselves might be influenced by *N*, *K*, and *Z*. For this more detailed inference, the equation could be expanded to ask about changes in ecosystem variables as a function of the above terms, plus terms representing effects that act through community structure. At the risk of getting too messy, here is an example (the dependencies that make up this equation are shown in the top right panel of fig. 9.1):

$$\frac{dE}{dt} = \frac{\partial E}{\partial C}\frac{\partial C}{\partial N}\frac{\partial N}{\partial K}\frac{dK}{dt} + \frac{\partial E}{\partial C}\frac{\partial C}{\partial N}\frac{\partial N}{\partial Z}\frac{dZ}{dt} + \frac{\partial E}{\partial C}\frac{\partial C}{\partial Z}\frac{dZ}{dt}$$
$$+ \frac{\partial E}{\partial N}\frac{\partial N}{\partial K}\frac{dK}{dt} + \frac{\partial E}{\partial N}\frac{\partial N}{\partial Z}\frac{dZ}{dt} + \frac{\partial E}{\partial Z}\frac{dZ}{dt}$$

In this equation, the change in ecosystem function (*E*) is partitioned into effects due to:

1. first term = changes in community structure (*C*) that result from changes in the population size (*N*) of a focal species that result from changes in an external environmental driver (*K*),
2. second term = changes in community structure that result from changes in the population size of a focal species that result from changes in a phenotypic trait (*Z*) in that species,
3. third term = changes in community structure that result from changes in a phenotypic trait of the focal species,
4. fourth term = changes in the population size of a focal species that result from changes in an external environmental driver,
5. fifth term = changes in the population size of a focal species that result from changes in a phenotypic trait in that species, and
6. sixth term = changes in a phenotypic trait of a focal species.

Equations such as the above can be summarized with path diagrams, as shown in figure 9.1. The top-left panel shows the original Hairston Jr. et al. (2005) model, where population dynamics are influenced by changes in phenotype and an external environmental driver. The top-middle panel shows the model presented for community dynamics (chapter 8), where community variables are influenced directly through changes in phenotype or indirectly through changes in population dynamics that are themselves influenced by changes in phenotype or an external environmental driver. The top-right panel shows the model given just above. The bottom panels show more complicated sets of influences, such as when the same environmental driver directly influences population dynamics, community structure, and ecosystem function (bottom-left); or when the environmental drivers are different (K_1, K_2, K_3) at each of those levels (bottom-right). Limitations of the Hairston Jr. et al. (2005) approach were previously discussed in chapter 7 and chapter 8. An additional difficulty, here made stark by the explosion of terms, is that empirical fitting will require extensive replication in space (e.g., numbers of plots in experiments or field surveys) or time (e.g., numbers of years of population monitoring).

Extending the Price equation: just as the Price equation can be used to explain changes in traits (chapter 3), population parameters (chapter 7), and community structure (chapter 8),

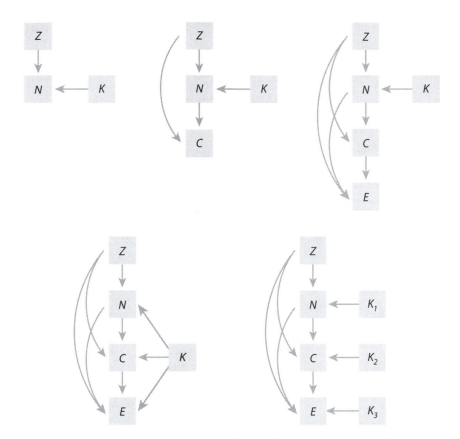

Fig 9.1. Path diagram representations of various effects of phenotypes (*Z*) and external environmental drivers (*K*) on population dynamics (*N*), community structure (*C*), and ecosystem function (*E*). Conceptually, a change at any level could cause a change (arrows) at another level. Note that a single *k* means that one external driver is important, whereas *k* with a subscript implies a specific environmental driver

it can do the same for changes in ecosystem function. For example, Collins and Gardner (2009) used the Price equation to consider how climate change would alter carbon uptake in communities of marine phytoplankton as a result of changes in "physiology" (plastic changes in uptake by a given clonal lineage), "evolution" (selection-driven changes in the frequency of clones within a species), and "community" (changes in the relative abundance of different species within a community). Limitations of the Price equation have been discussed earlier.

The Bailey et al. *(2009) path model*: as described above, path diagrams can be used to depict alternative scenarios in the Hairston Jr. et al. (2005) framework. Another way of using path diagrams to consider eco-evolutionary effects was presented by Bailey et al. (2009). The goal in this latter case was not to assess associations among *dynamics* at the different levels, but rather among different genetic/phenotypic variants and community/ecosystem properties. In essence, the Hairston Jr. et al. (2005) framework has been used for examining *dynamics*, whereas the Bailey et al. (2009) framework has been used for examining effects of standing variation (i.e., the "community genetics" perspective). In the initial formulation, Bailey et al. (2009) suggested that different alleles at a particular gene could influence a particular phenotypic trait (e.g., foliar condensed tannin), which could then influence a particular community variable (e.g., soil microbial community composition), which could then influence a particular ecosystem variable (e.g., soil nutrient cycling). All other effects influencing each level of variation (trait, community, ecosystem) were considered together as error terms (fig. 9.2). This basic model thus postulated that each "higher" level of organization was influenced only by direct effects from the immediately "lower" level of organization. This structure predicts that the effects of genes should decrease from lower to higher levels of organization (traits to communities to ecosystems) because the error terms are compounded as one moves through these levels. The same "compounding" has been argued to explain why heritabilities are lower for life history traits than for morphological traits (Price and Schluter 1991).

Expectations are less straightforward when multiple genes, multiple traits, and multiple ecological variables are involved (fig. 9.2). For instance, a single phenotypic trait might be influenced by multiple genes and a single ecological variable might be influenced by multiple phenotypic traits. In such cases, the strength of a particular genetic/phenotypic effect does not necessarily decay as one moves from "lower" to "higher" levels of organization. The reason is that a single gene could influence two phenotypic traits, which could then both influence a given community variable; in which case, the gene

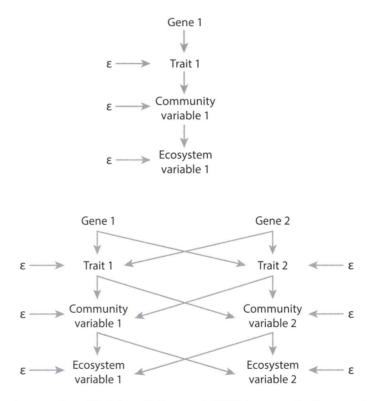

Fig 9.2. Another way (modified from Bailey et al. 2009) to use path diagrams to represent effects of genetic change on ecosystem variables. The top panel suggests that gene frequency changes can cause trait changes that can cause community changes that can cause ecosystem changes (ε corresponds to "error"). The bottom panel shows an extension to multiple genes, traits, community variables, and ecosystem variables. Direct effects from traits to ecosystems could also occur but are not illustrated, nor are effects acting through population dynamics

could more strongly influence the community variable than either of the two traits. Of course, even this more complicated model (fig. 9.2) is still oversimplified. For instance, opposing direct versus indirect effects, as well as any feedbacks (details below), can dramatically alter the outcome.

Biological (including ecological, metabolic, and organismal) stoichiometry is "the study of the balance of energy and multiple chemical elements in living systems" (Elser et al. 2000, see also Elser 2006, Matthews et al. 2011b, Crutsinger et al. 2014b, Jeyasingh et al. 2014). The uptake rate of nutrients (typically carbon, nitrogen, and phosphorus) by a species is expected to evolve in response to environmental conditions and should then influence those conditions. The uptake rates of nutrients might also evolve in response to changes in nutrient demand imposed by changes in organismal traits (which could have nothing to do with environmental conditions). As a specific example (Elser 2006), selection for increased growth rate should increase phosphorous demand because phosphorous is a key component of the ribosomal RNA needed for protein synthesis. Increased growth rate of a particular species therefore might decrease phosphorous levels in a lower trophic level and make more phosphorous available to a higher trophic level. These changes might then alter nutrient limitation and hence productivity and biomass at different trophic levels, as well as energy and nutrient transfer among levels (Schmitz 2013). Indeed, any limiting nutrient might lead to the evolution of increased uptake or utilization efficiency of that nutrient (or switches to other nutrients), which could then influence the availability and distribution of those nutrients in the food web. Eco-evolutionary *feedbacks* would seem particularly likely in such cases. For instance, evolution that leads to increased uptake efficiency of a particular nutrient by an ecologically influential species will likely alter the availability of that nutrient, and thereby influence selection related to uptake of that nutrient in multiple species.

Modeling efforts demonstrate a number of the above effects (Boudsocq et al. 2011, Matthews et al. 2011b), and nutrient-centered models are among the few to formally consider the evolution of overall ecosystems. Such models typically ask: what sorts of ecosystems will emerge through evolution (review: Fussmann et al. 2007)? Will they maximize overall productivity? Will they maximize stability? Will they lead to a "tragedy of the commons"? Results of these models are highly diverse and depend on particular assumptions. Also, laboratory studies are beginning to consider such effects. For instance, Declerk et al. (2015) showed that experimental adaptation to phosphorus-limited conditions caused rotifers (*Brachionus calyciflorus*) to evolve better performance under those conditions, which increased feeding rates and rotifer abundances, and decreased algal abundance. However, rotifer elemental composition did not evolve.

Relatively little empirical work has examined interactions between biological stoichiometry and evolution on contemporary time scales; yet developments in several systems illustrate the potential. As one example, high-predation and low-predation guppies differ in some aspects of their stoichiometry, with the former being more stoichiometrically balanced in their diets, being less susceptible to phosphorus limitation, and excreting less phosphorus and more nitrogen (El-Sabaawi et al. 2012, 2015b). In addition, high-predation guppies raised in the presence of predation risk plastically reduced nitrogen excretion by about 40% (Dalton et al. 2014). However, although guppy stoichiometry varies dramatically in nature, encompassing about 70% of the range across all freshwater fishes, guppy ecotype (predation regime) does not explain much of this variation, with (unspecified) local environmental conditions instead seeming to be much more important (El-Sabaawi et al.

2012, El-Sabaawi et al. 2014). Variation in guppy life history traits and behavior thus influences guppy stoichiometry; yet whether or not this variation is important to ecosystem dynamics in nature will depend on many other factors, such as how important guppies are for nutrient cycling and whether or not other organisms are limited by that nutrient. As another example, stickleback populations show the evolution of reduced armor in less than a decade when marine ancestors colonize fresh water (Bell et al. 2004, Bell and Aguirre 2013). Given that large amounts of armor (bone) will require more phosphorus, biological stoichiometry in stickleback should change dramatically during this period of evolution—indeed, phosphorous certainly differs among stickleback populations with high versus low armor (El-Sabaawi et al. 2016). Given that stickleback are an abundant and influential part of lake food webs, this evolution could have noteworthy influences on phosphorous dynamics (El-Sabaawi et al. 2016).

MORE ABOUT FEEDBACKS

Feedbacks and their implications have been extensively discussed within ecology (e.g., DeAngelis et al. 1986, Bever et al. 1997) and evolutionary biology (e.g., Crespi 2004). Feedbacks are also likely in interactions *between* ecology and evolution (Pimentel 1968, Odling-Smee et al. 2003, Kinnison and Hairston Jr. 2007, Haloin and Strauss 2008, Post and Palkovacs 2009, Schweitzer et al. 2014, Strauss 2014). Although I noted in chapter 1 that feedbacks were likely (fig. 1.7), I only occassionally noted them them in the following discussions of eco-to-evo effects (chapters 2–6) and evo-to-eco effects (chapters 7–9); yet some exceptions occurred. In chapter 4, I noted how ecological differences drive adaptive divergence in functional traits, which should thereby generate the divergent ecological effects of populations/ecotypes. In chapter 6, I noted how this feedback between ecological differences and adaptive divergence can really only have dramatic effects once reproductive isolation evolves, and how reproductive isolation is itself a part of the feedback. In chapter 7, I discussed feedbacks between the evolution of life history traits and density dependence in the context of population dynamics. In chapter 8, I discussed the possibility of feedbacks between community structure and evolutionary change in a focal species. My goal in the present section (and in question 6 of this chapter) is to more formally consider feedbacks between ecology and evolution. In essence, I take the above presentation of evo-to-eco effects encapsulated by the arrows in figure 9.1 and ask how such effects are modified by arrows also going in the reverse direction.

I first need to return to the distinction drawn in chapter 1 between eco-evolutionary feedbacks in the broad versus narrow (or strict) sense. By *feedbacks in the narrow sense*, I mean that the evolution of *a particular trait* in a population causes a change in an ecological variable that influences selection on *the same trait* in the same population. Alternatively, change in *a particular ecological variable* causes a trait to evolve in a way that causes further change in *the same ecological variable*. *Feedbacks in the broad sense* also include situations where the evolution of a trait causes a change in an ecological variable that then influences selection on some other trait. Alternatively, a change in an ecological variable causes a trait to evolve in a way that causes change in some other ecological variable. The distinction between these two types of feedback is that, in the narrow sense, the arrows between ecology and evolution must pass through the same traits and ecological variables; whereas, in the broad sense, they can pass through different traits and ecological variables. However, the narrow-sense criterion also can be satisfied even if intermediate traits or ecological variables are involved: that is, indirect eco-evolutionary feedbacks. For instance, the evolution of one trait could cause change in an ecological variable that imposes selection on a second trait, which thereby generates selection on the first trait.

As previously suggested, the way in which feedbacks play out will have important implications for eco-evolutionary dynamics (Schweitzer et al. 2014, Strauss 2014, Kinnison et al. 2015). In particular, feedbacks can be (1) reinforcing (positive), such as when an increase in trait value causes an ecological change that selects *for* a further increase in trait value, or (2) opposing (negative), such as when an increase in trait value causes an ecological change that selects *against* a further increase in trait value. Positive feedbacks are likely to accelerate changes in traits and ecological variables (Crespi 2004), whereas negative feedbacks are likely to promote stability or cycles. Of course, positive feedbacks are unlikely to continue indefinitely, so an environmental change might first lead to a positive feedback that eventually transitions to a negative feedback as the system stabilizes or cycles around a new equilibrium (Becks et al. 2012, Strauss 2014, Kinnison et al. 2015).

Feedbacks, especially in the narrow sense, will be less common than one-way eco-evolutionary effects: that is, ecology to evolution *or* evolution to ecology. The reason is that both of the pathways have to be occurring and they must be tied together by the same traits and ecological variables. Thus, several conditions must be met for feedbacks to occur. First, trait changes have to take place on contemporary time frames, which is common but not ubiquitous (chapter 3). (Remember that stability also could be a strong indicator of ongoing eco-evolutionary dynamics.) Second, those trait changes must have ecological effects on similar time frames, which is not inevitable (chapters 7–9). Third, those ecological effects must impose selection on the same trait. Such feedbacks seem most likely to be important when particular organismal traits are especially important to both fitness and ecological effects. Strong candidates include body size, feeding traits, and stoichiometry.

EXPECTATIONS

In the previous chapter, I outlined seven expectations that emerged from conceptual frameworks for evaluating the effects of contemporary evolution on community structure. Those expectations hold in the present context of ecosystem function, and I repeat them below by way of reminder (details were provided in the previous chapter). In this new context, some modifications and additional expectations emerge, which I here indicate with italics.

Evolutionary (or, more generally, phenotypic) change is most likely to influence ecosystem function on contemporary time scales when . . .

1. . . . phenotypic change is greatest.
2. . . . phenotypic trait variation has strong direct effects *on ecosystem function.*
3. . . . the focal organisms are very abundant.
4. . . . evolution influences population dynamics *or community structure. I am here referring to an indirect effect of evolution acting through population dynamics or community structure. For the latter, ecosystem function is clearly shaped by community composition and so changes in that composition can modify ecosystem function.*
5. . . . direct and indirect effects do not oppose each other.
6. . . . feedbacks are present—and positive.

7. . . . external drivers are not overpowering.
8. . . . *(a) individual phenotypic traits have influences on multiple ecological variables, or (b) multiple phenotypic traits influence the same ecological variable. Such complexity can increase (a) the total contribution of a given trait across all ecological variables, or (b) the total contribution across all traits to a given ecological variable (Bailey et al. 2009). Although this prediction was not presented in the previous chapter, it should hold in that context too.*
9. . . . *it leads to changes in biological stoichiometry for limiting nutrients.*

As before, these expectations are based on measureable evolutionary and ecological *change*. Yet a *lack of change* in traits, population dynamics, community structure, and ecosystem function might—and likely does—reflect underlying "cryptic" eco-evolutionary dynamics (Kinnison et al. 2015).

Common approaches

In the previous chapter, I described the common empirical approaches for studying "focal-species composite-response" eco-evolutionary dynamics at the community level. The same methods—and, indeed, sometimes the same experiments—are used for inferences at the ecosystem level: that is, the community-level variable is simply now an ecosystem-level variable. Given that these approaches were detailed previously (chapter 8), I will here simply list them. Two approaches manipulate the diversity or identity of genotypes of a focal species in experimental plots or treatments and then monitors ecosystem-level response variables. The third approach uses common-gardening experiments to compare divergent populations of a known age for their current ecosystem-level effects. The fourth approach relates real-time phenotypic/genetic changes within populations to concurrent changes in ecosystem-level variables.

Eco-evolutionary ecosystem function in nature

Many of the studies of phenotypic/genetic effects on community structure that were described in the previous chapter also measured ecosystem variables, and so I will first briefly update the questions of the previous chapter in this new context of ecosystem function. The relevant questions include: How does intraspecific diversity/identity influence ecosystem function (Q1)?, How strong are the ecosystem effects of intraspecific diversity/identity (Q2)?, On what time scales do evolutionary changes have ecosystem effects (Q3)?, and Are eco-evolutionary effects direct or indirect (Q4)? I will then turn to additional questions that emerge when adding this "higher" level of ecological complexity: Do the effects of genotypes decrease with increasingly complex levels of organization (Q5)?, and To what extent are feedbacks between evolution and ecosystem function evident (Q6)? Finally, I ask how ecosystem function can be shaped by selection within a generation (Q7).

QUESTION 1: DOES GENETIC DIVERSITY/IDENTITY MATTER FOR ECOSYSTEM FUNCTION?

As described in the previous chapter, an important topic in ecology has been the extent to which species diversity influences ecological properties and processes, including resistance/resilience, stability, productivity, and various "ecosystem services" (Loreau et al. 2001, Hooper et al. 2005, Lefcheck et al. 2012). The biodiversity-ecosystem function (BEF) studies addressing this question have recently progressed from considering only interspecific diversity to also considering intraspecific diversity (Haloin and Strauss 2008, Hughes et al. 2008, Bailey et al. 2009). In the previous chapter, I summarized some examples for community-level variables, especially for plants with regard to the diversity of arthropods and the resistance of communities to invasion by competitors. In the present chapter, I consider these experiments as they inform ecosystem-level variables.

BEF studies at the interspecific level frequently report positive associations between plant diversity and community productivity (Tilman et al. 1996, Loreau et al. 2001, Hooper et al. 2005, Lefcheck et al. 2015), with the same result also emerging for intraspecific diversity (Crutsinger et al. 2006, Hughes et al. 2008, Kotowska et al. 2010). The latter studies usually measure productivity of only the focal species, rather than an entire community, and so they can be considered population-level responses (Hughes et al. 2008). Indeed, I reviewed in the chapter on population dynamics (chapter 7) a number of cases where genetic diversity influenced aspects of population productivity. Variables that are less ambiguously assigned to the ecosystem level include fluxes of energy or nutrients. In this regard, litter mixtures that are more genetically diverse show greater nutrient fluxes and decomposition rates in some studies (e.g., Schweitzer et al. 2005) but not so much in others (e.g., Madritch et al. 2006). In yet other cases, ecosystem function is maximized at intermediate levels of genetic diversity, with an example being nitrogen in the soil below cottonwood stands (Schweitzer et al. 2011). As for whether such effects are strictly additive, or whether they also include some nonadditive contributions (see chapter 8 for details), results vary among studies (Schweitzer et al. 2005, Madritch et al. 2006, Zuppinger-Dingley et al. 2014).

Moving from genetic diversity to genetic identity, a number of studies have shown that different plant genotypes can have differential effects on ecosystem processes (reviews: Whitham et al. 2006, Bailey et al. 2009). For example, different cottonwood genotypes produce leaf litter that has different rates of nitrogen mineralization (Schweitzer et al. 2004). Moreover, placement of different black cottonwood (*Populus trichocarpa*) leaf genotypes into cattle tanks demonstrated effects on phytoplankton abundance, nutrient dynamics, and light availability (Crutsinger et al. 2014b). Recent studies of both community and ecosystem variables are particularly emphasizing interactions between genetic identity in a given species and various other factors, as a few examples will serve to illustrate. First, some studies have revealed interactions among genotype identities in different plant species (Genung et al. 2012a, 2013) or between a plant (cottonwood) and a fish (stickleback) species (Rudman et al. 2015). Second, aquatic decomposers appear adapted (at the community level) to local genotypes of the riparian red alder (*Alnus rubra*) (Jackrel and Wootton 2014; Jackrel et al. 2016). Third, genetic differences among coastal willow (*Salix hookeriana*) trees shape the complexity of food web interactions (Barbour et al. 2016). In some cases, these interactive effects are larger than either of the individual effects and many more such interactions are sure to be discovered.

Another major context for investigating effects of genetic identity is the use of fish ecotypes in mesocosms, as previously discussed for community structure (chapter 8) and as summarized here for ecosystem function (fig. 9.3). For guppies, mesocosms containing the high-predation ecotype had (relative to mesocosms containing the low-predation ecotype) higher algal standing stocks, lower biomass-specific gross primary productivity, and lower rates of leaf decomposition (Palkovacs et al. 2009, Bassar et al. 2010, 2012). In addition, interactions were seen between effects of guppy ecotype and guppy density for PO_4 flux, total nitrogen ($NH_4 + NO_3$) flux, and net production. These differences arose at least partly because different guppy ecotypes forage on different foods and have different excretion rates (Palkovacs et al. 2009, Bassar et al. 2010, 2012). For threespine stickleback, mesocosms with different ecotypes (benthic, limnetic, or generalist) had different gross primary productivity, respiration, algal biomass, dissolved organic carbon composition, and light transmission (Harmon et al. 2009), although ecosystem effects were often weaker in a subsequent study (Des Roches et al. 2013). For whitefish, benthic versus limnetic ecotypes had differential influences on dissolved organic carbon and phytoplankton concentrations (Lundsgaard-Hansen et al. 2014). Importantly, this last study and one on stickleback (Matthews et al. 2016) first reared the two ecotypes in a common-garden environment, thus more directly informing the effects of *genetic* divergence on ecosystems.

As was the case for community-level variables (chapter 8), **intraspecific diversity/ identity influences a wide variety of ecosystem-level variables**. Of course, most of the experiments have been conducted in controlled and artificial environments (e.g., cattle tanks), for which the label "ecosystem" is perhaps optimistic. Many more studies are needed—particularly in nature—if we are to reveal general patterns.

QUESTION 2: HOW STRONG ARE THE ECOSYSTEM EFFECTS OF INTRASPECIFIC DIVERSITY?

Documenting the existence of effects (question 1) does not say much about their *importance*, an additional inference for which some frame of reference is needed (Hairston Jr. et al. 2005, Hersch-Green et al. 2011, Matthews et al. 2014). In the previous chapter, I described how the effects of intraspecific variation on community structure were often (although not always) strong in relation to the effects of interspecific variation, species presence/absence, and variation in the density of a focal species; whereas intraspecific effects were often weak in relation to spatial variation and habitats. I here attempt the same comparisons at the ecosystem level, using some of the same studies discussed earlier at the community level.

The interspecific frame of reference has been applied in several studies. Crutsinger et al. (2009) conducted a common-garden experiment with multiple genotypes of goldenrod, as well as representatives of three closely related species. Leaf-litter decomposition rates and nitrogen dynamics were mostly influenced by species identity, followed by lesser effects of genotype identity, and genotype diversity. Conversely, Genung et al. (2013) showed that nitrogen and phosphorus dynamics during decomposition were more strongly influenced by genotypic variation within two species of *Solidago* than they were by differences between the two species. Taking a different approach, Wardle et al. (2009) examined how changes in leaf litter nutrients and decomposition along temporal sequences of ecosystem "retrogression" were determined by shifts in plant species composition versus changes in the properties of individual species. Although results varied among temporal sequences and response variables, interspecific variation was generally more important than intraspecific variation. A useful approach to implement in such studies might be the Price equation (Collins and Gardner 2009). That is, the total change in some ecosystem level variable could be partitioned into the relative contributions of plasticity within species, genetic change within species, and changes in species composition. Finally, in the mesocosm study of Palkovacs et al. (2009), the effects of intraspecific variation (relative to interspecific

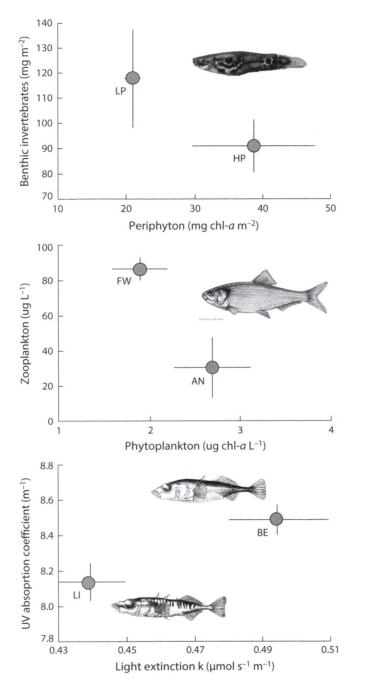

Fig 9.3. Examples of the effects of fish ecotypes on ecosystem function in mesocosms. The top panel shows some effects of low-predation (LP) versus high-predation (HP) guppies in experimental stream channels in Trinidad. The data are from Palkovacs et al. (2009). The middle panel shows some effects of fresh-water (FW) and anadromous (AN) alewife in experimental bags in a lake. The data are from the first sampling date in Palkovacs and Post (2009—zooplankton) and from Palkovacs and Dalton (2012), all provided by E. Palkovacs. The lower panel shows some effects of benthic (BE) and limnetic (LI) stickleback in large tanks. The data are from Harmon et al. (2009) and were provided by L. Harmon. In all panels, the data are means and standard errors. The figure is modified from Hendry et al. (2011)

variation) were stronger for algal accrual, weaker for nitrogen excretion, and similar (both nonsignificant) for phosphorous excretion and litter decomposition (fig. 9.4). Post et al. (unpubl. data) subjected these and related studies to meta-analysis and found that ecosystem function was influenced to a similar extent by intraspecific variation (different genotypes or different ecotypes) and interspecific variation (different species). However, the effects of species richness were greater than the effects of genotype richness in BEF studies.

The species presence/absence and density frames of reference have been applied to mesocosm studies of guppies and stickleback, as first discussed in the context of community structure (chapter 8). In mesocosm experiments with guppies, algal biomass was similarly influenced by guppy ecotype and by a doubling of guppy density, whereas the effect of guppy presence/absence was much greater (Bassar et al. 2010, 2012). In the same study, however, guppy ecotype had a much greater influence on leaf decomposition than did guppy density or presence/absence (results were mixed for benthic organic matter). In another study of the same system, El-Sabaawi et al. (2015a) found that guppy ecotype influenced several ecosystem properties (nutrient recycling, nutrient fluxes, ecosystem metabolism, leaf litter decay) as much as did a threefold difference in light levels. Although

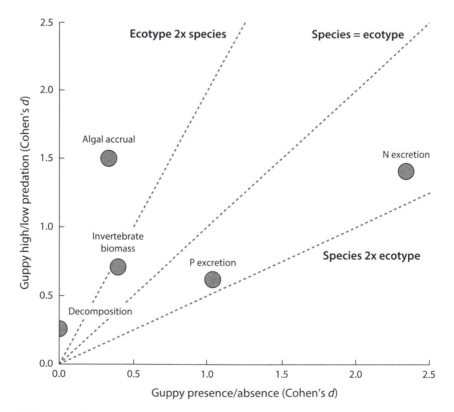

Fig 9.4. Relative effects of species (guppy presence/absence) versus ecotype (high-predation versus low-predation guppies) on five community and ecosystem variables in experimental mesocosms. Cohen's d is the difference in standard deviation units. The dashed lines indicate where ecotype effects are twice as large as species effects (Ecotype 2× species), species effects are twice as large as ecotype effects (Species 2× ecotype), and effects at the two levels are similar (Species = ecotype). The data are from Palkovacs et al. (2009)

these results were from mesocosms, they might also apply in nature given that several key parameters—such as ecotype-differences in nitrogen excretion—are similar for guppies in mesocosms and in nature (El-Sabaawi et al. 2015b). Thus, in a single system, intraspecific variation (guppy ecotype) has been shown to be as important for many ecosystem properties as are guppy presence/absence, guppy density, and light levels. In a mesocosm experiment with stickleback, by contrast, ecosystem properties (biomass of different trophic levels, dissolved oxygen) were generally more strongly influenced by stickleback presence/absence or density than by stickleback ecotype (Des Roches et al. 2013). However, Rudman et al. (2015) found that the effects of cottonwood genotype and stickleback ecotype were similar to those of stickleback presence/absence. By contrast, aphid (*Myzus persicae*) presence/absence and aphid density had greater effects on host plant biomass in field plots than did aphid genotype (Turley and Johnson 2015).

The spatial frame of reference, which is often leveraged for community variables (chapter 8), has been less often employed for ecosystem variables. Certainly, some studies have shown that spatial effects tend to be greater than intraspecific effects when it comes to the productivity of a focal species (e.g., Lojewski et al. 2009), but—as noted earlier—this variable is perhaps better considered at the population level. At the ecosystem level, Pregitzer et al. (2013) found that soil properties (pH, C, N) were more strongly influenced by spatial effects (different common gardens along an elevational gradient) than by different cottonwood genotypes, although the latter effects were certainly present and interacted with spatial variation. Of course, the spatial variation also could be due to evolutionary effects: (1) species composition (including microbes) likely differs among locations (chapter 8) as a result of species sorting based on prior adaptation to different conditions, and (2) the same species could differ genetically among locations as a result of local adaptation. Beyond space, several other frames of reference are informative. Schweitzer et al. (2005) found that the inclusion/exclusion of herbivores had a somewhat greater influence on the cottonwood leaf decomposition than did the diversity of genotypes making up the litter. Similarly, Madritch et al. (2006) found that nutrient treatments had a much greater influence on leaf decomposition than did the identity or diversity of cottonwood genotypes making up the litter. In contrast to these studies, C. Fitzpatrick et al. (2015) used experimental plots and laboratory experiments of evening primrose to show that ecosystem variables (leaf decay, soil respiration) were often more strongly influenced by plant genotype than by spatial variation (among plots, albeit on a small scale) or by the presence versus absence of herbivores. Stronger effects of plant genotype than nutrient additions (salmon carcasses) were also observed by LeRoy et al. (2016).

The importance of intraspecific variation is highly variable at the ecosystem level: sometimes being large and sometimes small in relation to various frames of reference, such as species presence/absence, density, space, and nutrient levels. An informal comparison between community variables (question 5 in chapter 8) and ecosystem variables (present question) seems to hint that effects of intraspecific diversity are stronger for the former than the latter. This point will be addressed more formally in question 5 below.

QUESTION 3: ON WHAT TIME SCALES DO EVOLUTIONARY CHANGES HAVE ECOSYSTEM EFFECTS?

Evolutionary influences on ecosystem variables are, as noted earlier, quite obvious on long time frames. At the other extreme, laboratory studies have repeatedly shown that evolution during the course of an experiment can alter a number of ecological variables in microcosms, including nutrient cycling and productivity (e.g., Lennon and Martiny 2008, Gravel et al. 2011, Lawrence et al. 2012). Yet the critical question we need to address is: how rapidly do such effects emerge in nature?

In the previous chapter on community structure, this topic was first addressed by reference to mesocosm studies of landlocked versus anadromous alewife that had diverged over several hundred years. The two ecotypes had different effects on zooplankton community structure (question 5 in chapter 8), and the same experiment revealed ecosystem-level effects on the total biomass of zooplankton and phytoplankton (Palkovacs and Post 2009, Palkovacs and Dalton 2012, fig. 9.3). Ecosystem-level effects were also seen in the study of recently divergent Swiss lake versus stream stickleback (Matthews et al. 2016). Further studies examining the ecosystem effects of contemporary evolution in fish seem likely to emerge given the many instances of introduction to (or colonization of) new environments, including for stickleback (Bell and Aguirre 2013) and guppies (review: Magurran 2005). For instance, guppies evolve changes in foraging traits and behavior when introduced to new environments, so it certainly seems likely that contemporary evolution will have ecological consequences (Palkovacs et al. 2011). However, the previously described caveats remain: (1) if the fish are wild-caught, the ecological effects might not reflect genetic difference; and (2) results in mesocosms might not reflect those in nature.

Also in the previous chapter, I described several studies that demonstrated community-level effects of contemporary evolution in wild plants (Lankau and Nodurft 2013) and in experimental genotype mixtures (Agrawal et al. 2012, 2013); such studies would ideally also consider ecosystem-level variables. One BEF experiment with relevant data was conducted at Jena, Germany, where experimental plots were established with monocultures or polycultures. Zuppinger-Dingley et al. (2014) postulated that selection in the polycultures should lead to character displacement (more about character displacement in chapters 6 and 8) that would decrease competition and enhance complementarity. If so, relationships between species diversity and ecosystem function (community-wide productivity) should be greater for genotypes from the polycultures than from the monocultures. The authors confirmed their prediction by employing new mini-BEF experiments that planted genotypes of a given species from each selection history (monoculture or polyculture) into pots with or without the other species. Also fitting the hypothesis, the authors documented character displacement in plant functional traits, although the relative contributions of selection with a generation versus evolution across generations have not been assessed. A laboratory experimental study with remarkably similar results is Lawrence et al. (2012). In addition, the C. Fitzpatrick et al. (2015) evening primrose study mentioned above showed considerable, but context-specific, effects of contemporary evolution of genotypic mixtures over five years on ecosystem variables. Context-dependent effects were also evident in a field experiment (Turley and Johnson 2015) of aphid evolution over five generations, where shifts in aphid genotypic composition influenced the biomass of one host plant species (*Brassica napus*) but not another (*Solanum nigrum*). These results, and those reported in question 6 of the previous chapter, confirm that community/ecosystem effects of evolution can emerge on very short time scales.

Contrast this conclusion with that reported for the previous question (and question 5 of the previous chapter) where interspecific effects on ecosystems/communities were not much greater than intraspecific effects. The intersection of these two results suggests a paradox: large ecological effects evolve on short time frames but do not accumulate very much over longer time frames. This apparent paradox parallels the one presented in chapter 3 (question 4) where it was seen that phenotypic traits can change rapidly on very short time scales but do not show much cumulative change on long time scales. Parallel outcomes for these two types of data (ecological effects and functional traits) makes sense because we expect the community/ecosystem effects of organisms to be related to their traits, although the links might not always be obvious (e.g., Matthews et al. 2011a). Indeed, ecological effects are often considered as phenotypes, albeit "extended" (Bailey 2012). To explore this seeming paradox, we can turn to studies that relate species differences in community/ecosystem effects to species differences in evolutionary time (e.g., phylogenetic distance). Suitable data come from the many BEF experiments that used experimental plots to assess the effects of species richness on community biomass. Based on 29 BEF experiments, Cadotte et al. (2008) found that phylogenetic diversity in a plot was a good predictor of total biomass (fig. 9.5), even after controlling for the diversity of species and functional groups. This first analysis pooled data from many experiments, whereas more precise relationships are often evident within single experiments. For example, Cadotte et al. (2012) showed that phylogenetic diversity was strongly predictive of aboveground biomass production in the Cedar Creek grasslands BEF experiment (fig. 9.5). These results (see also Cadotte 2013) show that increasing evolutionary time does indeed lead to increasing differences in ecosystem effects. A useful next step would be to combine these interspecific results with the aforementioned intraspecific results.

Contemporary evolution can influence ecosystem function on time scales at least as short as hundreds (alewife) to dozens (BEF experiments) of generations. At present, however, too few studies have been conducted to say how often and under what conditions such results emerge. As a second conclusion, **increasing evolutionary divergence does generate increasingly divergent effects on ecosystem function.** This result is not surprising on long enough time scales given that highly divergent effects often come from very divergent lineages, such as angiosperms versus gymnosperms, C4 versus C3 plants, and mammalian carnivores versus herbivores. Importantly, however, the relationship between divergence in time and divergence in function is very weak, indicating that the latter can be very high or very low for a given length of evolutionary time. A resulting postulate is that the hugely different effects of highly divergent groups, such as those listed above, might have arisen very quickly. Finally, perhaps the "adaptive zone" explanation applied to the trait pattern (question 4 in chapter 3) also applies here: that is, the evolution of ecosystem effects can play out rapidly within a given adaptive zone but can only rarely move outside that zone, at which time the change can be dramatic.

QUESTION 4: ARE ECO-EVOLUTIONARY EFFECTS DIRECT OR INDIRECT?

For Question 7 in the previous chapter, I discussed the classic contrast between direct and indirect effects in ecology (Abrams 1995, Werner and Peacor 2003). For example, the influence of one species on another species might be direct (predator eats prey) or it might work indirectly through a third species (predator eats one prey species and thus benefits that prey's competitor, "apparent competition" *sensu* Holt 1977, Estes et al. 2013). My focus in that chapter was on a different sort of direct versus indirect contrast: does variation in a focal species directly influence a community variable (phenotype-to-community) or does it act indirectly through the focal species' population

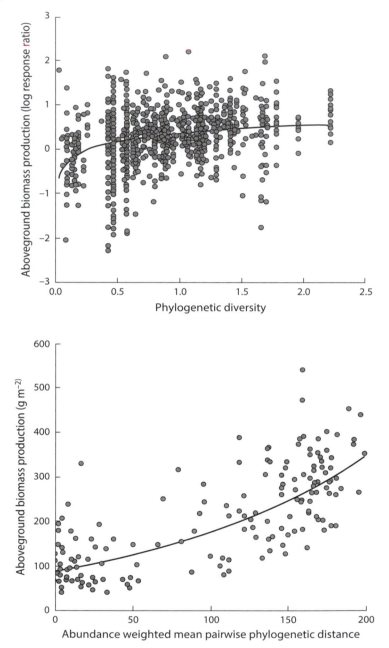

Fig 9.5. Phylogenetic effects on ecosystem function. The top panel combines data from BEF studies to illustrate how phylogenetic diversity of plants within experimental plots is positively related to aboveground biomass production. The curve is logarithmic. Phylogenetic diversity is the sum of branch lengths from a Bayesian ultrametric tree leading to all members of the community. Aboveground biomass production is the log response ratio of production for the polyculture plot relative to the average across monocultures of the same species. The data are from Cadotte et al. (2008) and were provided by M. Cadotte. The bottom panel shows data for the Cedar Creek grassland biodiversity experiment, illustrating that plots with greater phylogenetic diversity (abundance-weighted mean pairwise distance) have greater aboveground biomass production. Both biomass production and phylogenetic distance are averaged across 2001–2010 and the curve is exponential. The data are from Cadotte et al. (2012) and were provided by M. Cadotte

dynamics (phenotype-to-population-to-community)? Considering this contrast in the present context (phenotype-to-ecosystem versus phenotype-to-population-to-ecosystem), every mesocosm study that maintains ecotypes at the same densities and every community genetics study that measures the effects of individual plants (see above) documents effects that do not flow through population dynamics. Thus, indirect effects on ecosystem function that act through population dynamics are rarely studied; indeed, the few studies that showed such indirect effects on community variables (question 7 in chapter 8) did not examine ecosystem variables. Of course, such effects seem likely, such as in the case of rabbits introduced to Australia.

The increase in complexity that attends our current consideration of ecosystem function generates more types of potential indirect effects, starting with those shown in figure 1.7. For instance, variation (or change) in a focal species could directly influence an ecosystem variable (phenotype-to-ecosystem) or it could act indirectly through community structure (phenotype-to-community-to-ecosystem). One instance of the latter occurs when intraspecific variation in a focal species influences aquatic communities, which then influences light extinction (light availability at depth), which will then influence many other community and ecosystem parameters. Indeed, such effects have been demonstrated when the focal species are both plants and animals. For plants, Crutsinger et al. (2014b) used mesocosms to show that cottonwood leaf litter genotypes with more carbon and less condensed tannin generate higher phytoplankton abundance, which then increases light extinction. For animals, Harmon et al. (2009) argued that different stickleback ecotypes influenced aquatic communities, which influenced dissolved organic carbon, which influenced light extinction. Another instance of phenotype-to-community-to-ecosystem indirect effects is seen in eco-evolutionary trophic cascades (Estes et al. 2013, Moya-Laraño et al. 2014). For instance, alewife ecotypes change zooplankton communities (Palkovacs and Post 2009), which then "top-down" influence phytoplankton community structure (Weis and Post 2013) and "bottom-up" influence the diversification of piscivores (Brodersen et al. 2015). Other examples of indirect effects come from plant-soil interactions. As an example, Madritch and Lindroth (2011) showed how the effects of tree genotypes on soil processes were driven by "adaptation" (a combination of evolution and changes in community structure) of soil microbial communities to chemical characteristics that differed among plant genotypes. These different microbial communities then differentially influenced soil processes (see also question 4).

Other direct-indirect contrasts have been explored (review: Estes et al. 2013) and I here provide a few examples from studies of fish ecotypes in mesocosms. In the study of Lundsgaard-Hansen et al. (2014), genetically based whitefish ecotypes in mesocosms had larger effects on the invertebrates they eat (presumably a direct effect) than on the invertebrates they do not eat (presumably an indirect effect). Upping the level of complexity, Bassar et al. (2012) found that algal biomass was 17% greater in mesocosms with high-predation guppies than in mesocosms with low-predation guppies (fig. 9.6). The authors used electricity to exclude guppies from certain areas of the mesocosms and could thus separate some direct and indirect effects. The positive direct effect of high-predation guppies on algal biomass (they eat less algae than do low-predation guppies) was 218% greater than the total negative indirect effect (fig. 9.6). These results were used to parameterize a mechanistic model that revealed two large but opposing indirect effects: high-predation guppies consume more invertebrates, which reduces invertebrate consumption of algae but also reduces the contribution of invertebrate nutrient excretion to algal growth. These indirect effects oppose each other to yield the only a minor net indirect effect, leaving the total net effect of guppies to be mostly explained by the direct effect noted above. Although these examples seem to suggest stronger direct than indirect effects, it is important to remember that they did not consider numerous other potential indirect effects, including population dynamics. In this way, they are akin to inferences about "direct" and "indirect" selection (chapter 2) where, in reality, "direct" simply means "independent of the particular other indirect effects explicitly examined." As a result, "direct" effects might—and likely do—include a number of indirect effects that are not considered in a given study.

Indirect effects can lead to additional eco-evolutionary cascades. Again, further examples come from the alewife system. We have already seen how the evolution of foraging traits in alewife leads not only to changes in zooplankton community structure (Palkovacs and Post 2009) but, through a trophic cascade, the biomass and composition of phytoplankton (Palkovacs and Dalton 2012, Weis and Post 2013). It turns out that the evolution of these alewife traits also leads to the *evolution* of life history traits in *Daphnia* (Walsh and Post 2011), which should then have additional effects on phytoplankton. One can only imagine that the phytoplankton must also be evolving in response. Similarly, guppy evolution influences not only a variety of ecological processes at different levels but also the *evolution* of intraguild predators (Walsh et al. 2011) and parasites (Perez-Jvostov et al. 2015), which presumably then has additional ecological effects. Overall, ecology and evolution are almost certainly interacting in a complex web of interactions where the evolution of one species causes evolutionary and ecological effects that then cause additional evolutionary and ecological effects.

Eco-evolutionary effects will play out through a variety of direct and indirect pathways, which is not surprising given that the same statement is true for non-evolutionary ecological effects. I have here emphasized indirect pathways that link different levels of ecological complexity (populations, communities, ecosystems); yet indirect links certainly also occur within each of these levels, with complex food webs being an obvious case in point (Estes et al. 2013, Barbour et al. 2016). The existence of these multiple pathways means that the net effect of a given evolutionary change might be much less than the individual effects acting through each pathway. This realization once again reinforces the point that eco-evolutionary *stability* might reflect extensive cryptic eco-evolutionary *dynamics* (Kinnison et al. 2015). As to the relative frequency and strength of direct versus indirect effects, and as to the types of indirect effects that are strongest, we currently have little insight. Indeed, intensive work on model systems finds a great diversity of effects despite examining only a few of the possibilities; the real world is certainly much more complicated. Thus, it is likely impractical to attempt a holistic comparison of the strength of direct versus all indirect effects. However, it does seem profitable to compare and contrast particular types of direct and indirect effects, such as those acting through population dynamics or particular types of communities (e.g., soil or gut microbes). As is so often the case, more data are needed.

QUESTION 5: DO GENETIC EFFECTS DECREASE WITH INCREASINGLY COMPLEX LEVELS OF ORGANIZATION?

It is often expected that the importance of a given causal driver will decline with an increasing number of intermediate steps needed to reach the response variable, as

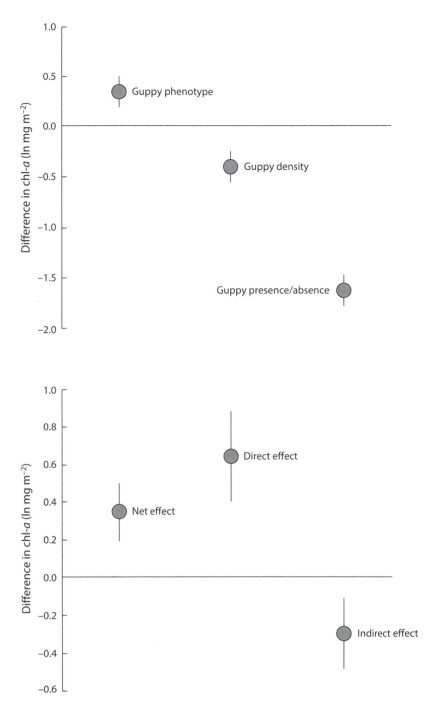

Fig 9.6. Ecosystem effects (chlorophyll-*a* standing crop: mg per m² of substrate) of guppies in mesocosms. The top panel shows the total net effects of guppy phenotype (high-predation versus low-predation), guppy density (two-fold change), and guppy presence/absence. The lower panel shows the net effect of guppy phenotype (as in the top panel) and how this net effect is separated into direct effects (consumption of algae) versus indirect effects (e.g., through consumption of invertebrates). The data (means and standard errors) are from Bassar et al. (2012) and were provided by R. Bassar

exemplified by results for trophic cascades (Micheli 1999). In the present context, we might expect the ecological effects of genotypes to decline as ecological variables become further removed from those genotypes (Bailey et al. 2009). For instance, the effects of genotypes might decline from phenotypes to population dynamics to community structure to ecosystem function. As noted in question 2, this prediction seems likely to hold from a crude qualitative comparison of results from the present chapter versus the previous chapter: in relation to various frames of reference, effects of intraspecific variation appear greater at community than ecosystem levels. At the same time, however, the answer isn't straightforward because ecosystem-level effects appear to be just as strong as community-level effects in at least some systems.

Bailey et al. (2009) performed two meta-analyses of the strength of associations among genes, plant phenotypes, community structure, and ecosystem function (population dynamics were not examined). The first analysis examined the relationship in community genetics common-garden studies between the proportion of *P. fremontii* genes that introgressed into *P. augustiflora* and plant phenotypes (e.g., phytochemistry, architecture, and physiology), community structure (arthropod species richness, total abundance, and community composition), and ecosystem function (energy flow and energy transformation). The second meta-analysis examined the relationship in seven plant systems between genotypic diversity in BEF experiments and each of the above three levels of organization. In both cases, the average effect of these plant "genetic factors" decreased from plant traits to community variables to ecosystem variables (fig. 9.7), thus supporting the original hypothesis. The amount of ecological variation explained by genes was relatively low in the first analysis ($r_{phenotype} = 0.42$, $r_{community} = 0.28$, and $r_{ecosystem} = 0.20$) and the second analysis ($r_{phenotype} = 0.30$, $r_{community} = 0.25$, and $r_{ecosystem} = 0.13$). These results are not surprising given that many factors extrinsic to the genetics of a single (albeit "foundation") species will influence variation in phenotypes, communities, and ecosystems. At the same time, the above values are averages, whereas the effects were sometimes very strong indeed—even at "higher" ecological levels (Bailey et al. 2009). Indeed, Crutsinger et al. (2014) later reported that *Populus* leaf litter genotypes placed in mesocosms have their greatest effects on individual plant traits and then ecosystem variables (mass loss and light extinction), followed by considerably weaker effects on community variables (various aquatic invertebrates).

Finally, we can merge our consideration of the relative strength of interspecific and intraspecific effects (question 5 in chapter 8 and question 2 in the present chapter) with our current consideration of whether genetic effects decline from communities to ecosystems. Specifically, we can ask: Does the relative importance of intraspecific versus interspecific effects change from community to ecosystem response variables? In this regard, the meta-analysis of Post et al. (unpubl. data) found that the magnitudes of intraspecific effects (genotype or ecotype replacement) and interspecific effects (species replacement/removal) were roughly similar at the ecosystem level for both plants and animals, whereas the interspecific effect was greater than the intraspecific effect at the community level for animals but not plants. Interestingly, the reason for the shift in animals was that interspecific effects weakened from communities to ecosystems, whereas intraspecific effects actually increased somewhat.

The influence of intraspecific variation on ecological dynamics sometimes decreases from less complex (e.g., phenotypes) to more complex (e.g., ecosystems) levels. However, such dampening is not always the case. Moreover, the meta-analyses performed thus far

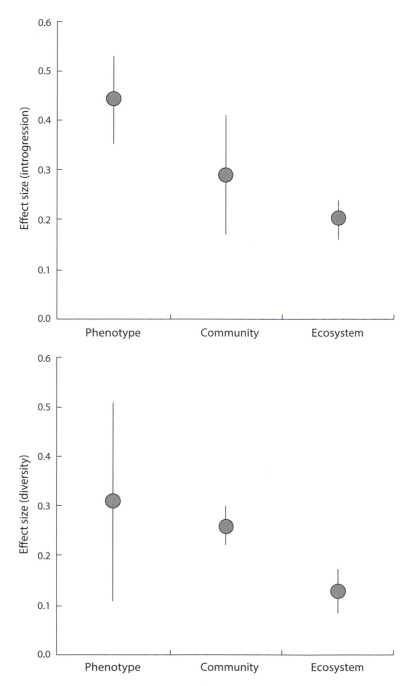

Fig 9.7. Effect sizes from a meta-analysis across plant studies relating genetic introgression (top panel) and genetic diversity (bottom panel) to phenotypic traits (e.g., phytochemistry, architectural, and physiological traits), community variables (e.g., abundance and richness of arthropods), and ecosystem variables (e.g., decomposition rate, productivity, carbon accumulation in the soil). Effect sizes in the top panel are the correlation (transformed *r*) between the proportion of *Populus fremontsii* genes introgressed into a genetic background of *Populus angustifolia* and ecological variables. Effect sizes in the bottom panel are the correlation (transformed *r*) between the number of genotypes in a plot and ecological variables. The data (means and 95% confidence intervals) are from Bailey et al. (2009) and were provided by J. Bailey

did not consider population dynamics, which might alter the situation by providing a powerful indirect route for genotypes/phenotypes to influence higher ecological levels. In short, the answer to this question is still open, and I suspect that extensive variation among systems and ecological variables will make generalities elusive.

QUESTION 6: HOW COMMON ARE FEEDBACKS?

Feedbacks in the broad sense (evolution of a trait causes an ecological response that influences the evolution of other traits) have been much discussed in the context of niche construction. For instance, Odling-Smee et al. (2003, see also Matthews et al. 2014) provide a table of 40 putative instances where changes to the environment imposed by a particular organism appear to have selected for additional adaptations by that organism. A number of these instances might also qualify as feedbacks in the narrow sense, particularly when considering indirect effects. For instance, the traits of beavers (chiseling teeth, broad tails, and webbed feet) dramatically influence the environment, which then presumably directly or indirectly influences selection on those same traits. However, such narrow-sense eco-evolutionary feedbacks are rarely documented and will be my main focus here.

Feedbacks—even in the narrow sense—must be very common in nature. The beaver example makes this clear: the evolution of traits in beavers that influence dam building then influence the environment in ways that must influence further evolution of the same traits. Similarly obvious logic applies to other niche constructing organisms, such as birds that build nests, animals that dig burrows, and so on (Odling-Smee et al. 2003). In these cases, however, feedbacks are inferred rather than demonstrated and the time scale over which they played out is not certain. Direct demonstrations of contemporary feedbacks have been provided in laboratory studies, such as in chemostats, where rotifer predators cause the evolution of alga prey traits (cell clumping) that influenced predator population growth that then influenced further prey evolution (Becks et al. 2012). To what extent can such feedbacks be seen on contemporary time scales in nature?

The best evidence for eco-evolutionary feedbacks in nature occur at the level of population dynamics, particularly in relation to population density and life history traits. Examples were provided in chapter 7, including that of side-blotched lizards, where changes in population density cause evolution along the egg size/number tradeoff, which then influences density and causes further evolution along that tradeoff (Sinervo et al. 2000). In that case, the feedback was negative such that the effect of increasing (or decreasing) density was to impose selection for traits that then reduced (or increased) density. In the present chapter, however, our focus is on whether eco-evolutionary feedbacks are also evident at higher levels, such as in communities and ecosystems.[2]

Post and Palkovacs (2009) provided several putative examples of eco-evolutionary feedbacks at the community level. One example was based on work by Grant and Grant (2002, 2006), where Darwin's finches on Daphne Major deplete seeds during droughts (when most plants don't reproduce) in ways that change the relative availability of seeds of different size. This change in the seed distribution then selects for birds with beak sizes best suited to exploit the remaining seeds. This selection causes the evolution of beak size,

---

[2]Although this chapter emphasizes ecosystem variables, I have here merged them with community variables as the evidence is too sparse to warrant sections in both chapters.

which should then (in principle) change the distribution of seed types in the soil. So far, however, no study has formally tested whether different distributions of finch beaks have differential influences that can be seen in plant communities. Another putative example comes from the work on alewife (Post et al. 2008, Palkovacs and Post 2009, Palkovacs et al. 2014): landlocked alewife cause zooplankton communities to shift to smaller sizes, which then selects on the alewife to have longer and more closely spaced gill rakers, which then influences zooplankton size structure, which then further selects on gill rakers.

Several other potential examples of eco-evolutionary feedbacks can be advanced. Continuing with guppies, Bassar et al. (2012) established that changes in the phenotype of guppies toward less herbivory (high-predation fish eat less algae than do low-predation fish) would influence algal biomass in ways that cause selection for further reductions in herbivory. So far, however, these were predictions of a mechanistic model parameterized for the system rather than direct measures of selection on phenotypes. More directly, lake ecotypes of adult threespine stickleback structure mesocosm environments (relative to stream ecotypes) so as to reduce the growth of lake juveniles (relative to stream juveniles) (Matthews et al. 2016). In the context of plant-soil feedbacks (van der Putten et al. 2013, Schweitzer et al. 2014), several studies have demonstrated that particular genotypes of plants lead to soil microbial communities that are best suited to decompose the litter of that particular genotype (Madritch and Lindroth 2011) and that, in at least some cases, this association can increase the success of seedlings of the same genotype as the parent (Pregitzer et al. 2010). It isn't certain, however, if the same plant traits/genes are both driving and being driven by the ecological effect: that is, a *narrow-sense* feedback. Getting closer to this goal, the evolution of allelopathic secondary compounds in plants can alter the surrounding community of competitors, which can then feedback to influence selection on those same compounds. Demonstrating these effects in *Brassica nigra* with different sinigrin levels and several competing plants, Lankau and Strauss (2007) argued that genetic diversity with species and species diversity would simultaneously enhance each other. As a final example, the fermentation of fruit caused by *S. cerevisiae* yeast produces "a hot, anaerobic, alcoholic environment that is toxic to the multifarious competing microbes" (Buser et al. 2014). The resulting alteration of the microbial community then reduces competition and thus feeds back to increase the fitness of *S. cerevisiae* (Goddard 2008).

Eco-evolutionary feedbacks at the community and ecosystem levels must be common, yet relatively few studies have formally tested for their action in the context of contemporary evolution. At present, then, the strength of feedbacks, the extent to which they involve direct versus indirect effects of phenotypes, and the commonality of positive versus negative feedbacks is unknown.

QUESTION 7: HOW DOES SELECTION WITHIN A GENERATION INFLUENCE ECOLOGICAL PROCESSES?

Eco-evolutionary dynamics resulting from contemporary evolution are mainly driven by natural selection. If this selection is strong enough in any particular generation, it could influence ecological dynamics within that generation, instead of only after that selection is subsequently translated into evolution. Stated another way, the phenotypic shift that results from selection within a generation can have ecological consequences within that generation. Moreover, these selective effects could be stronger than per-generation evolutionary effects because only part of the phenotypic shift that results from selection in a given generation will be translated into an evolutionary response by the next generation (owing to imperfect trait heritability—chapter 3). In one of the few studies to consider such effects in any context, Johnson et al. (2014) showed that much of the mortality in coral reef fishes that influences population dynamics (specifically recruitment) was selective mortality. Could similar effects be seen at the ecosystem level?

Carlson et al. (2011) used sockeye salmon returning from the ocean to breed in a freshwater stream (Hansen Creek) to consider how selection within a generation (henceforth just "selection") might influence ecological processes. They analyzed two episodes of selection: "stranding" (falling over) when the fish attempt to pass from the lake into the creek (the creek mouth is very shallow), and bear predation that removes fish from the creek and deposits them in the riparian zone (fig. 9.8). Both episodes induced substantial mortality (annual average = 16% for stranding and 42.6% for bear predation) that is size-selective against larger fish (Carlson and Quinn 2007). Carlson et al. (2011) describe how this selection can have ecological effects at the population, community, and ecosystem levels, and I here concentrate on the last by considering changes in biomass (and thus nutrient) flux.

Pacific salmon are a critical source of nutrients in temperate freshwater ecosystems and their surrounding riparian zones. The reason is that approximately 99% of a salmon's body mass is gained at sea and these nutrients are released in fresh water because the salmon die within weeks of breeding. These marine-derived nutrients supplement every level of the food web and thereby dramatically influence the productivity of aquatic and riparian communities (Gende et al. 2002). (Also, the presence or absence of salmon carcasses has been shown to interact with riparian tree genotypes to shape several community and ecosystem parameters in streams: LeRoy et al. 2015.) To consider how selection within a generation would influence this nutrient subsidy provided by salmon, Carlson et al. (2011) first calculated that total mortality due to stranding reduced nutrient flux into the creek by an average annual amount of 2913 kg, which represents 15% of the total flux of salmon that return to the creek. Of this total flux loss, an annual average of 6% could be attributed to natural selection (larger fish are more likely to "strand" in the shallow water), with the rest being determined by mortality independent of fish size. The authors next calculated that bear predation increased the flux of salmon biomass into the riparian zone by an annual average of 3300 kg. Of this flux gain to the riparian zone, an average of 5% could be attributed to size selection by the bears (bears kill larger fish), with the rest being determined by predation independent of fish size. These effects are summarized in figure 9.8.

Although the above effects of selection (the shift in average size owing to stranding or predation) might seem small here relative to the effects of "ecology" (numbers of fish owing to stranding or predation), the total effects of selection are certainly underestimated. First, the authors ignored other important selective episodes, such as size-selective mortality in the fishery (Kendall et al. 2009). Second, the authors considered only one trait under selection (body size), when multiple traits (e.g., timing and body shape) are likely under selection (Quinn et al. 2007). Moreover, the seemingly small selective effects have driven adaptive body size differences among populations. In particular, Hansen Creek fish are much smaller on average than are fish in creeks subject to lower levels of stranding and predation (Carlson et al. 2009). This adaptive divergence among populations surely has ecological effects. For instance, Carlson et al. (2011) calculated that flux into the creek would change by 30–32% if the fish in Hansen Creek were as large as those in an adjacent

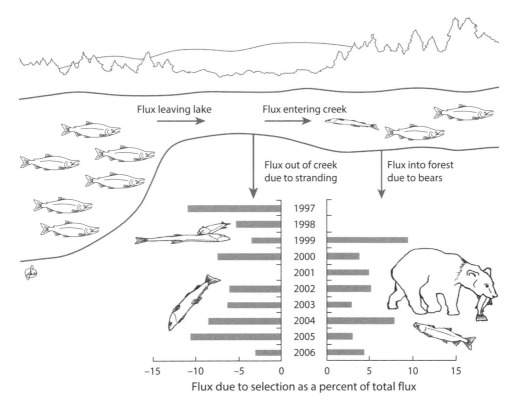

Fig 9.8. As salmon leave the lake to spawn, they transport biomass (flux) out of the lake. Some of that biomass is lost due to "stranding" in the very shallow water at the creek mouth. The rest of the biomass enters the creek proper. Some of that biomass is then removed from the creek and transported into the riparian zone forest due to predation by bears. The graphs show the proportion of these two fluxes (due to stranding and bears) that is the result of the *selective* stranding and the *selective* predation of larger fish. The data and images are from Carlson et al. (2011) and were provided by S. Carlson

creek not subject to stranding at the river mouth. Thus, local adaptation has had a very large effect on the average nutrient flux provided by an individual salmon. In addition, the above calculations ignore the fact that more fish would strand at Hansen Creek if they had not evolved smaller body size, and so the numerical effect of "nonselective" mortality attributed above to "ecology" probably ignores an evolutionary effect of average trait size on overall mortality rates.

Selection within a single generation can have a noticeable effect on ecosystem variables. This selection is expected to cause adaptive evolution, which should lead to even greater ecosystem-level effects, with the latter being emphasized elsewhere in this chapter.

Conclusions, significance, and implications

Many studies have demonstrated that genetic diversity/identity/variation matters for ecosystem function, and that such influences can be strong in relation to some benchmarks or standards of comparison, such as species presence/absence and density, yet weak in relation to others, such as spatial variation and nutrient addition. Most of the demonstrated effects are direct in the sense that they do not depend on changes in population density; yet indirect effects acting through population density are often important when they are explicitly considered. Intraspecific effects sometimes seem weaker on ecosystems than on communities, perhaps because phenotypic effects dampen along the cascade; yet counter examples exist and so generality remains uncertain. Eco-evolutionary feedbacks are likely common; yet narrow-sense feedbacks have only rarely been subject to serious investigation. Finally, selection within a generation (even without evolution across generations) can strongly influence ecosystem variables in at least some systems. With this summary of answers to the key questions addressed in the chapter, it now seems appropriate to return to the predictions made at the outset of this and the previous chapter as to when evo-to-eco effects should be most common and strong.

Evolutionary (or, more generally, phenotypic) change is most likely to influence community structure and ecosystem function on contemporary time scales when . . .

1. . . . *phenotypic change is greatest.* In spite of (or perhaps because of) the seeming obviousness of this prediction, it seems that no study has formally tested whether greater phenotypic changes lead to greater community/ecosystem effects. However, it is certainly true that many such effects can be attributed to particular phenotypic changes, such as gill rakers in alewife and whitefish influencing zooplankton communities (Palkovacs and Post 2009, Lundsgaard-Hasen et al. 2014), sinigrin levels in *Brassica* influencing plant communities (Lankau and Strauss 2007), and tannin levels in cottonwoods influencing decomposition (Whitham et al. 2006) and water clarity (Crutsinger et al. 2014b). Thus, although the prediction might not yet have been formally investigated, it seems very likely to hold.

2. . . . *phenotypic trait variation has strong direct effects on communities/ecosystems.* All of the effects mentioned immediately above, and many others described earlier, reflect direct effects (i.e., independent of changes in population density) because the experiments examined the effects of individuals or groups of constant density. And, in nearly every case, the trait was already known to have strong ecological effects. Thus, this prediction seems likely to hold true; yet its importance in nature relative to indirect effects awaits further analysis.

3. . . . *the focal organisms are very abundant.* The point here is that—orthogonal to the above prediction—individual organisms (and therefore variation in their traits) might not have large effects but, together with all other individuals in the focal population/species, they might have large effects in aggregate. One putative example is the evo-to-eco effects of introduced rabbits and myxoma that propagated through the ecosystem in Australia. In addition, the direct effects documented for fish (e.g., guppies, alewife, stickleback, whitefish) in mesocosms are expected to be important in nature only because the fish can be very abundant (El-Sabaawi et al. 2015b). By contrast, other species can have large effects as individuals, such as cottonwoods or beavers.

4. . . . *evolution has indirect effects moving through population dynamics to communities or ecosystems.* The finding that genotypes/ecotypes can have direct ecological effects at least as large as variation in density (e.g., Bassar et al. 2010, 2012) might suggest that indirect phenotype-to-population-to-community and

phenotype-to-population-ecosystem effects might not be strong. However, the density manipulations thus far employed might be less dramatic than the effects evolution can accomplish, with evolutionary rescue (and thereby community rescue) being a particularly dramatic manifestation (Kovach-Orr and Fussmann 2013). Moreover, some studies of plants have found that ecological effects are mostly due to indirect effects of changes in density (Hughes and Stachowicz 2004, Crutsinger et al. 2008b). In addition, a phenotype-to-community-to-ecosystem indirect effect has been shown for cottonwood leaf genotypes in mesocosms (Crutsinger et al. 2014b). Thus, indirect effects seem important in at least some systems and more examples will surely emerge from targeted studies.

5. . . . *direct and indirect effects do not offset each other.* I am not aware of any studies formally testing whether direct and indirect (through population dynamics) effects complement or offset each other in shaping community/ecosystem variables. However, studies examining other sorts of direct and indirect effects (i.e., not related to population dynamics) do find at least some opposing effects, such as for guppy ecotypes (e.g., Bassar et al. 2012) and pale chub (*Zacco platypus*) behavior types (Katano 2011) influencing periphyton. In addition, Palkovacs et al. (2015) argued that the evo-to-eco effects of consumers will be greatest when the effects of consumption act in the same direction as the effects of excretion.

6. . . . *feedbacks are present—and positive.* Feedbacks acting through community structure and ecosystem function have been demonstrated in the laboratory (Becks et al. 2012) and feedbacks acting through population dynamics have been demonstrated in nature (chapter 7). Yet the former feedbacks have not been demonstrated in the latter context (Post and Palkovacs 2009). Such feedbacks are easily postulated and seem likely to be common, such as in the interaction between finches and their seeds and between plants and their soils. Interestingly, most effects thus far postulated are positive, including interactions between alewife and zooplankton (Palkovacs and Post 2009, Palkovacs et al. 2014), plants and soils (Pregitzer et al. 2010), and genetic diversity and species diversity (Lankau and Strauss 2007, Vellend et al. 2014).

7. . . . *external drivers are not overpowering.* Most of the existing evo-to-eco studies have employed controlled environments, such as chemostats or common gardens or mesocosms. Some of these arenas allow for some external drivers (rainfall, temperature, predators), but to an extent that is always less than in nature. Thus, we should worry that external drivers in nature could be overwhelming, as suggested by studies showing that spatial effects generally outweigh genotype effects in community genetics studies of plants (Tack et al. 2012). This topic remains one of the most important and pressing ones (that is, Do eco-evolutionary dynamics *really matter* on contemporary time scales?) and it demands more studies of natural populations in nature.

8. . . . *(a) individual phenotypic traits have influences on multiple ecological variables, or (b) multiple phenotypic traits influence the same ecological variable.* Many studies demonstrate that particular phenotypic traits have effects that echo across multiple ecological variables, such as tannin concentrations in cottonwood (Whitham et al. 2006, Crutsinger et al. 2014b) and the morphology of coyote bush (Crutsinger et al. 2013, 2014a). In addition, different fish ecotypes often

have a wide variety of community/ecosystem effects, as demonstrated for guppies (Palkovacs et al. 2009 Bassar et al. 2010, 2012), stickleback (Harmon et al. 2009, Des Roches et al. 2013, Matthews et al. 2016, Rudman et al. 2016), alewife (Palkovacs and Post 2009), and whitefish (Lundsgaard-Hansen 2014). Yet all such studies have difficulty partitioning the various effects of each trait to each ecological variable, and so the general question remains unresolved.

9. . . . *it leads to changes in biological stoichiometry for limiting nutrients.* To my knowledge, no study has formally demonstrated such effects in nature, yet they certainly make logical sense (Elser et al. 2000, Elser 2006, Matthews et al. 2011b, Crutsinger et al. 2014b, Jeyasingh et al. 2014).

Considering all of the above points, the main message emerging from this part of the book is that eco-evolutionary dynamics are likely important for communities and ecosystems in nature. Yet our knowledge, even with respect to several key questions and critical predictions, is incredibly fragmentary. This realization is simultaneously depressing (we know so little) and exciting (we have so much to learn).

Chapter 10

Genetics and Genomics

Although I have argued that plasticity needs to be an integral part of any eco-evolutionary story (see also chapter 11), the *evolutionary* part of the term implies genetic change. That is, we are often interested specifically in how *genetic* changes in phenotypes are driven by ecological forces and how those genetic changes then shape ecological forces. The specifics of these genetic changes are expected to strongly influence how evolution proceeds in any given instance and how the resulting ecological effects play out. In theoretical studies, for example, eco-evolutionary dynamics are strongly influenced by the details of how genetic variation influences adaptive traits (e.g., Orr and Unckless 2008, Johnson et al. 2009, Botero et al. 2015). As an empirical example, the dynamics of predators (rotifers) and prey (algae) in chemostats can vary dramatically depending on the specific form of the genetic trade-off between defense and growth rate in prey (Kasada et al. 2014). As another example, the strength of genetic correlations in adaptive responses to two different enemies will shape the evolutionary dynamics of communities (O'Reilly-Wapstra et al. 2014). The present chapter therefore provides some of the genetic details that I have thus far ignored.

As repeatedly emphasized, it is phenotypes—not genotypes—that have ecological effects, and so what we are primarily interested in here is the genetics and genomics of phenotypic change. Stated another way, the genetics and genomics of eco-evolutionary dynamics are largely the genetics and genomics of phenotypic adaptation, adaptive divergence, and ecological speciation.[1] Existing literature on these topics is vast and has been reviewed in many previous articles (e.g., Orr 2005, Hoekstra and Coyne 2007, Stern and Orgogozo 2008, Barrett and Schluter 2008, Wolf et al. 2010, Barrett and Hendry 2012, Feder et al. 2012, Rockman 2012, Olson-Manning et al. 2012). I haven't the space to cover all of the ideas in a comprehensive fashion and have therefore chosen to focus on nine key questions that are particularly relevant to eco-evolutionary dynamics and that are often debated. In each case, I muster empirical data in a attempt to pass tentative judgment on some of those controversies. A version of this chapter was previously published as Hendry (2013).

---

[1] An exception occurs when asking about the "heritability" of the ecological effects of a given organism (Fritz and Price 1988, Johnson and Agrawal 2005, Shuster et al. 2006, Keith et al. 2010) or when asking how variation in a particular gene might have ecological effects (e.g., Zheng et al. 2009).

Approaches

I will start with a brief overview of some of the common methods for dissecting the genetic basis for adaptation, adaptive divergence, and ecological speciation. My goal here is not to be comprehensive but rather to provide a simple summary before getting into the juicier debates that these methods can inform. Analyses of the genetics and genomics of phenotypic adaptation can be conducted at two levels: variation within populations or divergence among populations/species. Methods at each of these levels are diverse and ever expanding. In particular, rapidly advancing sequencing technologies mean that many of the current methods will change dramatically in the near future, and new methods will become available (Gilad et al. 2009, Stapley et al. 2010). Owing to this fluidity, I will not here go into any detail on specific methods. I will instead merely introduce the main categories so that the reader can understand the basic underpinnings of the data presented later. Readers hoping for details on the specific methods can turn to the cited articles and books.

1. *Quantitative genetics* involves the statistical partitioning of phenotypic variance into different components, such as additive genetic variance, dominance variance, epistatic variance, maternal effects variance, and environmental variance (Falconer and Mackay 1996, Roff 1997, Lynch and Walsh 1998). Of these components, additive genetic variance is generally considered the most relevant to evolution because it determines the resemblance between parents and offspring. (The ratio of the additive genetic variance to the total variance is the narrow-sense heritability.) For investigations within populations, additive genetic variance is generally estimated through parent-offspring trait correlations, relationships among siblings, or pedigree-based analyses in natural or artificial populations (Falconer and Mackay 1996, Roff 1997, Lynch and Walsh 1998, Wilson et al. 2010). For investigations among populations, the starting point is often to raise individuals from the different populations in a common-garden environment to remove environmental (plastic) effects, and these experiments sometimes continue through multiple generations to also remove maternal effects. A common subsequent step is to perform "line-cross" analyses based on the comparison of pure crosses, F1 and F2 hybrids, and backcrosses to parental forms (Lynch and Walsh 1998, Roff and Emerson 2006).

2. *Linkage mapping* involves the detection of quantitative trait loci (QTL): regions of the genome that contribute to differences in phenotypic traits. The standard approach is to generate F1 hybrids (among genotypes, lines, populations, or species) and to then intercross those F1 hybrids to generate an F2 population. These F2s are then measured for phenotypic traits of interest and genotyped at genetic markers (e.g., microsatellites or single nucleotide polymorphisms [SNPs]) spread as evenly and densely as possible across the genome. Statistical analyses are then performed to test for associations between particular alleles at the marker loci and trait values for the phenotypes of interest (Lynch and Walsh 1998). Crosses beyond the F1 hybrids are important for this endeavor because recombination will have occurred between the two parental genomes, thus reducing associations between genetic markers and QTL to which they are not linked. The result of such an analysis is a linkage map showing the positions of marker loci on

different linkage groups (often chromosomes), along with information on the markers that are associated with loci influencing phenotypic trait values. Evaluation of the strength of these influences can be made by calculating the percentage of variation in the trait explained by each locus.

3. *Association mapping* can localize QTL in a way qualitatively similar to linkage mapping as described above. Association mapping proceeds by screening a large number of individuals from natural populations and relating among-individual variation in alleles at marker loci to phenotypic trait values (Slate 2005). The idea is that recombination in natural populations will allow persistent statistical associations only between phenotypic traits and marker loci that are close together on a chromosome. The known location of the marker on a linkage map can then be used to infer the locations of QTL. Association mapping can be performed in large outbred populations (Kruglyak 2008, Flint and Mackay 2009) or in situations of natural interbreeding among divergent populations, such as in hybrid zones (Buerkle and Lexer 2008). The former approach generally reveals QTL influencing trait variation within populations and the later approach, which is often called admixture mapping, reveals QTL influencing trait divergence among populations.

4. *Genome scans* involve genotyping large numbers of genetic markers, now usually SNPs, in multiple individuals from multiple populations or species and then searching statistically for markers that show the smallest or largest differences among those groups (Stinchcombe and Hoekstra 2008, Nosil et al. 2009a). These markers are expected to be located in, or close to, QTL that are under similar (stabilizing) or different (divergent) selection among those populations. The goal is often to infer the action of selection, but genetic markers can also differ among populations owing to drift. For this reason, the action of selection is usually invoked only for "outlier loci": that is, loci where divergence deviates substantially from the distribution based on most of the other loci (Stinchcombe and Hoekstra 2008, Nosil et al. 2009a); or for loci that show the strongest correlations with environmental variables (Hancock et al. 2011, Fournier-Level et al. 2011). The best way to infer these outliers is still in active discussion (Cruickshank and Hahn 2014, Whitlock and Lotterhos 2015).

5. *Gene expression* Phenotypic differences among populations can depend not only on divergence at protein-coding loci but also in regulatory sequences that influence gene expression. Genes influencing divergence thus can be inferred by examining variation in gene expression among populations in a given environment (genetic differences in expression) or among environments for a given population (environmental influences on expression). A variety of methods exist for measuring gene expression, including quantitative PCR, microarrays, and next-generation sequencing methods like RNA-Seq (Gilad et al. 2009, Pavey et al. 2010). Determinations are usually made as to which genes are up-regulated or down-regulated in one environment or population in relation to another environment or population. These genes are likely to play a role in phenotypic differences between populations and/or environments, whether the variation is present in those specific genes or in upstream regulatory genes. Of course, it is important to note that gene expression differences among natural populations can reflect

either sequence differences among those populations or environmental (plastic) responses to local environments (e.g., Stutz et al. 2015).

6. *Candidate genes* are often genes of known function in model organisms, and these same genes are sometimes targeted for study in nonmodel organisms (Stinchcombe and Hoekstra 2008). In particular, targeted sequencing of a gene in individuals with different phenotypes, whether within or among populations, can implicate that gene in the generation of those differences. The specific role of that gene can then be confirmed through a variety of methods, including gene knockout studies or transgenics (e.g., Colosimo et al. 2005, Abzhanov et al. 2006, Chan et al. 2010, Manceau et al. 2011).

Each of the above methods has strengths and weaknesses that have been extensively discussed (Lynch and Walsh 1998, Stinchcombe and Hoekstra 2008, Buerkle and Lexer 2008, Mackay et al. 2009, Stapley et al. 2010, Rockman 2012). I will not here detail these issues except when they bear directly on the key questions described below. In general, however, it is important to recognize that the best genetic studies incorporate multiple methods. For example, the adaptive divergence of dwarf and normal ecotypes of whitefish has been examined through quantitative genetics, linkage mapping, genome scans, gene expression analyses, and a variety of other methods (review: Bernatchez et al. 2010).

Genetics and genomics in nature

My goal in the remainder of this chapter is to address several key questions surrounding the genetics and genomics of adaptation. Because this book is about eco-evolutionary dynamics, which work through phenotypes, I will focus on the genetics of phenotypic and fitness variation through time, among individuals, and among populations. For this reason, I will be much less concerned with the role of specific genes than I will be with more basic properties of genetic architecture, such as (Q1) how much genetic variation is out there?, (Q2) to what extent is important genetic variation additive versus nonadditive, (Q3 and Q4) how many genes of what effect size are involved in adaptation, (Q5) how predictable (parallel/convergent) is the genomic basis of adaptation?, (Q6) is adaptation mainly driven by standing variation or new mutations?, (Q7) what do "adaptive walks" to new fitness peaks look like?, (Q8) to what extent is adaptation driven by regulatory versus structural genetic changes?, and (Q9) how heritable are the ecological effects of organisms (considered as "extended phenotypes")? Whenever possible, I will refer to meta-analyses but such analyses are available for only some of the questions and so I will often simply provide key examples and offer my general impressions, which can be thought of as predictions. When it comes to specific examples, I tend to rely mostly on vertebrates, with threespine stickleback having a particularly prominent position. This taxonomic focus is mainly because it is an especially well studied natural system, and also because I am reasonably familiar with some its nuances.

QUESTION 1: HOW MUCH GENETIC VARIATION IS OUT THERE?
The evolution of phenotypic traits will be heavily influenced by the amount of additive genetic variation—so how much of it is out there? The classic review is the survey by Mousseau and Roff (1987) of narrow-sense heritabilities (ratio of additive genetic

variance to total phenotypic variance) in wild, outbred animal populations. Based on 1120 estimates from the literature, the authors calculated mean heritabilities of 0.46 for morphological traits, 0.26 for life history traits, and 0.30 for behavioral traits. Figure 10.1 shows the results of a more recent survey by Hansen et al. (2011). Compilations of this sort universally show that most traits in most populations of most species show substantial evolutionary potential. For example, multiplying the median absolute value of bias-corrected selection gradients for morphology (0.15: Hereford et al. 2004) by the median heritability (0.43) yields an expected evolutionary response of 0.065 standard deviations per generation. If sustained, this response would shift the mean trait value by one standard deviation in only 16 generations. Also, heritabilities for a given trait in a given population vary depending on conditions (GxE) (Charmantier and Garant 2005) and are sometimes higher when selection is stronger (Husby et al. 2011), which should speed evolution.

Some authors have argued that heritabilities are not the best measure of evolutionary potential, and have instead argued for the use of "evolvability"—additive genetic variance

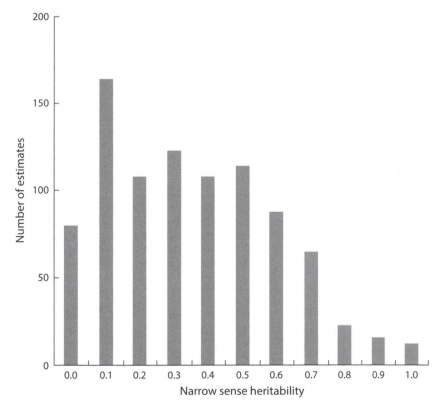

Fig 10.1. Frequency distribution of 901 narrow sense heritability estimates compiled by Hansen et al. (2011). Estimates less than zero are included in the zero column and estimates greater than one are included in the one column. Although other compilations have presented more heritability estimates, the set shown here is based on the same estimates for which "evolvabilities" are reported in figure 10.2. Regardless, the distribution looks roughly the same in all compilations of heritabilities

divided by the square of the mean trait value (Houle 1992, Hansen et al. 2011). Multiplying an estimate of evolvability by a mean-standardized selection gradient (Hereford et al. 2004, Matsumura et al. 2012) then gives the expected proportional change in the mean trait value per generation. Hansen et al. (2011) reviewed 1,465 estimates of evolvability and found that the median value was 0.26; a subset of these estimates is shown in figure 10.2 (although it seems that many authors have miscalculated evolvability—Garcia-Gonzalez et al. 2012). Multiplying this estimate by the median bias-corrected mean-standardized selection gradient in natural populations (0.28: Hereford et al. 2004) yields an expected 0.073% per-generation change in mean trait value. If sustained, this response would shift the mean trait value by 5% in 68 generations.

The main message here is simply that evolutionary potential—whatever the metric— is high for most traits in most populations, but the qualifier "most" is critical. For instance, Hoffmann et al. (2003) reported that desiccation resistance, an important fitness-related trait influencing adaptation and species distributions in insects, showed zero heritability and zero additive genetic variance in a rainforest population of *D. birchii*. Kellermann et al. (2006) then showed that this finding generalized to other populations of *D. birchii*—although

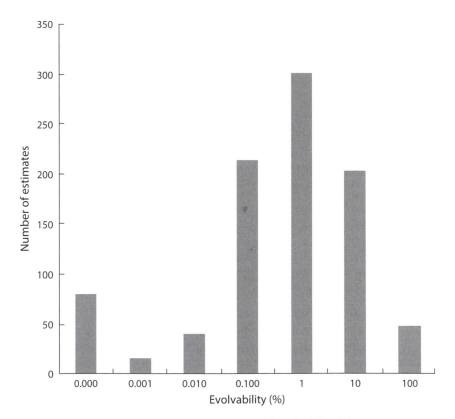

Fig 10.2. Frequency distribution of 901 estimates of "evolvability" (mean-scaled additive genetic variance) compiled by Hansen et al. (2011). Estimates less than zero are included in the zero column and estimates greater than 100 are included in the 100 column. I here show only evolvability estimates that correspond to the same studies/traits as those for which narrow sense heritabilities were reported in figure 10.1

other traits in the same populations were not so limited. Finally, Kellermann et al. (2009) showed that several other *Drosophila* species also lacked heritability and additive genetic variance for desiccation resistance. In particular, specialist rainforest species have lost most of the genetic variation in desiccation resistance. Although these are perhaps the best-known instances of low heritable variation in natural populations, other studies have reported analogous situations (Bradshaw and McNeilly 1991, Futuyma et al. 1995). But even in such cases, the relevance isn't clear because, for example, van Heerwaarden and Sgró (2014) later found that the *Drosophila* studied by Kellermann et al. (2009) actually had lots of genetic variation for adapting to more realistic levels of desiccation stress.

Interestingly, fitness itself seems often to have low additive genetic variance and/or heritability (Kruuk et al. 2000, Merilä and Sheldon 2000, Ellegren and Sheldon 2008, Teplitsky et al. 2009), suggesting that although traits usually have high evolutionary potential (as above), fitness itself may not. This difference is not surprising in some respects because (1) traits will not be perfectly correlated with fitness and so a given proportional increase in a trait should lead to a small proportional increase in fitness (Hereford et al. 2004), and (2) selection is expected to erode genetic variance in traits closely related to fitness (Fisher 1930). However, the apparently low evolvability of fitness does seem at odds with suggestions that selection needs to improve fitness by 1–10% per generation if it is to counteract the negative consequences of gene flow, mutation, and drift (Burt 1995). Whether or not this is a real discrepancy or simply the result of biased (and certainly imprecise) estimates remains to be seen. In reality, although traits more closely related with fitness do generally have lower heritability, they often do not have lower additive genetic variance (Price and Schluter 1991, Houle 1992, Merilä and Sheldon 1999, Teplitsky et al. 2009).

Most populations of most species harbor substantial additive genetic variance in fitness-related traits, which therefore should be able to evolve when exposed to altered selection pressures. However, the amount of this variation differs widely among populations and species, meaning that **the rate of evolutionary response to a given selective pressure also will be highly variable.** Of course, this question has taken a single-trait approach, whereas evolutionary potential can also be influenced by trait correlations, a topic considered in detail in chapters 3 and 4.

QUESTION 2: WHAT ABOUT NONADDITIVE EFFECTS?

I have thus far focused on additive genetic variation because it allows a relatively straightforward interpretation for how selection should influence phenotypic evolution. However, this focus begs the question: to what extent is evolution driven by additive genetic effects, as opposed to dominance or epistasis (Roff and Emerson 2006, Hill et al. 2008, Phillips 2008, Mäki-Tanila and Hill 2014)? This question is important because nonadditive effects can substantially alter evolutionary trajectories, as well as the magnitude and effects of gene flow (Wolf et al. 2000, Wade 2002).

Work on soapberry bugs (*Jadera heamatoloma*) adapting to different host plants provides a concrete example of nonadditive effects and how quickly they can contribute to adaptation. Specifically, the genetic basis of trait differences between two recently (fewer than 100 generations) diverged host races was examined by performing line cross analyses that compared mean phenotypes of parental types, F1 and F2 hybrids, and backcrosses (Carroll et al. 2001, 2003, Carroll 2007). Results differed among traits (fig. 10.3), ranging from almost perfect additivity for host plant preference to a diverse range of dominance and epistatic effects for other traits—and these effects depended on the rearing environment (host plant type). Overall, dominance appears generally to be important among host races of phytophagous insects, particularly with respect to performance on the different hosts (Matsubayashi et al. 2010). Mixtures of additive and nonadditive effects on adaptive divergence also have been described for many other groups discussed earlier, such as lake versus stream threespine stickleback (Berner et al. 2011, Moser et al. 2016) and dwarf versus normal lake whitefish (*Coregonus clupeaformis*) (Renaut et al. 2009, Bernatchez et al. 2010). Beyond these examples, the potential

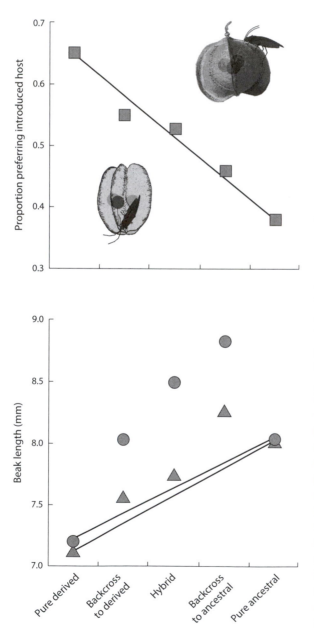

Fig 10.3. Results of line cross analyses between the "ancestral" host race of soapberry bugs adapted to native plants and the "derived" host race of soapberry bugs that recently adapted to an introduced plant. The top panel shows how the proportion of hatchling bugs preferring the introduced host plant has nearly perfect additive variation between the two host races (the additive expectation is shown by the line). The bottom panel has how beak length divergence between the host races includes substantial dominance and epistatic components (deviations from the additive expectation) and that these contributions differ depending on which host plant the experimental bugs are raised on: native plants in triangles and introduced plants in circles. The data are from Carroll et al. (2007) and were provided by S. Carroll. Images (ancestral bug on balloon vine, derived bug on *Koeleuteria elegans*) by T. Fowles of soapberrybug.org

generality of nonadditive effects was considered by Roff and Emerson (2006) in their meta-analysis of line cross analyses. They found that dominance made a significant contribution to population differentiation in nearly all cases: 96.5% of life history traits and 97.4% of morphological traits—and the effects were large: the ratio of dominance to additive effects was 1.57 for life history traits and 1.28 for morphological traits. Epistasis was also common, contributing to 79.4% of life history traits and 67.1% of morphological traits. Seemingly in contrast to the above line cross analyses between distinct groups (inbred lines, populations, species), genetic variation within groups has been argued to be predominantly additive (Hill et al. 2008, Mäki-Tanila and Hill 2014). It remains to be seen whether or not this is a real difference and, if so, what the reason might be.

It seems that evolutionary adaptation, including that on short (contemporary) time frames, often involves nonadditive genetic changes. Theory and empirical analysis needs to more carefully consider this phenomenon.

QUESTION 3: MANY SMALL OR FEW LARGE?

This question is a classic one, all the way back to debates between "biometricians" and "Mendelians" (review: Provine 1971). Those two schools of thought were successfully merged during the modern synthesis when it became clear that even polygenic traits were based on Mendelian genes, but the question remained open as to just how many genes of what effect size were important in shaping phenotypes. Classic analyses continued to reveal an apparent dichotomy. On the one hand, many continuous traits, such as body size, clearly involve many genes of small effect. On the other hand, many discrete traits, going all the way back to Mendel's peas, clearly have a mostly single-gene basis. I here ask how these alternatives have fared in the modern era of genomics. At first glance, evidence might seem to be growing for the importance of large-effect genes, given a number of high-profile examples, but I will use three points to argue that most adaptation is the result of many genes of small to modest effect.

First, current genomic methods are strongly biased against genes of small effect. This bias is particularly obvious in candidate gene approaches, which deliberately target just the opposite. It is also true of linkage and association mapping, where estimation problems arise when relevant alleles are found at low frequency, when not enough recombination has occurred to break up large linkage blocks, when the number of individuals is few, when the number of loci is few, when the effect size of alleles is small, and from the need to assume a high threshold effect size to reduce study-wide type-I errors (Buerkle and Lexer 2008, Mackay et al. 2009, Yeaman 2015). The aggregate extent of these biases can be illustrated by reference to the so-called "missing heritability paradox," in which genome-wide association studies (GWAS) can explain very little of the heritable variation in most human traits (Manolio et al. 2009, Mackay et al. 2009, Hill 2010). To address this problem, Yang et al. (2010) estimated the proportion of variance in human height explained by 294,831 SNPs (single nucleotide polymorphisms) genotyped on 3,925 unrelated individuals. Using a novel approach, the authors found that 88% of the variation due to SNPs had been undetected in previous GWAS because "the effects of the SNPs are too small to be statistically significant" (Yang et al. 2010; see also Yang et al. 2015). A scarcity of genes of large effect also appears to be the case for many other human traits, including susceptibility to numerous diseases (Manolio et al. 2009). In addition, artificial selection studies clearly show that evolutionary changes are often driven by many genes

(Hill and Kirkpatrick 2010), with classic examples including oil and protein content in maize (*Zea mays*) (Moose et al. 2004) and body weight in chickens (*Gallus domesticus*) (Johansson et al. 2010).

Second, nearly all studies have sought to explain variance in specific *traits*, rather than overall adaptation (or "fitness"). This distinction is critical because overall adaptation to a given environment will be influenced by many traits. As a result, even the genes explaining high levels of variation in a particular trait might contribute little to overall fitness differences. For example, studies in freshwater versus marine stickleback of the genes *EDA* (lateral plates) and *Pitx1* (pelvis) are often cited as evidence that adaptation is influenced by few genes of large effect. However, three points need to be kept in mind. First, these genes/traits are exceptions, with most other stickleback traits having no single QTL (quantitative trait locus) that explains more than 50% of the variance (Peichel et al. 2001, Albert et al. 2008, Rogers et al. 2012). This rarity of large-effect genes seems to be general: a meta-analysis of effect sizes in QTL studies comparing phenotypically divergent populations found that the most important QTL typically explained only 14.4% of the variation (Morjan and Rieseberg 2004). Second, freshwater and marine stickleback differ not only in lateral plates and pelvic structures but also in many other traits, with a partial listing including body size, gill raker number and length, dorsal and anal fin rays, body color, jaw size, spine length, salinity tolerance, swimming performance, reproductive behavior, and cold tolerance. Although single (or closely linked) genes might well have effects on several of these traits (Albert et al. 2008, Barrett et al. 2008, Kitano et al. 2010), the great diversity of traits is likely to involve a great diversity of genes. I expect this phenomenon—many traits, and so multiple genes, are involved in adaptation—to be general across organisms and environments.

Third, genome scans typically reveal that high-differentiation outliers, presumably influenced by divergent selection, represent 5–10% of the genome—and the distribution of these loci clearly implicates multiple unlinked genes (Nosil et al. 2009a). To continue with stickleback, Hohenlohe et al. (2010) examined 45,000 SNPs in each of 100 individuals from each of two marine and three freshwater populations. The authors found nine outlier genomic regions that showed elevated divergence in all marine versus freshwater comparisons—and so a minimum of nine genes must be involved. The actual number of genes, however, will be much higher because (1) multiple genes are present in each genomic region, and (2) the focus was only on regions showing parallel divergence across all comparisons. Not surprisingly, then, other work has found many more genes (or genomic regions) that differ between freshwater and marine stickleback (Shimada et al. 2011, Jones et al. 2012b), and similar results have been obtained for benthic versus limnetic stickleback (Jones et al. 2012a), dwarf versus normal whitefish (Renaut et al. 2011), and several independent studies of lake versus stream stickleback (Deagle et al. 2012, Roesti et al. 2012, 2015, Feulner et al. 2015). In short, genome scans typically implicate divergence in many genes, and yet these scans still remain biased against genes of small effect and the best method to infer outliers is still under debate (Cruickshank and Hahn 2014, Whitlock and Lotterhos 2015). Indeed, other approaches, such as reciprocal transplants and correlations with environmental variables, typically reveal many more loci under selection (Michel et al. 2010, Hancock et al. 2011, Fournier-Level et al. 2011, Bergland et al. 2014). Importantly, however, the vast majority of these loci appear to be under relatively weak selection (Thurman and Barrett 2016).

Although variation in some ecologically important traits is clearly influenced by a few genes of large effect ("keystone genes"), **the clear conclusion from genomic studies is that adaptation to a given environment will generally involve many genes, each of small to modest effect.** Current methods are poorly positioned to detect all, or even a substantial fraction, of these genes. More studies should examine the genetic basis not only of selected traits but also of overall adaptation or fitness, which integrates across all relevant traits. More sensitive analytical methods also need to be developed.

QUESTION 4: WHAT IS THE DISTRIBUTION OF EFFECT SIZES?

Even if we can be confident now that adaptation usually involves many genes of small to modest effect, the specific number of genes and their effect size distribution remains an open question. Several possibilities have been suggested (Fisher 1930, Orr 1999, 2005). One is the "infinitesimal model," where—stated in realistic form (Rockman 2012)—many genes are involved and all are of very small effect. Another is Fisher's "geometric model," which predicts an exponential distribution ranging from many loci with small effects to a few loci with large effects. After accounting for the difficulty of detecting loci of very small effect (as described above), the geometric model predicts a gamma distribution of effect sizes with a shape parameter greater than unity (Otto and Jones 2000).

Albert et al. (2008) explicitly tested the above alternatives through a QTL mapping study that compared the body shape of marine stickleback from Japan to that of derived freshwater benthic stickleback from Paxton Lake, British Columbia. The authors found that effects of particular QTLs on the morphological difference approximately followed a gamma distribution—consistent with the geometric model. The largest effect QTL explained about 22% of the difference, showing yet again that most of the variation is due to genes of small to modest effect. The mapping cross in this study was between two very different populations found on opposite sides of the Pacific, and it involved only a single family. The first point is of concern because the differences do not reflect a true ancestor-descendent scenario, although marine populations are considered nearly panmictic across their range. The second point is of concern because it can miss differences that are not fixed among the populations, and so will underestimate the number of genes involved in adaptation, particularly those of small effect.

Rogers et al. (2012) considered the same topic in the same study system but avoided some of the above concerns by crossing stickleback from a marine population to stickleback from each of four nearby lake populations (although only a single cross was performed in each case). Of additional interest, the phenotypic optimum for stickleback was expected to be farther from the ancestral marine optimum in the two lakes (Cranby and Hoggan) that lacked a predator (prickly sculpin, *Cottus asper*) versus the two lakes (Graham and Paq) that contained the predator. The resulting difference in the expected magnitude of adaptive evolution allowed testing another of the geometric model's predictions: larger mutations should contribute when adaptation is to more distant fitness peaks. The authors found that (1) most genes were again of small effect (only three QTL explained more than 20% of the variation in particular landmark coordinates), and (2) the lake populations adapting to the more distant fitness peak were more likely to have the larger effect QTL (two of the above three QTL were in Cranby Lake and one was in Hoggan Lake) (fig. 10.4). An experimental evolution study with *Aspergillus nidulans*

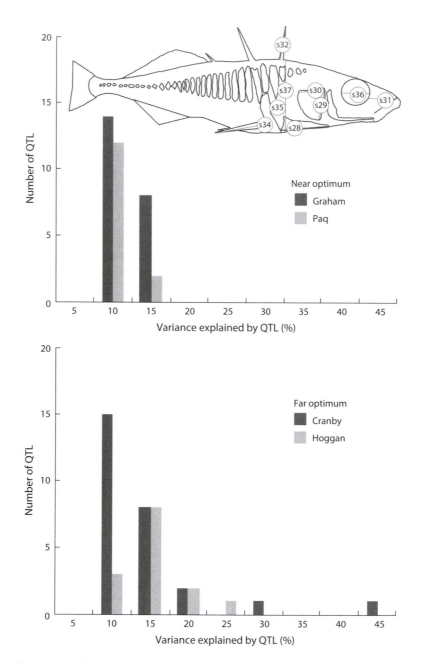

Fig 10.4. Frequency distributions of QTL effect sizes in crosses between marine (anadromous) stickleback from the Little Campbell River (British Columbia, Canada) and freshwater stickleback from each of four lakes in the same region. Graham and Paq lakes also contain predatory sculpins, whereas Cranby and Hoggan lakes do not. The latter lakes are expected to have a phenotypic optimum farther from the ancestral marine population. Estimates of variance explained are for particular geometric morphometric landmarks or univariate measurements (the latter shown on the fish image, which was provided by S. Rogers). Note that the lack of QTL of very small effect (<5% variance explained) reflects estimation limitations rather than the absence of such QTL. the data are from Rogers et al. (2012) and were provided by S. Rogers

found roughly the same pattern: populations farther from fitness optima tended to fix larger mutations during the course of adaptation (Gifford et al. 2011).

The above studies are just examples, and most other studies generate similar findings in that **the vast majority of QTLs for divergence between selective environments are of very small effect but that a few are of larger effect than the others.** (This pattern is also seen in selection differentials at the genomic level—Thurman and Barrett 2016.) Although determining the distribution of effect sizes is interesting in its own right, it is important to remember that studies of the sort reported above do not provide a validation or rejection of the geometric model. In particular, the theory is based on new mutations, whereas standing genetic variation is clearly important in adaptation to freshwater by stickleback (see also Lescak et al. 2015) and probably in most other instances of adaptation in nature (Q6). In addition, the theory assumes that the traits under study are the only traits involved in adaptation to the new environment, whereas QTL studies generally map just a few such traits. It would therefore also be useful to do QTL mapping studies for fitness, thus integrating variation across all phenotypic traits.

QUESTION 5: HOW REPEATABLE IS GENETIC DIVERGENCE?

Many studies report the independent (repeated) evolution of similar phenotypes in similar environments, either from similar or different ancestors. This repeatability (often couched as "predictability") of phenotypic patterns implies, with caveats, a strong deterministic role for environmentally determined natural selection (Simpson 1953b, Mayr 1963, Endler 1986, Schluter 2000a, Arendt and Reznick 2008, Losos 2011, Rosenblum et al. 2014). In chapter 4, I considered repeatability from a phenotypic perspective in the context of parallel evolution. Here I consider repeatability at the genetic level. At one extreme, adaptation by independent populations to similar environments could be driven by the same frequency changes in the same alleles (and nucleotides) at the same loci, with the relevant alleles at each locus having arisen only once (i.e., identical by descent). We might call this end of the spectrum, parallelism at the genomic level. Moving away from this extreme, the same allele might have had multiple origins, the alleles might be different but have similar effects, the alleles might be different and have different effects, and different genes might be involved (Arendt and Reznick 2008, Manceau et al. 2010, Elmer and Meyer 2011, Linnen et al. 2013, Rosenblum et al. 2014). We might call some patterns at this end of the spectrum convergence at the genomic level.

All of the above alternatives seem important in nature—even within the same study systems. For instance, stickleback provide a nice example of a single allele involved in adaptation to a similar environment in multiple independent instances. In particular, the low-plate *EDA* allele favored (and almost universally found) in fresh water arose once through mutation, is retained in marine stickleback because it is recessive, and increases toward fixation whenever marine stickleback colonize fresh water (Colosimo et al. 2005). Stickleback also provide a nice example of different alleles of the same gene (*Pitx1*) having the same phenotypic effect (pelvic reduction) in multiple independent instances (Chan et al. 2010). Divergence in *Pitx1* is thus parallel at the level of the gene but is convergent at the level of the allele. Additional examples of independent mutations at the same gene having similar phenotypic effects include *FRI* and flowering time in *Arabidopsis* (Shindo et al. 2005) and *VNR1* and season growth in cereal plants (Cockram et al. 2007). Importantly, parallelism even at the level of the gene is not universal even in the above case: for instance, some freshwater stickleback populations show lateral plate reduction not due to *EDA* (Lucek et al. 2012, Leinonen et al. 2012). Also in stickleback, about half of the effects detectable in QTL mapping studies are shared between two benthic-limnetic pairs found in lakes a few km apart (Conte et al. 2015). Similarly diverse results are common in other organisms, with a well-described example being the evolution of color in animals (review: Manceau et al. 2010).

As the preceding examples illustrate, the genetics of adaptation run the gamut of possibilities from very high to very low parallelism (see also Flint and Mackay 2009), but can any generalities be drawn? Conte et al. (2012) reviewed genetic mapping and candidate gene studies for the extent to which the same genes were shared during adaptation by different lineages. By their calculations, mean probabilities of gene reuse were "0.32 for genetic mapping studies and 0.55 for candidate gene studies" (Conte et al. 2012). They consider these estimates to be "surprisingly high" and conclude that "Frequent reuse of the same genes during repeated phenotypic evolution suggests that strong biases and constraints affect adaptive evolution, resulting in changes at a relatively small subset of available genes" (Conte et al. 2012, p. 5039). Without questioning the data itself (fig. 10.5), I find it easier to draw the opposite conclusion: gene reuse is low. For starters, a number of biases increase the estimated probabilities of gene reuse: publication bias against nonparallel patterns (particularly in candidate gene studies), the difficulty of detecting small effect genes in mapping studies (see above), the focus on traits as opposed to overall adaptation (see above), the explicit a priori focus on parallel phenotypic change (Conte et al. 2012), and the exclusion of unexplained phenotypic variance (which is often very high) from the calculations (Conte et al. 2012). (The authors do discuss some of these biases, as well as others that might act in the opposite direction.) Even ignoring any biases, the genetic mapping results find that the probability of nonreuse (68%) is more than twice the probability of reuse (32%). Evolution is thus considerably more likely to involve different genes in different instances than it is to involve the same genes.

Similar results have been obtained based on genome-wide comparisons of natural populations. For example, in three pairs of sunflower species, much of the divergence is nonparallel but some parallelism is observed, especially for regions under strong selection and when environmental gradients separating the species are very similar (Renaut et al. 2014). And, in guppies, Fraser et al. (2015) found parallel signatures of adaptation by different guppy populations to similar predation environments only in the case of very recent experimental introductions, as opposed to long-standing natural populations. (As an interesting aside, guppy gut microbiota show minimal predation-associated repeatability across watersheds: Sullam et al. 2015.) In this case, hitchhiking during initial divergence may have enhanced genomic parallelism that later decayed owing to recombination. Further, polygenic adaptation from genes of small effect is expected to be transitory, with different genes being involved at different points in the process (Yeaman 2015). More generally, theoretical models suggest that parallelism is likely rare and most adaptation will be driven by convergent genomic responses involving different genes and alleles (Ralph and Coop 2015).

Based on the above considerations and data, **I suggest that most of the genetic divergence among populations in different environments will be nonparallel, certainly at the allele level and probably also at the gene level.** What we need now are more

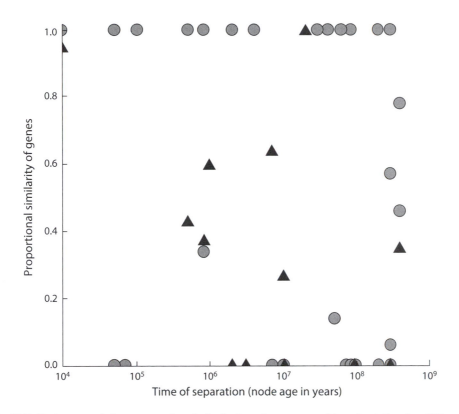

Fig 10.5. Estimates of the proportional similarity of genes used in adaptation by different species pairs according to the age of their separation. A value of zero means that no genes were shared and a value of one means that the sharing was perfect (e.g., the same gene was responsible in the two taxa). Triangles represent data from genetic crosses and circles represent data from candidate gene studies. A single point at 300 years (candidate gene study with 100% probability of gene reuse) is omitted to increase visibility in the rest of the data range. The data are from Conte et al. (2012)

objective inferential approaches that allow us to determine the portion of the genome that diverges between populations in different environments and the proportion of that divergence that is or is not parallel or convergent at a given level (see also Rosenblum et al. 2014).

QUESTION 6: STANDING GENETIC VARIATION OR NEW MUTATIONS?

Is adaptation to new conditions driven primarily by genetic variation already present in the population (standing) or is it primarily the result of new mutations (Barrett and Schluter 2008, Berg and Coop 2015)? At the most basic level, mutations are obviously important because standing variation started from new mutations—and these might have arisen and spread during previous adaptation. So the more important question is: how much of the adaptation to a particular selective event is the result of mutations that arose during that event? Theoretical predictions are diverse (Hermisson and Pennings 2005, Barrett and Schluter 2008, Orr and Unckless 2008, Rockman 2012), but a common perception is that standing variation often will be most important; because the waiting time for adaptation is shorter, existing alleles are less likely to be lost through drift, and existing variation is more likely to have been tested by past selection. Exceptions occur when standing variation is limited (e.g., inbreeding, strong past selection), mutational inputs are very high (e.g., large populations, short generation times, high mutation rates), and the new condition has not been previously experienced (e.g., pollutants, pesticides, herbicides, antibiotics, antivirals).

Empirically, the importance of standing variation is first implied by the earlier-described studies that report nearly ubiquitous additive genetic variance for fitness-related traits in natural populations (question 1). Evolution would seem likely to start with this variation, as long as it is relevant to selection and not unduly constrained by correlations with other traits. Fitting this expectation, the immediate and dramatic evolutionary responses often seen in artificial selection experiments suggest that plenty of relevant variation is present (Moose et al. 2004, Lango Allen et al. 2010, Johansson et al. 2010, Hill and Kirkpatrick 2010). Although the results of these studies might seem of questionable relevance because the selection was not "natural," similarly rapid responses have been observed in natural populations experiencing environmental change (Hendry and Kinnison 1999, Reznick and Ghalambor 2001, Hendry et al. 2008). These arguments, however, are indirect: they point to standing variation only because intuition suggests the changes were too fast to result from new mutations.

Additional evidence for the role of standing variation can be gained through three other approaches: signatures of selective sweeps in the genome, evidence that adaptive alleles in new populations were present in the ancestral population, and phylogenetic analyses that establish whether adaptive alleles arose before or after the environmental change (Barrett and Schluter 2008). Application of these approaches to the afore-mentioned selection experiments on maize (Moose et al. 2004) and chickens (Johansson et al. 2010) strongly implicates standing genetic variation. Application to natural populations often yields a similar conclusion, with examples including lactose tolerance in humans (Myles et al. 2005), warfarin resistance in brown rats (*Rattus norvegicus*) (Pelz et al. 2005), malathion resistance in blowflies (*Lucilia* spp.) (Hartley et al. 2006), local adaptation in *A. thaliana* (Fournier-Level et al. 2011), and several traits in stickleback (Colosimo et al. 2005, Miller et al. 2007, Kitano et al. 2008, Jones et al. 2012b, Terekhanova et al. 2015). Of particular relevance, the evolutionary changes observed in many of these cases took place over relatively short time frames (decades to centuries).

Although many studies thus point to standing variation, evidence for the role of new mutations is also growing. Interestingly, this evidence often comes from some of the same systems and traits that were discussed above, including *Pitx1* in stickleback, lactose tolerance in some human populations (Tishkoff et al. 2007), diazinon resistance in blowflies (Hartley et al. 2006), acetylcholinesterase resistance in *Drosophila* (Karasov et al. 2010), and local adaptation in *Arabidopsis* (Hancock et al. 2011). These findings suggest that new mutations can be important even when populations contain lots of standing variation, although not necessarily for the same traits. At present, not enough data exist to state more generally the relative importance of new mutations versus standing variation. However, one body of work with some tantalizing results is that considering the population dynamics of predators and prey in laboratory chemostats (e.g., Yoshida et al. 2003, Becks et al. 2010, Ellner 2013). The key design element of this work is that some chemostats start with only a single clone of prey, whereas others started with multiple clones. Evolution in the single-clone chemostats will require new mutations, whereas evolution in the multiclone chemostats can proceed through standing variation. The observed dynamics are very different among these cases: the single-clone chemostats show a general lack of evolutionary change whereas the multiclone chemostats show dramatic and ongoing evolutionary change that has population dynamic consequences. Here, at least, standing genetic variation dramatically shapes eco-evolutionary dynamics, whereas new mutations seemingly do not.

Adaptation to new conditions is brought about by a combination of standing genetic variation and new mutations. Unfortunately, the relative contributions of these two factors is difficult to infer from point-in-time surveys of genomic variation (Barrett and Schluter 2008, Roesti et al. 2014, 2015, Berg and Coop 2015). Although no complete analysis exists, their relative importance would seem likely to differ dramatically depending on the conditions. **Standing variation should be especially important for outbred populations (standing genetic variation will be higher) with moderate-to-long generation times (fewer new mutations can arise per unit time) facing environmental changes that are not entirely novel (previous conditions will have shaped standing genetic variation). Mutations should be increasingly important when populations are more inbred or are otherwise depleted in standing variation, as generation lengths become shorter, and as environmental disturbances are increasingly novel (e.g., certain pollution or pesticides).** Even under this last set of conditions, however, the contribution of standing genetic variation could still be greater than that from new mutations. This topic thus remains an open question.

QUESTION 7: WHAT DO ADAPTIVE WALKS LOOK LIKE?

I earlier mentioned Fisher's geometric model in the context of the distribution of genetic effect sizes expected to characterize adaptive differences among populations. Fisher's model also makes predictions about the temporal progression of mutational effect sizes fixed during the course of adaptation (Orr 1998). The set-up is that a multidimensional trait optimum exists some distance away from current trait values. New mutations then arise and can vary in direction (toward or away from the optimum) and size (small to large). Assuming the direction of mutations is random with respect to the optimum, large effect mutations are expected to be fixed earlier in the sequence than later because, as the population approaches the optimum, large effect mutations start to overshoot the optimum and so become maladaptive (fig. 10.6). Thus, the "adaptive walk" to an optimum should involve the early fixation of relatively few, but large effect, mutations and the late fixation of relatively numerous, but small effect, mutations (fig. 10.6).

Studies empirically testing this prediction come mainly from laboratory-based experimental evolution, often where replicate lines of a single clonal lineage are exposed to a new environment and the effects of sequential mutations are recorded during the subsequent adaptive walk to the new optimum. A general pattern in these studies is that evolution is initially rapid and then slows down, consistent with approach to a new adaptive peak (Dettman et al. 2012). In addition, large effect mutations are more likely to fix early in the process and the overall adaptive walk (number of mutations fixed) can be quite short, often only two to four steps (e.g., Betancourt 2009, Rokyta et al. 2009, Schoustra et al. 2009, Dettman et al. 2012). Also, adaptive walks can be quite short

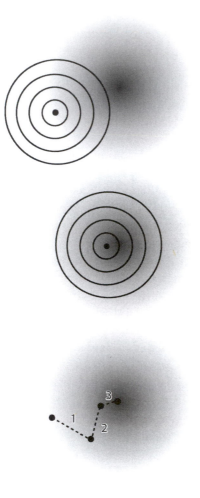

Fig 10.6. Illustration of Fisher's geometric model and some of its predictions. The filled circle shows a fitness surface in two-dimensional trait space, with the darker region toward the center indicating higher fitness. The dark dot shows the position of a population's mean phenotype in this space. Concentric circles around the dot show the distribution of potential mutational effects of a given size. The top panel shows that any mutation that takes the population closer to the fitness peak (i.e., into the gray circle) will be favored and the population should evolve in that direction. The middle panel shows a population that has evolved through this process but hasn't yet reached the peak. Thus, large effect mutations are more likely to provide fitness improvements when the population is far from the peak (top panel) than when it is near the peak (middle panel). The bottom panel shows a hypothetical distribution of mutations fixed during the course of a three-step adaptive walk to the fitness peak—larger effect mutations are expected to contribute earlier in the process

regardless of whether the population starts near or far from an adaptive peak (Gifford et al. 2011). This last study caused different lines to deviate to different degrees from adaptive optima by giving them different deleterious mutations. As the lines were only one deleterious mutation away from the optimum, the adaptive walks might be expected to especially short (Gifford et al. 2011). Overall, these studies suggest that adaptation to a new environment is often the result of relatively few mutations of large effect that are fixed early in the course of adaptation (Orr 2005). But we need to be careful just what we infer from this work. In particular, epistatic effects can dictate that a given mutation has a larger positive effect on fitness if it fixes earlier than later (Martin et al. 2007, Betancourt 2009, Chou et al. 2011, Dettman et al. 2012). For instance, later mutations might have smaller effects simply because part of the walk is over and so less scope exists to further improve fitness. Thus, it isn't clear whether mutations fix early because they have the largest effects or whether they have the largest effects because they fix early.

Several other reasons exist to be skeptical regarding the relevance of these laboratory studies for understanding natural populations. First, laboratory studies often focus on new mutations, which is consistent with the theory. Adaptation in natural populations, however, is likely often to involve standing genetic variation—and this variation is more likely to be of small effect, as described above. In this case, the earliest-fixing mutations could well be those of small effect—simply because they are already segregating in the population. (Of course, this doesn't mean that large effect mutations will contribute later—because little further adaptation may be necessary following the accomplishments of standing genetic variation.) Second, new environments in the laboratory often represent a single-dimensional manipulation, such as a new substrate or toxin. It does seem reasonable that adaptation would here involve fewer traits—and therefore fewer mutations—than adaptation to more natural environments that typically differ in all sorts of dimensions. Third, some laboratory studies expose organisms to stressors that they have never experienced in their evolutionary history—such as some new contaminant. Adaptation is here more likely to involve new and large mutations than in the case of quantitative changes in the conditions already experienced in nature: for example, more or less predation, more or less oxygen, hotter or colder conditions, etc. Finally, laboratory studies of adaptive walks often involve relatively simple organisms, for which the genetics of adaptation is expected to be simpler (Orr 2000, Welch and Waxman 2003).

A somewhat more relevant context might be the evolution of resistance to herbicides, fungicides, or pesticides in natural populations. This evolution often involves rapid increases in resistance consistent with the spread of a large effect mutation, sometimes one already segregating in the population and sometimes a new one (Raymond et al. 2001). A number of other studies show more gradual increases consistent with a polygenic mode of inheritance (Gressel 2009). However, these situations still represent very strong, uni-dimensional, novel selective pressures that might not be representative of most cases of adaptation in nature. For this last context, it might be most informative to ask how many genes are involved in adaptation to different environments—and these studies generally find dozens to hundreds of genes (see above). Real adaptive walks in nature are thus likely to have many more steps than those observed in the above laboratory experimental evolution studies. However, the current data from natural populations do not yet give us any indication of which types of mutations fix first because they don't track the time course of adaptation. And yet this tracking could be done through experimental evolution studies in nature, such as marine stickleback colonizing new freshwater environments (Bell et al. 2004, Le Rouzic et al. 2011), high-predation guppies introduced into low-predation environments (Reznick et al. 1997), or *Anolis* lizards introduced to islands with different predators or habitat conditions (Losos et al. 1997).

Overall, we still have no idea of the distribution of allelic effects fixed during adaptation to new environments. Theoretical deliberations and laboratory studies suggest that the largest effect mutations might fix first, whereas there are several reasons to doubt the applicability of these results to natural populations. Regardless, just because the first mutations to fix have the largest effect, doesn't mean that those effects are large relative to the overall amount of adaptation that is necessary. **The conditions under which large effect mutations might be most likely to contribute early in the course of adaptation include an absence of relevant standing genetic variation, a very novel new environment, and simple organisms.** It has also been argued that large effect mutations will be particularly critical in the case of mimicry, where a large phenotypic effect may be necessary to get two unrelated taxa to converge enough for selection to then favor increasing similarity (Baxter et al. 2009).

QUESTION 8: ARE THE GENETIC CHANGES REGULATORY OR STRUCTURAL?

An ongoing debate, here characterized in simple form, is whether evolution occurs mostly as a result of genetic changes that alter the amino acid sequence of proteins (structural) versus genetic changes that alter the amount, timing, or location of protein production (regulatory). Answering this question is important because the two types of changes can have very different evolutionary effects: for example, they represent different targets for mutation and they have different expectations for pleiotropic effects and selection (Hoekstra and Coyne 2007, Wray 2007, Carroll 2008, Stern and Orgogozo 2008). Although regulatory changes can occur in several ways, much of the current debate has focused on *cis*-regulatory regions: short, noncoding sequences that influence the expression of a nearby gene.[2]

Stern and Ozgogozo (2008) compiled a database of individual genetic mutations influencing phenotypic traits in "domesticated species (99 cases), intraspecific variation in wild species (157 cases), and interspecific differences (75 cases)." Of these mutations, 22% involved *cis*-regulatory regions and the authors argue that this is a major underestimate owing to investigator bias. Their survey additionally suggested that *cis*-regulatory mutations are more important for morphological traits (as opposed to physiological traits) and in interspecific comparisons (as opposed to domesticated species and intraspecific comparisons) (fig. 10.7). The first observation was interpreted as support for the importance of pleiotropy. Specifically, genes embedded more deeply in regulatory networks (hypothesized to be the case for morphology as opposed to physiology) are more likely to evolve through *cis*-regulatory changes, which are less likely to disrupt the entire network. The second observation was suggested to imply that the more subtle and local changes that can result from *cis*-regulatory mutations are more likely to be fixed during evolution over longer time periods.

Objectively obtained, whole-genome data on structural versus regulatory variation are rare, but a study of stickleback provides an exemplar. Jones et al. (2012b) performed full genome sequencing of 21 individual stickleback representing freshwater and marine forms across the northern hemisphere. Of the 64 genomic regions showing the strongest evidence of parallel habitat-associated divergence (i.e., the same genetic changes at different places in the world), 17% were in coding regions, 41% were in noncoding regions, and 42% included both coding and noncoding sequences. All of the latter regions involved alleles that did not cause protein-coding changes. The authors interpreted these results to imply that at least 41%, and perhaps as much as 83%, of the most important and parallel genomic regions influencing adaptation involved regulatory mutations.

Both coding and noncoding (regulatory) genetic changes clearly contribute to adaptation to new environments, but a relative rarity of objective studies and the existence of major ascertainment biases mean that their relative contribution is not yet clear. Moreover, this might not be the most interesting question anyway. Perhaps we should be instead asking "what kinds of phenotypic changes [e.g., morphological versus physiological] are expected under particular coding versus *cis*-regulatory changes" (Stern and Orgogozo 2008) or, almost equivalently, "whether *cis*-regulatory mutations

---

[2]"The *cis*-regulatory region of a gene encompasses all of the DNA elements (enhancer, promoter, 5'UTR, 3'UTR, introns, etc.) that regulate its expression in *cis*, in other words that act directly on the gene-coding region located on the same DNA strand, without encoding intermediary factors." (Stern and Orgogozo 2008).

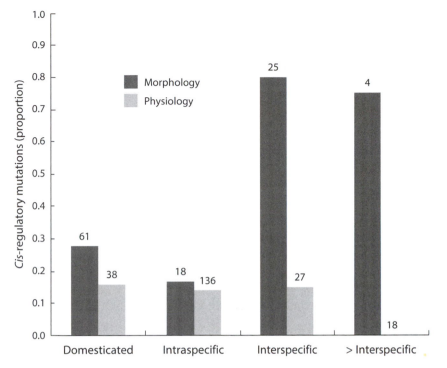

Fig 10.7. Proportion of *cis*-regulatory mutations in studies providing "compelling" evidence for individual mutations influencing either morphological or physiological traits. The data are divided into those from domesticated species, intraspecific comparisons, interspecific comparisons, or comparisons above the species level (> Interspecific). Numbers above each bar give the total number of mutations per category. The data are from Stern and Ozgozogo (2008)

have a qualitatively distinct role [as opposed to other types of mutations] in phenotypic evolution" (Wray 2007). Returning to an earlier point, it will be important to consider these effects in the context of overall adaptation (fitness) and also to find more sensitive ways to detect nonparallel effects.

QUESTION 9: HOW HERITABLE ARE ECOLOGICAL EFFECTS?

A premise of this chapter has been that the genetics and genomics of eco-evolutionary dynamics are largely equivalent to the genetics and genomics of phenotypes, at least to the extent that studying the latter provides a good foundation for understanding the former. For instance, a change in some ecological variable at the population, community, or ecosystem level might be predicted from information about selection acting on traits, genetic (co)variances for traits, and the ecological effects of traits (Johnson et al. 2009, Collins and Gardner 2009, Ellner et al. 2011, Moya-Laraño et al. 2014). A limitation of this approach is that the traits having large effects on ecological variables are often unknown. An alternative approach is to consider the ecological effects of individuals as "extended phenotypes" (or "interspecific indirect genetic effects") and directly estimating their heritability or evolvability (Fritz and Price 1988, Johnson and Agrawal 2005, Shuster et al. 2006). That is, standard quantitative genetic methods can be used to relate the ecological effects of individuals to the genetic relationships among them (parent-off-spring, half-sibs, etc.) and to thereby estimate the heritability of the ecological effects of organisms, which I will here call "ecological heritabilities."

A number of studies have estimated ecological heritabilities—usually for community-level variables and usually based on broad-sense estimates (proportion of the total variance due to genetic effects). For example, Shuster et al. (2006) studied cottonwood (*Populus* spp.) trees planted in a common-garden experiment in nature with multiple replicate clones of each of multiple tree genotypes. Arthropod communities were assessed on each individual tree and the proportion of the total variance in arthropods attributed to tree genotypes was estimated. The resulting broad-sense heritabilities were 56–68% for arthropod community composition, 30–34% for arthropod species richness, and 31–43% for arthropod abundance (Shuster et al. 2006, Keith et al. 2010). In addition, Keith et al. (2010) showed that these estimates were remarkably constant across the same trees in three different years, such that the broad-sense heritability of community *similarity* was 32%. This last estimate indicates that arthropod communities were more consistent among years on some genotypes than on others: the highest similarity for a genotype was 61% and the lowest was 24%. Studies of other plant systems have estimated heritabilities of arthropod community variables at 9–51% (Fritz and Price 1988) and 0–43% (Johnson and Agrawal 2005). Future work would ideally estimate ecological heritabilities in the narrow sense (proportion of the total variance due to *additive* genetic effects) and for more types of ecological variables (e.g., Lojewski et al. 2009, Crutsinger et al. 2014b).

With such information, one could theoretically estimate "selection" on an ecological variable as though it were an organismal phenotype and multiply this selection by the ecological heritability to predict at least short-term changes in the ecological variable. Although this thinking stretches the traditional meaning of "selection," it could provide a trait-independent, whole-organism approach to predicting how selection on organisms might drive evolutionary changes that alter ecological variables. Even more directly (but with more difficulty), predictions of this sort can be obtained by measuring genetic covariances between the fitness of individuals and their ecological effects (Johnson et al. 2009). This is the Robertson-Price Identity approach described in chapter 3, where the traits described there are replaced here with ecological effects. Of course, the accuracy of such predictions in nature will depend on the extent to which other factors also influence the ecological variables. For instance, ecological effects owing to the evolution of one species might be washed out by other factors influencing the same variable.

The ecological "extended phenotypes" of organisms are heritable in at least some instances, perhaps just as heritable as traits themselves. Given that traits can evolve on contemporary time scales (Hendry and Kinnison 1999, Reznick and Ghalambor 2001), the same would also seem likely for ecological variables. **However, the degree to which different ecological variables are heritable is highly variable** (Fritz and Price 1988), and so we can't assume that ecological variables will always be so responsive to selection on organisms.

Conclusions, significance, and implications

Some conceptualizations of eco-evolutionary dynamics are couched in terms of "genes to ecosystems" (Elser et al. 2000, Whitham et al. 2006, Bailey et al. 2009). Some readers might interpret this phrasing to mean that we should be searching for particular genes that have large ecological effects and, indeed, a few such genes have been found. We might call them "keystone genes." For example, *Pgi* influences population dynamics in Glanville fritillary butterflies (Hanski and Saccheri 2006). Overall, however, I suggest that searching for particular genes of large ecological effect, while perhaps flashy and more likely to be rewarded by publication in fancy journals, will only rarely be the best approach to the genetics and genomics of eco-evolutionary dynamics. A key reason is that *phenotypes* are the interface between ecology and evolution, with genes only being important indirectly through their effects on phenotypes. Eco-evolutionary investigations therefore should be concerned with the genetics and genomics of phenotypic adaptation, which I have here argued is based mostly on standing genetic variation at many genes of small to modest effect. Thus, although the search for particular genes that have ecological effects will sometimes be successful, it will miss the majority of the important links between evolution and ecology. Thus, we should implement approaches that can examine and quantify the polygenic basis of eco-evolutionary dynamics, including through estimates of the heritability of ecological effects considered as extended phenotypes. Such work will involve a mixture of quantitative genetics (including nonadditive effects), whole genome analyses and genome scans (with improved inferential methods), and gene expression (because many phenotypic differences among populations appear to be regulatory). Of additional importance will be the incorporation of microbiome studies because most of the genetic variation in organisms is found in their microbiome, and changes in that variation likely play a critical role in adaptive responses to many environmental changes.

Chapter 11

Plasticity

Eco-evolutionary dynamics are driven by interactions between environmental features and organismal *phenotypes*. Although it can be tempting, or at least convenient, to assume that phenotypic change is the result of genetic evolution, a likely alternative is that environmental conditions influence the expression of phenotypic traits for a given genotype. In its various guises, this phenomenon is discussed as phenotypic plasticity, developmental plasticity, environmental induction, acclimation, epigenetics, induced defenses, maternal effects, genotype-by-environment interaction (GxE), and indirect genetic effects (West-Eberhard 2003, Wolf and Wade 2009). Plasticity in these various manifestations can influence ecological dynamics through several effects (Miner et al. 2005, Yamamichi et al. 2011, Kovach-Orr and Fussmann 2013, Fischer et al. 2014), which I here briefly list and later discuss in more detail. First, current levels of plasticity are expected to have evolved as a result of past selection, and so plastic changes expressed by individuals in the present can be adaptive and have a genetic basis. In such cases, plasticity can be viewed as a contemporary expression of past evolution. Second, plasticity can evolve on contemporary time scales, and so phenotypic changes in a population can reflect the ongoing evolution of plasticity. Third, plasticity modifies selection on genotypes, and thereby influences evolutionary responses to ecological change and ecological responses to evolutionary change. In short, plasticity must be an integral part of any general framework for eco-evolutionary dynamics.

Plasticity has been the focus of considerable interest, with summaries of its theoretical and empirical development appearing in several books (Schlichting and Pigliucci 1998, West-Eberhard 2003). Of particularly recent popularity has been the concept of epigenetics, whereby genetic changes (e.g., DNA methylation or stable chromatin modifications) that do not alter the DNA sequence have phenotypic effects that can be heritable (Feil and Fraga 2012, Duncan et al. 2014, Schlichting and Wund 2014). Stated another way, epigenetics provides a genetic mechanism for plasticity and its propagation across generations. My focus in the present chapter will be on interactions between environment and phenotype, and so I will not be concerned with the precise genetic mechanisms. In addition, my specific focus will be on key features of plasticity that are especially relevant for eco-evolutionary dynamics. Another version of this chapter was published as Hendry (2016).

What is plasticity?

I here introduce some key features of plasticity by outlining a specific example: behavioral responses of *Daphnia* to fish predators (De Meester 1996, Boersma et al. 1998, Cousyn et al. 2001). *Daphnia* feed most effectively near the lake surface in daylight but doing so can expose them to very strong predation by fish. Thus, when planktivorous fish are present, we expect *Daphnia* to migrate downward (away from the lake surface) near dawn but upward (toward the lake surface) near dusk, behaviors mediated by responses to light (phototaxis). Even when fish are present in a lake, however, the intensity of predation will vary through time and space, in which case it pays *Daphnia* to remain near the surface even during the day as long as they do not detect proximate fish predators but to move away from the surface when they do detect predators. A useful cue for inferring the likelihood of predation is the presence of fish chemicals (kairomones) that can be sensed by *Daphnia*. We therefore expect *Daphnia* in lakes with fish to show greater plasticity in phototactic responses to fish kairomones (move away from the light when fish kairomones are present but not when they are absent) than *Daphnia* in lakes without fish.

De Meester (1996) tested this hypothesis in a laboratory common-garden experiment by recording the phototactic responses (on exposure to fish kairomones) of *Daphnia* clones (multiple individuals per clone) from lakes with different levels of fish predation. He found that the plastic response of *Daphnia* (phototaxis in relation to fish kairomones) was greatest for clones from the lake that had the highest natural level of fish predation. Later, Cousyn et al. (2001) showed that these plastic responses evolved over a 30-year period in a single lake in response to temporal changes in fish predation. This evolution was demonstrated by using cores from lake sediments to isolate *Daphnia* resting eggs from different layers (and therefore different time periods), which then could be hatched and tested for phototactic responses (fig. 11.1). Beyond phototaxis, a number of other *Daphnia* traits also respond plastically to fish kairomones and differ as expected among lakes with different levels of fish predation (Boersma et al. 1998, Stoks et al. 2016).

These results illustrate several key features of plasticity. First, plasticity occurs when a single genotype (a *Daphnia* clone) produces different phenotypes (different levels of phototaxis) depending on the environmental context (kairomone present versus absent). Second, adaptive plastic responses are often triggered by environmental cues (fish kairomones) that reliably indicate the appropriate phenotype. Third, different genotypes within populations can differ in plasticity (different clones show different responses), indicating genetic variation in plasticity (see also Dingemanse and Wolf 2013 for behavioral plasticity). Fourth, the evolutionary potential provided by this variation generates adaptive differences in plasticity among populations (*Daphnia* show greater plasticity in situations with fish than in situations without fish). Fifth, adaptive plasticity can evolve over short time scales (30 years), including reductions in plasticity when the selective force favoring it (fish predation) is lost. This last observation implies that plasticity has costs. We will return to these concepts after first considering some methods for studying plasticity.

How to infer plasticity

By far the most common approach to studying phenotypic plasticity is to implement experimental manipulations under otherwise controlled conditions. To follow up the

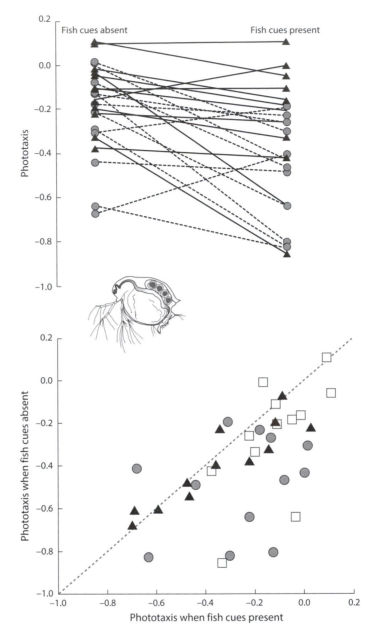

Fig 11.1. Empirical example illustrating phenotypic plasticity, genetic variation in plasticity, and the evolution of plasticity. Phototaxis is the ratio of *D. magna* individuals in a compartment near a light source versus a compartment away from the light source, either in the presence or absence of fish cues (kariomones). The top panel shows norms of reaction for the responses of each of 24 clones (lines) from two different time periods in a single lake. The triangles and solid lines show clones collected from sediments from a period without fish predation, whereas the circles and dashed lines show clones collected from sediments from a period with fish predation. The bottom panel shows the same data in a different way, where each clone is a single point, and deviations from the 1:1 line correspond to plasticity. The lower panel also shows data from a third time period (squares) after fish predation declined in the lake. The data are from Stoks *et al.* (2016) and were provided by K. Pauwels and L. De Meester

above *Daphnia* example, different individuals of each clone were exposed to different kairomone conditions in the laboratory. These experiments thus yield information on the phenotype produced by a given genotype under different conditions, a relationship termed the "reaction norm" (Scheiner 1993, Schlichting and Pigliucci 1998). Often, the goal is to compare plasticity among different groups, such as different time periods or species or populations, which can be accomplished by comparing slopes and elevations of their reaction norms. In figure 11.1, for example, we see that (1) plasticity is present in many *Daphnia* clones (reaction norm slope), (2) genetic variation in plasticity is present (slopes differ among clones), and (3) genotypes differ irrespective of plasticity (reaction norm elevations). Using these clone-specific slopes and elevations as units of replication, reaction norms can be statistically compared among groups.

Such experiments work most effectively for species, like *Daphnia*, that allow the use of inbred lines or clones—because a single genotype can be examined in multiple environments. This approach is not possible for many organisms, where different "genotypes" must instead be represented by different full-sibling families (different individuals from a given family are split among the different conditions) or by unrelated individuals randomly sampled from the different groups (usually different populations). The assumption in these cases is that substantial genetic differences are not present among the individuals from a given family/population exposed to the different conditions. If this assumption is correct, the resulting family/population-level reaction norm should be representative of the average reaction norm of genotypes from those families/populations.

Once average reaction norms are estimated for different groups (e.g., clones, populations, or species), a number of outcomes might emerge, as shown in figure 11.2. In panel A, environmental conditions do not have a plastic effect on the trait (flat reaction norms) and the groups do not differ genetically for that trait (identical reaction norms). In panel B, environmental conditions have a plastic effect on the trait, but the groups do not differ genetically. In panel C, environmental conditions do not have a plastic effect, but the two groups differ genetically in trait expression (elevations differ). In panel D, environmental conditions have a similar plastic effect on both populations (similar slopes) but the populations differ genetically in trait expression for a given environment (elevations differ), with the genetic and plastic influences here reinforcing each other (i.e., cogradient). In panel E, environmental conditions have a plastic effect that differs in direction between the two groups (slopes differ in sign). In panel F, environmental conditions have a similar plastic effect on both populations (similar slopes) but the populations differ genetically in trait expression for a given environmental condition (elevations differ), with the genetic and plastic influences this time opposing each other (i.e., counter-gradient).

Several considerations attend reaction norms and their estimation (Schlichting and Pigliucci 1998). First, I have illustrated the concept with only two environments, but reaction norms can be quantified for any number of environments and for continuous environmental variables. Second, although reaction norms are typically shown as linear, they can take any shape. Third, reaction norms can be quantified for phenotypes of any sort. Traditional phenotypes include behavior, physiology, color, morphology, life history, and various fitness metrics; but reaction norms also can be evaluated for variables such as gene expression (e.g., Swindell et al. 2007, McCairns and Bernatchez 2010, Stutz et al. 2015) or protein expression (e.g., Tomanek 2008, Martínez-Fernández et al. 2010).

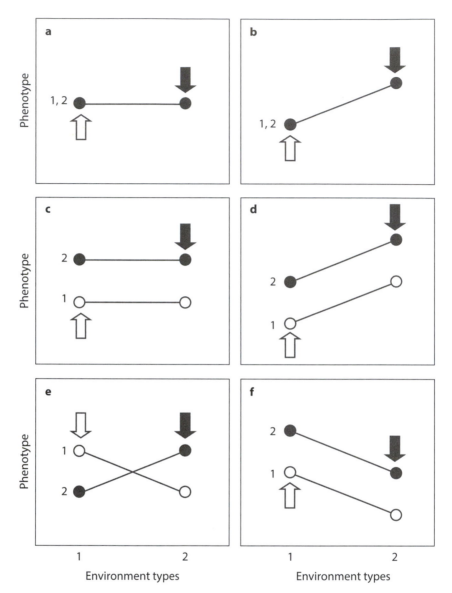

Fig 11.2. Hypothetical reaction norms for two populations (or genotypes). The two lines represent phenotypes (*y*-axis) for two different populations (1 and 2) in two different environment types (1 and 2). The arrows point to phenotypes expressed by each population in the wild: that is, population 1 in environment 1 (open arrow) and population 2 in environment 2 (filled arrow). Redrafted from Morbey and Hendry (2008), from which I also repeat some of the description in the text

In addition, phenotypes can be continuous variables, such as growth curves, which are also called "function-valued traits" (Kingsolver et al. 2001, Stinchcombe et al. 2012). Fourth, additive genetic (co)variances and heritabilities can be calculated for the slopes and elevations of reaction norms, just as they can for trait values in a single environment (chapter 3). Fifth, it is sometimes possible to study the specific genes or gene regions (QTLs) underlying some aspects of plasticity (e.g., Gutteling et al. 2007), with prime examples being heat shock proteins (e.g., Gibert et al. 2007) and DNA methylation patterns (e.g., Herrera and Bazaga 2010). Sixth, reaction norms induced by a particular environmental variable likely depend on levels of other environmental variables (GxE becomes GxExE), and so results will be context-dependent.

Another approach to the study of plasticity is observational: follow individuals/populations through time to quantify the relationship between environmental conditions and trait expression. At the individual level, this approach relies on the same individuals experiencing different environments at multiple discrete episodes in their life (Nussey et al. 2007). In long-lived birds or mammals, for example, the breeding time of an individual in each of several years can be related to temperature in those years, allowing the estimation of individual-level reaction norms for breeding time in relation to temperature (Nussey et al. 2005, Charmantier et al. 2008, Husby et al. 2010, Porlier et al. 2012, Charmantier and Gienapp 2014). Such analyses are most useful for traits that can be adjusted by an individual on an episode-by-episode basis; most obviously various aspects of behavior and physiology. By contrast, it is not informative for traits that are developmentally plastic, such as many morphological and life history traits. At the population level, the observational approach relates average trait values to average environmental conditions through time (Quinn and Adams 1996, Phillimore et al. 2010). Although such population-level analyses are extremely common (Parmesan and Yohe 2003), they have severe limitations (Merilä and Hendry 2014): for instance, factors other than plasticity can cause temporal changes, and unmeasured correlated traits or environmental variables might influence observed trends. The observational approach at either level (individual or population) also can be applied in a spatial context by relating trait values in different individuals/populations to environmental conditions experienced by those individuals/populations during development (Urban et al. 2014). This approach is particularly likely to run afoul of the limitations listed just above, especially the fact that spatial variation among populations could be shaped by genetically based adaptive divergence (chapter 4).

Another set of approaches for inferring plasticity seeks to rule out (or partition out) genetic contributions to observed differences, thus leaving plasticity as the default explanation. First, groups that differ phenotypically in nature can be raised under common-garden conditions to see if those differences vanish, which thus implies their plastic basis, with a classic example being James (1983). Second, estimates of selection and genetic variation can be used in the breeder's equation or Robertson-Price Identity (chapter 3) to estimate the likely contribution of evolution, leaving plasticity as the remaining explanation for change not explained thereby (Crozier et al. 2011, Anderson et al. 2012). Third, animal model analyses can be used to infer genetic change based on breeding values and thus, by the remainder, any plastic contributions (Merilä et al. 2001b, Charmantier and Gienapp 2014). Fourth, the Price equation (chapter 2) can be used to partition phenotypic changes into those due to selection versus various forms of plasticity (Ozgul et al. 2009, 2010, Ellner et al. 2011). Each of these approaches can be informative but each is also attended by a number of inferential caveats (Merilä and Hendry 2014).

Plasticity in nature

I now address key questions surrounding phenotypic plasticity and its interactions with ecology and evolution. I start with the basic question: (Q1) Is plasticity adaptive (i.e., increases fitness), as opposed to a nonadaptive or maladaptive response to (for example) stressful conditions. I then ask (Q2) whether plasticity has limits or costs—because such constraints will influence whether or not plasticity is a sufficient solution to environmental change. If plasticity has adaptive consequences, it should vary among populations in response to selection. In particular, plasticity should be greater in more variable environments, a prediction I address in the next question (Q3). I then consider (Q4) to what extent plasticity can aid the colonization of new environments and persistence in the face of environmental change. Given that plasticity and genetic change can be alternative responses, the next questions consider how plasticity (Q5) promotes or constrains genetic evolution, including (Q6) ecological speciation. At the same time, plasticity and evolution are not mutually exclusive, and so I will also consider (Q7) how quickly and frequently plasticity can evolve. Finally, I will (Q8) evaluate the extent to which plasticity influences community or ecosystem variables. As in most chapters, but even more so here, the questions are hard to answer definitively because strong generalities are elusive. In such cases, I seek to outline the conditions under which different outcomes are most likely to result.

QUESTION 1: TO WHAT EXTENT IS PLASTICITY ADAPTIVE?

Some types of plasticity are clearly adaptive, such as immune responses to parasites or behavioral avoidance of predators, whereas other types of plasticity are clearly not adaptive (Grether 2005, Ghalambor et al. 2007). For instance, resource limitation can cause developmental problems that generate phenotypes of no benefit to the organism. Given these alternative possibilities, it is important to not only quantify plasticity but to also evaluate its adaptive significance. One way to do so is through experiments where plastic responses are induced and changes in fitness are monitored. For instance, defensive responses to a particular enemy often decrease vulnerability to that enemy. As a specific example, herbivory on plants decreases following herbivore-induced increases in volatile chemicals (Kessler and Baldwin 2001), setose trichome density (Agrawal 1999), and spine length (Milewski et al. 1991). Similarly, predation decreases following predator-induced increases in body depth in *Carassius* carp (Bronmark and Miner 1992) and shell thickness in *Physa acuta* snails (Auld and Relyea 2011). Further, animals that evolved on islands without predators often lack adaptive antipredator behaviors (e.g., Cooper Jr. et al. 2014), and so suffer major declines when a predator is introduced (Sih et al. 2010).

Even in cases where plasticity is seemingly adaptive, caveats and nuances exist. For instance, defenses induced by exposure to one enemy might be disadvantageous in the presence of a different enemy (Dewitt et al. 2000, Decaestecker et al. 2002), and induced defenses can be costly in general. Moreover, the above examples were targeted investigations of specific changes expected a priori to be adaptive, whereas more diverse results are obtained when traits are chosen more objectively. For example, Caruso et al. (2006) exposed two wildflowers (*Lobelia cardinalis* and *Lobelia siphilitica*) to wet or dry conditions, measured a series of phenotypic traits related to photosynthesis, and used above-ground biomass as a surrogate for fitness. The two species showed different levels of plasticity in different traits, and the consequences ran the gamut from adaptive to maladaptive to neutral. In a meta-analysis of reciprocal transplant experiments in plants, Palacio-López et al. (2015) reported that plasticity was observed in about half of the cases and was adaptive in 65% of those cases. The adaptive (or maladaptive) nature of such responses is expected to influence population dynamics (Duputié et al.

2015). As will be considered further in the next question, a series of similar studies have been performed with other organisms and the results are highly variable with respect to the adaptive significance of plasticity (van Kleunen and Fischer 2005, Auld and Relyea 2011).

Most studies of the adaptive significance of plasticity are conducted under controlled experimental conditions, such as common gardens or mesocosms. Given that these arenas do not include all potential selective forces, the overall adaptive significance of plasticity often remains uncertain. The alternative is to evaluate plasticity and its consequences in natural populations (Nussey et al. 2007). This approach is rarely implemented owing to logistical constraints, but we do have some informative cases studies. I would especially like to highlight a contrast between studies of individual plasticity in two populations of great tits, one in the Netherlands (Nussey et al. 2005) and one in the United Kingdom (Charmantier et al. 2008). In each case, plasticity was quantified as the extent to which individual birds changed in their breeding date between years as a function of changes in temperature, and this individual plasticity was related to lifetime reproductive success. In the Dutch population, individuals differed dramatically in plasticity, selection favored increased plasticity, and current levels of plasticity were insufficient for fully adaptive responses to climate change (Nussey et al. 2005). Results were opposite in the UK population: individuals did not differ strongly in plasticity, plasticity was not under selection, and the current levels of plasticity were sufficient for fully adaptive responses (Charmantier et al. 2008). More recent work has formally compared these two studies and, although some conclusions change, the different populations (and traits) clearly differ dramatically in individual plasticity, its genetic basis, and its adaptive significance (Husby et al. 2010). Dramatic intraspecific variation in these properties is also present on small spatial scales, as demonstrated by work on blue tits (*Cyanistes caeruleus*) (Porlier et al. 2012).

Plasticity is sometimes adaptive, sometimes maladaptive, and sometimes neutral. The set of conditions under which each result emerges is not yet clear, not the least because plasticity and its consequences vary dramatically within species. However, it seems likely that plasticity will be most adaptive for traits that have different optimal values under different environmental conditions that the population has routinely experienced in the past, particularly when reliable environmental cues allow appropriate and timely plastic changes. These expectations will be considered further in question 3.

QUESTION 2: IS PLASTICITY COSTLY OR LIMITED?

Organisms faced with variable environments might evolve genetically based adaptive divergence or might instead use plasticity to mold individual phenotypes to current conditions; or both, including adaptive divergence in plasticity. In the absence of constraints, plasticity would seem the best of these alternatives because it should be the most immediately responsive to environmental change. Yet meta-analyses of reciprocal transplants suggest that adaptive plasticity is "perfect" only about 25% of the time (Palacio-López et al. 2015) and adaptive genetic divergence is common (chapter 4). Both of these observations suggest that plasticity must have constraints in the form of costs or limits. I here outline these constraints according to DeWitt et al. (1998), although other frameworks exist (e.g., Auld et al. 2010, Murren et al. 2015). *Costs* of plasticity could include "maintenance costs" associated with the sensory and regulatory machinery that enables plasticity, "production costs" associated with implementing a plastic response, "information acquisition costs" associated with acquiring information about the best plastic responses, "developmental instability" associated with plastic traits, and "genetic costs" such as linkage, pleiotropy, or epistasis. *Limits* of plasticity could include "information reliability limits" stemming from an inability to determine appropriate phenotypes, "lag-time limits" due to the time it takes to implement a plastic change, "developmental range" limits wherein some plastic changes aren't possible, and the "epiphenotype problem" whereby plasticity in one trait might negatively affect other traits.

One suggested method for assessing costs of plasticity is to relate fitness in a given environment to trait values in that environment and plasticity among environments, with the data coming from sib-ships, clonal genotypes, or inbred lines in multiple environments (van Tienderen 1991, DeWitt 1998, Scheiner and Berrigan 1998). The data are then analyzed in a Lande style multiple regression model (chapter 2), where one predictor is the trait value in an environment and the other predictor is the difference in trait value between environments. The partial regression coefficient for the latter term provides an estimate of the net cost or benefit of plasticity while controlling for mean trait value. When this coefficient is negative (selection against plasticity), a cost is inferred. When this coefficient is positive (selection for plasticity), a "cost of canalization" (benefit of plasticity) is inferred. Van Buskirk and Steiner (2009) performed a meta-analysis of 27 studies reporting 536 separate estimates. Costs of canalization were found to be as common as costs of plasticity and both types of costs were relatively weak and rarely significant (see also van Kleunen and Fischer 2005, Auld et al. 2010) (fig. 11.3). At face value, these results might be taken to mean that costs of plasticity are not strong (see also Auld et al. 2010). In reality, however, these analyses test for *selection* on plasticity (i.e., question 1), which will reflect a combination of costs and benefits. Thus, a lack of selection against plasticity doesn't imply an absence of costs, merely that any such costs are offset by compensatory benefits.

Plasticity certainly has limits—both ultimate and proximate. In an ultimate sense, some phenotypic changes will be forever impossible through plasticity, just as some phenotypic changes will be forever impossible through evolution. In a proximate sense, the plasticity currently present within a population is often insufficient for fully adaptive responses to environmental change. For example, although some birds can plastically match their breeding time to appropriate conditions, such as the timing of peak caterpillar abundance, migratory birds cannot breed before they arrive. Migratory timing, which is often genetically based, thus places a limit on what can be achieved through plasticity of breeding time (Both and Visser 2001, Gill et al. 2014). Further examples of limits to plasticity are legion: current plasticity appears insufficient for responding to climate change in British frogs (Phillimore et al. 2010), a number of birds (Nussey et al. 2005, Gill et al. 2014), and many plants (Willis et al. 2008, Wolkovich et al. 2012, Van Buskirk et al. 2012, Duputié et al. 2015). What remains uncertain is just how prevalent and important are these limitations (Murren et al. 2015).

Another context for considering limits to plasticity is the idea of behavioral "types" and "syndromes" suites of "correlated behaviors expressed either within a given behavioral context (e.g., correlations between foraging behaviors in different habitats) or across different contexts (e.g., correlations among feeding, antipredator, mating, aggressive, and dispersal behaviors)" (Sih et al. 2004). The basic idea is that

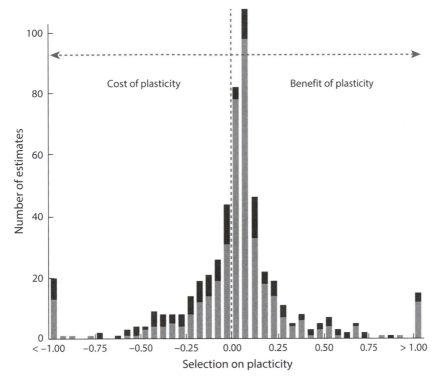

Fig 11.3. Frequency distribution of 540 estimates from 27 studies of selection on plasticity, with significant estimates shown in the dark portions of the bars. Negative values imply a cost of plasticity, whereas positive values imply a benefit of plasticity (or a cost of canalization). The data are from Van Buskirk and Steiner (2009) and were provided by J. Van Buskirk

different individuals fall at different positions along behavioral or "personality" axes, which makes it difficult to alter behaviors from one context to another (Sih et al. 2004). For example, selection might favor boldness in the presence of potential mates but shyness in the presence of potential predators (Smith and Blumstein 2008), and yet bold individuals might remain bold in both contexts as a result of limited moment-to-moment flexibility. Such axes could have important consequences for a variety of ecological and evolutionary processes (Wolf and Weissing 2012). At present, however, the relative frequency and importance of syndromes in causing maladaptive context-dependent behavior is unknown (Brommer 2014). Another uncertainty is the extent to which behavioral types and syndromes reflect hard limits to behavioral plasticity, as opposed to adaptive responses to past selection resulting from, for example, high costs of excessive plasticity. What is known from meta-analyses is that (1) the behavioral repeatability of individuals is highly variable (Bell et al. 2009)—that is, behaviors are sometimes very repeatable and sometimes not (fig. 11.4), and (2) different personality axes (boldness, exploration, aggression) can influence fitness components (reproductive success and survival) in a variety of ways (Smith and Blumstein 2008).

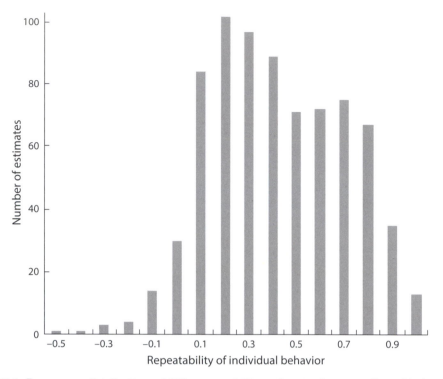

Fig 11.4. Frequency distribution of 659 repeatability estimates from studies of behavior. Repeatability is the variance in behavior among individuals divided by the sum of the variance among individuals and the variance across repeated measurements within individuals. For improved presentation, one very low repeatability value (−0.95) is not shown. The data are from Bell et al. (2009)

Costs and limits of plasticity should be context dependent: for example, costs might be strong only when plastic responses are large and environmental conditions are stressful. The first possibility (large responses) was considered by Lind and Johansson (2009) through a comparison of common frog (*Rana temporaria*) populations that showed large versus small plastic changes in developmental timing in response to simulated pond drying. Costs of plasticity were found only in populations that showed the largest plastic responses (see also Merilä et al. 2004). The second possibility (stressful conditions) was considered in the meta-analysis of Van Buskirk and Steiner (2009). Specifically, costs of plasticity were highest when environmental stress was greatest, at least for animals (fig. 11.5)—although this result is not universal (Steiner and Van Buskirk 2008). Not surprisingly, then, costs of plasticity depend on properties of organisms, traits, and environments.

Plasticity must have costs and limits but these constraints are highly variable, often weak, and hard to detect. In particular, it has proven difficult for studies to reliably separate limits, costs, and benefits, all of which might interact and be context-dependent (Auld et al. 2010). Overall, it seems likely that costs of plasticity will be highest when plastic changes are greatest, when environmental conditions are more stressful, and in

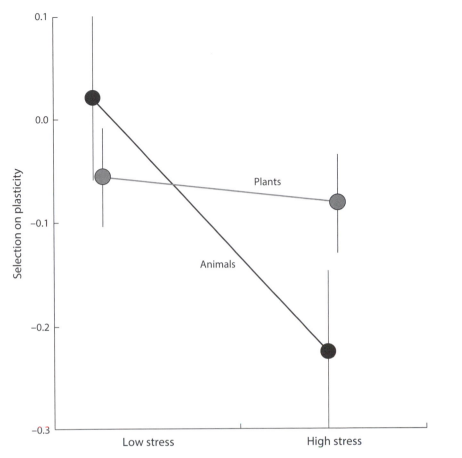

Fig 11.5. Selection on plasticity for plants and animals in low stress versus high stress conditions. Negative values imply a cost of plasticity, whereas positive values imply a benefit of plasticity (or a cost of canalization). The data (means and standard errors) are from a mixed model analyzed by Van Buskirk and Steiner (2009)

rarely experienced environmental conditions (because past selection will have had less opportunity to reduce costs). In addition, it has been suggested that limits to plasticity are most likely in cases of relaxed selection and variable selection intensities (Murren et al. 2015).

QUESTION 3: TO WHAT EXTENT DO VARIABLE ENVIRONMENTS FAVOR THE EVOLUTION OF PLASTICITY?

Given that the adaptive benefits and costs/limits of plasticity vary among traits, organisms, and environments, the evolution of plasticity should vary at these same levels. In particular, theoretical models have shown that adaptive phenotypic plasticity readily evolves when selective conditions are variable. Some of these models have altered optimal phenotypes/genotypes through time for a single population (Gabriel 2005, Stomp et al. 2008, Lande 2009, Svanbäck et al. 2009, Gomez-Mestre and Jovani 2013, Ezard et al. 2014, Botero et al. 2015, Tufto 2015), whereas others have analyzed meta-populations where optimal phenotypes/genotypes vary across space (Levins 1968, Via and Lande

1985, van Tienderen 1997, Sultan and Spencer 2002, Thibert-Plante and Hendry 2011b, Scheiner and Holt 2012). These models consistently suggest that greater plasticity is favored when (1) spatial variation is greater, (2) dispersal is higher, (3) temporal variation is greater, (4) environmental cues are more reliable, (5) genetic variation for plasticity is higher, and (6) costs/limits of plasticity are lower. I now summarize studies testing the first four of these predictions.

1. A number of empirical studies have tested the prediction that higher plasticity should evolve when environments are more spatially heterogeneous. Lind and Johansson (2007) examined plasticity in common frog populations from 14 islands off the coast of Sweden. Islands with more spatial variation in pond-drying regimes (some ponds dry quickly and others slowly) were found to have frogs with greater plasticity in developmental timing when exposed to simulated drying regimes (water volume changes) (fig. 11.6). Along the same lines, Baythavong (2011) showed that plasticity for the plant *Erodium cicutarium* was higher in environments with more fine-grained spatial variation. Although a number of other such studies further support the above expectation, too few have been conducted to warrant sweeping generalizations.

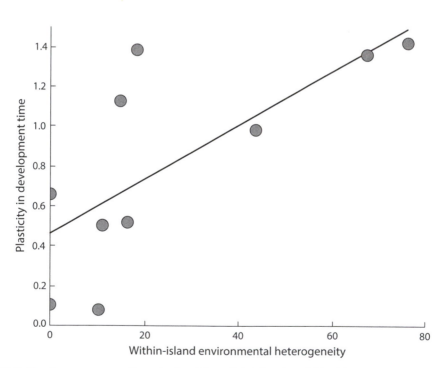

Fig 11.6. For frogs on recently colonized Swedish islands, within-island heterogeneity of pond drying regimes is positively correlated (across islands) with the degree of plasticity frogs show in development time. Plasticity is the mean development time in constant water level minus the mean development time under artificial drying. Environmental heterogeneity is the coefficient of variance in pool drying. The data are from Lind et al. (2011) and were provided by M. Lind

2. If spatial variation favors the evolution of plasticity (as above), greater plasticity is expected to evolve under higher dispersal rates—because this condition increases the spatial variation experienced by a given lineage. Lind et al. (2011) used the frog system described just above to suggest (statistical significance was lacking) that phenotypic plasticity was greater when gene flow (based on microsatellites) was higher among islands with different drying regimes. A potential uncertainty in such analyses is ascertaining whether higher gene flow is the cause or the consequence of higher plasticity (Crispo 2008). The role of dispersal was evaluated more generally in a meta-analysis of 258 experiments on plasticity in marine invertebrates (Hollander 2008). In accordance with the expectation, species with low dispersal (viviparous/ovoviviparous development or direct development from benthic egg masses) showed lower plasticity than did species with high dispersal (planktonic) (fig. 11.7). The idea is that planktonic species have limited control over the conditions they experience and should therefore evolve higher plasticity.

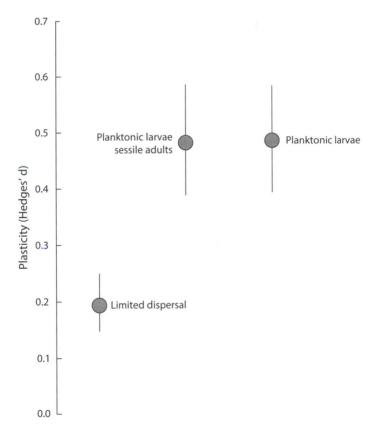

Fig 11.7. Plasticity in marine invertebrates is highest for species with planktonic larvae. Plasticity estimates were for many traits in many environmental contrasts. Shown are estimates based on all data—similar results are obtained in reduced analyses (one study per species). The "planktonic larvae" category includes all estimates from the "planktonic larvae, sessile adults" category plus estimates from species where both larvae and adults are planktonic. The data (means and 95% confidence intervals) are from Hollander (2008)

3. Several already-mentioned studies inform the expectation that populations experiencing greater temporal variation will evolve greater plasticity. For example, temporal variation in fish predation on *Daphnia* is present in lakes with fish but not in lakes without fish—and plasticity in kairomone-induced phototactic responses of *Daphnia* is correspondingly higher in the former than the latter (De Meester 1996). Interestingly, an opposite result is seen for some traits in Trinidadian guppies, where the population not experiencing the predator shows greater plastic responses to predator cues (Torres-Dowdall et al. 2012)—probably as a result of correlated responses to selection on the mean phenotype. Returning to supportive examples, Gianoli and González-Teuber (2005) compared three populations of the plant *Convolvulus chilensis* that experience dramatically different interannual variation in precipitation and therefore drought stress. Four traits showed plastic responses to simulated drought conditions in the laboratory and, in each case, plasticity was greatest for the population that experienced the greatest temporal variation in nature. However, the adaptive significance of plasticity could be confirmed for only one of the traits: foliar trichome density.

4. Adaptive responses should evolve only when an environmental cue provides a reliable and timely indicator of appropriate phenotypes. This topic has been studied extensively in plants that respond to crowding conditions by elongating internodes and accelerating flowering, with the first response helping to escape competition for light and the second response helping to increase reproduction before death. The environmental cues that initiate this response are overall irradiance and the ratio of red to far red wavelengths, both of which are indicators of vegetation-generated shade. However, the same cues will not reliably indicate local competitive conditions in woodland habitats, where shade is mostly determined by larger trees. As expected, plants from nonwoodland habitats show greater responses to light cues than do populations from woodland habitats (Morgan and Smith 1979), and reciprocal transplant experiments in *Impatiens capensis* have confirmed the adaptive significance of these differences (Donohue et al. 2000, 2001). Evidence that plasticity is stronger under more predictive conditions has also been reported for animals (e.g., Porlier et al. 2012).

Multiple lines of evidence support the expectation that greater trait plasticity evolves in more variable environments. Importantly, the plasticity can buffer performance and fitness across a range of environments (Lynch and Gabriel 1987, Chevin et al. 2010, Lande 2014, Reusch 2014). Yet counter-examples exist, such as the maintenance of high plasticity in isolated populations experiencing relatively stable environments (Torres-Dowdall et al. 2012, Wiens et al. 2014) and the failure of generalists to evolve in variable environments in some laboratory experimental evolution studies (Condon et al. 2014). Thus, while the general expectations are often upheld, numerous exceptions point to the importance of additional interacting factors (Angilletta 2009, Condon et al. 2014).

QUESTION 4: DOES PLASTICITY AID COLONIZATION AND RESPONSES TO ENVIRONMENTAL CHANGE?

Large environmental shifts should pose problems for populations because existing phenotypes no longer will be well suited for the new conditions. In such cases, organisms are expected to shift their phenotypes in an adaptive direction, which might then make the difference between persistence versus extirpation. This "phenotypic rescue" can occur if populations undergo adaptive genetic change ("evolutionary rescue"), if individuals move to more appropriate locations, or if individuals manifest plasticity ("plastic rescue") (Chevin and Lande 2010, Chevin et al. 2010, Yamamichi et al. 2011, Barrett and Hendry 2012, Gomez-Mestre and Jovani 2013, Kovach-Orr and Fussmann 2013, Ezard et al. 2014, Duputié et al. 2015). I have previously discussed how populations can show contemporary evolution of important phenotypic traits (chapter 3) and how such changes might contribute to evolutionary rescue (chapter 7). In the present question, I focus on how phenotypic rescue might be achieved through plasticity in two contexts: in situ environmental disturbance (e.g., climate change) and the introduction of populations to new environments.

The basic idea behind plastic rescue is that individuals might be able to evaluate altered conditions and adjust their phenotypes appropriately, which might then increase mean population fitness and thereby enhance persistence and colonization of new environments. This phenomenon has been called the "Baldwin Effect" (Simpson 1953a, Price et al. 2003, Ghalambor et al. 2007, Crispo 2007) following its exposition by James Baldwin (1896, 1902). Baldwin further suggested that, once adaptive plasticity occurred, genetic change would be expected in the direction of the plastic response. This second step has been termed "genetic accommodation" (West-Eberhard 2003, Schlichting and Wund 2014). Waddington (1953, 1961) argued that the specific type of post-plasticity genetic change would be canalization of the trait such that the new phenotypes would no longer require environmental induction, a phenomenon he called "genetic assimilation" (West-Eberhard 2003, Crispo 2007, Schlichting and Wund 2014). Spalding (1873) had a similar idea, as described by Price (2008, p. 133). Despite the appeal of these ideas, it has been argued that concrete evidence is lacking (de Jong 2005) and the opposite sequence (evolution first, then plasticity) is also possible (Scheiner and Holt 2012). Here I consider evidence for the first part of the idea: plasticity aids persistence, colonization, and invasiveness.

Perhaps the best evidence for the importance of plasticity in responding to environmental change comes from studies of phenological responses to climate warming. Many organisms have advanced the timing of spring life history events (e.g., flowering, breeding, migration) as temperatures have increased and winters have shortened over the past 50 years (Parmesan and Yohe 2003). It is hard to ascertain whether these changes are the result of genetic evolution versus phenotypic plasticity (or both), mainly because some common methods for confirming a genetic basis for phenotypic change (e.g., common-garden experiments) are difficult to apply in a temporal (allochronic) context (Gienapp et al. 2008, Merilä and Hendry 2014). Without disputing the importance of evolution in at least some phenological changes (Bradshaw and Holzapfel 2006, Merilä and Hendry 2014, chapter 7), plasticity must also often be important. For instance, the aforementioned study of UK great tits (Charmantier et al. 2008) suggested that plasticity was entirely sufficient for adaptive responses of reproductive timing to climate change. Similar arguments have been made for other species, including collared flycatchers (*Ficedula albicollis*) in Gotland (Przybylo et al. 2000). By contrast, phenotypic plasticity seems insufficient for fully adaptive responses to climate change in other instances (see question 2). The next step is transition from trait changes to fitness consequences.

The populations referenced in the above paragraph all persisted in the face of environmental change, and perhaps adaptive plasticity was the reason, although explicit confirmation is not available. A more informative comparison, however, would be to consider the role of plasticity in populations showing alternative demographic responses to climate change. For example, Willis et al. (2008) recorded changes in the flowering time and abundance of plant species over 150 years in "Thoreau's Woods," Concord, Massachusetts. In this location, the species that were extirpated were those that showed low plasticity in flowering time in relation to temperature. The implication is that persistence of the remaining species, whose flowering time advanced by an average of 7 days, was at least partly due to plastic rescue. As another example, the bird studies discussed in chapter 7 (Q1) represent a situation where limits to plasticity have prevented sufficient change in breeding time, which has caused population declines (Both and Visser 2001, Both et al. 2006, Møller et al. 2008). It thus seems that plasticity will be sufficient for phenotypic rescue in some instances, whereas evolutionary changes will be needed in others (see also Phillimore et al. 2010, Duputié et al. 2015). With this recognition, plasticity has been increasingly incorporated into population viability and evolutionary rescue models for specific taxa (Baskett et al. 2009, Gienapp et al. 2012, Benton 2012, Vedder et al. 2013). The upshot of these analyses is that plasticity should often have a positive effect on population persistence, although negative effects are also possible (Duputié et al. 2015).

When organisms are introduced into new environments, adaptive plasticity might play a key role in colonization, persistence, and invasiveness (Baker 1965, Richards et al. 2006, Hulme 2008), perhaps even on macro-evolutionary scales (Standen et al. 2014). One way to inform this possibility is to compare levels of plasticity in fitness-related traits between invasive and noninvasive species, with the latter being either native species with which the invasive species is interacting or noninvasive species from the invasive species' native range. Early qualitative reviews for plants yielded inconclusive results, with greater plasticity found as commonly for noninvasive species as for invasive species (Bossdorf et al. 2005, Richards et al. 2006). However, a subsequent quantitative meta-analysis that examined 75 invasive/noninvasive plant species pairs came down decisively in favor of greater plasticity in invaders (Davidson et al. 2011). A related, but independent, line of inquiry asks whether behavioral plasticity promotes invasion success (Wright et al. 2010). Sol and colleagues (2002, 2005a, 2008) found that brain size (expected to be correlated with behavioral flexibility) and foraging innovation (a measure of behavioral flexibility) were positively associated with the probability that introduced birds and mammals became invasive. At the same time, however, some species with modest brain sizes become invasive and some species with large brain sizes don't (fig. 11.8): that is, the variance explained isn't very high. Overall, then, although behavioral plasticity in animals (and trait plasticity in plants) might sometimes aid responses to new environments, it certainly isn't a universal solution.

Plasticity sometimes aids colonization of new environments and responses to in situ environmental change. However, plastic responses aren't always necessary or sufficient in these contexts. Plasticity seems likely to be most beneficial when (1) the trait is particularly important for fitness, (2) the new conditions are similar to those previously experienced by a lineage (plasticity is then more likely to have been shaped by past selection), (3) it can accomplish large phenotypic changes, (4) it isn't very costly, and (5) the traits are behavioral or physiological as these traits should be the most malleable traits on short time scales. Future work would do well to focus on the population dynamic consequences of plasticity, as this key topic is currently understudied.

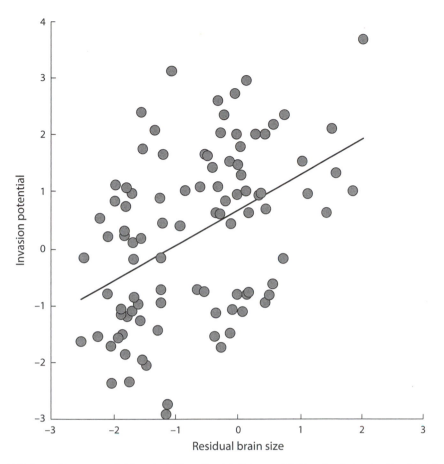

Fig 11.8. Invasion potential (probability of establishment following introduction) in birds is positively related to residual brain size (brain size corrected for body size). Each data point represents a different species and similar results are obtained if phylogeny is controlled with independent contrasts. The data are from Sol et al. (2012) and were provided by D. Sol

QUESTION 5: DOES PLASTICITY PROMOTE OR CONSTRAIN
GENETIC EVOLUTION?[1]

A number of arguments have been advanced for how plasticity might promote or constrain adaptive genetic change. On the constraining side, the main argument is that plasticity shields the genotype from selection, thereby slowing adaptive *genetic* change (Linhart and Grant 1996, Huey et al. 2003, Ghalambor et al. 2007). This argument applies mainly to *adaptive* plasticity, whereas maladaptive plasticity would be expected to increase selection for adaptive "genetic compensation" (West-Eberhard 2003, Grether 2005, Ghalambor et al. 2007). On the promoting side, a common argument

---

[1] This question and the one that follows overlap with debates about "genes as leaders" versus "genes as followers," with the first perspective being that trait differences start as genetic polymorphisms and the second being that trait differences start as environmentally induced plasticity (West-Eberhard 2003). The reality is that genes are sometimes leaders and sometimes followers (and, I suppose, they often walk together), and the relative importance of each cannot be determined with most existing data (Schwander and Leimar 2011).

is that plasticity allows colonization of, or persistence in, extreme environments (question 4), which thereby increases selection on that trait or other traits (West-Eberhard 2003, Schlichting and Wund 2014). Another suggested positive influence is that plasticity can expose otherwise cryptic genetic variation (not expressed under normal conditions) to selection (West-Eberhard 2003, Suzuki and Nijhout 2006, Pfennig et al. 2010, Moczek et al. 2011, Gomez-Mestre and Jovani 2013, Wund and Schlichting 2014). In addition, simulations have suggested that plasticity can alter genetic architecture so as to increase the production of adaptive phenotypes (Fierst 2011). Given that nearly everything seems possible in theory (Paenke et al. 2007),[2] I will here focus on empirical observations relevant to the key ideas.

The phenomenon of counter-gradient variation, where genetic effects are in the opposite direction to plastic effects (fig. 11.2), provides a nice example of how maladaptive plasticity can promote, indeed necessitate, compensatory adaptive genetic change (Levins 1968, Conover and Schultz 1995, Conover et al. 2009). A well-established example is growth rate in Atlantic silversides. Northern and southern populations of this fish have similar body sizes in nature despite better growing conditions in the south. When raised in a common garden, however, fish from the northern population grow faster and to a larger size than do fish from the southern population. In this case, plastic effects on growth that result from environmental differences have led to the evolution of compensating genetic differences in intrinsic growth rate (Conover and Present 1990, Present and Conover 1992). The contrasting pattern of cogradient variation, where plastic effects are in the same direction as genetic effects, can imply that plasticity has reduced genetic divergence (e.g., Byars et al. 2007). As both patterns are known to exist in nature, the important question becomes: how common is each? Although a formal meta-analysis has not been conducted, Conover et al. (2009) summarized more than 60 examples of counter-gradient variation, while finding many fewer examples of cogradient variation.

Recent work has examined associations between plastic and evolutionary gene expression differences in introduced populations. Pointing toward countergradient effects, Ghalambor et al. (2015) showed that guppies introduced from high-predation to low-predation environments in nature documented evolutionary divergence in gene expression that was in the opposite direction to the plastic effect on gene expression that arose when captive guppies were experimentally exposed to predators. Pointing toward cogradient effects, Mäkinen et al. (2016) found that plastic and evolutionary gene expression differences in response to temperature differences were positively correlated in introduced European grayling (*Thymallus thymallus*). Clearly, no universal pattern will be found and future work should focus on delineating the conditions that determine when cogradient versus counter-gradient patterns emerge.

---

[2] Many theoretical studies have considered whether plasticity will accelerate or decelerate genetic evolution. These have been reviewed and "unified" by Paenke et al. (2007), who conclude that "if the change of the phenotype due to plasticity is adaptive (i.e., toward higher fitness) and its magnitude is similar for all genotypes, evolution is predicted to be accelerated if the logarithm of the fitness function is convex and decelerated when it is concave. More generally, a fitness landscape with a convex logarithm is more likely than one with a concave logarithm to result in evolution being accelerated due to directional plasticity, even if the effect of (plasticity) differs among genotypes." Additional theoretical treatments examining interactions between genetic change and plasticity (and the evolution of plasticity and bet hedging) include Botero et al. (2015) and Tufto (2015).

Counter-gradient variation provides a particularly obvious situation where plasticity can promote genetic change—because the plasticity is maladaptive and thus imposes selection for genetic compensation. Another situation occurs when plastic change in one trait necessitates genetic change in other traits: that is, altering one aspect of the phenotype should require compensatory changes in other aspects of the phenotype. As a clear example, the introduction of a predator (curly tailed lizards, *Leiocephalus carinatus*) caused a prey species (*Anolis sagrei* lizards) to plastically shift their habitat to narrow perches in trees, which imposed selection for shorter legs (Losos et al. 2006). Of course, the opposite effect is also known: behavioral thermoregulation (plasticity) reduces exposure to extreme temperatures and thereby reduces selection for physiological temperature adaptation (Huey et al. 2003).

For the remainder of this question, I focus on the more complicated possibility that *adaptive* plasticity in a trait *promotes* genetic change in the same trait. The best evidence here would come from experiments showing that populations with greater adaptive plasticity in a trait also show faster adaptive evolution of that trait. At least one such study has been performed. Schaum and Collins (2014) conducted a laboratory experimental evolution study using 16 lineages of *Ostreococcus* (a marine green algae microbe) that initially differed in CO_2-related plasticity for "oxygen evolution rates" (generating oxygen through chemical reaction). During 400 generations of rearing under constant or fluctuating CO_2 conditions, lineages with higher ancestral plasticity showed faster evolution of population growth rates (a measure of fitness). This positive relationship between plasticity and evolution was, as expected, strongest in the treatments with fluctuating CO_2. Evidence for the converse—plasticity in a trait *constrains* genetic divergence in that trait—also could be generated through the above-suggested experiments. In the only example I know of, Samani and Bell (2016) found a negative association between plasticity and evolutionary responses to stress in laboratory evolutionary rescue studies with yeast.

With so few experiments of this sort, all of them in the laboratory, we also need to consider correlative support from natural populations. Two sorts of comparisons—one based on populations and one based on traits—are useful. For the first, populations showing greater plasticity in a trait should show lower genetic divergence in that trait. For the second, traits showing greater plasticity should show lower genetic divergence. Exemplifying a population-based comparison, Misty Lake versus Misty Inlet stream stickleback show strong genetic divergence in a number of adaptive traits, whereas Misty Lake versus Misty Outlet stream stickleback show no genetic divergence but rather plastic differences (Hendry et al. 2002, Sharpe et al. 2008). Of course, cause and effect is here difficult to establish given that high gene flow between the lake and outlet populations (Moore et al. 2007, Roesti et al. 2012) could prevent genetic divergence, leaving plasticity as the only recourse (as opposed to plasticity evolving and then limiting genetic divergence). Exemplifying a trait-based comparison, we have the adaptive responses of fish to low dissolved oxygen. In many species, fish from low-oxygen environments have larger gills so as to extract more oxygen, and also have smaller brains due to the resulting limitations on cranial space. Crispo and Chapman (2010) collected populations of the cichlid fish *Pseudocrenilabrus multicolor victoriae* from different oxygen environments in nature and raised their offspring under high and low oxygen conditions in the laboratory. Essentially all of the resulting variation in gill size was plastic, with no apparent genetic differences among populations (fig. 11.9). By contrast, brain size was less plastic and showed more

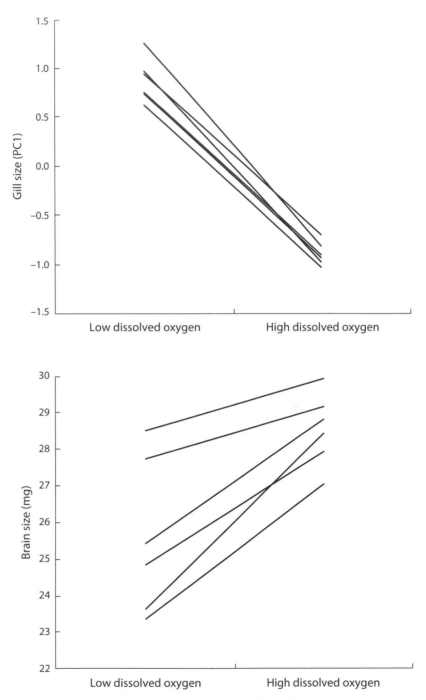

Fig 11.9. Population-level reaction norms for gill size (PC1 of measurements standardized to a common body size) and brain size (standardized to a common body size) for six populations of *Pseudocrenilabrus multicolor victoriae* raised under different oxygen conditions in the laboratory. The data are from Crispo and Chapman (2010)

genetic variation among populations (fig. 11.9). Similar findings (more plastic traits—and gene constructs—show lower genetic divergence) have emerged in studies of the effects of predators on guppies (Torres-Dowdall et al. 2012, Ghalambor et al. 2015).

Although the above observations are consistent with the idea that plasticity constrains genetic divergence, causation is hard to establish and, regardless, too few studies have been conducted to invite generalization. Moreover, the effects of plasticity could be—indeed they are often expected to be—transient during the course of evolution, such as in the case of genetic assimilation. Detecting such effects requires the tracking of genetic and plastic contributions during the course of environmental change or in controlled experiments (e.g., Schaum and Collins 2014, Ghalambor et al. 2015).

Plasticity will sometimes promote and sometimes constrain genetic evolution. Some possible predictions are that promoting effects will be most likely when (1) plasticity enables colonization/persistence where it would not otherwise be possible, (2) plasticity in one trait (e.g., behavioral flexibility that alters resource use) results in altered selection on other traits, (3) selection is on plasticity itself, (4) plasticity exposes otherwise cryptic genetic variation, and (5) plasticity is maladaptive. Although correlative tests of these hypotheses will be useful, particularly informative approaches would be experimental. In particular, more versus less plastic genotypes could be introduced into new environments and subsequent adaptive evolution could be monitored—as Schaum and Collins (2014) did in the laboratory.

QUESTION 6: DOES PLASTICITY HELP OR HINDER ECOLOGICAL SPECIATION?

The previous question focused on variation within species, whereas the present question considers the same issues with respect to species formation. The debate has crystallized around two opposing schools of thought, which I here dichotomize for the sake of argument. (Both perspectives, and various intermediates, are acknowledged in most publications.) The first perspective is an extension of the "Baldwin Effect" described in question 4. It argues that plasticity facilitates colonization of new environments, or the use of new resources, after which phenotypes are exposed to divergent selection that causes adaptive genetic divergence and hence ecological speciation (Skúlason and Smith 1995, Smith and Skúlason 1996, Robinson and Parsons 2002, West-Eberhard 2003, Pfennig et al. 2010). One branch of this argument specifically emphasizes behavioral flexibility that results in the use of new resources, which can then enhance speciation through a process sometimes called "behavioral drive" (Wyles et al. 1983). The opposing school of thought is an extension of question 5, arguing that plasticity shields the genotype from selection and thereby reduces genetic divergence and hampers speciation (Price et al. 2003, Ghalambor et al. 2007, Crispo 2008, Svanbäck et al. 2009, Thibert-Plante and Hendry 2011b). Behavior also could play into this constraining effect—by allowing organisms to use similar resources, thus reducing divergent selection, even in different environments (Duckworth 2009). I start by summarizing and evaluating three empirical observations suggested to indicate that plasticity promotes ecological speciation.

1. Plasticity within species, ideally demonstrated in ancestral forms, is sometimes in the same direction as genetic differences among species. One clear example is trophic morphology in fishes, where "limnetic" versus "benthic" diets cause plastic divergence in trophic morphology in a direction that parallels genetically based

divergence among closely related species (Day et al. 1994, Robinson and Parsons 2002, Adams and Huntingford 2004, Wund et al. 2008). Another example comes from color in ecomorphs of Hawaiian spiders (Brewer et al. 2015). The common inference therefrom is that ancestral plasticity initiated and promoted the subsequent genetic divergence.

2. Character displacement among species is sometimes facilitated by polyphenism (different, *discrete* phenotypes emerge when the same genotype is exposed to different environments), which can then sharpen reproductive barriers (Pfennig and Pfennig 2009). In spadefoot toads, for example, two species (*Spea bombifrons* and *Spea multiplicata*) can develop either herbivorous or carnivorous tadpoles: but, when reared together, *S. multiplicata* produces many fewer carnivores than does *S. bombifrons*. In addition, *S. multiplicata* from ponds with more *S. bombifrons* in nature are genetically less likely to produce carnivores in a common-garden environment (Pfennig and Murphy 2002). The common inference therefrom is that selection in sympatry has enhanced ancestral polyphenism and thereby exaggerated species divergence.

Both of these arguments are *consistent* with the idea that plasticity promotes speciation, yet neither provides strong evidence. One reason is that no meta-analysis has quantified the extent to which plasticity within species is similar to divergence among species, nor the extent to which sympatry exaggerates polyphenic differences. In addition, observed plastic effects within species are often much smaller than observed differences among species (e.g., Losos et al. 2000). Another limitation is that a low-plasticity "control" comparison is not normally considered: that is, speciation might have been even more likely/rapid/dramatic if plasticity wasn't present. Most critically, the identical prediction (plastic and genetic differences are in the same direction) also emerges from arguments that plasticity *constrains* divergence (see above).

3. Plasticity is higher in taxonomic groups that are more speciose, as a number of studies have shown. Nicolakakis et al. (2003) reported that innovation rate, a proxy for behavioral flexibility, is positively related to the number of species within bird taxa. Sol et al. (2005b) showed that relative brain size, which is correlated with behavioral flexibility, is positively related to the number of subspecies in Holarctic passerines. Tebbich et al. (2010) pointed out that the bird group that has diversified most in Galápagos, Darwin's finches, shows very high levels of behavioral flexibility. Pfennig and McGee (2010) used sister group comparisons to show that fish and amphibian lineages that include polyphenic species are more speciose than those that do not. The common inference from such findings is that plasticity generally promotes diversification, speciation, and adaptive radiation.

Again, such observations are consistent with the hypothesis, but are attended by important caveats. In particular, the level of plasticity is often unknown in the ancestral species. As a result, it is difficult to establish whether plasticity was the cause or the consequence of high diversification. In addition, the number of polyphenic species is often very low even in the speciose/polyphenic groups; and, if polyphenism is rare, it would be more likely to occur simply by chance in groups with more species. Thus, although numerous studies argue that plasticity enhances speciation, remaining ambiguities leave the question still very much open.

One process by which plasticity is particularly likely to promote speciation occurs when juveniles imprint on parents, conspecifics, environments, or resources. Such imprinting can lead to assortative mating that allows genetic divergence and the evolution of reproductive barriers. In birds, nestlings sometimes imprint on the songs of their fathers (Price 2008), with male offspring later singing—and female offspring later preferring—similar songs. The result can be mating isolation among groups whose songs have diverged for whatever reason (Price 2008). Remarkably, male and female nestlings of brood-parasitic *Vidua* finches imprint in a similar way on the songs of their host species, which leads to assortative mating among finches parasitizing different hosts (Payne et al. 2000). In insects, larvae sometimes imprint on the plant on which they feed ("conditioning") and then preferentially select those plants during mating and oviposition, thus generating mating isolation among groups using different host plants (Funk et al. 2002). Moreover, Rebar and Rodríguez (2014) have shown that treehoppers raised on different plant genotypes have different sexual signals, which could promote reproductive isolation. In salmonid fishes, juveniles often imprint on chemical properties of their natal site and then strongly "home" back to that site for reproduction, which reduces gene flow among populations (Hendry et al. 2004b). Considering these examples as instances of "positive" imprinting, "negative" imprinting can also reduce gene flow: for example, exposure to heterospecifics can strengthen preferences against them (Price 2008, Delbarco-Trillo et al. 2010). In each of these cases, reproductive isolation depends on individuals being exposed to different environments, with another example being the commensal bacteria in *Drosophila* that seemingly determine the chemical signals that drive mating isolation (Sharon et al. 2010).

Interestingly, such effects of plasticity reverse the causal pathway typically assumed in ecological speciation: that is, adaptive divergence commences and causes reproductive barriers to evolve as a consequence (Schluter 2000a, Rundle and Nosil 2005, Räsänen and Hendry 2008). Instead, plasticity can cause phenotypic differences in traits or preferences as soon as divergent environments are colonized. Those plasticity-driven phenotypic differences could then cause reproductive barriers, such as through imprinting, before any genetic divergence has taken place. These plastic phenotypic differences, if adaptive, could then reduce gene flow among environments to the point that adaptive genetic divergence can take place, which could then strengthen reproductive barriers (Thibert-Plante and Hendry 2011b). How often this process occurs is unknown.

What of the opposing school of thought—that plasticity retards speciation? Empirical support might be provided through evidence that groups with *lower* plasticity speciate more often, or that the traits determining reproductive isolation among species are not particularly plastic (especially in the ancestor). Formal tests of these predictions have not been performed; however, many populations in different environments show strong plastic differences and yet minor—if any—reproductive isolation. Following up on examples from question 5, gene flow is high among populations where phenotypic divergence has a primarily plastic basis in *Pseudocrenilabrus* from different oxygen environments (Crispo and Chapman 2008) and between the Misty Lake and Outlet stickleback populations (Hendry et al. 2002, Moore et al. 2007, Roesti et al. 2012). Despite these and other suggestive examples, the idea that plasticity hampers speciation has not yet been subject to rigorous testing.

Plasticity will sometimes help and sometimes hinder ecological speciation (see also Duckworth 2009) but, at present, empirical tests are insufficient to allow general conclusions as to how often and when each result emerges. Plasticity would seem most likely to have positive effects in the various manifestations of imprinting, when mating cues depend on environmental exposure, and when dispersal occurs after (rather than before) plastic changes occur. Specifically, reproductive barriers among populations in different environments should be greatest when developmental plasticity occurs before dispersal than when it occurs afterward—because, in the latter case, individuals can adaptively adjust their phenotypes to the new conditions and thereby increase their survival and reproductive success (Thibert-Plante and Hendry 2011b).

QUESTION 7: HOW FAST DOES PLASTICITY EVOLVE?

Many studies dichotomize phenotypic change into that caused by genetic change versus that caused by plasticity. In reality, both effects can occur at the same time and can influence each other—as described above. Moreover, plasticity can evolve and this evolution can have important consequences for population dynamics, including evolutionary rescue (question 4). It is therefore important to ask how quickly plasticity can evolve and what factors increase or decrease this rate. These questions are parallel to those asked previously with regard to rates of evolution of mean phenotypes (chapter 3), reproductive isolation (chapter 6), and ecological effects (chapters 7–9). In each of these previous cases, rates of evolution were highly variable but substantial effects were sometimes seen on contemporary time scales, such as years to centuries.

A first point is that many studies have documented the evolution of reaction norms on the time scale of decades, including the above-described phototactic behavior of *Daphnia* in response to changing fish predation (fig. 11.1). Many other examples exist—and I will here mention only two. The Asian shade annual plant *Polygonum cespitosum* colonized North America in the early 1900s and has recently spread into more open habitats. In the ten years following this niche expansion, the plant has evolved increased plasticity in root allocation and physiological traits in response to open versus shaded conditions (Sultan et al. 2013). The Asian shore crab *Hemigrapsus sanguineus* was first reported in North America in 1988 and feeds on native marine mussels (*Mytilus edulis*). At present, mussels in areas where the crab has invaded (southern New England) show inducible shell thickening in response to waterborne *H. sanguineus* cues, whereas mussels in areas where the crab has not invaded (northern New England) do not (Freeman and Byers 2006). At the same time, however, a number of other studies have documented instances where plasticity did not evolve even on long time scales despite a change in selection pressure. A particularly obvious example is the retention of antipredator behavior long after the predator is no longer present (Lahti et al. 2009). These latter cases likely reflect relaxed selection (the trait is not expressed in the absence of the cue), in which case trait evolution would occur only through the relatively slow processes of drift and mutation, and perhaps also correlated selection on other traits.

As always, selected examples can only take us so far and any hope of generality must come from meta-analyses. With this goal, Crispo et al. (2010) analyzed 20 studies that measured plasticity in two or more populations, at least one of which was subject to recent human disturbance and at least one of which was not. The authors calculated rates of change for plasticity in darwins and haldanes, the same metrics encountered earlier with regard to the evolution of mean trait values (chapter 3). Results showed that disturbed plant populations often evolved changes in plasticity and that different taxa and traits showed different responses. Based on a qualitative comparison between Crispo et al. (2010) and Hendry et al. (2008), rates of evolution of plasticity were about the same (on average) as the rates of evolution of mean phenotypes.

Plasticity can show considerable evolutionary change on contemporary time scales, although the rates of this evolution are highly variable among taxa and traits. These results confirm theoretical expectations that plasticity can evolve quickly and that it is likely to have important fitness consequences (Lande 2009, Chevin et al. 2010, 2013, Chevin and Lande 2011). What we now need are more studies assessing the population dynamic consequences of plasticity—and its evolution—in natural populations.

QUESTION 8: DOES PLASTICITY HAVE COMMUNITY/ECOSYSTEM EFFECTS?

To the extent that organismal traits have community/ecosystem effects, plastic changes in traits should alter those effects (Collins and Gardner 2009, Chevin et al. 2013, Kovach-Orr and Fussmann 2013). Very few empirical studies have directly assessed this question but a few examples will illustrate some possibilities, starting with community influences and then moving to ecosystem influences.

Many foraging traits of fishes are phenotypically plastic in response to diet. For instance, fish fed on zooplankton diets (as opposed to benthic diets) tend to have longer gill rakers and changes in jaw morphology that increase foraging efficiency on those food items (Day and McPhail 1996, Lundsgaard-Hansen et al. 2013). Because these plastic changes influence foraging efficiency on the different food types, and because fish traits have dramatic influences on aquatic prey communities (Brooks and Dodson 1965, Post et al. 2008), diet-induced trophic plasticity should influence prey communities. Indeed, mesocosm experiments have demonstrated that rearing European whitefish (Lundsgaard-Hansen et al. 2014) or threespine stickleback (Matthews et al. 2016) on different diets can influence aquatic communities and some ecosystem processes. Such effects have not yet been demonstrated formally in nature but they would be fascinating to explore, not the least because they show a strong chance of feedbacks. That is, plastic changes in foraging traits that influence foraging success on a given food type should reduce the availability of that food type (and induce its evolution), which should then influence further plasticity and selection.

Rates of feeding, metabolism, and growth dramatically influence biological stoichiometry (chapter 10) by altering the consumption and excretion of various elements. As a result, plastic changes in these rates could have dramatic effects on the availability and transfer of elements within communities and ecosystems. As one example, Schmitz (2013) argued that increasing animal metabolic rates with increasing temperature owing to climate change should cause "phenotypically plastic shifts in animal elemental demand, from nitrogen-rich proteins that support production to carbon-rich soluble carbohydrates that support elevated energy demands." The resulting change in diets should then have important consequences for carbon cycling (Schmitz 2013, for related ideas see Moya-Loraño et al. 2014). As another example, Dalton and Flecker (2014) showed that the presence of dangerous predators (simulated with predator cues in the laboratory) decreased N excretion rates of guppies by 39%, which could have important consequences for this limiting nutrient in their stream ecosystems.

Another likely arena for ecosystem effects of plasticity is for organisms that produce chemical resources that are used by many other organisms, such as plants producing CO_2 or fixing nitrogen. Such effects seem particularly likely given the great plasticity in these processes depending on environmental conditions, such as ambient levels of CO_2 or nitrogen, as well as temperature and humidity. For instance, Collins and Gardner (2009) provide a "worked example" of how to calculate the potential contribution of plasticity, evolution, and community change to carbon uptake by marine phytoplankton experiencing elevated CO_2 levels. At present, however, nearly all such applications (for other examples, see Chevin et al. 2013) are theoretical, hypothetical, or lab-based. Sorely needed are formal assessments in natural systems. As an example, Jackrel and Wootton (2015) showed in nature that plastic responses in red alder to herbivore stress altered leaf nutrients and, hence, ecosystem function through the suppression of aquatic decomposition.

Plasticity seems likely to have considerable influences on ecological dynamics at the community and ecosystem levels, with stoichiometry and foraging traits being just a few likely direct effects. In addition, any plasticity-induced effect on population dynamics (e.g., plastic rescue) should also have indirect effects on community and ecosystem parameters. For instance, Derry et al. (2013) describe how plasticity, coupled with dispersal, can aid native-invasive species coexistence when native species have an uninvaded "refuge" habitat.

Conclusions, significance, and implications

Plasticity is the prodigal son of evolutionary biology. Ignored for many years owing to a primary focus on the role of natural selection in shaping genetic differences among populations, plasticity has recently returned with a vengeance—perhaps too much so for the liking of some (de Jong 2005). The reality is that essentially all traits will be influenced by plasticity, although the extent, type, and consequences will be highly variable among traits and taxa. Importantly, evolution and plasticity are not strict alternatives, but rather both contribute to divergence in many instances. Of particular note, many trait differences are likely due to the *evolution of* plastic differences (i.e., divergent reaction norms), which can occur on contemporary time scales. With regard to its evolution, plasticity is often (but not always) adaptive, often (but not always) costly, and often (but not always) an evolved response to spatiotemporal environmental variation. Given these numerous and diverse possibilities, it is not surprising that valid arguments can be raised that plasticity will sometimes promote and sometimes constrain adaptive genetic divergence and speciation. At present, however, the data supporting either alternative is indirect and often unconvincing.

Given that plasticity will influence traits in nearly every instance, it is also likely to have an important role in shaping the ecological effects of those traits: i.e., the evo-to-eco of eco-evolutionary dynamics. For instance, adaptive plasticity will often (but not always) aid the ability of populations to persist in changing environments and to colonize and spread in new environments. These population dynamics effects likely then cascade to indirect influences on communities and ecosystems. Plasticity likely also has direct influences on communities and ecosystems. As one example, trophic traits are often strongly influenced by plasticity, which will then have many community and ecosystem consequences as outlined in chapters 8 and 9. Similarly, plant traits are often very plastic and so many of the ecological effects of those traits (most obviously individual plant biomass) also will be fundamentally altered by plasticity. For all of these reasons, plasticity needs to be an integral part of any conceptual framework and empirical investigation of eco-evolutionary dynamics.

Chapter 12

What We Do and Don't Know

After an introductory chapter (conceptual framework), five chapters on the eco-to-evo side of the story (selection, adaptation, adaptive divergence, gene flow, ecological speciation), three chapters on the evo-to-eco side of the story (population dynamics, community structure, ecosystem function), and two chapters on underpinnings (genetics/genomics, plasticity), the time has come to attempt a wrap-up. I will do so by describing what I think we now know about eco-evolutionary dynamics and, more importantly, what we don't know.

Eco-to-evo: what we know

As I wrote the first part of the book, a general hypothesis—that is, a particular model of the way eco-to-evo works—began to coalesce for me. The hypothesis is not entirely novel but instead builds on previous views of the phenotypic adaptive landscape (Simpson 1944, Schluter 2000a, Arnold et al. 2001). Although metaphors can be a dangerous way of presenting an hypothesis, the basics of one kept springing to mind. Yet it took some time to settle on the specific metaphor that best captured the essence—and the nuances—of the hypothesis. It was on a recent visit to Galápagos that the appropriate metaphor finally came to mind, which I will here call the *Española-Isabela Hypothesis*.

Española is one of the oldest islands in Galápagos. It is located in the extreme southeast of the archipelago (fig. 12.1), to whence it drifted on the Nazca plate after forming over a geological hotspot more than 250 km to the northwest. The island is now rather low and flat (maximum elevation = 206 m), in sharp contrast to its probable topography when it formed via lava extrusion from the hotspot. At that time, its shape may have been more akin to Fernandina (elevation = 1476 m), which is the most recently formed island and still sits on the edge of the hotspot. Thus, between its initial formation and its current condition, a time span of approximately 3.2 million years, Española has weathered and sunk to its current low and flat topography.

To develop the metaphor, we can imagine Española as a peak on an adaptive landscape, with elevation equivalent to mean fitness for a population with two phenotypic traits having mean values corresponding to (in our metaphor) latitude and longitude. Sea level then represents a mean absolute population fitness of unity for a terrestrial organism, such as a colonizing ancestral finch. On land, population growth is positive at a rate that increases toward the peak. In the sea, it is negative at a rate that increases with distance from the island.

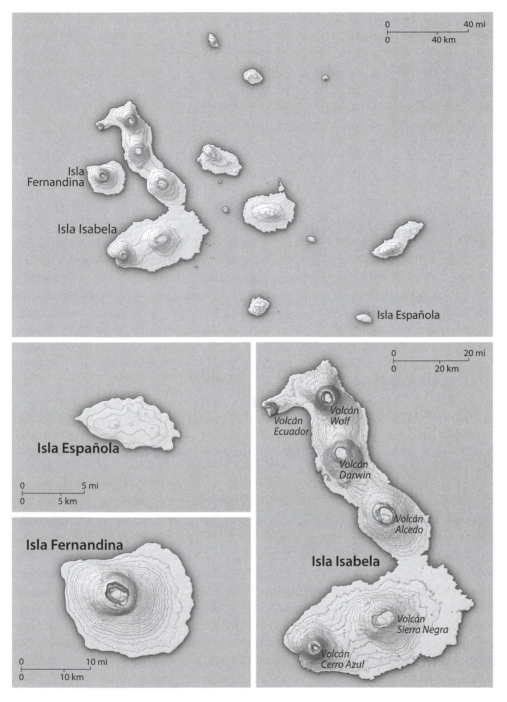

Fig 12.1. Galápagos Islands and their topography as a metaphor for the Española-Isabela Hypothesis

When the fitness peak is unoccupied by any population (metaphorically, right after volcanic activity ceases), it is quite high (high elevation) and its slopes are steep. Successful colonization is likely to be difficult because it requires the colonizing organisms to coincidentally have phenotypes that fall somewhere above the water level, allowing positive population growth. And, when colonization does happen, it seems most likely that the colonists will find themselves some distance from the peak, perhaps close to the shoreline. The closer the colonists are to the shoreline, the more vulnerable they will be to changing conditions of demographic stochasticity, making expansion relatively unlikely. For all of these reasons, many adaptive peaks in nature are likely unoccupied (chapter 4, question 4: i.e., C4-Q4) and introduced species usually fail to establish themselves in new locations (C7-Q5).

For new colonizing populations that do persist, selection will be very strong toward the peak, which represents the observation that populations not well adapted for the local environment are likely to experience strong directional selection (C2-Q1). The evolutionary response to this selection will next be shaped by, metaphorically, the spatial spread of the organism across the landscape on Española, representing the amount (heritability/evolvability: C3-Q1, C10-Q1) and shape (correlated traits: C3-Q5) of the genetic (co)variance matrix. Unless these constraints are severe, the population will rapidly evolve up the steep slope toward the peak, which represents the observation that contemporary evolution can be very rapid when current phenotypes don't match optimal phenotypes (C3-Q1). As the population approaches the peak, selection will decrease and adaptation will slow (C3-Q4). Eventually, the population's phenotypic distribution will rest on—and immediately around—the peak and selection will become stabilizing, although the population phenotypes may well bounce around the peak a bit depending on stochastic effects (Haller and Hendry 2014). As a result of this process, most established populations will be reasonably well adapted to their local environments (C3-Q6) and generally under weak selection (C2-Q1, C2-Q4).

As the population was approaching the fitness peak, its abundance was expected to increase as a result of improving adaptation (C7-Q2, C7-Q3) and, once the peak was reached, population size would be expected to approach or exceed carrying capacity. Competition would thus increase and, accordingly, reduce per capita resource availability. Per capita growth rate would thus decrease and, as a result, the peak would start to sink—metaphorically like the weathering that has caused Española to lower and flatten with time. Once the peak flattens, not only is directional selection absent but so too is stabilizing selection (C2-Q4). Stated another way, nearly all phenotypes are now present on the (now relatively flat) fitness peak and are rare on the steep slopes that extend off the island and underwater.

Importantly, the peak has drifted through phenotypic space (perhaps more than 250 km for Española) as it has lowered and flattened, which represents environmentally determined movement in the optimum value for mean phenotype. This movement, representing environmental change, can maintain directional selection and adaptive evolution acting on a population. Of course, if the tape of life were to replay all over again, as we might envision the different Galápagos islands, the initial peak likely (1) would be in a somewhat different location, (2) would be formed into a different size and shape, (3) would be colonized by a somewhat different population with different phenotypic/genetic properties, and (4) would drift in a different direction and with a potentially different speed. As a result, adaptation by the same sort of organism to the same sort of environment will likely lead to different phenotypic outcomes (nonparallel evolution, C3-Q2, C4-Q5) underlain by different genetic architectures (C10-Q5).

To summarize some of the key aspects of the Española-Isabela Hypothesis as they relate most directly to chapters 2 and 3: (1) Colonization of a new fitness peak can be difficult; (2) When colonization is successful, directional selection will be strong and adaptation rapid; (3) As the population mean phenotype approaches the peak, selection gradually shifts from directional to stabilizing; (4) Once the population is well adapted, selection (directional and stabilizing) mostly disappears apart from responses to peak movement; and (5) this process is likely to lead to somewhat different evolutionary outcomes in different replicates. Overall, most populations in relatively stable environments in nature should be well-adapted and currently experience weak, if any, selection—even though directional and stabilizing selection were what got them to where they are in the first place: that is, selection erases its traces (Haller and Hendry 2014).

Isabela, another Galápagos island, is quite different from Española. It has six high-elevation volcanoes (Alcedo = 1130 m, Cerro Azul = 1689 m, Darwin = 1330 m, Ecuador = 790 m, Sierra Negra = 1124 m, and Wolf =1707 m; fig. 12.1) separated to varying degrees by low-elevation, often bare-lava, areas that are inhospitable to most organisms. As a result, Isabela is often thought of as a collection of six islands separated by seas of lava. In the metaphor, Isabela is another sort of adaptive landscape: one with multiple high-fitness peaks separated by low-fitness valleys or moats—as is likely the case in nature (C4-Q2). The process of adaptation to any one of the six peaks might not be any different from the process described above in the context of Española. Yet Isabela allows the metaphor to better illustrate how evolution might or might not generate multiple forms, such as adaptively divergent populations or species.

To a crude approximation, diversification on an Isabela-style adaptive landscape might proceed in two ways. First, it could play out on a single fitness peak, metaphorically a single volcano. Just as the peak on Española lowered and flattened through time (due to competition) so too will the Isabela-style adaptive peaks lower and flatten. This process can, under the right circumstances, lead to disruptive selection that generates two species on the same initial peak (Dieckmann and Doebeli 1999, Doebeli et al. 2007, Rettelbach et al. 2013). Stated another way, competition generates two peaks on an underlying resource distribution that was initially unimodal.

Second, and likely much more common (C6-Q5), is diversification among existing resource peaks, such as those different Isabela volcanoes. A first important point is that, once a population has adapted to one peak (one Isabela volcano in the metaphor), it becomes hard to bud off a new population onto another peak (another volcano), which is likely why many potential fitness peaks (niches for potential species) are likely currently unoccupied (C4-Q3). The main reason is that switching peaks requires a subset of the original population to evolve lower fitness as it moves down the fitness peak and across the barren lava of low fitness to get to the new peak.

Although potentially difficult, the split of a population adapted to one fitness peak into separate populations adapted to different fitness peaks can be facilitated by several processes of which I here mention two. The first process emerges from the above-described effect of increasing competition for an existing adaptive peak that lowers and flattens the peak (C6-Q5). In such cases, the valley between the peaks, while not changing in absolute fitness, increases in relative fitness because the adjacent peak occupied by the population is sinking. In this case, a population on one peak doesn't have to evolve down so steep a fitness slope and doesn't suffer such a disadvantaged when traveling through the adjacent valley toward a (potentially higher) alternative peak. That is, competition can shallow the valleys between peaks and make the colonization of new peaks much more likely. The second process invokes the emergence of a new peak or ridge (new volcanic activity between existing volcanoes) between the existing peaks, which also shallows the fitness valley between peaks and enables adaptive divergence (Schluter 2000a).

Although the shallowing of fitness valleys facilitates adaptive divergence, it can also generate constraints that limit how far divergence proceeds. In particular, shallow valleys between peaks can enhance gene flow between populations on the two peaks, which can prevent each from embarking on a separate evolutionary trajectory (C5-Q1-4). The various outcomes can be (1) collapse back to the original peak (no population permanently colonizes the new peak), (2) movement of a single population from one peak to the other (extinction on the old peak coinciding with spread on the new peak), or a persistently intermediate generalist that uses both peaks without splitting into separate forms (Rueffler et al. 2006).

If, however, circumstances reduce—perhaps only temporarily—gene flow between the ancestral population on the old peak and the diverging subpopulation on the new peak, the latter can climb the new peak and phenotypic divergence from the ancestral population can become substantial. In short, limited gene flow among populations in different environments should lead to adaptive divergence (C4-Q1). (Of course, gene flow can sometimes have considerable positive consequences for adaptation (C5-Q2).) As a result of adaptive divergence, the fitness of adapting residents on the new peak rises in relation to less well-adapted immigrants from the ancestral population (Schluter 2000a, Rundle and Nosil 2005, Nosil 2012). The outcome is increasingly effective selection against migrants and hybrids, as well as selection for habitat preference, assortative mating, and other reproductive barriers. These barriers to gene flow can arise as a byproduct of adaptive divergence or as a result of selection to avoid maladaptive interbreeding (C6-Q6) and to avoid competition with the other populations (C6-Q5). Eventually, the ancestral and descendent populations, now on different fitness peaks, become so divergent as to have strong reproductive barriers and are said to have undergone substantial ecological speciation (C6-Q1, C6-Q2). Of course, these species remain susceptible to merging if environmental conditions change to further shallow fitness valleys or bring fitness peaks closer together (C6-Q4).

To summarize some of the key aspects of the Española-Isabela Hypothesis as they relate most directly to chapters 4–6: (1) evolutionary diversification often proceeds through adaptation to different fitness peaks; (2) colonization of new peaks is likely promoted by competition that depresses occupied peaks and/or by the emergence of ridges between peaks; (3) these same properties (shallow valleys and ridges) that favor adaptive divergence through peak shifts can constrain the magnitude of adaptive divergence by enhancing gene flow; and (4) if adaptive divergence proceeds far enough, it becomes self-reinforcing and can lead to the formation of distinct species each adapted to distinct peaks.

Eco-to-evo: what we don't know

The above Española-Isabela Hypothesis, which is—at its essence—another way of framing the "ecological theory of adaptive radiation" (Schluter 2000a), has considerable empirical support as outlined in chapters 2–6. However, ambiguities and uncertainties exist regarding its details. I here list and discuss a few of these gaps that seem most critical for future work to address.

1. *Just how important are genetic constraints?* Genetic constraints on evolution clear-ly exist but controversy remains as to just how important they are and on what time scales. I have argued that genetic (co)variances can influence short-term evo-lution (C3-Q1) but are less likely to be a major constraint on longer time frames (C3-Q5, C4-Q7, C10-Q1). (Of course, physical and functional limits certainly do ultimately constrain the realm of what is possible.) Yet other authors (Blows and Hoffmann 2005, Hansen and Houle 2008, Kirkpatrick 2009, Walsh and Blows 2009) feel strongly that genetic constraints can be strong on any time frame. At present, then, all we can state with assurance is that considerable disagreement exists about the prevalence and severity of genetic constraints on evolution.

2. *To what extent does sexual selection facilitate or constrain adaptation and specia-tion?* Sexual selection clearly drives the evolution of secondary sexual traits, but how does sexual selection interact with natural selection? Classic opinion and logic dictate that natural and sexual selection should often oppose each other; yet, although supportive examples do exist, meta-analyses have not yielded strong general support (C2-Q6). What are we missing from our logic or analyses? As another manifestation of the uncertainty, one set of arguments and analyses sug-gests that sexual selection enhances speciation, whereas a different set of argu-ments and analyses suggests that sexual selection constrains speciation (C6-Q6). Perhaps the contribution of sexual selection to each of these phenomena is so context-specific that attempts at generality will be unproductive. If so, can we define a set of conditions under which each effect emerges?

3. *What is the general role of phenotypic plasticity?* Phenotypic plasticity has long been recognized as important in the adaptive fitting of organisms to their envi-ronment, and it seems inevitable that plasticity will often aid colonization of new environments and persistence in the face of environmental change (C11-Q4). Classically, this ability of plasticity to bring phenotypes closer to the optimum has been thought to diminish natural selection and slow adaptive genetic change and speciation. Recently, however, the pendulum has swung toward the idea that plasticity promotes adaptive evolution, including speciation. Although specific examples of both constraining and promoting effects are known, the evidence is too fragmentary for any general conclusions about which effect is more common (C11-Q5, C11-Q6). This uncertainty may stem from a rarity of definitive tests, an insufficiency of existing meta-analyses, or the fact that—as just suggested for sexual selection—the roles of plasticity are highly variable and context-specific. Perhaps the way forward is to seek subsets of conditions that determine when and how plasticity either promotes or constrains evolution.

4. *What is the general role of gene flow?* Following the format of the last two points, we here have another continuum bounded by two extremes, each of which has vocal proponents. Classically, gene flow is thought to constrain adaptation by preventing the independent evolution of populations in different environments (C5-Q1), which thereby reduces population sizes (C5-Q3), constrains range ex-pansion (C5-Q4), and opposes speciation (C5-Q2). Yet work on coevolutionary dynamics and various forms of "rescue" (review: Carlson et al. 2014) has empha-sized the positive roles of gene flow (C5-Q5). I suspect that increasing gene flow is beneficial only so long as it remains relatively low: that is, some gene flow is better

than none but a lot is worse than a little (Garant et al. 2007). This "optimal" level of gene flow is likely higher in cases of antagonistic coevolution and when environments aren't too divergent—but these predictions have yet to be adequately tested.

5. *How much gene flow is typical during speciation?* This statement is a more productive way of framing the old sympatry-allopatry speciation debate (Gavrilets 2004, Nosil 2012). At the one extreme (classic allopatry), populations become separated by a barrier to dispersal and then coincidentally evolve reproductive isolation (Coyne and Orr 2004). At the other extreme (the new sympatry), competition for shared resources leads to disruptive selection that splits populations into reproductive isolated forms (C6-Q5). Probably most speciation events occur near the allopatric end of the gene flow continuum, yet many examples of putative allopatric speciation (such as diversification on island archipelagos) surely involved some gene flow. Moreover, some speciation events, especially for phytophagous insects (Berlocher and Feder 2002), fishes in small lakes (Schluter 1996b, Siwertsson et al. 2010, Gordeeva et al. 2015), and plants on small islands (Papadopulos et al. 2011), are clearly toward the sympatric end of the continuum, although many of these examples could well involve some micro-allopatry. So, the appropriate question becomes: how are speciation events distributed across the gene flow continuum in different contexts and taxa? Answering this question will have to take into consideration the reality that gene flow is not a single genome-wide quantity but rather varies considerably in relation to selection and recombination (C5-Q7).

6. *To what extent do peaks multiply?* For this debate, the extremes range from (1) the number of adaptive peaks is set by the "environment" and species radiate to occupy those peaks (fill those niches), at which time diversification ceases (diversity constrains diversification), to (2) a species adapting to one peak generates additional peaks that can lead to the evolution of more species (diversity begets diversification); The reality must surely be somewhere between these two extremes and likely differs within trophic levels (perhaps closer to the first extreme) versus among trophic levels (presumably closer to the second extreme). Yet where any environment or taxon falls on this spectrum is generally unknown (C4-Q4), and yet the answer is critical to the way we conceptualize adaptive radiation.

7. *How important are ecological differences to speciation?* A major premise of this book, and others (Schluter 2000a, Nosil 2012), is that adaptation to different environments is a major—perhaps the main—driver of speciation, adaptive radiation, and evolutionary diversification. Many outstanding examples of this process exist, yet two basic questions need to be answered before its general importance is known. First, we need some way of determining how many of the existing speciation events were the result of adaptation to different environments. At present, the only safe general assertion is that various metrics of reproductive isolation are correlated with ecological/adaptive differences, yet these effects are often weak and thus leave much of the existing reproductive isolation unexplained (C6-Q1). Second, we need to ascertain just how often the colonization of different ecological environments leads to speciation. I expect the general answer here will be "very rarely" given the extensive diversity of habitats and adaptations that occur among populations within species (C6-Q2). Perhaps, ecological differences rarely

drive speciation but when speciation occurs it is (often? sometimes?) the result of those ecological differences.

8. *What is the role of the microbiome?* A relatively recent realization is that by far the most genetic variation in the body of an organism is not in the DNA of that organism but rather in its attendant fungi, bacteria, parasites, viruses, and so on. This variation is known to be important in shaping adaptive responses to a great diversity of environmental challenges experienced many organisms. Yet traditional evolutionary theory has, with very few exceptions, ignored this component of "adaptation." Future work should rectify this deficiency as it seems likely that changes in an organism's microbiome will dramatically alter adaptive responses of that organism to environmental change and thus the selection on—and evolution of—that organism's own genetic variation.

Evo-to-eco: what we know

The evo-to-eco side of eco-evolutionary dynamics is much less well developed than the just-described eco-to-evo side. For this reason, it has not yet coalesced—for me anyway—into a coherent overall hypothesis that might warrant a metaphor to match the eco-to-evo *Española-Isabela Hypothesis*. Thus, I will here focus instead on a set of observations that we can safely assert. The first three correspond to different ways to consider evo-to-eco effects (genetic identity, genetic diversity, contemporary evolution) and the second three are organized around ecological "levels" (population dynamics, community structure, ecosystem function).

1. *Intraspecific identity influences ecological dynamics.* A number of studies have now shown that different individuals (or genotypes or phenotypes) or populations (or ecotypes) have different ecological effects at the population (C7-Q6), community (C8-Q4), and ecosystem (C9-Q1) levels. Moreover, these effects of differences within species are often as large as are the effects of differences among species (C8-Q5, C9-Q2). This striking—and perhaps initially surprising—set of observations is reminiscent of the finding that phenotypic changes are often similarly large on short and long time scales (C3-Q4). That is, evolution—and its ecological consequences—can be very rapid on short time scales but will only rarely accumulate over longer time scales. This pattern would be expected if ecological differences are the main source of phenotypic variation within species, as well as the driver of speciation (C4-Q1-3, C6-Q1, C6-Q2), whereas the differences that accumulate after speciation have less to do with ecological differences.

2. *Intraspecific diversity influences ecological dynamics.* Extending the above point about intraspecific *identity*, the ecological effects of a given population or species also depend on the *diversity* of genotypes/phenotypes within that population or species (C7-Q6, C8-Q4, C9-Q1). For instance, various population dynamic (e.g., stability), community structure (e.g., species diversity), and ecosystem function (e.g., productivity) parameters increase as genetic variation within a population (or an experimental plot) increases. These effects parallel—but seem to be somewhat weaker than (C8-Q5, C9-Q2)—the well-documented effects of increasing species/ functional/phylogenetic diversity in communities (or experimental plots). Some of

these effects of diversity at the community and ecosystem levels are additive whereas others are nonadditive (C8-Q4, C9-Q1), and some of the effects are direct (effects of traits independent of changes in population dynamics) whereas others are indirect (effects of traits that act through changes in population dynamics) (C8-Q7, C9-Q4).

3. *Contemporary evolution influences ecological dynamics.* The above two assertions relate to existing (standing) variation within or among populations. Other studies have shown that short-term changes in this variation, such as allele frequencies or phenotypes, can influence ecological dynamics at the population (C7-Q2-5), community (C8-Q6), and ecosystem levels (C9-Q3). Based on existing analyses, these evolutionary effects can be as large as various ecological drivers, such as rainfall or species differences, but they seem to explain only small-to-modest amounts of the total ecological variation in nature (C7-Q3, C8-Q5, C9-Q2, C9-Q7). Yet more sophisticated analyses are necessary to definitively describe the ecological importance of contemporary evolutionary/phenotypic change—as will be described in more detail below.

4. *Intraspecific diversity and contemporary evolution generally benefit populations.* No surprise here: most evolution will be the result of natural selection and will therefore be adaptive (chapters 2-4), which should thereby enhance population fitness (C7-Q2), stabilize population fluctuations (C7-Q6), facilitate range expansion (C7-Q5), and slow/reverse population declines (evolutionary rescue: C7-Q4). Yet beyond this generalization lurk many ambiguities and uncertainties, a number of which will be discussed below with respect to what we don't know. Opposite effects are also likely in some instances, wherein adaptive evolution can decrease population size and increase fluctuations (Matsuda and Abrams 1994, Gyllenberg and Parvinen 2001, Webb 2003). Such effects can arise when, for example, carrying capacity is exceeded, per capita resource demand increases, life history traits evolve in response to density, or environments change rapidly (chapter 7).

5. *Intraspecific variation and contemporary evolution generally enhance community diversity.* Such effects are, again, not surprising—at least not across trophic levels. Most obviously, intraspecific variation at a lower trophic level (e.g., plants) will enhance species diversity at a higher trophic level (e.g., herbivores), with effects potentially then cascading to even higher levels (e.g., predators). This point takes the "diversity begets diversification" idea that originated at the interspecific level (C4-Q4) and extends it to include the intraspecific level. Consistent with this idea, spatial variation in genetic diversity within key species is often correlated with spatial variation in species diversity (Lankau and Strauss 2007, Vellend et al. 2014). Further, the contemporary evolution of intraspecific variation likely enhances interspecific diversity by maintaining and stabilizing population dynamics of each of the component species, as introduced in the above point. Indeed, evolution seems (often, but not always) to enhance the coexistence of predators and prey (C8-Q1) and of competitors (C8-Q2). Importantly, however, some clear exceptions exist, perhaps most obviously the fact that high genetic variation in introduced species can increase their invasiveness (C7-Q5), which can then negatively impact native species.

6. *Intraspecific variation and contemporary evolution generally enhance ecosystem function.* This point is an extension of the one above and relies on similar arguments, with a few different twists. First, the aspects of ecosystem function that are

enhanced by genetic variation include many of the parameters of interest from the anthropocentric perspective of ecosystem *services*, such as productivity, decomposition, and rates of nutrient cycling (C9-Q1-3). Combining this observation with the realization that even the effects of different species have their origins in evolution, we can see that ecosystem services are really EVOsystem services (Faith et al. 2010, Hendry et al. 2010). Second, effects at the ecosystem level might be weaker than those at the community level (Bailey et al. 2009), which might be weaker than those at the population level (C9-Q5). However, a number of exceptions have been described (C9-Q5). Third, ecosystem parameters are the arena where strong effects of selection within a generation (as opposed to evolution across generations) have been conclusively demonstrated (C9-Q7). Such effects also seem likely at other ecological levels (Johnson et al. 2014).

Evo-to-eco: what we don't know

1. *How important are evo-to-eco effects in nature?* Most studies of the ecological effects of intraspecific identity, intraspecific diversity, and contemporary evolution are conducted in controlled environments, such as the laboratories, greenhouses, mesocosms, or experimental gardens. As just described, these studies have revealed noteworthy effects, suggesting that genetic variation and contemporary evolution are important drivers of ecological processes and patterns. However, the controlled conditions that prevail in even the most "natural" of these experiments dictate that we currently remain uncertain as to just how important evo-to-eco effects would be in the complicated and messy natural world (C7-Q3, C8-Q5, C9-Q2, C9-Q7). My suspicion is that effects documented under controlled conditions will be comparatively weaker than effects documented under uncontrolled conditions, simply because the natural environment has so many other external effects (floods, droughts, temperature variation) that can swamp, or at least obscure, the ecological effects of intraspecific variation and its evolution. Yet this assertion doesn't mean that evo-to-eco effects are trivial: instead, it means that we need more information on how they play out in uncontrolled contexts. In addition, current approaches to eco-evolutionary dynamics, including the focal-species composite-response emphasis and the focus on *dynamic*, likely misses many important effects of evolution.

2. *How does evolutionary rescue work?* One of the hottest topic in eco-evolutionary dynamics is the potential role of evolution in saving populations that would otherwise go extinct when environments change (Gonzalez et al. 2013, Martin et al. 2013, Schiffers et al. 2013, Carlson et al. 2014). Theoretical models have been numerous and diverse, demonstrations in the laboratory have been elegant, and putative examples from nature have been advanced (C7-Q4). However, formal demonstrations of evolutionary rescue in nature are virtually nonexistent. This paucity of confirmation probably has a lot to do with the difficulty of conclusively demonstrating evolutionary rescue in nature, simply because the process must often occur when environments change and yet populations persist. Yet many factors besides evolution can enhance population growth, such as a relaxation of

density dependence (as soon as population size declines), an increasing availability of resources (e.g., insects that can use crop plants), or enemy release (such as in invasive species). Moreover, many populations clearly persist and grow even when they are not well-adapted to a new environment (C7-Q2), probably partly because competition is initially low. In addition, it seems likely that evolutionary rescue often will be cryptic (details below). In the end, however, it seems likely that every population currently persisting in nature must be continually undergoing evolutionary rescue because—if they were to stop evolving—the Red Queen dictates that extirpation would certainly result and probably sooner rather than later.

3. *How do we study cryptic eco-evolutionary dynamics?* As discussed many times in this book, methods for assessing eco-evolutionary dynamics focus—not surprisingly—on dynamic changes. In essence, investigators related ecological and evolutionary differences/changes to one another so as to infer how changes in the one influence changes in the other. In addition to the classic problem of inferring cause versus effect in such two-way situations, it also seems likely that ecological *stability* will reflect ongoing evolutionary change. Stated another way; if evolution were to cease, ecological changes would be much more dramatic—as introduced above in the context of evolutionary rescue. This realization means that contemporary evolution may be making much greater contributions to eco-evolutionary *stability* than it is to eco-evolutionary *dynamics*. In short, we need a set of approaches for uncovering and quantifying eco-evolutionary stability as this phenomenon is likely to be the next important, but almost entirely unstudied, frontier of the field (Hiltunen et al. 2014, Strauss 2014, Kinnison et al. 2015).

4. *How do we move beyond single-species composite-response approaches?* Studies of the evo-to-eco side of evolutionary dynamics mostly emphasize the focal-species composite-response approach, where one quantifies the effects of variation/change in one species on a set of ecological variables in the surrounding environment (chapters 8 and 9). This approach emerged as a way of being more inclusive when considering community and ecosystem effects than was possible in the classic species-by-species approaches. Yet all members of a community could be having eco-evolutionary effects that might or might not be additive and interactive. Similarly, all aspects of the community and ecosystem have the potential to respond to those effects, and cascades of eco-evolutionary dynamics seem likely (C8-Q4). As a result, we currently have little understanding of the sum-total of eco-evolutionary effects in any given location; nor have effective methods been developed for their quantification. Developing those methods won't be trivial but it will be necessary if we are to understand just how important intraspecific variation and its contemporary evolution are to ongoing ecological dynamics.

5. *How do eco-evolutionary feedbacks play out?* Eco-evolutionary feedbacks, such as from ecology to evolution and back again, must be extremely common, and a number of putative examples have been suggested for nature or demonstrated in the laboratory (C9-Q6). Yet clear demonstrations of these feedbacks have largely been absent in nature, because the confirmation of causality usually requires experimental approaches. As a result, we don't know how strong eco-evolutionary feedbacks are, whether they flow through the same or different traits or the same or different ecological parameters (i.e., narrow-sense or broad-sense), to what

extent they are positive (reinforcing) or negative (opposing), and how they shape evolutionary and ecological trajectories.

6. *To what extent are evo-to-eco effects direct or indirect?* Intraspecific variation and contemporary evolution should strongly influence population dynamics. Given that the population size/structure of key species is known to influence community structure and ecosystem function, we here have a particularly obvious indirect pathway from evolution to ecology (C8-Q7, C9-Q4). Yet the trait values of individuals can also influence communities and ecosystems directly—that is, intraspecific variation and contemporary evolution of a focal species can have per capita effects on communities and ecosystems (C8-Q7, C9-Q4). More studies should be designed to reveal direct and indirect effects in relation to population dynamics, as well as in relation to other intermediate pathways.

Underpinnings: what we do and don't know

Eco-evolutionary dynamics will be influenced by both genetic and plastic effects, as well as their combination and interaction. The above sections have already discussed some of these underpinnings, including the possibility of genetic constraints on evolution, and the role of plasticity in enhancing or constraining evolution. I will here highlight a few more topics, mixing together what we do and don't know.

1. *Genetic and plastic variation are both important.* As I have repeatedly emphasized, eco-evolutionary dynamics are driven by phenotypes, which can vary as a result of genetic differences (chapter 10) and environmental (plastic) effects (chapter 11). At present, it remains unclear how these two sources of phenotypic variation interact to shape population dynamics, community structure, and ecosystem function. Most studies conducted to date cannot address this question because they tend to fall into two mutually exclusive categories: comparisons of different "genotypes" (usually in plants or insects) or different phenotypes/ecotypes (usually in fish or mammals). The former studies focus on different clones and thus reveal genetic effects—but they rarely also consider plasticity. The latter studies focus on phenotypes and thus cannot separate genetic from plastic effects. The few studies that have explicitly separated such effects (e.g., Lundsgaard-Hansen et al. 2014) have revealed that both can be important and more such studies are needed before we can attempt any general statements.

2. *It's a many-small world after all.* Eco-evolutionary dynamics are driven by phenotypes, as just noted above. Thus, the *genetic* basis of eco-evolutionary dynamics will be—in the main—the genetic basis of phenotypes. Although some traits are strongly shaped by genes of large effect, most phenotypic variation is clearly the result of many genes, each of small-to-modest effect (C10-Q3-4). Moreover, overall fitness and presumably many ecological effects of organisms will be driven by the combined effects of multiple traits. Combine these two points (many traits and many genes for each trait) and it is clear that eco-evolutionary dynamics will be generally driven by many genes of small effect (Hendry et al. 2013). Although some specific genes with large ecological effects will be documented and described ("keystone genes"), such instances will not be representative

of the typical situation, as is already clear for the genetics of phenotypic traits (C10-Q3-4).

3. *What are the most important aspects of genetic variation?* Beyond the reasonably safe assertion that eco-evolutionary dynamics usually will be driven by many genes of small effect, little else is known with surety. First, trait variation is now known to be shaped, perhaps in equal measure, by structural and regulatory genetic changes (C10-Q8). The same will be true of eco-evolutionary dynamics, although the relative contribution of each driver—and how these contributions might differ among contexts—is entirely unknown. Second, most studies of adaptation focus on additive genetic variation; whereas nonadditive variation, including dominance, epistasis, and pleiotropic effects, is clearly also important (C10-Q2). Indeed, even the broad-sense heritability of genotypic effects on community and ecosystem parameters is low enough (C10-Q9) to leave plenty of room for nonadditive contributions. Third, heritable changes in traits can be driven by sequence changes or by epigenetic changes, such as DNA methylation or histone modifications, yet the contribution of epigenetics to eco-evolutionary dynamics is entirely unexplored.

4. *Standing variation versus new mutations.* Three observations make clear that eco-evolutionary dynamics will be driven mainly standing genetic variation, as opposed to new mutations. First, many studies have shown that standing genetic variation within populations has important ecological effects (C7-Q6, C8-Q4, C9-Q1). Second, most short-term adaptation is driven by selection acting on standing variation (C10-Q6). Third, most eco-evolutionary dynamics will be driven by many genes of small effect (as above). However, new mutations could well be important for organisms with larger population sizes and short generation times, most obviously micro-organisms and some insects, and when completely new selective conditions are experienced, such as human attempts to control pests and pathogens (C10-Q6).

5. *Eco-evolutionary dynamics typically will be nonparallel (and nonconvergent) at the genetic level.* Despite some examples of the same allele at the same gene having the same phenotypic effect in multiple independent instances of adaptation to similar environments, most such adaptation involves different genes and different alleles (C10-Q5). Eco-evolutionary dynamics are likely to be even less parallel/convergent given the above described expectation of many genes being involved.

The End

Stripped down to its barest essence, what we are striving to explain is the relationship between ecology and evolution. We might seek to elucidate these relationships for a specific taxonomic group in a specific environment, such as Darwin's finches in Galápagos. In such cases, we need—by definition—detailed studies of that particular group in that particular environment, ideally supplemented with theoretical models and laboratory studies that help to elucidate mechanisms. Alternatively, or additionally, we might seek to discover generalities of relationships between ecology and evolution that transcend specific groups and environments. In such cases, we might want to start with theory and then move to laboratory experiments (Ellner 2013); yet we ultimately still need to work on real populations in real environments. Thus, no matter the ultimate goal, we need detailed studies of specific taxonomic groups in specific environments, where we have two general questions: (1) how has the diversity of organisms and their traits been shaped by the environment?, and (2) how has this evolved organismal diversity fed-back to influence the environment? Although these two questions have been explored in many taxonomic groups, our understanding of their interplay is still rudimentary. Thus, the main message of the book is simply that we need to get out in nature and collectively work to figure out what is going on; because, stated baldly, we really are still shining a dim flashlight around a vast cavern. We need more and better flashlights—and some new batteries!

References

Abdo, Z., K. A. Crandall, and P. Joyce. 2004. Evaluating the performance of likelihood methods for detecting population structure and migration. Molecular Ecology 13:837–851.

Abrams, P. A. 1987. Alternative models of character displacement and niche shift. I. Adaptive shifts in resource use when there is competition for nutritionally nonsubstitutable resources. Evolution 41:651–661.

Abrams, P. A. 1995. Implications of dynamically variable traits for identifying, classifying, and measuring direct and indirect effects in ecological communities. American Naturalist 146:112–134.

Abrams, P. A. 2000. The evolution of predator-prey interactions: theory and evidence. Annual Review of Ecology and Systematics 31:79–105.

Abzhanov, A., W. P. Kuo, C. Hartmann, B. R. Grant, P. R. Grant, and C. J. Tabin. 2006. The calmodulin pathway and evolution of elongated beak morphology in Darwin's finches. Nature 442:563–567.

Adams, C. E., and F. A. Huntingford. 2004. Incipient speciation driven by phenotypic plasticity? Evidence from sympatric populations of arctic charr. Biological Journal of the Linnean Society 81:611–618.

Adler, P. B., J. HilleRisLambers, and J. M. Levine. 2007. A niche for neutrality. Ecology Letters 10:95–104.

Agashe, D. 2009. The stabilizing effect of intraspecific genetic variation on population dynamics in novel and ancestral habitats. American Naturalist 174:255–267.

Agrawal, A. A. 1999. Induced responses to herbivory in wild radish: effects on several herbivores and plant fitness. Ecology 80:1713–1723.

Agrawal, A. A., A. P. Hastings, M. T. J. Johnson, J. L. Maron, and J.-P. Salminen. 2012. Insect herbivores drive real-time ecological and evolutionary change in plant populations. Science 338:113–116.

Agrawal, A. A., M. T. J. Johnson, A. P. Hastings, and J. L. Maron. 2013. A field experiment demonstrating plant life-history evolution and its eco-evolutionary feedback to seed predator populations. American Naturalist 181:S35–S45.

Agrawal, A. F., J. L. Feder, and P. Nosil. 2011. Ecological divergence and the origins of intrinsic post-mating isolation with gene flow. International Journal of Ecology Article ID 435357.

Agrawal, A. F., and J. R. Stinchcombe. 2009. How much do genetic covariances alter the rate of adaptation? Proceedings of the Royal Society B. Biological Sciences 276:1183–1191.

Agrawal, A. F., E. D. Brodie III, and L. H. Rieseberg. 2001. Possible consequences of genes of major effect: transient changes in the G-matrix. Genetica 112:33–43.

Aguilée, R., D. Claessen, and A. Lambert. 2013. Adaptive radiation driven by the interplay of eco-evolutionary and landscape dynamics. Evolution 67:1291–1306.

Albert, A. Y. K., and D. Schluter. 2004. Reproductive character displacement of male stickleback mate preference: reinforcement or direct selection? Evolution 58:1099–1107.

Albert, A. Y. K., S. Sawaya, T. H. Vines, A. K. Knecht, C. T. Miller, B. R. Summers, S. Balabhadra, D. M. Kingsley, and D. Schluter. 2008. The genetics of adaptive shape shift in stickleback: pleiotropy and effect size. Evolution 62:76–85.

Albert, C. H., F. de Bello, I. Boulangeat, G. Pellet, S. Lavorel, and W. Thuiller. 2012. On the importance of intraspecific variability for the quantification of functional diversity. Oikos 121:116–126.

Albert, C. H., W. Thuiller, N. G. Yoccoz, R. Douzet, S. Aubert, and S. Lavorel. 2010. A multi-trait approach reveals the structure and the relative importance of intra vs. interspecific variability in plant traits. Functional Ecology 24:1192–1201.

Alexander, R. D., and R. S. Bigelow. 1960. Allochronic speciation in field crickets, and a new species, *Acheta veletis*. Evolution 14:334–346.

Alexandrou, M. A., C. Oliveira, M. Maillard, R. A. R. McGill, J. Newton, S. Creer, and M. I. Taylor. 2011. Competition and phylogeny determine community structure in Müllerian co-mimics. Nature 469:84–88. Nature 469:84–88.

Alfaro, M. E., D. I. Bolnick, and P. C. Wainwright. 2005. Evolutionary consequences of many-to-one mapping of jaw morphology to mechanics in labrid fishes. American Naturalist 165:E140–E154.

Alleaume-Benharira, M., I. R. Pen, and O. Ronce. 2006. Geographical patterns of adaptation within a species' range: interactions between drift and gene flow. Journal of Evolutionary Biology 19:203–215.

Alphey, N., P. G. Coleman, C. A. Donnelly, and L. Alphey. 2007. Managing insecticide resistance by mass release of engineered insects. Journal of Economic Entomology 100:1642–1649.

Alroy, J. 1998. Cope's rule and the dynamics of body mass evolution in North American fossil mammals. Science 280:731–734.

Anderson, E. C., and E. A. Thompson. 2002. A model-based method for identifying species hybrids using multilocus genetic data. Genetics 160:1217–1229.

Anderson, J. T., D. W. Inouye, A. M. McKinney, R. I. Colautti, and T. Mitchell-Olds. 2012. Phenotypic plasticity and adaptive evoltuion contribute to advancing flowering phenology in response to climate change. Proceedings of the Royal Society B. Biological Sciences 279:3843–3852.

Andersson, M. 1982. Female choice selects for extreme tail length in a widowbird. Nature 299:818–820.

Andersson, M. 1994. *Sexual selection*. Princeton University Press, Princeton, NJ.

Angeler, D. G. 2007. Resurrection ecology and global climate change research in freshwater ecosystems. Journal of the North American Benthological Society 26:12–22.

Angilletta, M. G. 2009. *Thermal adaptation: a theoretical and empirical synthesis*. Oxford University Press, Oxford, UK.

Antonovics, J. 1976. The nature of limits to natural selection. Annals of the Missouri Botanical Garden 63:224–247.

Antonovics, J., and A. D. Bradshaw. 1970. Evolution in closely adjacent plant populations. VIII. Clinal patterns at a mine boundary. Heredity 25:349–362.

Arendt, J. D. 1997. Adaptive intrinsic growth rates: an integration across taxa. Quarterly Review of Biology 72:149–177.

Arendt, J., and D. Reznick. 2008. Convergence and parallelism reconsidered: what have we learned about the genetics of adaptation? Trends in Ecology and Evolution 23:26–32.

Arnegard, M. E., M. D. McGee, B. Matthews, K. B. Marchinko, G. L. Conte, S. Kabir, N. Bedford, S. Bergek, Y. F. Chan, F. C. Jones, D. M. Kingsley, C. L. Peichel, and D. Schluter. 2014. Genetics of ecological divergence during speciation. Nature 511:307–311.

Arnold, M. L. 1997. *Natural hybridization and evolution*. Oxford University Press, Oxford, UK.

Arnold, M. L., and S. A. Hodges. 1995. Are natural hybrids fit or unfit relative to their parents? Trends in Ecology and Evolution 10:67–71.

Arnold, S. J. 1983. Morphology, performance and fitness. American Zoologist 23:347–361.

Arnold, S. J., and M. J. Wade. 1984a. On the measurement of natural and sexual selection: theory. Evolution 38:709–719.

Arnold, S. J., and M. J. Wade. 1984b. On the measurement of natural and sexual selection applications. Evolution 38:720–734.

Arnold, S. J., M. E. Pfrender, and A. G. Jones. 2001. The adaptive landscape as a conceptual bridge between micro- and macroevolution. Genetica 112–113:9–32.

Arnold, S. J., R. Bürger, P. A. Hohenlohe, B. C. Ajie, and A. G. Jones. 2008. Understanding the evolution and stability of the G-matrix. Evolution 62:2451–2461.

Ashton, K. G., M. C. Tracy, and A. de Queiroz. 2000. Is Bergmann's Rule valid for mammals? American Naturalist 156:390–415.

Atkins, K. E., and J. M. J. Travis. 2010. Local adaptation and the evolution of species' ranges under climate change. Journal of Theoretical Biology 266:449–457.

Auld, J. R., A. A. Agrawal, and R. A. Relyea. 2010. Re-evaluating the costs and limits of adaptive phenotypic plasticity. Proceedings of the Royal Society B. Biological Sciences 277:503–511.

Auld, J. R., and R. A. Relyea. 2011. Adaptive plasticity in predator-induced defenses in a common freshwater snail: altered selection and mode of predation due to prey phenotype. Evolutionary Ecology 25:189–202.

Bauer, S., and B. J. Hoye. 2014. Migratory animals couple biodiversity and ecosystem functioning worldwide. Science 344:54–62.

Bailey, J. K., J. A. Schweitzer, B. J. Rehill, R. L. Lindroth, G. D. Martinsen, and T. G. Whitham. 2004. Beavers as molecular genticists: a genetic basis of the foraging of an ecosystem engineer. Ecology 85:603–608.

Bailey, J. K., J. A. Schweitzer, F. Úbeda, J. Koricheva, C. J. LeRoy, M. D. Madritch, B. J. Rehill, R. K. Bangert, D. G. Fischer, G. J. Allan, and T. G. Whitham. 2009. From genes to ecosystems: a synthesis of the effects of plant genetic factors across levels of organization. Philosophical Transactions of the Royal Society B. Biological Sciences 364:1607–1616.

Bailey, N. W. 2012. Evolutionary models of extended phenotypes. Trends in Ecology and Evolution 27:561–569.

Bailey, J. K., M. A. Genung, I. Ware, C. Gorman, M. E. Van Nuland, H. Long, and J. A. Schweitzer. 2014. Indirect genetic effects: an evolutionary mechanism linking feedbacks, genotypic diversity and coadaptation in a climate change context. Functional Ecology 28:87–95.

Baker, H. G. 1965. Characteristics and modes of origin of weeds. Pages 147–169 in H. G. Baker and G. L. Stebbins, Editors. The genetics of colonizing species. Academic Press, New York.

Bakker, J., M. E. C. van Rijswijk, F. J. Weissing, and R. Bijlsma. 2010. Consequences of fragmentation for the ability to adapt to novel environments in experimental Drosophila metapopulations. Conservation Genetics 11:435–448.

Baldwin, J. M. 1896. A new factor in evolution. American Naturalist 30:441–451,536–553.

Bank, C., J. Hermisson, and M. Kirkpatrick. 2012. Can reinforcement comlete speciation? Evolution 66:229–239.

Baldwin, J. M. 1902. Development and Evolution. The McMillan Company, New York.

Barbour, M. A., M. A. Fortuna, J. Bascompte, J. R. Nicholson, R. Julkunen-Tiitto, E. S. Jules, and G. M. Crutsinger. 2016. Genetic specificity of a plant-insect food web: Implications for linking genetic variation to network complexity. Proceedings of the National Academy of Sciences USA 113: 2128–2133.

Barbour, R. C., J. M. O'Reilly-Wapstra, D. W. De Little, G. J. Jordan, D. A. Steane, J. R. Humphreys, J. K. Bailey, T. G. Whitham, and B. M. Potts. 2009b. A geographic mosaic of genetic variation within a foundation tree species and its community-level consequences. Ecology 90:1762–1772.

Barbour, R. C., L. G. Forster, S. C. Baker, D. A. Steane, and B. M. Potts. 2009a. Biodiversity consequences of genetic variation in bark characteristics within a foundation tree species. Conservation Biology 23:1146–1155.

Barluenga, M., K. N. Stölting, W. Salzburger, M. Muschick, and A. Meyer. 2006. Sympatric speciation in Nicaraguan crater lake cichlid fish. Nature 439:719–723.

Barraclough, T. G., and A. P. Vogler. 2000. Detecting the geographical pattern of speciation from species-level phylogenies. American Naturalist 155:419–434.

Barraclough, T. G., P. H. Harvey, and S. Nee. 1995. Sexual selection and taxonomic diversity in passerine birds. Proceedings of the Royal Society B. Biological Sciences 259:211–215.

Barrett, R. D. H., A. Paccard, T. M. Healy, S. Bergek, P. M. Schulte, D. Schluter, and S. M. Rogers. 2011. Rapid evolution of cold tolerance in stickleback. Proceedings of the Royal Society B. Biological Sciences 278:233–238.

Barrett, R. D. H., and A. P. Hendry. 2012. Evolutionary rescue under environmental change? Pages 216–233 in U. Candolin and B. B. M. Wong, Editors. Behavioural responses to a changing world: mechanisms and consequences. Oxford University Press, Oxford, UK.

Barrett, R. D. H., and D. Schluter. 2008. Adaptation from standing genetic variation. Trends in Ecology and Evolution 23:38–44.

Barrett, R. D. H., S. M. Rogers, and D. Schluter. 2008. Natural selection on a major armor gene in threespine stickleback. Science 322:255–257.

Barton, N. 2001a. Adaptation at the edge of a species' range. Page 365–392 in J. Silvertown and J. Antonovics, Editors. Integrating ecology and evolution in a spatial context. Blackwell Science, Oxford, UK.

Barton, N. H. 2000. Genetic hitchhiking. Philosophical Transactions of the Royal Society B. Biological Sciences 355:1553–1562.

Barton, N. H. 2001b. The role of hybridization in evolution. Molecular Ecology 10:551–568.

Barton, N., and B. O. Bengtsson. 1986. The barrier to genetic exchange between hybridising populations. Heredity 57:357–376.

Barton, N., and L. Partridge. 2000. Limits to natural selection. BioEssays 22:1075–1084.

Baskett, M. L., S. D. Gaines, and R. M. Nisbet. 2009. Symbiont diversity may help coral reefs survive moderate climate change. Ecological Applications 19:3–17.

Basolo, A. L. 1990. Female preference predates the evolution of the sword in swordtail fish. Science 250:808–810.

Bassar, R. D., A. Lopez-Sepulcre, D. N. Reznick, and J. Travis. 2013. Experimental evidence for density-dependent regulation and selection on Trinidadian guppy life histories. American Naturalist 181:25–38.

Bassar, R. D., M. C. Marshall, A. López-Sepulcre, E. Zandonà, S. K. Auer, J. Travis, C. M. Pringle, A. S. Flecker, S. A. Thomas, D. F. Fraser, and D. N. Reznick. 2010. Local adaptation in Trinidadian guppies alters ecosystem processes. Proceedings of the National Academy of Sciences USA 107:3616–3621.

Bassar, R. D., R. Ferriere, A. López-Sepulcre, M. C. Marshall, J. Travis, C. M. Pringle, and D. N. Reznick. 2012. Direct and indirect ecosystem effects of evolutionary adaptation in the Trinidadian guppy (*Poecilia reticulata*). American Naturalist 180:167–185.

Bassar, R. D., A. Lopez-Sepulcre, D. N. Reznick, and J. Travis. 2014. Experimental evidence for density-dependent regulation and selection on Trinidadian guppy life histories. American Naturalist 181:25–38.

Baxter, S. W., S. E. Johnston, and C. D. Jiggins. 2009. Butterfly speciation and the distribution of gene effect sizes fixed during adaptation. Heredity 102:57–65.

Baythavong, B. S. 2011. Linking the spatial scale of environmental variation and the evolution of phenotypic plasticity: selection favors adaptive plasticity in fine-grained environments. American Naturalist 178:75–87.

Beall, C. M. 2007. Two routes to functional adaptation: Tibetan and Andean high-altitude natives. Proceedings of the National Academy of Sciences USA 104:8655–8660.

Beall, C. M., M. J. Decker, G. M. Brittenham, I. Kushner, A. Gebremedhin, and K. P. Strohl. 2002. An Ethiopian pattern of human adaptation to high-altitude hypoxia. Proceedings of the National Academy of Sciences USA 99:17215–17218.

Beans, C. M. 2014. The case for character displacement in plants. Ecology and Evolution 4: 852–865.

Bearhop, S., W. Fiedler, R. W. Furness, S. C. Votier, S. Waldron, J. Newton, G. J. Bowen, P. Berthold, and K. Farnsworth. 2005. Assortative mating as a mechanism for rapid evolution of a migratory divide. Science 310:502–504.

Becks, L., S. P. Ellner, L. E. Jones, and N. G. Hairston Jr. 2010. Reduction of adaptive genetic diversity radically alters eco-evolutionary community dynamics. Ecology Letters 13:989–997.

Becks, L., S. P. Ellner, L. E. Jones, and N. G. Hairston Jr. 2012. The functional genomics of an eco-evolutionary feedback loop linking gene expression, trait evolution, and community dynamics. Ecology Letters 15:492–501.

Becquet, C., and M. Przeworski. 2009. Learning about modes of speciation by computational approaches. Evolution 63:2547–2562.

Beerli, P., and J. Felsenstein. 1999. Maximum-likelihood estimation of migration rates and effective population numbers in two populations using a coalescent approach. Genetics 152:763–773.

Beerli, P., and J. Felsenstein. 2001. Maximum likelihood estimation of a migration matrix and effective population sizes in n subpopulations by using a coalescent approach. Proceedings of the National Academy of Sciences USA 98:4563–4568.

Behm, J. E., A. R. Ives, and J. W. Boughman. 2010. Breakdown in postmating isolation and the collapse of a species pair through hybridization. American Naturalist 175:11–26.

Bell, A. M., S. J. Hankison, and K. L. Laskowski. 2009. The repeatability of behaviour: a meta-analysis. Animal Behaviour 77:771–783.

Bell, G. 2008. *Selection: the mechanism of evolution*. Oxford University Press, Oxford, UK.

Bell, G. 2010. Fluctuating selection: the perpetual renewal of adaptation in variable environments. Philosophical Transactions of the Royal Society B. Biological Sciences 365:87–97.

Bell, G., and A. Gonzalez. 2009. Evolutionary rescue can prevent extinction following environmental change. Ecology Letters 12:942–948.

Bell, G., and A. Gonzalez. 2011. Adaptation and evolutionary rescue in metapopulations experiencing environmental deterioration. Science 332:1327–1330.

Bell, M. A., M. P. Travis, and D. M. Blouw. 2006. Inferring natural selection in a fossil threespine stickleback. Paleobiology 32:562–577.

Bell, M. A., W. E. Aguirre, and N. J. Buck. 2004. Twelve years of contemporary armor evolution in a threespine stickleback population. Evolution 58:814–824.

Bell, M. A., and W. E. Aguirre. 2013. Contemporary evolution, allelic recycling, and adaptive radiation of the threespine stickleback. Evolutionary Ecology Research 15:377–411.

Benkman, C. W. 2003. Divergent selection drives the adaptive radiation of crossbills. Evolution 57:1176–1181.

Bensch, S., D. Hasselquist, B. Nielsen, and B. Hansson. 1998. Higher fitness for philopatric than for immigrant males in a semi-isolated population of great reed warblers. Evolution 52:877–883.

Benton, T. G. 2012. Individual variation and population dynamics: lessons from a simple system. Philosophical Transactions of the Royal Society B. Biological Sciences 367:200–210.

Benton, T. G., and A. Grant. 2000. Evolutionary fitness in ecology: Comparing measures of fitness in stochastic, density-dependent environments. Evolutionary Ecology Research 2:769–789.

Berg, J. J., and G. Coop. 2015. A coalescent model for a sweep of a unique standing variant. Genetics 201:707–725.

Bergland, A. O., E. L. Berhman, K. R. O'Brien, P. S. Schmidt, and D. A. Petrov. 2014. Genomic evidence of rapid and stable adaptive oscillations over seasonal time scales in *Drosophila*. PLoS Genetics 10:e1004775.

Bergstrom, C. T., and M. Feldgarden. 2008. The ecology and evolution of antibiotic-resistant bacteria. Pages 124–138 *in* S. C. Stearns and J. C. Koella, Editors. *Evolution in health and disease, 2nd ed.* Oxford University Press, Oxford, UK.

Berlocher, S. H., and J. L. Feder. 2002. Sympatric speciation in phytophagous insects: moving beyond controversy? Annual Review of Entomology 47:773–815.

Bernardo, J. 1996. Maternal effects in animal ecology. American Zoologist 36:83–105.

Bernatchez, L., A. Chouinard, and G. Lu. 1999. Integrating molecular genetics and ecology in studies of adaptive radiation: whitefish, *Coregonus* sp., as a case study. Biological Journal of the Linnean Society 68:173–194.

Bernatchez, L., S. Renaut, A. R. Whiteley, N. Derome, J. Jeukens, L. Landry, G. Lu, A. W. Nolte, K. Østbye, S. M. Rogers, and J. St-Cyr. 2010. On the origin of species: insights from the ecological genomics of lake whitefish. Philosophical Transactions of the Royal Society B. Biological Sciences 365:1783–1800.

Berner, D., A.-C. Grandchamp, and A. P. Hendry. 2009. Variable progress toward ecological speciation in parapatry: stickleback across eight lake-stream transitions. Evolution 63:1740–1753.

Berner, D., D. C. Adams, A.-C. Grandchamp, and A. P. Hendry. 2008. Natural selection drives patterns of lake-stream divergence in stickleback foraging morphology. Journal of Evolutionary Biology 21:1653–1665.

Berner, D., R. Kaeuffer, A.-C. Grandchamp, J. A. M. Raeymaekers, K. Räsänen, and A. P. Hendry. 2011. Quantitative genetic inheritance of morphological divergence in a lake-stream stickleback ecotype pair: implications for reproductive isolation. Journal of Evolutionary Biology 24:1975–1983.

Berner, D., W. E. Stutz, and D. I. Bolnick. 2010. Foraging trait (co)variances in stickleback evolve deterministically and do not predict trajectories of adaptive diversification. Evolution 64:2265–2277.

Berry, O., M. D. Tocher, and S. D. Sarre. 2004. Can assignment tests measure dispersal? Molecular Ecology 13:551–561.

Berry, R. J. 1964. The evolution of an island population of the house mouse. Evolution 18:468–483.

Best, S. M., and P. J. Kerr. 2000. Coevolution of host and virus: the pathogenesis of virulent and attenuated strains of myxoma virus in resistant and susceptible European rabbits. Virology 267:36–48.

Betancourt, A. J. 2009. Genomewide patterns of substitution in adaptively evolving populations of the RNA bacteriophage MS2. Genetics 181:1535–1544.

Bever, J. D., K. M. Westover, and J. Antonovics. 1997. Incorporating the soil community into plant population dynamics: the utility of the feedback approach. Journal of Ecology 85:561–573.

Bijlsma, R., M. D. D. Westerhof, L. P. Roekx, and I. Pen. 2010. Dynamics of genetic rescue in inbred *Drosophila melanogaster* populations. Conservation Genetics 11:449–462.

Björklund, M., E. Ranta, V. Kaitala, L. A. Bach, P. Lundberg, and N. C. Stenseth. 2009. Quantitative trait evolution and environmental change. PLoS ONE 4:e4521.

Björklund, M., A. Husby, and L. Gustafsson. 2013. Rapid and unpredictable changes of the G-matrix in a natural bird population over 25 years. Journal of Evolutionary Biology 26:1–13.

Blanckenhorn, W. U. 2000. The evolution of body size: what keeps organisms small. Quarterly Review of Biology 75:385–407.

Blossey, B., and R. Nötzold. 1995. Evolution of increased competitive ability in invasive nonindigenous plants: a hypothesis. Journal of Ecology 83:887–889.

Blows, M. W., and A. A. Hoffmann. 2005. A reassessment of genetic limits to evolutionary change. Ecology 86:1371–1384.

Blows, M. W., and R. Brooks. 2003. Measuring nonlinear selection. American Naturalist 162:815–820.

Blows, M. W., S. F. Chenoweth, and E. Hine. 2004. Orientation of the genetic variance-covariance matrix and the fitness surface for multiple male sexually selected traits. American Naturalist 163:329–340.

Boag, P. T. 1983. The heritability of external morphology in Darwin's ground finches (*Geospiza*) on Isla Daphne Major, Galapagos. Evolution 37:877–894.

Boag, P. T., and P. R. Grant. 1981. Intense natural selection in a population of Darwin's finches (Geospizinae) in the Galápagos. Science 214:82–85.

Bobdyl Roels, S. A., and J. K. Kelly. 2011. Rapid evolution caused by pollinator loss in *Mimulus guttatus*. Evolution 65:2541–2552.

Bocedi, G., S. C. F. Palmer, G. Peér, R. K. Heikkinen, Y. G. Matsinos, K. Watts, and J. M. J. Travis. 2014. RangeShifter: a platform for modelling spatial eco-evolutionary dyanmics and species' responses to environmental changes. Methods in Ecology and Evolution 5:388–396.

Boersma, M., P. Spaak, and L. De Meester. 1998. Predator-mediated plasticity in morphology, life history, and behavior of *Daphnia*: the uncoupling of responses. American Naturalist 152:925–927.

Bohannan, B. J. M., and R. E. Lenski. 2000. Linking genetic change to community evolution: insights from studies of bacteria and bacteriophage. Ecology Letters 3:362–377.

Bohonak, A. J. 1999. Dispersal, gene flow, and population structure. Quarterly Review of Biology 74:21–45.

Bolnick, D. I. 2001. Intraspecific competition favours niche width expansion in *Drosophila melanogaster*. Nature 410:463–466.

Bolnick, D. I. 2004. Can intraspecific competition drive disruptive selection? An experimental test in natural populations of sticklebacks. Evolution 58:608–618.

Bolnick, D. I. 2011. Sympatric speciation in threespine stickleback: why not? International Journal of Ecology. Article ID 942847.

Bolnick, D. I., and B. M. Fitzpatrick. 2007. Sympatric speciation: models and empirical evidence. Annual Review of Ecology, Evolution, and Systematics 38:459–487.

Bolnick, D. I., and M. S. Araújo. 2011. Partitioning the relative fitness effects of diet and trophic morphology in the threespine stickleback. Evolutionary Ecology Research 13:439–459.

Bolnick, D. I., and O. L. Lau. 2008. Predictable patterns of disruptive selection in stickleback in postglacial lakes. American Naturalist 172:1–11.

Bolnick, D. I., and P. Nosil. 2007. Natural selection in populations subject to a migration load. Evolution 61:2229–2243.

Bolnick, D. I., L. K. Snowberg, C. Patenia, W. E. Stutz, T. Ingram, and O. L. Lau. 2009. Phenotype-dependent native habitat preference facilitates divergence between parapatric lake and stream stickleback. Evolution 63:2004–2016.

Bolnick, D. I., P. Amarasekare, M. S. Araújo, R. Bürger, J. M. Levine, M. Novak, V. H. W. Rudolf, S. J. Schreiber, M. C. Urban, and D. A. Vasseur. 2011. Why intraspecific trait variation matters in community ecology. Trends in Ecology and Evolution 26:183–192.

Bolnick, D. I., T. J. Near, and P. C. Wainwright. 2006. Body size divergence promotes post-zygotic reproductive isolation in centrarchids. Evolutionary Ecology Research 8:903–913.

Bonduriansky, R. 2011. Sexual selection and conflict as engines of ecological diversification. American Naturalist 178:729–745.

Booth, R. E., and J. P. Grime. 2003. Effects of genetic impoverishment on plant community diversity. Journal of Ecology 91:721–730.

Bossdorf, O., H. Auge, L. Lafuma, W. E. Rogers, E. Siemman, and D. Prati. 2005. Phenotypic and genetic differentiation between native and introduced plant populations. Oecologia 144:1–11.

Bossuyt, B. 2007. Genetic rescue in an isolated metapopulation of a naturally fragmented plant species, *Parnassia palustris*. Conservation Biology 21:832–841.

Botero, C. A., F. J. Weissing, J. Wright, and D. R. Rubenstein. 2015. Evolutionary tipping points in the capacity to adapt to environmental change. Proceedings of the National Academy of Sciences USA 112:184–189.

Both, C., and M. E. Visser. 2001. Adjustment to climate change is constrained by arrival date in a long-distance migrant bird. Nature 411:296–298.

Both, C., S. Bouwhuis, C. M. Lessells, and M. E. Visser. 2006. Climate change and population declines in a long-distance migratory bird. Nature 441:81–83.

Boudsocq, S., S. Barot, and N. Loeuille. 2011. Evolution of nutrient acquisition: when adaptation fills the gap between contrasting ecological theories. Proceedings of the Royal Society B. Biological Sciences 278:449–457.

Boughman, J. W. 2001. Divergent sexual selection enhances reproductive isolation in sticklebacks. Nature 411:944–948.

Boughman, J. W. 2002. How sensory drive can promote speciation. Trends in Ecology and Evolution 17:571–577.

Boughman, J. W., H. D. Rundle, and D. Schluter. 2005. Parallel evolution of sexual isolation in sticklebacks. Evolution 59:361–373.

Boulding, E. G., and T. Hay. 2001. Genetic and demographic parameters determining population persistence after a discrete change in the environment. Heredity 86:313–324.

Bourne, E. C., G. Bocedi, J. M. J. Travis, R. J. Pakeman, R. W. Brooker, and K. Schiffers. 2014. Between migration load and evolutionary rescue: dispersal, adaptation and the response of spatially structured populations to environmental change. Proceedings of the Royal Society B. Biological Sciences 281:20132795.

Bowler, D. E., and T. G. Benton. 2005. Causes and consequences of animal dispersal strategies: relating individual behaviour to spatial dynamics. Biological Reviews 80:205–225.

Bradburd, G. S., P. L. Ralph, and G. M. Coop. 2013. Disentangling the effects of geographic and ecological isolation on genetic differentiation. Evolution 67:3258–3273.

Bradshaw, A. D. 1984. Adaptation of plants to soils containing toxic metals—a test for conceit. CIBA Foundation Symposia 102:4–14.

Bradshaw, A. D., and T. McNeilly. 1991. Evolutionary response to global climate change. Annals of Botany 67:5–14.

Bradshaw, W. E., and C. M. Holzapfel. 2006. Evolutionary response to rapid climate change. Science 312:1477–1478.

Brady, K. U., A. R. Kruckeberg, and H. D. Bradshaw Jr. 2005. Evolutionary ecology of plant adaptation to serpentine soils. Annual Review of Ecology, Evolution, and Systematics 36:243–266.

Brewer, M. S., R. A. Carter, P. J. P. Croucher, and R. G. Gillespie. 2015. Shifting habitats, morphology, and selective pressures: Developmental polyphenism in an adaptive radiation of Hawaiian spiders. Evolution 69:162–178.

Bridle, J. R., and T. H. Vines. 2007. Limits to evolution at range margins: when and why does adaptation fail? Trends in Ecology and Evolution 22:140–147.

Bridle, J. R., J. Polechová, M. Kawata, and R. K. Butlin. 2010. Why is adaptation prevented at ecological margins? New insights from individual-based simulations. Ecology Letters 13:485–494.

Bridle, J. R., S. Gavaz, and W. J. Kennington. 2009. Testing limits to adaptation along altitudinal gradients in rainforest *Drosophila*. Proceedings of the Royal Society B. Biological Sciences 276:1507–1515.

Brodersen, J., J. G. Howeth, and D. M. Post. 2015. Emergence of a novel prey life history promotes contemporary sympatric diversification in a top predator. Nature Communications 6:8115.

Brodie III, E. D., A. J. Moore, and F. J. Janzen. 1995. Visualizing and quantifying natural selection. Trends in Ecology and Evolution 10:313–318.

Brommer, J. E. 2014. Using average autonomy to test whether behavioral syndromes constrain evolution. Behavioral Ecology and Sociobiology 68:691–700.

Brommer, J. E., L. Gustafsson, H. Pietiäinen, and J. Merilä. 2004. Single-generation estimates of individual fitness as proxies for long-term genetic contribution. American Naturalist 163:505–517.

Bronikowski, A. M., M. E. Clark, F. H. Rodd, and D. N. Reznick. 2002. Population-dynamic consequences of predator-induced life history variation in the guppy (*Poecilia reticulata*). Ecology 83:2194–2204.

Bronmark, C., and J. G. Miner. 1992. Predator-induced phenotypical change in body morphology in crucian carp. Science 258:1348–1350.

Brooks, J. L., and S. I. Dodson. 1965. Predation, body size, and composition of plankton. Science 150:28–35.

Brooks, T. M., R. A. Mittermeier, C. G. Mittermeier, G. A. B. da Fonseca, A. B. Rylands, W. R. Konstant, P. Flick, J. Pilgrim, S. Oldfield, G. Magin, and C. Hilton-Taylor. 2002. Habitat loss and extinction in the hotspots of biodiversity. Conservation Biology 16:909–923.

Broquet, T., J. Yearsley, A. H. Hirzel, J. Goudet, and N. Perrin. 2009. Inferring recent migration rates from individual genotypes. Molecular Ecology 18:1048–1060.

Brown Jr., W. L., and E. O. Wilson. 1956. Character displacement. Systematic Zoology 5:49–64.

Brown, J. S., and T. L. Vincent. 1992. Organization of predator-prey communities as an evolutionary game. Evolution 46:1269–1283.

Brown, M. B., and C. R. Brown. 2011. Intense natural selection on morphological of cliff swallows (*Petrochelidon pyrrhonota*) a decade later: did the population move between adaptive peaks? Auk 128:69–77.

Buerkle, C. A., and C. Lexer. 2008. Admixture as the basis for genetic mapping. Trends in Ecology and Evolution 23:686–694.

Bumpus, H. 1899. The elimination of the unfit as illustrated by the introduced sparrow, *Passer domesticus*. *Biological Lectures*, Marine Biology Laboratory, Woods Hole.

Burdon, J. J., R. H. Groves, and J. M. Cullen. 1981. The impact of biological control on the distribution and abundance of *Chonrilla juncea* in south-eastern Australia. Journal of Applied Ecology 18:957–966.

Bürger, R. 1986. Constraints for the evolution of functionally coupled characters: a nonlinear analysis of a phenotypic model. Evolution 40:182–193.

Bürger, R., and M. Lynch. 1995. Evolution and extinction in a changing environment: a quantitative-genetic analysis. Evolution 49:151–163.

Burns, J. H., and S. Y. Strauss. 2011. More closely related species are more ecologically similar in an experimental test. Proceedings of the National Academy of Sciences USA 108:5302–5307.

Burt, A. 1995. The evolution of fitness. Evolution 49:1–8.

Buser, C. C., R. D. Newcomb, A. C. Gaskett, and M. R. Goddard. 2014. Niche construction initiates the evolution of mutualistic interactions. Ecology Letters 17:1257–1264.

Bush, G. L. 1969. Sympatric host race formation and speciation in frugivorous flies of Genus *Rhagoletis* (Diptera, Tephritidae). Evolution 23:237–251.

Bush, G. L. 1994. Sympatric speciation in animals: new wine in old Bottles. Trends in Ecology and Evolution 9:285–288.

Bush, M., and R. Rivera. 1998. Pollen dispersal and representation in a neotropical rain forest. Global Ecology and Biogeography Letters 7:379–392.

Butlin, R. K., J. Galindo, and J. W. Grahame. 2008. Sympatric, parapatric or allopatric: the most important way to classify speciation? Philosophical Transactions of the Royal Society B. Biological Sciences 363:2997–3007.

Byars, S. G., W. Papst, and A. A. Hoffmann. 2007. Local adaptation and cogradient selection in the alpine plant, *Poa hiemata*, along a narrow altitudinal gradient. Evolution 61:2925–2941.

Byrne, K., and R. A. Nichols. 1999. *Culex pipiens* in London Underground tunnels: differentiation between surface and subterranean populations. Heredity 82:7–15.

Cadena, C. D., R. E. Ricklefs, I. Jiménez, and E. Bermingham. 2005. Ecology: is speciation driven by species diversity? Nature 438:E1–E2.

Cadotte, M. W. 2013. Experimental evidence that evolutionarily diverse assemblages result in higher productivity. Proceedings of the National Academy of Sciences USA 110:8996–9000.

Cadotte, M. W., B. J. Cardinale, and T. H. Oakley. 2008. Evolutionary history and the effect of biodiversity on plant productivity. Proceedings of the National Academy of Sciences USA 105:17012–17017.

Cadotte, M. W., R. Dinnage, and D. Tilman. 2012. Phylogenetic diversity promotes ecosystem stability. Ecology 93:S223–S233.

Cain, A. J. 1964. The perfection of animals. Biological Journal of the Linnean Society 36:3–29.

Cain, A. J., and P. M. Sheppard. 1950. Selection in the polymorphic land snail *Cepaea nemoralis*. Heredity 4:275–294.

Cameron, T. C., D. O'Sullivan, A. Reynolds, S. B. Piertney, and T. G. Benton. 2013. Eco-evolutionary dynamics in response to selection on life-history. Ecology Letters 16:754–763.

Calsbeek, B. 2012. Exploring variation in fitness surfaces over time or space. Evolution 66:1126–1137.

Calsbeek, R., and R. M. Cox. 2010. Experimentally assessing the relative importance of predation and competition as agents of selection. Nature 465:613–616.

Camacho, C., D. Canal, and J. Potti. 2015. Testing the matching habitat choice hypothesis in nature: phentoype-environment correlation and fitness in a songbird population. Evolutionary Ecology 29:873–886.

Carlson, S. M., C. J. Cunningham, and P. A. H. Westley. 2014. Evolutionary rescue in a changing world. Trends in Ecology and Evolution 29:521–530.

Carlson, S. M., A. P. Hendry, and B. H. Letcher. 2004. Natural selection acting on body size, growth rate and compensatory growth: an empirical test in a wild trout population. Evolutionary Ecology Research 6:955–973.

Carlson, S. M., and T. P. Quinn. 2007. Ten years of varying lake level and selection on size-at-maturity in sockeye salmon. Ecology 88:2620–2629.

Carlson, S. M., E. M. Olsen, and L. A. Vøllestad. 2008. Seasonal mortality and the effect of body size: a review and an empirical test using individual data on brown trout. Functional Ecology 22:663–673.

Carlson, S. M., H. B. Rich, and T. P. Quinn. 2009. Does variation in selection imposed by bears drive divergence among populations in the size and shape of sockeye salmon? Evolution 63:1244–1261.

Carlson, S. M., T. P. Quinn, and A. P. Hendry. 2011. Eco-evolutionary dynamics in pacific salmon. Heredity 106:438–447.

Carmona, D., C. R. Fitzpatrick, and M. T. J. Johnson. 2015. Fifty years of co-evolution and beyond: integrating co-evolution from molecules to species. Molecular Ecology 24:5315–5329.

Carneiro, M., N. Ferrand, and M. W. Nachman. 2009. Recombination and speciation: loci near centromeres are more differentiated than loci near telomeres between subspecies of the European rabbit (*Oryctolagus cuniculus*). Genetics 181:593–606.

Carrière, Y., and B. E. Tabashnik. 2001. Reversing insect adaptation to transgenic insecticidal plants. Proceedings of the Royal Society B. Biological Sciences 268:1475–1480.

Carrière, Y., D. W. Crowder, and B. E. Tabashnik. 2010. Evolutionary ecology of insect adaptation to Bt crops. Evolutionary Applications 3:561–573.

Carroll, S. B. 2008. Evo-devo and an expanding evolutionary synthesis: A genetic theory of morphological evolution. Cell 134:25–36.

Carroll, S. P. 2007. Brave New World: the epistatic foundations of natives adapting to invaders. Genetica 129:193–204.

Carroll, S. P., A. P. Hendry, D. N. Reznick, and C. W. Fox. 2007. Evolution on ecological time-scales. Functional Ecology 21:387–393.

Carroll, S. P., H. Dingle, and T. R. Famula. 2003. Rapid appearance of epistasis during adaptive divergence following colonization. Proceedings of the Royal Society B. Biological Sciences 270:S80–S83.

Carroll, S. P., H. Dingle, T. R. Famula, and C. W. Fox. 2001. Genetic architecture of adaptive differentiation in evolving host races of the soapberry bug, *Jadera haematoloma*. Genetica 112–113:257–272.

Carroll, S. P., J. E. Loye, H. Dingle, M. Mathieson, T. R. Famula, and M. P. Zalucki. 2005. And the beak shall inherit—evolution in response to invasion. Ecology Letters 8:944–951.

Caruso, C. M., H. Maherali, and M. Sherrard. 2006. Plasticity of physiology in *Lobelia*: testing for adaptation and constraint. Evolution 60:980–990.

Case, T. J., and M. L. Taper. 2000. Interspecific competition, environmental gradients, gene flow, and the coevolution of species' borders. American Naturalist 155:583–605.

Caswell, H. 2001. Matrix population models: construction, analysis, and interpretation. Sinauer Associates, Inc., Sunderland, MA.

Cavender-Bares, J., A. Keen, and B. Miles. 2006. Phylogenetic structure of Floridian plant communities depends on taxonomic and spatial scale. Ecology 87:S109–S122.

Cavender-Bares, J., D. D. Ackerly, D. A. Baum, and F. A. Bazzaz. 2004. Phylogenetic overdispersion in Floridian oak communities. American Naturalist 163:823–843.

Ceballos, G., and P. R. Ehrlich. 2002. Mammal population losses and the extinction. Science 296:904–907.

Chaine, A. S., and B. E. Lyon. 2008. Adaptive plasticity in female mate choice dampens sexual selection on male ornaments in the lark bunting. Science 319:459–462.

Chakraborty, R., and L. Jin. 1992. Heterozygote deficiency, population substructure and their implications in DNA fingerprinting. Human Genetics 88:267–272.

Chan, Y. F., M. E. Marks, F. C. Jones, G. Villarreal Jr, M. D. Shapiro, S. D. Brady, A. M. Southwick, D. M. Absher, J. Grimwood, J. Schmutz, R. M. Myers, D. Petrov, B. Jónsson, D. Schluter, M. A. Bell, and D. M. Kingsley. 2010. Adaptive evolution of pelvic reduction of a *Pitx1* enhancer. Science 327:302–305.

Charlesworth, B., M. Nordborg, and D. Charlesworth. 1997. The effects of local selection, balanced polymorphism and background selection on equilibrium patterns of genetic diversity in subdivided populations. Genetical Research 70:155–174.

Charlesworth, B., R. Lande, and M. Slatkin. 1982. A neo-Darwinian commentary on macroevolution. Evolution 36:474–498.

Charlesworth, D., and B. Charlesworth. 1987. Inbreeding depression and its evolutionary consequences. Annual Review of Ecology and Systematics 18:237–268.

Charmantier, A., and D. Garant. 2005. Environmental quality and evolutionary potential: lessons from wild populations. Proceedings of the Royal Society B. Biological Sciences 272:1415–1425.

Charmantier, A., and P. Gienapp. 2014. Climate change and timing of avian breeding and migration: evolutionary versus plastic changes. Evolutionary Applications 7: 15–28.

Charmantier, A., R. H. McCleery, L. R. Cole, C. Perrins, L. E. B. Kruuk, and B. C. Sheldon. 2008. Adaptive phenotypic plasticity in response to climate change in a wild bird population. Science 320:800–803.

Chenoweth, S. F., H. D. Rundle, and M. W. Blows. 2010. The contribution of selection and genetic constraints to phenotypic divergence. American Naturalist 175:186–196.

Cheptou, P.-O., O. Carrue, S. Rouifed, and A. Cantarel. 2008. Rapid evolution of seed dispersal in an urban environment in the weed *Crepis sancta*. Proceedings of the National Academy of Sciences USA 105:3796–3799.

Cheverud, J. M. 1982. Phenotypic, genetic, and environmental morphological integration in the cranium. Evolution 36:499–516.

Chevin, L.-M., and B. C. Haller. 2014. The temporal distribution of directional gradients under selection for an optimum. Evolution 68:3381–3394.

Chevin, L.-M., and R. Lande. 2010. When do adaptive plasticity and genetic evolution prevent extinction of a density-regulated population? Evolution 64:1143–1150.

Chevin, L.-M., and R. Lande. 2011. Adaptation to marginal habitats by evolution of increased phenotypic plasticity. Journal of Evolutionary Biology 24:1462–1476.

Chevin, L.-M., G. Decorzent, and T. Lenormand. 2014. Niche dimensionality and the genetics of ecological speciation. Evolution 68: 1244–1256.

Chevin, L.-M., R. Gallet, R. Gomulkiewicz, R. D. Holt, and S. Fellous. 2013. Phenotypic plasticity in evolutionary rescue experiments. Philosophical Transactions of the Royal Society B. Biological Sciences 368:20120089.

Chevin, L.-M., R. Lande, and G. M. Mace. 2010. Adaptation, plasticity, and extinction in a changing environment: towards a predictive theory. PLoS Biology 8:1–8.

Childs, D. Z., B. C. Sheldon, and M. Rees. 2016. The evolution of labile traits in sex- and age-structured populations. Journal of Animal Ecology 85:329–342.

Chitty, D. 1952. Mortality among voles (*Microtus agrestis*) at Lake Vyrnwy, Montgomeryshire in 1936–9. Philosophical Transactions of the Royal Society of London B. Biological Sciences 236:505–552.

Chitty, D. 1960. Population process in the vole and their relevance to general theory. Canadian Journal of Zoology 38:99–113.

Chou, H.-H., H.-C. Chiu, N. F. Delaney, D. Segrè, and C. J. Marx. 2011. Diminishing returns epistasis among beneficial mutations decelerates adaptation. Science 332:1190–1192.

Clark, J. S. 2010. Individuals and the variation needed for high species diversity in forest trees. Science 327:1129–1132.

Clark, J. S., D. M. Bell, M. H. Hersh, M. C. Kwit, E. Moran, C. Salk, A. Stine, D. Valle, and K. Zhu. 2011. Individual-scale variation, species-scale differences: inference needed to understand diversity. Ecology Letters 14:1273–1287.

Clausen, J., M. A. Nobs, O. Bjorkman, D. D. Keck, and W. M. Hiesey. 1940. Experimental studies on the nature of species. I. Effect of varied environments on western North American plants. Carnegie Institute of Washington, Washington, DC.

Clegg, S. M., F. D. Frentiu, J. Kikkawa, G. Tavecchia, and I. P. F. Owens. 2008. 4000 Years of phenotypic change in an island bird: heterogeneity of selection over three microevolutionary timescales. Evolution 62:2393–2410.

Clobert, J., E. Danchin, A. A. Dhondt, and J. D. Nichols (Editors). 2001. *Dispersal*. Oxford University Press, Oxford, UK.

Clobert, J., J. F. J.-F. Le Galliard, J. Cote, S. Meylan, and M. Massot. 2009. Informed dispersal, heterogeneity in animal dispersal syndromes and the dynamics of spatially structured populations. Ecology Letters 12:197–209.

Clutton-Brock, T. 2009. Sexual selection in females. Animal Behaviour 77:3–11.

Clutton-Brock, T., and B. C. Sheldon. 2010. Individuals and populations: the role of long-term, individual-based studies of animals in ecology and evolutionary biology. Trends in Ecology and Evolution 25:562–573.

Cockram, J., I. J. Mackay, and D. M. O'Sullivan. 2007. The role of double-stranded break repair in the creation of phenotypic diversity at cereal *VRN1* loci. Genetics 177:2535–2539.

Cody, M. L., and J. M. Diamond. 1975. Evolution and ecology of communities. Harvard University Press, Cambridge, MA.

Colautti, R. I., J. L. Maron, and S. C. H. Barrett. 2009. Common garden comparisons of native and introduced plant populations: latitudinal clines can obscure evolutionary inferences. Evolutionary Applications 2:187–199.

Colautti, R. I., and J. A. Lau. 2015. Contemporary evolution during invasion: evidence for differentiation, natural selection, and local adaptation. Molecular Ecology 24:1999–2017.

Collins, S., and A. Gardner. 2009. Integrating physiological, ecological and evolutionary change: a Price equation approach. Ecology Letters 12:744–757.

Colosimo, P. F., K. E. Hosemann, S. Balabhadra, G. Villarreal Jr., M. Dickson, J. Grimwood, J. Schmutz, R. M. Myers, D. Schluter, and D. M. Kingsley. 2005. Widespread parallel evolution in sticklebacks by repeated fixation of Ectodysplasin alleles. Science 307:1928–1933.

Comeault, A. A., S. M. Flaxman, R. Riesch, E. Curran, V. Soria-Carrasco, Z. Gompert, T. E. Farkas, M. Muschick, T. L. Parchman, T. Schwander, J. Slate, and P. Nosil. 2015. Selection on a genetic polymorphism counteracts ecological speciation in a stick insect. Current Biology 25:1975–1981.

Condon, C., B. S. Cooper, S. Yeaman, and M. J. Angilletta Jr. 2014. Temporal variation favors the evolution of generalists in experimental populations of *Drosophila melanogaster*. Evolution 68:720–728.

Connell, J. H. 1980. Diversity and the coevolution of competitors, or the ghost of competition past. Oikos 35:131–138.

Conover, D. O., and E. T. Schultz. 1995. Phenotypic similarity and the evolutionary significance of countergradient variation. Trends in Ecology and Evolution 10:248–252.

Conover, D. O., T. A. Duffy, and L. A. Hice. 2009. The covariance between genetic and environmental influences across ecological gradients: reassessing the evolutionary significance of countergradient and cogradient variation. Annals of the New York Academy of Sciences 1168:100–129.

Conover, D., and T. M. C. Present. 1990. Countergradient variation in growth rate: compensation for length of the growing season among Atlantic silversides from different latitudes. Oecologia 83:316–324.

Constanza, R., R. D'Arge, R. de Groot, S. Farber, M. Grasso, B. Hannon, K. Limburg, S. Naeem, R. V O'Neill, J. Paruelo, R. G. Raskin, P. Sutton, and M. van den Belt. 1997. The value of the world's ecosystem services and natural capital. Nature 387:253–260.

Conte, G. L., M. E. Arnegard, C. L. Peichel, and D. Schluter. 2012. The probability of genetic parallelism and convergence in natural populations. Proceedings of the Royal Society B. Biological Sciences 279:5039–5047.

Conte, G. L., M. E. Arnegard, J. Best, Y. F. Chan, F. C. Jones, D. M. Kingsley, D. Schluter, and C. L. Peichel. 2015. Extent of QTL reuse during repeated phenotypic divergence of sympatric threespine stickleback. Genetics 201:1189–1200.

Cooke, F., P. D. Taylor, C. M. Francis, and R. F. Rockwell. 1990. Directional selection and clutch size in birds. American Naturalist 136:261–267.

Cooper, N., and A. Purvis. 2010. Body size evolution in mammals: complexity in tempo and mode. American Naturalist 175:727–738.

Cooper Jr., W. E., R. A. Pyron, and T. Garland Jr. 2014. Island tameness: living on islands reduces flight initiation distance. Proceedings of the Royal Society B. Biological Sciences 281:20133019.

Cope, E. D. 1887. The origin of the fittest. Appelton, New York.

Cornell, H. V., and J. H. Lawton. 1992. Species interactions, local and regional processes, and limits to the richness of ecological communities: a theoretical perspective. Journal of Animal Ecology 61:1–12.

Coulson, T., and S. Tuljapurkar. 2008. The dynamics of a quantitative trait in an age-structured population living in a variable environment. American Naturalist 172:599–612.

Coulson, T., T. G. Benton, P. Lundberg, S. R. X. Dall, and B. E. Kendall. 2006. Putting evolutionary biology back in the ecological theatre: a demographic framework mapping genes to communities. Evolutionary Ecology Research 8:1155–1171.

Cousyn, C., L. De Meester, J. K. Colbourne, L. Brendonck, D. Verschuren, and F. Volckaert. 2001. Rapid, local adaptation of zooplankton behavior to changes in predation pressure in the absence of neutral genetic changes. Proceedings of the National Academy of Sciences USA 98:6256–6260.

Cox, G. W. 2004. Alien species and evolution: the evolutionary ecology of exotic plants, animals, microbes, and interacting native species. Island Press, Washington, DC.

Coyne, J. A., and H. A. Orr. 1997. "Patterns of speciation in *Drosophila*" revisited. Evolution 51:295–303.

Coyne, J. A., and H. A. Orr. 2004. Speciation. Sinauer Associates, Inc., Sunderland, MA.

Coyne, J. A., and T. D. Price. 2000. Little evidence for sympatric speciation in island birds. Evolution 54:2166–2171.

Coyne, J. A., N. H. Barton, and M. Turelli. 1997. A critique of Sewall Wright's shifting balance theory of evolution. Evolution 51:643–671.

Craig, T. P., J. K. Itami, and J. V Craig. 2007. Host plant genotype influences survival of hybrids between *Eurosta solidaginis* host races. Evolution 61:2607–2613.

Crawford, K. M., and K. D. Whitney. 2010. Population genetic diversity influences colonization success. Molecular Ecology 19:1253–1263.

Crespi, B. J. 2000. The evolution of maladaptation. Heredity 84:623–629.

Crespi, B. J. 2004. Vicious circles: positive feedback in major evolutionary and ecological transitions. Trends in Ecology and Evolution 19:627–633.

Crespi, B. 2007. The Baldwin effect and genetic assimilation: revisiting two mechanisms of evolutionary change mediated by phenotypic plasticity. Evolution 61:2469–2479.

Crispo, E. 2008. Modifying effects of phenotypic plasticity on interactions among natural selection, adaptation and gene flow. Journal of Evolutionary Biology 21:1460–1469.

Crispo, E., and L. J. Chapman. 2008. Population genetic structure across dissolved oxygen regimes in an African cichlid fish. Molecular Ecology 17:2134–2148.

Crispo, E., and L. J. Chapman. 2010. Geographic variation in phenotypic plasticity in response to dissolved oxygen in a cichlid fish. Journal of Evolutionary Biology 23:2091–2103.

Crispo, E., J. D. DiBattista, C. Correa, X. Thibert-Plante, A. E. McKellar, A. K. Schwartz, D. Berner, L. F. De Leon, and A. P. Hendry. 2010. The evolution of phenotypic plasticity in response to anthropogenic disturbance. Evolutionary Ecology Research 12:47–66.

Crispo, E., P. Bentzen, D. N. Reznick, M. T. Kinnison, and A. P. Hendry. 2006. The relative influence of natural selection and geography on gene flow in guppies. Molecular Ecology 15:49–62.

Crooks, J. A. 2005. Lag times and exotic species: The ecology and management of biological invasions in slow-motion. Ecoscience 12:316–329.

Crozier, L. G., M. D. Scheuerell, and R. W. Zabel. 2011. Using time series analysis to characterize evolutionary and plastic responses to environmental change: a case stduy of a shift toward earlier migration date in sockeye salmon. American Naturalist 178: 755–773.

Cruickshank, T. E., and M. W. Hahn. 2014. Reanalysis suggests that genomic islands of speciation are due to reduced diversity not reduced gene flow. Molecular Ecology 23:3133–3157.

Crutsinger, G. M., L. Souza, and N. J. Sanders. 2008b. Intraspecific diversity and dominant genotypes resist plant invasions. Ecology Letters 11:16–23.

Crutsinger, G. M., M. D. Collins, J. A. Fordyce, Z. Gompert, C. C. Nice, and N. J. Sanders. 2006. Plant genotypic diversity predicts community structure and governs an ecosystem process. Science 313:966–968.

Crutsinger, G. M., N. J. Sanders, and A. T. Classen. 2009. Comparing intra- and interspecific effects on litter decomposition in an old-field ecosystem. Basic and Applied Ecology 10:535–543.

Crutsinger, G. M., W. N. Reynolds, A. T. Classen, and N. J. Sanders. 2008a. Disparate effects of plant genotypic diversity on foliage and litter arthropod communities. Oecologia 158:65–75.

Crutsinger, G. M., B. E. Carter, and J. A. Rudgers. 2013. Soil nutrients trump intraspecific effects on understory plant communities. Oecologia 173: 1531–1538.

Crutsinger, G. M., M. A. Rodriguez-Cabal, A. B. Roddy, K. G. Peay, J. L. Bastow, A. G. Kidder, T. E. Dawson, P. V. A. Fine, and J. A Rudgers. 2014a. Genetic variation wihtin a dominant shrub structures green and brown community assemblages. Ecology 95:387–398.

Crutsinger, G. M., S. M. Rudman, M. A. Rodriguez-Cabal, A. D. McKown, T. Sato, A. M. MacDonald, J. Heavyside, A. Geraldes, E. M. Hart, C. J. Leroy, and R. W. El-Sabaawi. 2014b. Testing a 'genes-to-eco-systems' approach to understanding aquatic-terrestrial linkages. Molecular Ecology 23:5888–5903.

Cruz, R., M. Carballo, P. Conde-Padín, and E. Rolán-Alvarez. 2004. Testing alternative models for sexual isolation in natural populations of *Littorina saxatilis*: indirect support for by-product ecological speciation? Journal of Evolutionary Biology 17:288–293.

Cummings, M. E., G. G. Rosenthal, and M. J. Ryan. 2003. A private ultraviolet channel in visual communication. Proceedings of the Royal Society B. Biological Sciences 270:897–904.

Cunningham, C. J., G. T. Ruggerone, and T. P. Quinn. 2013. Size selectivity of predation by brown bears depends on the density of their sockeye salmon prey. American Naturalist 181:663–673.

Dalton, C. M., and A. S. Flecker. 2014. Metabolic stoichiometry and the ecology of fear in Trinidadian guppeis: consequences for life histories and stream ecosystems. Oecologia 176:691–701.

Dargent, F., M. E. Scott, A. P. Hendry, and G. F. Fussmann. 2013. Experimental elimination of parasites in nature leads to the evolution of increased resistance in hosts. Proceedings of the Royal Society B. Biological Sciences 280:20132371.

Darimont, C. T., S. M. Carlson, M. T. Kinnison, P. C. Paquet, T. E. Reimchen, and C. C. Wilmers. 2009. Human predators outpace other agents of trait change in the wild. Proceedings of the National Academy of Sciences USA 106:952–954.

Darwin, C. 1859. On the origin of species. John Murray, London.

Darwin, C. 1866. On the origin of species, 4th edition. John Murray, London.

Darwin, C. 1871. The descent of man, and selection in relation to sex. John Murray, London.

Davidson, A. M., M. Jennions, and A. B. Nicotra. 2011. Do invasive species show higher phenotypic plasticity than native species and, if so, is it adaptive? A meta-analysis. Ecology Letters 14:419–431.

Davies, M. S., and R. W. Snaydon. 1976. Rapid population differentiation in a mosaic environment. Heredity 36:59–66.

Davies, T. J., N. Cooper, J. A. F. Diniz-Filho, G. H. Thomas, and S. Meiri. 2012. Using phylogenetic trees to test for character displacement: a model and an example from a desert mammal community. Ecology 93:S44–S51.

Davis, M. A. 2003. Biotic globalization: does competition from introduced species threaten biodiversity? BioScience 53:481–489.

Davis, M. B., and R. G. Shaw. 2001. Range shifts and adaptive responses to Quaternary climate change. Science 292:673–679.

Dawkins, R. 1989. *The extended phenotype.* Oxford University Press, Oxford, UK.

Day, T., and J. D. McPhail. 1996. The effect of behavioural and morphological plasticity on foraging efficiency in the threespine stickleback (*Gasterosteus* sp.). Oecologia 108:380–388.

Day, T., J. Pritchard, and D. Schluter. 1994. A comparison of two sticklebacks. Evolution 48:1723–1734.

Dayan, T., and D. Simberloff. 2005. Ecological and community-wide character displacement: the next generation. Ecology Letters 8:875–894.

De Busschere, C., F. Hendrickx, S. M. Van Belleghem, T. Backeljau, L. Lens, and L. Baert. 2010. Parallel habitat specialization within the wolf spider genus *Hogna* from the Galápagos. Molecular Ecology 19:4029–4045.

de Jong, G. 2005. Evolution of phenotypic plasticity: patterns of plasticity and the emergence of ecotypes. New Phytologist 166:101–118.

De León, L. F., E. Bermingham, J. Podos, and A. P. Hendry. 2010. Divergence with gene flow as facilitated by ecological differences: within-island variation in Darwin's finches. Philosophical Transactions of the Royal Society B. Biological Sciences 365:1041–1052.

De León, L. F., J. A. M. Raeymaekers, E. Bermingham, J. Podos, A. Herrel, and A. P. Hendry. 2011. Exploring possible human influences on the evolution of Darwin's finches. Evolution 65:2258–2272.

DeLong, J. P., V. E. Forbes, N. Galic, J. P. Gibert, R. G. Laport, J. S. Phillips, and J. M. Vavra. 2016. How fast is fast? Eco-evolutionary dynamics and rates of change in populations and phenotypes. Ecology and Evolution 6:573–581.

de Mazancourt, C., E. Johnson, and T. G. Barraclough. 2008. Biodiversity inhibits species' evolutionary responses to changing environments. Ecology Letters 11:380–388.

De Meester, L. 1996. Evolutionary potential and local genetic differentiation in a phenotypically plastic trait of a cyclical parthenogen, *Daphnia magna*. Evolution 50:1293–1298.

De Meester, L., A. Gómez, B. Okamura, and K. Schwenk. 2002. The monopolization hypothesis and the dispersal–gene flow paradox in aquatic organisms. Acta Oecologica 23:121–135.

De Meester, L., J. Vanoverbeke, L. J. Kilsdonk, and M. C. Urban. 2016. Evolving perspectives on monopolization and priority effects. Trends in Ecology and Evolution 31:136–146.

De Roos, A. M., T. Schellekens, T. van Kooten, K. van de Wolfshaar, D. Claessen, and L. Persson. 2007. Food-dependent growth leads to overcompensation in stage-specific biomass when mortality increases: the influence of maturation versus reproduction regulation. American Naturalist 170:E59–E76.

Deagle, B. E., F. C. Jones, Y. F. Chan, D. M. Absher, D. M. Kingsley, and T. E. Reimchen. 2012. Population genomics of parallel phentoypic evolution in stickleback across stream-lake ecological transitions. Proceedings of the Royal Society of London. B. Biological Sciences 279:1277–1286.

DeAngelis, D. L., W. M. Post, and C. C. Travis. 1986. Positive feedback in natural systems. Springer-Verlag, Berlin, NY.

Débarre, F. 2012. Refining the conditions for sympatric ecological speciation. Journal of Evolutionary Biology 25:2651–2660.

Débarre, F., S. Yeaman, and F. Guillaume. 2015. Evolution of quantitative traits under a migration-selection balance: When does skew matter? American Naturalist 186:S37–S47.

Decaestecker, E., L. De Meester, and D. Ebert. 2002. In deep trouble: Habitat selection constrained by multiple enemies in zooplankton. Proceedings of the National Academy of Sciences USA 99:5481–5485.

Decaestecker, E., S. Gaba, J. A. M. Raeymaekers, R. Stoks, L. Van Kerckhoven, D. Ebert, and L. De Meester. 2007. Host-parasite "Red Queen" dynamics archived in pond sediment. Nature 450:870–873.

Declerk, S. A. J., A. R. Malo, S. Diehl, D. Waasdrop, K. D. Lemmen, K. Prioios, and S. Papakostas. 2015. Rapid adaptation of herbivore consumers to nutrient limitation: eco-evolutionary feedbacks to population demography and resource control. Ecology Letters 18:553–562.

Delbarco-Trillo, J., M. E. McPhee, and R. E. Johnston. 2010. Adult female hamsters avoid interspecific mating after exposure to heterospecific males. Behavioral Ecology and Sociobiology 64:1247–1253.

Denison, R. F., E. T. Kiers, and S. A. West. 2003. Darwinian agriculture: when can humans find solutions beyond the reach of natural selection. Quaterly Review of Biology 78:145–168.

Derry, A. M., and S. E. Arnott. 2007. Adaptive reversals in acid tolerance in copepods from lakes recovering from historical stress. Ecological Applications 17:1116–1126.

Derry, A. M., Å. M. Kestrup, and A. P. Hendry. 2013. Possible influences of plasticity and genetic/maternal effects on species coexistence: native Gammarus fasciatus facing exotic amphipods. Functional Ecology 27:1212–1223.

Des Roches, S., J. B. Shurin, D. Schluter, and L. J. Harmon 2013. Ecological and evolutionary effects of stickleback on community structure. PLoS ONE 8:e59644.

Dettman, J. R., C. Sirjusingh, L. M. Kohn, and J. B. Anderson. 2007. Incipient speciation by divergent adaptation and antagonistic epistasis in yeast. Nature 447:585–588.

Dettman, J. R., N. Rodrigue, A. H. Melnyk, A. Wong, S. T. Bailey, and R. Kassen. 2012. Evolutionary insight from whole-genome sequencing of experimentally evolved microbes. Molecular Ecology 21:2058–2077.

Deutsch, C. A., J. J. Tewksbury, R. B. Huey, K. S. Sheldon, C. K. Ghalambor, D. C. Haak, and P. R. Martin. 2008. Impacts of climate warming on terrestrial ectotherms across latitude. Proceedings of the National Academy of Sciences USA 105:6668–6672.

DeWitt, T. J. 1998. Costs and limits of phenotypic plasticity: tests with predator-induced morphology and life history in a freshwater snail. Journal of Evolutionary Biology 11:465–480.

DeWitt, T. J., A. Sih, and D. S. Wilson. 1998. Costs and limits of phenotypic plasticity. Trends in Ecology and Evolution 13:77–81.

DeWitt, T. J., B. W. Robinson, and D. S. Wilson. 2000. Functional diversity among predators of a freshwater snail imposes an adaptive trade-off for shell morphology. Evolutionary Ecology Research 2:129–148.

Dey, S., N. G. Prasad, M. Shakarad, and A. Joshi. 2008. Laboratory evolution of population stability in *Drosophila*: constancy and persistence do not necessarily coevolve. Journal of Animal Ecology 77:670–677.

Di Cesnola, A. P. 1907. A first study of natural selection in "*Helix arbustorum*" (Helicogena). Biometrika 5:387–399.

Diamond, J., S. L. Pimm, M. E. Gilpin, and M. LeCroy. 1989. Rapid evolution of character displacement in myzomelid honeyeaters. American Naturalist 134:675–708.

Díaz, S., and M. Cabido. 2001. Vive la différence: plant functional diversity matters to ecosystem processes. Trends in Ecology and Evolution 16:646–655.

Díaz, S., A. Purvis, J. H. C. Cornelissen, G. M. Mace, M. J. Donoghue, R. M. Ewers, P. Jordano, and W. D. Pearse. 2013. Functional traits, the phylogeny of function, and ecosystem service vulnerability. Ecology and Evolution 3:2958–2975.

DiBattista, J. D. 2008. Patterns of genetic variation in anthropogenically impacted populations. Conservation Genetics 9:141–156.

DiBattista, J. D., K. A. Feldheim, S. H. Gruber, and A. P. Hendry. 2007. When bigger is not better: selection against large size, high condition and fast growth in juvenile lemon sharks. Journal of Evolutionary Biology 20:201–212.

Didiano, T. J., N. E. Turley, G. Everwand, H. Schaefer, M. J. Crawley, and M. T. J. Johnson. 2014. Experimental test of plant defence evolution in four species using long-term rabbit exclosures. Journal of Ecology 102:584–594.

Dieckmann, U., and M. Doebeli. 1999. On the origin of species by sympatric speciation. Nature 400:354–357.

Dieckmann, U., M. Doebeli, J. A. J. Metz, and D. Tautz (Editors). 2004. *Adaptive speciation*. Cambridge University Press, Cambridge, UK.

Dingemanse, N. J., and M. Wolf. 2013. Between-individual differences in behavioural plasticity within populations: causes and consequences. Animal Behaviour 85:1031–1039.

Dingle, H. 1996. *Migration: the biology of life on the move*. Oxford University Press, Oxford, UK.

Dobzhansky, T. 1940. Speciation as a stage in evolutionary divergence. American Naturalist 74:312–321.

Doebeli, M. 1996. A quantitative genetic competition model for sympatric speciation. Journal of Evolutionary Biology 9:893–909.

Doebeli, M., and U. Dieckmann. 2003. Speciation along environmental gradients. Nature 421:259–264.

Doebeli, M., H. J. Blok, O. Leimar, and U. Dieckmann. 2007. Multimodal pattern formation in phenotype distributions of sexual populations. Proceedings of the Royal Society B. Biological Sciences 274:347–357.

Donohue, K., D. Messiqua, E. H. Pyle, M. S. Heschel, and J. Schmitt. 2000. Evidence of adaptive divergence in plasticity: density- and site-dependent selection on shade-avoidance responses in *Impatiens capensis*. Evolution 54:1956–1968.

Donohue, K., E. H. Pyle, D. Messiqua, M. S. Heschel, and J. Schmitt. 2001. Adaptive divergence in plasticity in natural populations of *Impatiens capensis* and its consequences for performance in novel habitats. Evolution 55:692–702.

Drès, M., and J. Mallet. 2002. Host races in plant-feeding insects and their importance in sympatric speciation. Philosophical Transactions of the Royal Society B. Biological Sciences 357:471–492.

Drummond, E. B. M., and M. Vellend. 2012. Genotypic diversity effects on the performance of *Taraxacum officinale* populations increase with time and environmental favorability. PLoS ONE 7:e30314.

Duckworth, R. A. 2009. The role of behavior in evolution: a search for mechanism. Evolutionary Ecology 23:513–531.

Duckworth, R. A., and S. M. Aguillon. 2015. Eco-evolutionary dynamics: investigating multiple causal pathways linking changes in behaivor, population density, and natural selection. Journal of Ornithology 156:S115–S124.

Duffy, M. A., and S. E. Forde. 2009. Ecological feedbacks and the evolution of resistance. Journal of Animal Ecology 78:1106–1112.

Duffy, M. A.,and L. Sivars-Becker. 2007. Rapid evolution and ecological host-parasite dynamics. Ecology Letters 10:44–53.

Dulvy, N. K., J. D. Metcalfe, J. Glanville, M. G. Pawson, and J. D. Reynolds. 2000. Fishery stability, local extinctions, and shifts in community structure in skates. Conservation Biology 14:283–293.

Duncan, E. J., P. D. Gluckman, and P. K. Dearden. 2014. Epigenetics, plasticity, and evolution: How do we link epigenetic chagne to phenotype? Journal of Experimetnal Zoology B: Molecular and Developmental Evolution 322:208–220.

Duputié, A., A. Rutschmann, O. Ronce, and I. Chuine. 2015. Phenotypic plasticity will not help all species adapt to climate change. Global Change Biology 21:3062–3073.

Dwyer, G., S. A. Levin, and L. Buttel. 1990. A simulation model of the population dynamics and evolution of myxomatosis. Ecological Monographs 60:423–447.

Dybdahl, M. F., and C. M. Lively. 1998. Host-parasite coevolution: evidence for rare advantage and time-lagged selection in a natural population. Evolution 52:1057–1066.

Easty, L. K., A. K. Schwartz, S. P. Gordon, and A. P. Hendry. 2011. Does sexual selection evolve following introduction to new environments? Animal Behaviour 82:1085–1095.

Ebert, D., C. Haag, M. Kirkpatrick, M. Riek, J. W. Hottinger, and V. I. Pajunen. 2002. A selective advantage to immigrant genes in a *Daphnia* metapopulation. Science 295:485–488.

Eckhart, V. M., M. A. Geber, W. F. Morris, E. S. Fabio, P. Tiffin, and D. A. Moeller. 2011. The geography of demography: long-term demographic studies and species distribution models reveal a species border limited by adaptation. American Naturalist 178:S26–S43.

Edelaar, P., A. M. Siepielski, and J. Clobert. 2008. Matching habitat choice causes directed gene flow: a neglected dimension in evolution and ecology. Evolution 62:2462–2472.

Edelaar, P., P. Burraco, and I. Gomez-Mestre. 2011. Comparisons between Q_{ST} and F_{ST} --how wrong have we been? Molecular Ecology 20:4830–4389.

Edmands, S. 1999. Heterosis and outbreeding depression in interpopulation crosses spanning a wide range of divergence. Evolution 53:1757–1768.

Edmands, S. 2007. Between a rock and a hard place: evaluating the relative risks of inbreeding and outbreeding for conservation and management. Molecular Ecology 16:463–475.

Egan, S. P., and D. J. Funk. 2009. Ecologically dependent postmating isolation between sympatric host forms of *Neochlamisus bebbianae* leaf beetles. Proceedings of the National Academy of Sciences USA 106:19426–19431.

Ehrlich, P. R., and P. H. Raven. 1969. Differentiation of populations. Science 165:1228–1232.

Eizaguirre, C., T. L. Lenz, A. Traulsen, and M. Milinksi. 2009. Speciation accelerated and stabilized by pleiotropic major histocompatibility complex immunogenes. Ecology Letters 12:5–12.

El-Sabaawi, R. W., R. D. Bassar, C. Rakowski, M. C. Marshall, B. L. Bryan, S. N. Thomas, C. Pringle, D. N. Reznick, and A. S. Flecker. 2015a. Intraspecific phenotypic differences in fish affect ecosystem processes as much as bottom-up factors. Oikos 9:1181–1191.

El-Sabaawi, R. W., M. C. Marshall, R. D. Bassar, A. López-Sepulcre, E. P. Palkovacs, and C. Dalton. 2015b. Assessing the effects of guppy life history evolution on nutrient recycling: from experiments to the field. Freshwater Biology 60:590–601.

El-Sabaawi, R. W., M. L. Warbanski, S. Rudman, R. Hovel, and B. Matthews. 2016. Investment in boney defensive traits alters organismal stoichiometry and excretion in fish. Oecologia. In press.

El-Sabaawi, R. W., J. Travis, E. Zadonà, P. B. McIntyre, D. N. Reznick, and A. Flecker. 2014. Intraspecific variability modulates interspecitic variation in animal organismal stoichiometry. Ecology and Evolution 4:1505–1515.

El-Sabaawi, R. W., E. Zadonà, T. J. Kohler, M. C. Marshall, J. M. Moslemi, J. Travis, A. López-Sepulcre, R. Ferriére, C. M. Pringle, S. A. Thomas, D. N. Reznick, and A. S. Flecker. 2012. Widespread intraspecific organismal stoichiometry among populations of the Trinidadian guppy. Functional Ecology 26:666–676.

Eldredge, N., and S. J. Gould. 1972. Punctuated equilibria: an alternative to phyletic gradualism. Pages 82–115 *in* T. J. M. Schopf, Editor. *Models in paleobiology*. Freeman, Cooper and Company, San Francisco, CA.

Ellegren, H., and B. C. Sheldon. 2008. Genetic basis of fitness differences in natural populations. Nature 452:169–175.

Ellers, J., and C. L. Boggs. 2003. The evolution of wing color: male mate choice opposes adaptive wing color divergence in *Colias* butterflies. Evolution 57:1100–1106.

Ellers, J., S. Rog, C. Braam, and M. P. Berg. 2011. Genotypic richness and phenotypic dissimilarity enhance population performance. Ecology 92:1605–1615.

Ellison, A. M., M. S. Bank, B. D. Clinton, E. A. Colburn, K. Elliott, C. R. Ford, D. R. Foster, B. D. Kloeppel, J. D. Knoepp, G. M. Lovett, J. Mohan, D. A. Orwig, N. L. Rodenhouse, W. V Sobczak, K. A. Stinson, J. K. Stone, C. M. Swan, J. Thompson, B. Von Holle, and J. R. Webster. 2005. Loss of foundation species: consequences for the structure and dynamics of forested ecosystems. Frontiers in Ecology and the Environment 3:479–486.

Ellner, S. P. 2013. Rapid evolution: from genes to communities, and back again? Functional Ecology 27:1087–1099.

Ellner, S. P., N. G. Hairston Jr, and M. A. Geber. 2011. Does rapid evolution matter? Measuring the rate of contemporary evolution and its impacts on ecological dynamics. Ecology Letters 14:603–614.

Ellstrand, N. C., R. Whitkus, and L. H. Rieseberg. 1996. Distribution of spontaneous plant hybrids. Proceedings of the National Academy of Sciences USA 93:5090–5093.

Elmer, K. R., T. K. Lehtonen, A. F. Kautt, C. Harrod, and A. Meyer. 2010. Rapid sympatric ecological differentiation of crater lake cichlid fishes within historic times. BMC Biology 8:60.

Elmerk, K. R., and A. Meyer. 2011. Adaptation in the age of ecological genomics: insights from parallelism and convergence. Trends in Ecology and Evolution 26:298–306.

Elser, J. 2006. Biological stoichiometry: a chemical bridge between ecosystem ecology and evolutionary biology. American Naturalist 168:S25–S35.

Elser, J. J., R. W. Sterner, E. Gorokhova, W. F. Fagan, T. A. Markow, J. B. Cotner, J. F. Harrison, S. E. Hobbie, G. M. Odell, and L. J. Weider. 2000. Biological stoichiometry from genes to ecosystems. Ecology Letters 3:540–550.

Elton, C. S. 1958. The ecology of invasions by plants and animals. Methuen & Co, London.

Emerson, B. C., and N. Kolm. 2005. Species diversity can drive speciation. Nature 434:1015–1017.

Emerson, B. C., and N. Kolm. 2007. Response to comments on Species diversity can drive speciation. Ecography 30:334–338.

Emerson, B. C., and R. G. Gillespie. 2008. Phylogenetic analysis of community assembly and structure over space and time. Trends in Ecology and Evolution 23:619–630.

Emery, S. M., and K. L. Gross. 2006. Dominant species identity regulates invasibility of old-field plant communities. Oikos 115:549–558.

Endler, J. A. 1977. *Geographic variation, speciation, and clines*. Princeton University Press, Princeton, NJ.

Endler, J. A. 1978. A predator's view of animal colour patterns. Evolutionary Biology 11:319–364.

Endler, J. A. 1980. Natural selection on color patterns in Poecilia reticulata. Evolution 34:76–91.

Endler, J. A. 1986. *Natural selection in the wild*. Princeton University Press, Princeton, NJ.

Endler, J. A. 1995. Multiple-trait coevolution and environmental gradients in guppies. Trends in Ecology and Evolution 10:22–29.

Endler, J. A., and A. E. Houde. 1995. Geographic variation in female preferences for male traits in *Poecilia reticulata*. Evolution 49:456–468.

Ennos, R. A. 1994. Estimating the relative rates of pollen and seed migration among plant populations. Heredity 72:250–259.

Eroukhmanoff, F., A. Hargeby, and E. I. Svensson. 2009. Rapid adaptive divergence between ecotypes of an aquatic isopod inferred from $F_{ST} - Q_{ST}$ analysis. Molecular Ecology 18:4912–4923.

Eroukhmanoff, F., A. Hargeby, and E. I. Svensson. 2011. The role of different reproductive barriers during phenotypic divergence of isopod ecotypes. Evolution 65:2631–2640.

Eroukhmanoff, F., and E. I. Svensson. 2009. Contemporary parallel diversification, antipredator adaptations and phenotypic integration in an aquatic isopod. PLoS ONE 4:e6173.

Eroukhmanoff, F., and E. I. Svensson. 2011. Evolution and stability of the G-matrix during the colonization of a novel environment. Journal of Evolutionary Biology 24:1363–1373.

Erwin, D. H. 2008. Macroevolution of ecosystem engineering, niche construction and diversity. Trends in Ecology and Evolution 23:304–310.

Essington, T. E., T. P. Quinn, and V. E. Ewert. 2000. Intra- and inter-specific competition and the reproductive success of sympatric Pacific salmon. Canadian Journal of Fisheries and Aquatic Sciences 57:205–213.

Estes, J. A., J. S. Brashares, and M. E. Power. 2013. Predicting and detecting reciprocity between indirect ecological interactions and evolution. American Naturalist 181:S76–S99.

Estes, S., and S. J. Arnold. 2007. Resolving the paradox of stasis: models with stabilizing selection explain evolutionary divergence on all timescales. American Naturalist 169:227–244.

Ezard, T. H. G., S. D. Côté, and F. Pelletier. 2009. Eco-evolutionary dynamics: disentangling phenotypic, environmental and population fluctuations. Philosophical Transactions of the Royal Society B. Biological Sciences 364:1491–1498.

Ezard, T. H. G., R. Prizak, and R. B. Hoyle. 2014. The fitness costs of adaptation via phenotypic plasticity and maternal effects. Functional Ecology 28:693–701.

Fahrig, L. 1997. Relative effects of habitat loss and fragmentation on population extinction. Journal of Wildlife Management 61:603–610.

Faith, D. P., S. Magallón, A. P. Hendry, E. Conti, T. Yahara, and M. J. Donoghue. 2010. Evosystem services: an evolutionary perspective on the links between biodiversity and human well-being. Current Opinion in Environmental Sustainability 2:66–74.

Falconer, D. S. 1989. Introduction to quantitative genetics, 3rd edition. Longman Scientific & Technical, New York.

Falconer, D. S., and T. F. C. Mackay. 1996. Introduction to quantitative genetics, 4th edition. Longman Science and Technology, Harlow, UK.

Faria, R., and A. Navarro. 2010. Chromosomal speciation revisited: rearranging theory with pieces of evidence. Trends in Ecology and Evolution 25:660–669.

Farkas, T. E., T. Mononen, A. A. Comeault, I. Hanski, and P. Nosil. 2013. Evolution of camouflage drives rapid ecological chagne in an insect community. Current Biology 23:1835–1843.

Farkas, T. E., and G. Montejo-Kovacevich. 2014. Density-dependent selection closes an eco-evolutionary feedback loop in the stick insect *Timema cristinae*. Biology Letters 10:20140896.

Farkas, T. E., A. P. Hendry, P. Nosil, and A. P. Beckerman. 2015. How maladaptation can structure biodiversity: Eco-evolutionary island biogeography. Trends in Ecology and Evolution 30:154–160.

Faubet, P., R. S. Waples, and O. E. Gaggiotti. 2007. Evaluating the performance of a multilocus Bayesian method for the estimation of migration rates. Molecular Ecology 16:1149–1166.

Fear, K. K., and T. Price. 1998. The adaptive surface in ecology. Oikos 82:440–448.

Feder, J. L., and P. Nosil. 2010. The efficacy of divergence hitchhiking in generating genomic islands during ecological speciation. Evolution 64:1729–1747.

Feder, J. L., S. P. Egan, and P. Nosil. 2012. The genomics of speciation-with-gene-flow. Trends in Genetics 28:342–350.

Feder, J. L., X. Xie, J. Rull, S. Velez, A. Forbes, B. Leung, H. Dambroski, K. E. Filchak, and M. Aluja. 2005. Mayr, Dobzhansky, and Bush and the complexities of sympatric speciation in *Rhagoletis*. Proceedings of the National Academy of Sciences USA 102:6573–6580.

Feil, R., and M. F. Fraga. 2012. Epigenetics and the environment: emerging patterns and implications. Nature Reviews Genetics 13:97–109.

Felker-Quinn, E., J. A. Schweitzer, and J. K. Bailey. 2013. Meta-analysis reveals evolution in invasive plant species but little support for Evolution of Increased Competitive Ability (EICA). Ecology and Evolution 3:739–751.

Felsenstein, J. 1976. The theoretical population genetics of variable selection and migration. Annual Review of Genetics 10:253–280.

Felsenstein, J. 1981. Skepticism towards Santa Rosalia, or why are there so few kinds of animals? Evolution 35:124–138.

Fenner, F. 1983. Biological control , as exemplified by smallpox eradication and myxomatosis. Proceedings of the Royal Society B. Biological Sciences 218:259–285.

Fenner, F., and R. N. Ratcliffe. 1965. *Myxomatosis*. Cambridge University Press, Cambridge, UK.

Fenster, C. B., and L. F. Galloway. 2000. Inbreeding and outbreeding depression in natural populations of *Chamaecrista fasciculata* (Fabaceae). Conservation Biology 14:1406–1412.

Ferrari, J., H. C. J. Godfray, A. S. Faulconbridge, K. Prior, and S. Via. 2006. Population differentiation and genetic variation in host choice among pea aphids from eight host plant genera. Evolution 60:1574–1584.

Ferriere, R., and S. Legendre. 2013. Eco-evolutionary feedbacks, adaptive dynamics and evolutionary rescue theory. Philosophical Transactions of the Royal Society B. Biological Sciences 368:20120081.

Feulner, P. G. D., F. J. J. Chain, M. Panchal, Y. Huang, C. Eizaguirre, M. Kalbe, T. L. Lenz, I. E. Samonte, M. Stoll, E. Bornberg-Bauer, T. B. H. Reusch, and M. Milinski. 2015. Genomics of divergence along a continuum of parapatric population differentiation. PLoS Genetics 11:e1004966.

Fierst, J. L. 2011. A history of phenotypic plasticity accelerates adaptation to a new environment. Journal of Experimental Biology 24:1992–2001.

Filchak, K. E., J. B. Roethele, and J. L. Feder. 2000. Natural selection and sympatric divergence in the apple maggot *Rhagoletis pomonella*. Nature 407:739–742.

Filin, I., R. D. Holt, and M. Barfield. 2008. The relation of density regulation to habitat specialization, evolution of a species' range, and the dynamics of biological invasions. American Naturalist 172:233–247.

Fischer, B. B., M. Kwiatkowski, M. Ackermann, J. Krismer, S. Roffler, M. J. F. Suter, R. I. L. Eggen, and B. Matthews. 2014. Phenotypic plasticity influences the eco-evolutionary dynamics of a predator-prey system. Ecology 95:3080–3092.

Fisher, R. A. 1930. *The genetical theory of natural selection*. Oxford University Press, Oxford, UK.

Fisk, D. L., L. C. Latta IV, R. A. Knapp, and M. E. Pfrender. 2007. Rapid evolution in response to introduced predators. I: rates and patterns of morphological and life-history trait divergence. BMC Evolutionary Biology 7:22.

Fitter, A. H., and R. S. R. Fitter. 2002. Rapid changes in flowering time in British plants. Science 296:1689–1691.

Fitzpatrick, C. R., A. A. Agrawal, N. Basiliko, A. P. Hastings, M. E. Isaac, M. Preston, and M. T. J. Johnson. 2015. The importance of plant genotype and contemporary evolution for terrestrial ecosystem processes. Ecology 96: 2632–2642.

Fitzpatrick, B. M., J. A. Fordyce, and S. Gavrilets. 2008. What, if anything, is sympatric speciation? Journal of Evolutionary Biology 21:1452–1459.

Fitzpatrick, S. W., J. Torres-Dowdall, D. N. Reznick, C. K. Ghalambor, and W. C. Funk. 2014. Parallelism isn't perfect: could disease and flooding drive a life-history anomaly in Trinidadian guppies. American Naturalist 183:290–300.

Fitzpatrick, S. W., J. C. Gerberich, J. A. Kronenberger, L. M. Angeloni, and W. C. Funk. 2015. Locally adapted traits maintained in the face of high gene flow. Ecology Letters 18:37–47.

Fitzpatrick, S. W., J. C. Gerberich, L. M. Angeloni, L. L. Bailey, E. D. Broder, J. Torres-Dowdall, C. A. Handelsman, A. López-Sepulcre, D. N. Reznick, C. K. Ghalambor, and W. C. Funk. 2016. Gene flow from an adaptively divergent source causes rescue through genetic and demographic factors in two wild populations of Trinidadian guppies. Evolutionary Applications. In press.

Flint, J., and T. F. C. Mackay. 2009. Genetic architecture of quantitative traits in mice, flies, and humans. Genome Research 19:723–733.

Flores-Moreno, H., E. S. García-Treviño, A. D. Letten, and A. T. Moles. 2015. In the beginning: phenotypic change in three invasive species through their first two centuries since introduction. Biological Invasions 17:1215–1225.

Forbes, A. A., T. H. Q. Powell, L. L. Stelinski, J. J. Smith, and J. L. Feder. 2009. Sequential sympatric speciation across trophic levels. Science 323:776–779.

Forde, S. E., J. N. Thompson, and B. J. M. Bohannan. 2004. Adaptation varies through space and time in a coevolving host—parasitoid interaction. Nature 431:841–844.

Forister, M. L. 2004. Oviposition preference and larval performance within a diverging lineage of lycaenid butterflies. Ecological Entomology 29:264–272.

Forsman, A. 2014. Effects of genotypic and phenotypic variation on establishment are important for conservation, invasion, and infection biology. Proceedings of the National Academy of Sciences USA 111:302–307.

Fournier-Level, A., A. Korte, M. D. Cooper, M. Nordborg, J. Schmitt, and A. M. Wilczek. 2011. A map of local adaptation in *Arabidopsis thaliana*. Science 334:86–89.

Fox, J. W., and D. A. Vasseur. 2008. Character convergence under competition for nutritionally essential resources. American Naturalist 172:667–680.

Frankham, R., J. D. Ballou, M. D. B. Eldridge, R. C. Lacy, K. Ralls, M. R. Dudash, and C. B. Fenster. 2011. Predicting the probability of outbreeding depression. Conservation Biology 25:465–475.

Franks, S. J., S. Sim, and A. E. Weis. 2007. Rapid evolution of flowering time by an annual plant in response to a climate fluctuation. Proceedings of the National Academy of Sciences 104:1278–1282.

Fraser, B. A., A. Kunstner, D. N. Reznick, C. Dreyer, and D. Weigel. 2015. Population genomics of natural and experimental populations of guppies *(Poecilia reticulata)*. Molecular Ecology 24:389–408.

Freeman, A. S., and J. E. Byers. 2006. Divergent induced responses to an invasive predator in marine mussel populations. Science 313:831–833.

Friesen, V. L., A. L. Smith, E. Gómez-Díaz, M. Bolton, R. W. Furness, J. González-Solís, and L. R. Monteiro. 2007. Sympatric speciation by allochrony in a seabird. Proceedings of the National Academy of Sciences USA 104:18589–18594.

Friman, V.-P., A. Jousset, and A. Buckling. 2014. Rapid prey evolution can alter the structure of predator-prey communities. Journal of Evolutionary Biology 27:374–380.

Fritz, R. S., and P. W. Price. 1988. Genetic variation among plants and insect community structure: willows and sawflies. Ecology 69:845–856.

Fry, J. D. 2003. Multilocus models of sympatric speciation: Bush versus Rice versus Felsenstein. Evolution 57:1735–1746.

Fryer, G., and T. D. Iles. 1972. The cichlid fishes of the great lakes of Africa: their biology and evolution. Oliver and Boyd, Edinburgh, UK.

Fukami, T., and D. A. Wardle. 2005. Long-term ecological dynamics: reciprocal insights from natural and anthropogenic gradients. Proceedings of the Royal Society of London B. Biological Sciences 272:2105–2115.

Fukushima, M., T. J. Quinn, and W. W. Smoker. 1998. Estimation of eggs lost from superimposed pink salmon (*Oncorhynchus gorbuscha*) redds. Canadian Journal of Fisheries and Aquatic Sciences 55:618–625.

Funk, D. J. 1998. Isolating a role for natural selection in speciation: host adaptation and sexual isolation in *Neochlamisus bebbianae* leaf beetles. Evolution 52:1744–1759.

Funk, D. J., K. E. Filchak, and J. L. Feder. 2002. Herbivorous insects: model systems for the comparative study of speciation ecology. Genetica 116:251–267.

Funk, D. J., P. Nosil, and W. J. Etges. 2006. Ecological divergence exhibits consistently positive associations with reproductive isolation across disparate taxa. Proceedings of the National Academy of Sciences USA 103:3209–3213.

Furin, C. G., F. A. von Hippel, and M. A. Bell. 2012. Partial reproductive isolation of a recently derived resident-freshwater population of threespine stickleback (*Gastosteus aculeatus*) from its putative anadromous ancestor. Evolution 66:3277–3286.

Fussmann, G. F., and A. Gonzalez. 2013. Evolutionary rescue can maintain an oscillating community undergoing environmental change. Interface Focus 3:20130036.

Fussmann, G. F., M. Loreau, and P. A. Abrams. 2007. Eco-evolutionary dynamics of communities and ecosystems. Functional Ecology 21:465–477.

Fussmann, G. F., S. P. Ellner, and N. G. Hairston Jr. 2003. Evolution as a critical component of plankton dynamics. Proceedings of the Royal Society B. Biological Sciences 270:1015–1022.

Futuyma, D. J. 1987. On the role of species in anagenesis. American Naturalist 130:465–473.

Futuyma, D. J. 2010. Evolutionary constraint and ecological consequences. Evolution 64:1865–84.

Futuyma, D. J., and G. Moreno. 1988. The evolution of ecological specialization. Annual Review of Ecology and Systematics 19:207–233.

Futuyma, D. J., M. C. Keese, and D. J. Funk. 1995. Genetic constraints on macroevolution—the evolution of host affiliation in the leaf beetle Genus *Ophraella*. Evolution 49:797–809.

Gabriel, W. 2005. How stress selects for reversible phenotypic plasticity. Journal of Evolutionary Biology 18:873–883.

Gandon, S., and S. L. Nuismer. 2009. Interactions between genetic drift, gene flow, and selection mosaics drive parasite local adaptation. American Naturalist 173:212–224.

Gandon, S., and Y. Michalakis. 2002. Local adaptation, evolutionary potential and host–parasite co-evolution: interactions between migration, mutation, population size and generation time. Journal of Evolutionary Biology 15:451–462.

Gandon, S., Y. Capowiez, Y. Dubois, Y. Michalakis, and I. Olivieri. 1996. Local adaptation and gene-for-gene coevolution in a metapopulation model. Proceedings of the Royal Society B. Biological Sciences 263:1003–1009.

Garant, D., L. E. B. Kruuk, R. H. McCleery, and B. C. Sheldon. 2007. The effects of environmental heterogeneity on multivariate selection on reproductive traits in female great tits. Evolution 61:1546–1559.

Garant, D., L. E. B. Kruuk, R. H. McCleery, and B. C. Sheldon. 2004. Evolution in a changing environment: a case study with great tit fledging mass. American Naturalist 164:E115–E129.

Garant, D., S. E. Forde, and A. P. Hendry. 2007. The multifarious effects of dispersal and gene flow on contemporary adaptation. Functional Ecology 21:434–443.

Garcia-Gonzalez, F., L. W. Simmons, J. L. Tomkins, J. S. Kotiaho, and J. P. Evans. 2012. Comparing evolvabilities: common errors surrounding the calculation and use of coefficients of additive genetic variation. Evolution 66:2341–2349.

García-Ramos, G., and D. Rodríguez. 2002. Evolutionary speed of species invasions. Evolution 56:661–668.

García-Ramos, G., and M. Kirkpatrick. 1997. Genetic models of adaptation and gene flow in peripheral populations. Evolution 51:21–28.

Gassmann, A. J., Y. Carrière, and B. E. Tabashnik. 2009. Fitness costs of insect resistance to *Bacillus thuringiensis*. Annual Review of Entomology 54:147–163.

Gause, G. F. 1934. The competitive exclusion principle. Williams & Wilkins, Baltimore, MD.

Gavrilets, S. 2003. Models of speciation: what have we learned in 40 years? Evolution 57:2197–2215.

Gavrilets, S. 2004. *Fitness landscapes and the origin of species*. Princeton University Press, Princeton, NJ.

Gavrilets, S., A. Vose, M. Barluenga, W. Salzburger, and A. Meyer. 2007. Case studies and mathematical models of ecological speciation. 1. Cichlids in a crater lake. Molecular Ecology 16:2893–909.

Gavrilets, S., and A. Vose. 2005. Dynamic patterns of adaptive radiation. Proceedings of the National Academy of Sciences USA 102:18040–18045.

Gavrilets, S., and A. Vose. 2007. Case studies and mathematical models of ecological speciation. 2. Palms on an oceanic island. Molecular Ecology 16:2910–2921.

Gavrilets, S., H. Li, and M. D. Vose. 2000. Patterns of parapatric speciation. Evolution 54:1126–1134.

Gende, S. M., R. T. Edwards, M. F. Willson, and M. S. Wipfli. 2002. Pacific Salmon in aquatic and terrestrial ecosystems. BioScience 52:917–928.

Genung, M. A., J. A. Schweitzer, F. Úbeda, B. M. Fitzpatrick, C. C. Pregitzer, E. Felker-Quinn, and J. K. Bailey. 2011. Genetic variation and community change—selection, evolution, and feedbacks. Functional Ecology 25:408–419.

Genung, M. A., J. K. Bailey, and J. A. Schweitzer. 2012a. Welcome ot the neighbourhood: interspecific genotype by genotype intearctions in *Solidago* influence above- and belowground biomass and associated communities. Ecology Letters 15:65–73.

Genung, M. A., G. M. Crutsinger, J. K. Bailey, J. A. Schweitzer, and N. J. Sanders. 2012b. Aphid and lady beetle abundance depend on the interaction of spatial effects and genotypic diversity. Oecologia 168:167–174.

Genung, M. A., J. K. Bailey, and J. A Schweitzer. 2013. The afterlife of interspecific indirect genetic effects: genotype interactions alter litter quality with consequences for decompositioni and nutrient dynamics. PLoS ONE 8:e53718.

Ghalambor, C. K., J. K. McKay, S. P. Carroll, and D. N. Reznick. 2007. Adaptive versus non-adaptive phenotypic plasticity and the potential for contemporary adaptation in new environments. Functional Ecology 21:394–407.

Ghalambor, C. K., K. L. Hoke, E. W. Ruell, E. K. Fischer, D. N. Reznick, and K. A. Hughes. 2015. Non-adaptive plasticity potentiates rapid adaptive evolution of gene expression in nature. Nature 525:372–375.

Gianoli, E., and M. González-Teuber. 2005. Environmental heterogeneity and population differentiation in plasticity to drought in *Convolvulus chilensis* (Convolvulaceae). Evolutionary Ecology 19:603–613.

Gibert, J.-M., F. Peronnet, and C. Schlötterer. 2007. Phenotypic plasticity in *Drosophila* pigmentation caused by temperature sensitivity of a chromatin regulator network. PLoS Genetics 3:e30.

Gienapp, P., M. Lof, T. E. Reed, J. McNamara, S. Verhulst, and M. E. Visser. 2012. Predicting demographically stable rates of adaptation: can great tit breeding time keep pace with climate change? Philosophical Transactions of the Royal Society. B. Biological Sciences 368:201202898.

Gienapp, P., C. Teplitsky, J. S. Alho, J. A. Mills, and J. Merilä. 2008. Climate change and evolution: disentangling environmental and genetic responses. Molecular Ecology 17:167–178.

Gifford, D. R., S. E. Schoustra, and R. Kassen. 2011. The length of adaptive walks is insensitive to starting fitness. Evolution 65:3070–3078.

Gilad, Y., J. K. Pritchard, and K. Thornton. 2009. Characterizing natural variation using next-generation sequencing technologies. Trends in Genetics 25:463–471.

Gilbert, B., and J. M. Levine. 2013. Plant invasions and extinction debts. Proceedings of the National Academy of Sciences USA 110:1744–1749.

Gilchrist, G. W., R. B. Huey, J. Balanyà, M. Pascual, and L. Serra. 2004. A time series of evolution in action: a latitudinal cline in wing size in South American *Drosophila subobscura*. Evolution 58:768–780.

Gill, J. A., J. A. Alves, W. J. Sutherland, G. F. Appleton, P. M. Potts, and T. G. Gunnarsson. 2014. Why is timing of bird migration advancing when individuals are not? Proceedings of the Royal Society of London. B. Biological Sciences 281:20132161.

Gillespie, R. 2004. Community assembly through adaptive radiation in Hawaiian spiders. Science 303:356–359.

Gilman, R. T., and J. E. Behm. 2011. Hybridization, species collapse, and species reemergence after disturbance to premating mechanisms of reproductive isolation. Evolution 65:2592–2605.

Gimenez, O., A. Grégoire, and T. Lenormand. 2009. Estimating and visualizing fitness surfaces using mark-recapture data. Evolution 63:3097–3105.

Gimenez, O., R. Covas, C. R. Brown, M. D. Anderson, M. B. Brown, and T. Lenormand. 2006. Nonparametric estimation of natural selection on a quantitative trait using mark-recapture data. Evolution 60:460–466.

Gingerich, P. D. 1983. Rates of evolution: effects of time and temporal scaling. Science 222:159–161.

Gingerich, P. D. 1993. Quantification and comparison of evolutionary rates. American Journal of Science 293:453–478.

Gingerich, P. D. 2001. Rates of evolution on the time scale of the evolutionary process. Genetica 112–113:127–144.

Gíslason, D., M. M. Ferguson, S. Skúlason, and S. S. Snorrason. 1999. Rapid and coupled phenotypic and genetic divergence in Icelandic Arctic char (*Salvelinus alpinus*). Canadian Journal of Fisheries and Aquatic Sciences 56:2229–2234.

Givnish, T. J. 2010. Ecology of plant speciation. Taxon 59:1326–1366.

Goddard, M. R. 2008. Quantifying the complexities of *Saccharomyces cerevisiae*'s ecosystem engineering via fermentation. Ecology 89:2077–2082.

Goldschmidt, R. 1940. The material basis of evolution. Yale University Press, New Haven, CT.

Gomez-Mestre, I., and R. Jovani. 2013. A heuristic model on the role of plasticity in adaptive evolution: plasticity increases adaptation, population viability and genetic variation. Proceedings of the Royal Society B. Biological Sciences 280:20131869.

Gomulkiewicz, R., and D. Houle. 2009. Demographic and genetic constraints on evolution. American Naturalist 174:E218–E229.

Gomulkiewicz, R., and R. D. Holt. 1995. Why does evolution by natural selection prevent extinction? Evolution 49:201–207.

Gomulkiewicz, R., and R. G. Shaw. 2013. Evolutionary rescue beyond the models. Philosophical Transactions of the Royal Society B. Biological Sciences 368:20120093.

Gomulkiewicz, R., R. D. Holt, and M. Barfield. 1999. The effects of density dependence and immigration on local adaptation and niche evolution in a black-hole sink environment. Theoretical Population Biology 55:283–296.

Gonzalez, A., J. H. Lawton, F. S. Gilbert, T. M. Blackburn, and I. Evans-Freke. 1998. Metapopulation dynamics, abundance, and distribution in a microecosystem. Science 281:2045–2047.

Gonzalez, A., O. Ronce, R. Ferriere, and M. E. Hochberg. 2013. Evolutionary rescue: an emerging focus at the intersection between ecology and evolution. Philosophical Transactions of the Royal Society B. Biological Sciences 368:20120404.

González-Suárez, M., and E. Revilla. 2013. Variability in life-history and ecological traits is a buffer against extinction in mammals. Ecology Letters 16:242–251.

Gordeeva, N. V., S. S. Alekseyev, A. N. Matveev, and V. P. Samusenok. 2015. Parallel evolutionary divergence in Arctic char *Salvelinus alpinus* complex from Transbaikalia: variation in differentiation degree and segregation of genetic diversity among sympatric forms. Canadian Journal of Fisheries and Aquatic Sciences 72:96–115.

Gordon, S. P., D. N. Reznick, M. T. Kinnison, M. J. Bryant, D. J. Weese, K. Räsänen, N. P. Millar, and A. P. Hendry. 2009. Adaptive changes in life history and survival following a new guppy introduction. American Naturalist 174:34–45.

Gosden, T. P., and E. I. Svensson. 2008. Spatial and temporal dynamics in a sexual selection mosaic. Evolution 62:845–856.

Gotanda, K. M., and A. P. Hendry. 2014. Using adaptive traits to consider potential consequences of temporal variation in selection: male guppy colour through time and space. Biological Journal of the Linnean Society 112:108–122.

Gotanda, K. M., C. Correa, M. M. Turcotte, G. Rolshausen, and A. P. Hendry. 2015. Linking macrotrends and microrates: Re-evaluating microevolutionary support for Cope's rule. Evolution 69:1345–1354.

Gotelli, N. J., and G. R. Graves. 1996. Null models in ecology. Smithsonian Institution Press, Washington, DC.

Goudet, J., N. Perrin, and P. Waser. 2002. Tests for sex-biased dispersal using bi-parentally inherited genetic markers. Molecular Ecology 11:1103–1114.

Gould, S. J., and N. Eldredge. 1977. Punctuated equilibria: the tempo and mode of evolution reconsidered. Paleobiology 3:115–151.

Gould, S. J., and R. C. Lewontin. 1979. The spandrels of San Marco and the Panglossian paradigm: a critique of the adaptationist program. Proceedings of the Royal Society B. Biological Sciences 205:581–598.

Goulson, D., and K. Jerrim. 1997. Maintenance of the species boundary between *Silene dioica* and *S. latifolia* (red and white campion). Oikos 79:115–126.

Gow, J. L., C. L. Peichel, and E. B. Taylor. 2006. Contrasting hybridization rates between sympatric three-spined sticklebacks highlight the fragility of reproductive barriers between evolutionarily young species. Molecular Ecology 15:739–752.

Gow, J. L., C. L. Peichel, and E. B. Taylor. 2007. Ecological selection against hybrids in natural populations of sympatric threespine sticklebacks. Journal of Evolutionary Biology 20:2173–2180.

Grant, B. R., and P. R. Grant. 1993. Evolution of Darwin's finches caused by a rare climatic event. Proceedings of the Royal Society B. Biological Sciences 251:111–117.

Grant, B. R., and P. R. Grant. 2003. What Darwin's finches can teach us about the evolutionary origin and regulation of biodiversity. BioScience 53:965–975.

Grant, B., and P. Grant. 1996. High survival of Darwin's finch hybrids: effects of beak morphology and diets. Ecology 77:500–509.

Grant, J. W. A., and D. L. Kramer. 1990. Territory size as a predictor of the upper limits to population density of juvenile salmonids in streams. Canadian Journal of Fisheries and Aquatic Sciences 47:1724–1737.

Grant, P. R. 1999. *Ecology and evolution of Darwin's finches, 2nd ed.* Princeton University Press, Princeton, NJ.

Grant, P. R., and B. R. Grant. 1995. Predicting microevolutionary responses to directional selection on heritable variation. Evolution 49:241–251.

Grant, P. R., and B. R. Grant. 2002. Unpredictable evolution in a 30-year study of Darwin's finches. Science 296:707–711.

Grant, P. R., and B. R. Grant. 2006. Evolution of character displacement in Darwin's finches. Science 313:224–226.

Grant, P. R., and B. R. Grant. 2008. *How and why species multiply*. Princeton University Press, Princeton, NJ.

Grant, P. R., and B. R. Grant. 2009. Sympatric speciation, immigration, and hybridization in island birds. Pages 326–357 *in* J. B. Losos and R. E. Ricklefs, Editors. *The theory of island biogeography revisited*. Princeton University Press, Princeton, NJ.

Grant, P. R., B. R. Grant, and K. Petren. 2005. Hybridization in the recent past. American Naturalist 166:56–67.

Grant, P. R., B. R. Grant, J. A. Markert, L. F. Keller, and K. Petren. 2004. Convergent evolution of Darwin's finches caused by introgressive hybridization and selection. Evolution 58:1588–1599.

Grant, V. 1949. Pollination systems as isolating mechanisms in angiosperms. Evolution 3:82–97.

Grant, V. 1994. Modes and origins of mechanical and ethological isolation in angiosperms. Proceedings of the National Academy of Science USA 91:3–10.

Gravel, D., F. Guichard, and M. E. Hochberg. 2011. Species coexistence in a variable world. Ecology Letters 14:828–839.

Greenwood, P. J. 1980. Mating systems, philopatry and dispersal in birds and mammals. Animal Behaviour 28:1140–1162.

Gressel, J. 2009. Evolving understanding of the evolution of herbicide resistance. Pest Management Science 65:1164–1173.

Grether, G. F. 2005. Environmental change, phenotypic plasticity, and genetic compensation. American Naturalist 166:E115–E123.

Griffith, T. M., and M. A. Watson. 2006. Is evolution necessary for range expansion? Manipulating reproductive timing of a weedy annual transplanted beyond its range. American Naturalist 167:153–164.

Gross, M. R. 1996. Alternative reproductive strategies and tactics: diversity within sexes. Trends in Ecology and Evolution 11:92–98.

Gruner, D. S., N. J. Gotelli, J. P. Price, and R. H. Cowie. 2008. Does species richness drive speciation? A reassessment with the Hawaiian biota. Ecography 31:279–285.

Guillaume, F. 2011. Migration-induced phenotypic divergence: the migration-selection balance of correlated traits. Evolution 65:1723–1738.

Gutteling, E. W., J. A. G. Riksen, J. Bakker, and J. E. Kammenga. 2007. Mapping phenotypic plasticity and genotype-environment interactions affecting life-history traits in *Caenorhabditis elegans*. Heredity 98:28–37.

Gyllenberg, M., and K. Parvinen. 2001. Necessary and sufficient conditions for evolutionary suicide. Bulletin of Mathematical Biology 63:981–993.

Hadfield, J. D. 2008. Estimating evolutionary parameters when viability selection is operating. Proceedings of the Royal Society B. Biological Sciences 275:723–734.

Hadfield, J. D., A. J. Wilson, and L. E. B. Kruuk. 2011. Cryptic evolution: does environmental deterioration have a genetic basis? Genetics 187:1099–1113.

Hadfield, J. D., A. J. Wilson, D. Garant, B. C. Sheldon, and L. E. B. Kruuk. 2010. The misuse of BLUP in ecology and evolution. American Naturalist 175:116–125.

Hairston, N. G., D. W. Tinkle, and H. M. Wilbur. 1970. Natural selection and the parameters of population growth. Journal of Wildlife Management 34:681–690.

Hairston Jr., N. G., S. P. Ellner, M. A. Geber, T. Yoshida, and J. A. Fox. 2005. Rapid evolution and the convergence of ecological and evolutionary time. Ecology Letters 8:1114–1127.

Haldane, J. B. S. 1948. The theory of a cline. Journal of Genetics 48:277–284.

Haldane, J. B. S. 1949. Suggestions as to quantitative measurement of rates of evolution. Evolution 3:51–56.

Haldane, J. B. S. 1956. The relation between density regulation and natural selection. Proceedings of the Royal Society B. Biological Sciences 145:306–308.

Hall, M. C., and J. H. Willis. 2006. Divergent selection on flowering time contributes to local adaptation in *Mimulus guttatus* populations. Evolution 60:2466–2477.

Haller, B. C., L. F. De León, G. Rolshausen, K. M. Gotanda, and A. P. Hendry. 2012. Magic traits: distinguishing the important from the trivial. Trends in Ecology and Evolution 27:4–5.

Haller, B. C., and A. P. Hendry. 2014. Solving the paradox of stasis: squashed stabilizing selection and the limits of detection. Evolution 68:483–500.

Haloin, J. R., and S. Y. Strauss. 2008. Interplay between ecological communities and evolution: review of feedbacks from microevolutionary to macroevolutionary scales. Annals of the New York Academy of Sciences 1133:87–125.

Hancock, A. M., B. Brachi, N. Faure, M. W. Horton, L. B. Jarymowycz, F. G. Sperone, C. Toomajian, F. Roux, and J. Bergelson. 2011. Adaptation to climate across the *Arabidopsis thaliana* genome. Science 334:83–86.

Hansen, T. F. 2003. Is modularity necessary for evolvability? Remarks on the relationship between pleiotropy and evolvability. Biosystems 69:83–94.

Hansen, T. F., A. J. R. Carter, and C. Pélabon. 2006. On adaptive accuracy and precision in natural populations. American Naturalist 168:168–181.

Hansen, T. F., and D. Houle. 2008. Measuring and comparing evolvability and constraint in multivariate characters. Journal of Evolutionary Biology 21:1201–1219.

Hansen, T. F., C. Pélabon, and D. Houle. 2011. Heritability is not evolvability. Evolutionary Biology 38:258–277.

Hanski, I. 2000. Estimating the parameters of survival and migration of individuals in metapopulations. Ecology 81:239–251.

Hanski, I. 2011. Eco-evolutionary spatial dynamics in the Glanville fritillary butterfly. Proceedings of the National Academy of Sciences USA 108:14397–14404.

Hanski, I., and I. Saccheri. 2006. Molecular-level variation affects population growth in a butterfly metapopulation. PLoS Biology 4:e129.

Hanski, I., and O. Ovaskainen. 2000. The metapopulation capacity of a fragmented landscape. Nature 404:755–758.

Hanski, I., T. Mononen, and O. Ovaskainen. 2011. Eco-evolutionary metapopulation dynamics and the spatial scale of adaptation. American Naturalist 177:29–43.

Hanson, D., R. D. H. Barrett, and A. P. Hendry. 2015. Testing for parallel allochronic isolation in lake-stream stickleback. *Journal of Evolutionary Biology* 29:47–57.

Hargrave, C. W., K. D. Hambright, and L. J. Weider. 2011. Variation in resource consumption across a gradient of increasing intra- and interspecific richness. Ecology 92:1226–1235.

Hargreaves, A. L., K. E. Samis, and C. G. Eckert. 2014. Are species' range limits simply niche limits writ large? A review of transplant experiments beyond the range. American Naturalist 183: 157–173.

Harman, O. 2010. The Price of altruism. W. W. Norton & Company, New York.

Harmon, L. J., B. Matthews, S. Des Roches, J. M. Chase, J. B. Shurin, and D. Schluter. 2009. Evolutionary diversification in stickleback affects ecosystem functioning. Nature 458:1167–1170.

Harmon, L. J., J. B. Losos, T. J. Davies, R. G. Gillespie, J. L. Gittleman, W. B. Jennings, K. H. Kozak, M. A. McPeek, F. Moreno-Roark, T. J. Near, A. Purvis, R. E. Ricklefs, D. Schluter, J. A. Schulte II, O. Seehausen, B. L. Sidlauskas, O. Torres-Carvajal, J. T. Weir, and A. Ø. Mooers. 2010. Early bursts of body size and shape evolution are rare in comparative data. Evolution 64:2385–2396.

Harmon, L. J., J. J. Kolbe, J. M. Cheverud, and J. B. Losos. 2005. Convergence and the multidimensional niche. Evolution 59:409–421.

Harper, J. L. 1977. Population biology of plants. Academic Press, London, UK.

Harrison, S., and A. Hastings. 1996. Genetic and evolutionary consequences of metapopulation structure. Trends in Ecology and Evolution 11:180–183.

Hartley, C. J., R. D. Newcomb, R. J. Russell, C. G. Yong, J. R. Stevens, D. K. Yeates, J. La Salle, and J. G. Oakeshott. 2006. Amplification of DNA from preserved specimens shows blowflies were preadapted for the rapid evolution of insecticide resistance. Proceedings of the National Academy of Sciences USA 103:8757–8762.

Hasselman, D. J., E. E. Argo, M. C. McBride, P. Bentzen, T. F. Schlutz, A. A. Perez-Umphrey, and E. C. Palkovacs. 2014. Human disturbance causes the formation of a hybrid swarm between two naturally sympatric fish species. Molecular Ecology 23:1137–1152.

Heap, I. M. 1997. The occurrence of herbicide-resistant weeds worldwide. Pesticide Science 51:235–243.

Hedrick, P. W. 1995. Gene flow and genetic restoration: the Florida panther as a case study. Conservation Biology 9:996–1007.

Hedrick, P. W., and R. Fredrickson. 2010. Genetic rescue guidelines with examples from Mexican wolves and Florida panthers. Conservation Genetics 11:615–626.

Heino, M., J. A. J. Metz, and V. Kaitala. 1998. The enigma of frequency-dependent selection. Trends in Ecology and Evolution 13:367–370.

Hellmann, J., and M. Pineda-Krch. 2007. Constraints and reinforcement on adaptation under climate change: Selection of genetically correlated traits. Biological Conservation 137:599–609.

Hendry, A. 2007. The Elvis paradox. Nature 446:147–149.

Hendry, A. P. 2001. Adaptive divergence and the evolution of reproductive isolation in the wild: an empirical demonstration using introduced sockeye salmon. Genetica 112–113:515–534.

Hendry, A. P. 2002. QST > = ≠ < FST? Trends in Ecology and Evolution 17:502.

Hendry, A. P. 2004. Selection against migrants contributes to the rapid evolution of ecologically dependent reproductive isolation. Evolutionary Ecology Research 6:1219–1236.

Hendry, A. P. 2009. Ecological speciation! Or the lack thereof? Canadian Journal of Fisheries and Aquatic Sciences 66:1383–1398.

Hendry, A. P. 2013. Key questions in the genetics and genomics of eco-evolutionary dynamics. Heredity 111:456–466.

Hendry, A. P. 2016. Key questions on the role of phenotypic plasticity in eco-evolutionary dynamics. Journal of Heredity 107:25–41.

Hendry, A. P., and A. Gonzalez. 2008. Whither adaptation? Biology & Philosophy 23:673–699.

Hendry, A. P., and E. B. Taylor. 2004. How much of the variation in adaptive divergence can be explained by gene flow? An evaluation using lake-stream stickleback pairs. Evolution 58:2319–2331.

Hendry, A. P., and M. T. Kinnison. 1999. The pace of modern life: measuring rates of contemporary microevolution. Evolution 53:1637–1653.

Hendry, A. P., and M. T. Kinnison. 2001. An introduction to microevolution: rate, pattern, process. Genetica 112–113:1–8.

Hendry, A. P., and T. Day. 2005. Population structure attributable to reproductive time: isolation by time and adaptation by time. Molecular Ecology 14:901–916.

Hendry, A. P., D. I. Bolnick, D. Berner, and C. L. Peichel. 2009a. Along the speciation continuum in sticklebacks. Journal of Fish Biology 75:2000–2036.

Hendry, A. P., L. G. Lohmann, E. Conti, J. Cracraft, K. A. Crandall, D. P. Faith, C. Häuser, C. A. Joly, K. Kogure, A. Larigauderie, S. Magallón, C. Moritz, S. Tillier, R. Zardoya, A.-H. Prieur-Richard, B. A. Walther, T. Yahara, and M. J. Donoghue. 2010. Evolutionary biology in biodiversity science, conservation, and policy: a call to action. Evolution 64:1517–1528.

Hendry, A. P., E. B. Taylor, and J. D. McPhail. 2002. Adaptive divergence and the balance between selection and gene flow: lake and stream stickleback in the Misty system. Evolution 56:1199–1216.

Hendry, A. P., J. K. Wenburg, P. Bentzen, E. C. Volk, and T. P. Quinn. 2000b. Rapid evolution of reproductive isolation in the wild: evidence from introduced salmon. Science 290:516–518.

Hendry, A. P., K. Hudson, J. A. Walker, K. Räsänen, and L. J. Chapman. 2011. Genetic divergence in morphology-performance mapping between Misty Lake and inlet stickleback. Journal of Evolutionary Biology 24:23–35.

Hendry, A. P., P. Nosil, and L. H. Rieseberg. 2007. The speed of ecological speciation. Functional Ecology 21:455–464.

Hendry, A. P., P. R. Grant, B. R. Grant, H. A. Ford, M. J. Brewer, and J. Podos. 2006. Possible human impacts on adaptive radiation: beak size bimodality in Darwin's finches. Proceedings of the Royal Society B. Biological Sciences 273:1887–1894.

Hendry, A. P., S. K. Huber, L. F. De León, A. Herrel, and J. Podos. 2009b. Disruptive selection in a bimodal population of Darwin's finches. Proceedings of the Royal Society B. Biological Sciences 276:753–759.

Hendry, A. P., S. M. Vamosi, S. J. Latham, J. C. Heilbuth, and T. Day. 2000a. Questioning species realities. Conservation Genetics 1:67–76.

Hendry, A. P., T. Day, and E. B. Taylor. 2001. Population mixing and the adaptive divergence of quantitative traits in discrete populations: a theoretical framework for empirical tests. Evolution 55:459–466.

Hendry, A. P., T. J. Farrugia, and M. T. Kinnison. 2008. Human influences on rates of phenotypic change in wild animal populations. Molecular Ecology 17:20–29.

Hendry, A. P., V. Castric, M. T. Kinnison, and T. P. Quinn. 2004b. The evolution of philopatry and dispersal: homing versus straying in salmonids. Pages 52–91 *in* A. P. Hendry and S. C. Stearns, Editors. *Evolution illuminated: salmon and their relatives.* Oxford University Press, Oxford, UK.

Hendry, A. P., V. Millien, A. Gonzalez, and H. C. E. Larsson. 2012. How humans influence evolution on adaptive landscapes. Pages 180–202 *in* E. Svensson and R. Calsbeek, Editors. *The adaptive landscape in evolutionary biology.* Oxford University Press, Oxford, UK.

Hendry, A. P., Y. E. Morbey, O. K. Berg, and J. K. Wenburg. 2004a. Adaptive variation in senescence: reproductive lifespan in a wild salmon population. Proceedings of the Royal Society of London B. Biological Sciences 271:259–266.

Herder, F., A. W. Nolte, J. Pfaender, J. Schwarzer, R. K. Hadiaty, and U. K. Schliewen. 2006. Adaptive radiation and hybridization in Wallace's Dreamponds: evidence from sailfin silversides in the Malili Lakes of Sulawesi. Proceedings of the Royal Society B. Biological Sciences 273:2209–2217.

Hereford, J. 2009. A quantitative survey of local adaptation and fitness trade-offs. American Naturalist 173:579–588.

Hereford, J., T. F. Hansen, and D. Houle. 2004. Comparing strengths of directional selection: how strong is strong? Evolution 58:2133–2143.

Hermisson, J., and P. S. Pennings. 2005. Soft sweeps: molecular population genetics of adaptation from standing genetic variation. Genetics 169:2335–2352.

Herrel, A., J. Podos, B. Vanhooydonck, and A. P. Hendry. 2009. Force-velocity trade-off in Darwin's finch jaw function: a biomechanical basis for ecological speciation? Functional Ecology 23:119–125.

Herrera, C. M., and P. Bazaga. 2010. Epigenetic differentiation and relationship to adaptive genetic divergence in discrete populations of the violet *Viola cazorlensis.* New Phytologist 187:867–876.

Hersch, E. I., and P. C. Phillips. 2004. Power and potential bias in field studies of natural selection. Evolution 58:479–485.

Hersch-Green, E. I., N. E. Turley, and M. T. J. Johnson. 2011. Community genetics: what have we accomplished and where should we be going. Philosophical Transactions of the Royal Society B. 366:1453–1460.

Hewitt, G. 2000. The genetic legacy of the Quaternary ice ages. Nature 405:907–913.

Hey, J., and R. Nielsen. 2004. Multilocus methods for estimating population sizes, migration rates and divergence time, with applications to the divergence of *Drosophila pseudoobscura* and *D. persimilis.* Genetics 167:747–60.

Heywood, J. S. 2005. An exact form of the breeder's equation for the evolution of a quantitative trait under natural selection. Evolution 59:2287–2298.

Higgie, M., S. Chenoweth, and M. W. Blows. 2000. Natural selection and the reinforcement of mate recognition. Science 290:519–521.

Hilborn, R., T. P. Quinn, D. E. Schindler, and D. E. Rogers. 2003. Biocomplexity and fisheries sustainability. Proceedings of the National Academy of Sciences USA 100:6564–6568.

Hill, W. G. 2010. Understanding and using quantitative genetic variation. Philosophical Transactions of the Royal Society B. Biological Sciences 365:73–85.

Hill, W. G., and A. Caballero. 1992. Artificial selection experiments. Annual Review of Ecology and Systematics 23:287–310.

Hill, W. G., and M. Kirkpatrick. 2010. What animal breeding has taught us about evolution. Annual Review of Ecology, Evolution, and Systematics 41:1–19.

Hill, W. G., M. E. Goddard, and P. M. Visscher. 2008. Data and theory point to mainly additive genetic variance for complex traits. PLoS Genetics 4:e1000008.

Hiltunen, T., N. G. Hairston Jr., G. Hooker, L. E. Jones, and S. P. Ellner. 2014. A newly discovered role of evolution ion previously published consumer-resource dynamics. Ecology Letters 17:915–923.

Hine, E., K. McGuigan, and M. W. Blows. 2011. Natural selection stops the evolution of male attractiveness. Proceedings of the National Academy of Sciences USA 108:3659–3664.

Hobson, K. A. 2005. Using stable isotopes to trace long-distance dispersal in birds and other taxa. Diversity and Distributions 11:157–164.

Hochberg, M. E., and R. D. Holt. 1995. Refuge evolution and the population dynamics of coupled host-parasitoid associations. Evolutionary Ecology 9:633–661.

Hodges, S. A., and M. L. Arnold. 1995. Spurring plant diversification: are floral nectar spurs a key innovation? Proceedings of the Royal Society B. Biological Sciences 262:343–348.

Hoeksema, J. D., and S. E. Forde. 2008. A meta-analysis of factors affecting local adaptation between interacting species. American Naturalist 171:275–290.

Hoekstra, H. E., and J. A. Coyne. 2007. The locus of evolution: evo devo and the genetics of adaptation. Evolution 61:995–1016.

Hoekstra, H. E., J. M. Hoekstra, D. Berrigan, S. N. Vignieri, A. Hoang, C. E. Hill, P. Beerli, and J. G. Kingsolver. 2001. Strength and tempo of directional selection in the wild. Proceedings of the National Academy of Sciences USA 98:9157–9160.

Hoffmann, A. A., and M. W. Blows. 1994. Species borders: ecological and evolutionary perspectives. Trends in Ecology and Evolution 9:223–227.

Hoffmann, A. A., R. J. Hallas, J. A. Dean, and M. Schiffer. 2003. Low potential for climatic stress adaptation in a rainforest *Drosophila* species. Science 301:100–102.

Hogg, J. T., S. H. Forbes, B. M. Steele, and G. Luikart. 2006. Genetic rescue of an insular population of large mammals. Proceedings of the Royal Society B. Biological Sciences 273:1491–1499.

Hohenlohe, P. A., S. Bassham, P. D. Etter, N. Stiffler, E. A. Johnson, and W. A. Cresko. 2010. Population genomics of parallel adaptation in threespine stickleback using sequenced RAD tags. PLoS Genetics 6:e1000862.

Hollander, J. 2008. Testing the grain-size model for the evolution of phenotypic plasticity. Evolution 62:1381–1389.

Holsinger, K. E., and B. S. Weir. 2009. Genetics in geographically structured populations: defining, estimating and interpreting F_{ST}. Nature Reviews Genetics 10:639–650.

Holt, R. D. 1977. Predation, apparent competition, and structure of prey communities. Theoretical Population Biology 12:197–229.

Holt, R. D., and M. Barfield. 2011. Theoretical perspectives on the statics and dynamics of species' borders in patchy environments. American Naturalist 178:S6–S25.

Holt, R. D., and M. E. Hochberg. 1997. When is biological control evolutionarily stable (or is it)? Ecology 78:1673–1683.

Holt, R. D., and R. Gomulkiewicz. 1997. How does immigration influence local adaptation? A reexamination of a familiar paradigm. American Naturalist 149:563–572.

Holt, R. D., R. Gomulkiewicz, and M. Barfield. 2003. The phenomenology of niche evolution via quantitative traits in a "black-hole" sink. Proceedings of the Royal Society B. Biological Sciences 270:215–224.

Holt, R. D., T. M. Knight, and M. Barfield. 2004. Allee effects, immigration, and the evolution of species' niches. American Naturalist 163:253–262.

Hooper, D. U., F. S. Chapin III, J. J. Ewel, A. Hector, P. Inchausti, S. Lavorel, J. H. Lawton, D. M. Lodge, M. Loreau, S. Naeem, B. Schmid, H. Setälä, A. J. Symstad, J. Vandermeer, and D. A. Wardle. 2005. Effects of biodiversity on ecosystem functioning: a consensus of current knowledge. Ecological Monographs 75:3–35.

Hopkins, R., and M. D. Rausher. 2012. Pollinator-mediated selection on flower color allele drives reinforcement. Science 335:1090–1092.

Hori, M. 1993. Frequency-dependent natural selection in the handedness of scale-eating cichlid fish. Science 260:216–219.

Hoso, M., Y. Kameda, S.-P. Wu, T. Asami, M. Kato, and M. Hori. 2010. A speciation gene for left–right reversal in snails results in anti-predator adaptation. Nature Communications 1:133.

Houde, A. E. 1987. Mate choice based upon naturally occurring color-pattern variation in a guppy population. Evolution 41:1–10.

Houde, A. E., and J. A. Endler. 1990. Correlated evolution of female mating preferences and male color patterns in the guppy *Poecilia reticulata*. Science 248:1405–1408.

Houle, D. 1992. Comparing evolvability and variability of quantitative traits. Genetics 130:195–204.

Howeth, J. G., J. J. Weis, J. Bordersen, E. C. Hatton, and D. M. Post. 2013. Intraspecific phenotypic variation in a fish predator affects multitrophic lake metacommunity structure. Ecology and Evolution 3:5031–5044.

Hubbell, S. P. 2001. *The unified neutral theory of biodiversity and biogeography*. Princeton University Press, Princeton, NJ.

Hubbell, S. P. 2006. Neutral theory and the evolution of ecological equivalence. Ecology 87:1387–1398.

Huber, S. K., L. F. De León, A. P. Hendry, E. Bermingham, and J. Podos. 2007. Reproductive isolation of sympatric morphs in a population of Darwin's finches. Proceedings of the Royal Society B. Biological Sciences 274:1709–1714.

Huey, R. B., and J. G. Kingsolver. 1993. Evolution of resistance to high temperature in ecotherms. American Naturalist 142:S21–S46.

Huey, R. B., P. E. Hertz, and B. Sinervo. 2003. Behavioral drive versus behavioral inertia in evolution: a null model approach. American Naturalist 161:357–366.

Hufbauer, R. A., and G. K. Roderick. 2005. Microevolution in biological control: Mechanisms, patterns, and processes. Biological Control 35:227–239.

Hughes, A. R., and J. J. Stachowicz. 2004. Genetic diversity enhances the resistance of a seagrass ecosystem to disturbance. Proceedings of the National Academy of Sciences USA 101:8998–9002.

Hughes, A. R., and J. J. Stachowicz. 2011. Seagrass genotypic diversity increases disturbance response via complementarity and dominance. Journal of Ecology 99:445–453.

Hughes, A. R., B. D. Inouye, M. T. J. Johnson, N. Underwood, and M. Vellend. 2008. Ecological consequences of genetic diversity. Ecology Letters 11:609–623.

Hughes, A. R., R. J. Best, and J. J. Stachowicz. 2010. Genotypic diversity and grazer identity interactively influence seagrass and grazer biomass. Marine Ecology Progress Series 403:43–51.

Hughes, C. L., J. K. Hill, and C. Dytham. 2003. Evolutionary trade-offs between reproduction and dispersal in populations at expanding range boundaries. Proceedings of the Royal Society of London Series B. Biological Sciences 270:S147–S150.

Hughes, J. B., G. C. Daily, and P. R. Ehrlich. 1997. Population diversity: its extent and extinction. Science 278:689–692.

Hughes, K. A., L. Du, F. H. Rodd, and D. N. Reznick. 1999. Familiarity leads to female mate preference for novel males in the guppy, Poecilia reticulata. Animal Behaviour 58:907–916.

Hughes, K. A., A. E. Houde, A. C. Price, and F. H. Rodd. 2013. Mating advantage for rare males in wild guppy populations. Nature 503:108–110.

Hulme, P. E. 2008. Phenotypic plasticity and plant invasions: is it all Jack? Functional Ecology 22:3–7.

Hunt, G. 2007. Evolutionary divergence in directions of high phenotypic variance in the ostracode genus Poseidonamicus. Evolution 61:1560–1576.

Hunt, G., M. A. Bell, and M. P. Travis. 2008. Evolution toward a new adaptive optimum: phenotypic evolution in a fossil stickleback lineage. Evolution 62:700–710.

Hunt, J., R. Brooks, M. D. Jennions, M. J. Smith, C. L. Bentsen, and L. F. Bussière. 2004. High-quality male field crickets invest heavily in sexual display but die young. Nature 432:1024–1027.

Husband, B. C., and D. W. Schemske. 1996. Evolution of the magnitude and timing of inbreeding depression in plants. Evolution 50:54–70.

Husby, A., D. H. Nussey, M. E. Visser, A. J. Wilson, B. C. Sheldon, and L. E. B. Kruuk. 2010. Contrasting patterns of phenotypic plasticity in reproductive traits in two great tit (Parus major) populations. Evolution 64:2221–2237.

Husby, A., M. E. Visser, and L. E. B. Kruuk. 2011. Speeding up microevolution: the effects of increasing temperature on selection and genetic variance in a wild bird population. PLoS Biology 9:e1000585.

Hutchinson, G. E. 1959. Homage to Santa Rosalia or why are there so many kinds of animals? American Naturalist 93:145–159.

Iason, G. R., J. J. Lennon, R. J. Pakeman, V. Thoss, J. K. Beaton, D. A. Sim, and D. A. Elston. 2005. Does chemical composition of individual Scots pine trees determine the biodiversity of their associated ground vegetation? Ecology Letters 8:364–369.

Ingram, T., R. Svanbäck, N. J. B. Kraft, P. Kratina, L. Southcott, and D. Schluter. 2012. Intraguild predation drives evolutionary shift in threespine stickleback. Evolution 66:1819–1832.

Irschick, D. J., J. J. Meyers, J. F. Husak, and J. Le Galliard. 2008. How does selection operate on whole-organism functional performance capacities? A review and synthesis. Evolutionary Ecology Research 10:177–196.

Jackrel, S. L., and J. T. Wootton. 2014. Local adaptation of stream communities to intraspecific variation in a terrestrial ecosystem subsidy. Ecology 95:37–43.

Jackrel, S. L., and J. T. Wootton. 2015. Cascading effects of induced terrestrial plant defenses on aquatic and terrestrial ecosystem function. Proceedings of the Royal Society B: Biological Sciences 282:20142522.

Jackrel, S. L., T. C. Morton, and J. T. Wootton. 2016. Intraspecific leaf chemistry drives locally accelerated ecosystem function in aquatic and terrestrial communities. Ecology. In press.

Jaenike, J. 1990. Host specialization in phytophagous insects. Annual Review of Ecology and Systematics 21:243–273.

Jain, S. K., and A. D. Bradshaw. 1966. Evolutionary divergence among adjacent plant populations. I. The evidence and its theoretical analysis. Heredity 21:407–441.

James, F. C. 1983. Environmental component of morphological differentiation in birds. Science 221:184–186.

Janzen, F. J., and H. S. Stern. 1998. Logistic regression for empirical studies of multivariate selection. Evolution 52:1564–1571.

Jennions, M. D., A. P. Møller, and M. Petrie. 2001. Sexually selected traits and adult survival: a meta-analysis. Quaterly Review of Biology 76:3–36.

Jennions, M. D., and A. P. Møller. 2002. How much variance can be explained by ecologists and evolutionary biologists? Oecologia 132:492–500.

Jeyasingh, P. D., R. D. Cothran, and M. Tobler. 2014. Testing the ecological consequences of evolutionary change using elements. Ecology and Evolution 4:528–538.

Jetz, W., K. G. Ashton, and F. A. La Sorte. 2009. Phenotypic population divergence in terrestrial vertebrates at macro scales. Ecology Letters 12:1137–1146.

Jiang, L., J. Tan, and Z. Pu. 2010. An experimental test of Darwin's naturalization hypothesis. American Naturalist 175:415–423.

Jiang, Y., D. I. Bolnick, and M. Kirkpatrick. 2013. Assortative mating in animals. American Naturalist 181: E125–E138.

Jiggins, C. D. 2008. Ecological speciation in mimetic butterflies. BioScience 58:541–548.

Jiggins, C. D., C. Salazar, M. Linares, and J. Mavarez. 2008. Hybrid trait speciation and *Heliconius* butterflies. Philosophical Transactions of the Royal Society B. Biological Sciences 363:3047–3054.

Johansson, F., M. I. Lind, P. K. Ingvarsson, and F. Bokma. 2012. Evolution of the G-matric in life history traits in the common frog during a recent colonisation of an island systems. Evolutionary Ecology 26:863–878.

Johansson, A. M., M. E. Pettersson, P. B. Siegel, and Ö. Carlborg. 2010. Genome-wide effects of long-term divergent selection. PLoS Genetics 6:e1001188.

Johnson, D. W., K. Grorud-Colvert, S. Sponaugle, and B. X. Semmens. 2014. Phenotypic variation and selective mortality as major drivers of recruitment variability in fishes. Ecology Letters 17:743–755.

Johnson, J. B. 2002. Divergent life histories among populations of the fish *Brachyrhaphis rhabdophora*: detecting putative agents of selection by candidate model analysis. Oikos 96:82–91.

Johnson, M. L., and M. S. Gaines. 1990. Evolution of dispersal: theoretical models and empirical tests using birds and mammals. Annual Review of Ecology and Systematics 21:449–480.

Johnson, M. S. 2011. Thirty-four years of climatic selection in the land snail *Theba pisana*. Heredity 106:741–748.

Johnson, M. T. J. 2008. Bottom-up effects of plant genotype on aphids, ants, and predators. Ecology 89:145–154.

Johnson, M. T. J., and A. A. Agrawal. 2005. Plant genotype and environment interact to shape a diverse arthropod community on evening primrose (*Oenothera biennis*). Ecology 86:874–885.

Johnson, M. T., and J. R. Stinchcombe. 2007. An emerging synthesis between community ecology and evolutionary biology. Trends in Ecology and Evolution 22:250–257.

Johnson, M. T. J., R. Dinnage, A. Y. Zhou, and M. D. Hunter. 2008. Environmental variation has stronger effects than plant genotype on competition among plant species. Journal of Ecology 96:947–955.

Johnson, M. T. J., M. J. Lajeunesse, and A. A. Agrawal. 2006. Additive and interactive effects of plant genotypic diversity on arthropod communities and plant fitness. Ecology Letters 9:24–34.

Johnson, M. T. J., M. Vellend, and J. R. Stinchcombe. 2009. Evolution in plant populations as a driver of ecological changes in arthropod communities. Philosophical Transactions of the Royal Society B. Biological Sciences 364:1593–1605.

Johnston, R. F., and R. K. Selander. 1964. House sparrows: rapid evolution of races in North America. Science 144:548–550.

Jones, A. G. 2008. A theoretical quantitative genetic study of negative ecological interactions and extinction times in changing environments. BMC Evolutionary Biology 8:119.

Jones, A. G., S. J. Arnold, and R. Bürger. 2003. Stability of the G-matrix in a population experiencing pleiotropic mutation, stabilizing selection, and genetic drift. Evolution 57:1747–1760.

Jones, A. G., S. J. Arnold, and R. Bürger. 2004. Evolution and stability of the G-matrix on a landscape with a moving optimum. Evolution 58:1639–1654.

Jones, A. G., S. J. Arnold, and R. Bürger. 2007. The mutation matrix and the evolution of evolvability. Evolution 61:727–745.

Jones, C. G., J. H. Lawton, and M. Shachak. 1994. Organisms as ecosystem engineers. Oikos 69:373–386.

Jones, E. I., and R. Gomulkiewicz. 2012. Biotic interactions, rapid evolution, and the establishment of introduced species. American Naturalist 179:E28–E36.

Jones, F. C., C. Brown, and V. Braithwaite. 2008. Lack of assortative mating between incipient species of stickleback from a hybrid zone. Behaviour 145:463–484.

Jones, F. C., C. Brown, J. M. Pemberton, and V. A. Braithwaite. 2006. Reproductive isolation in a threespine stickleback hybrid zone. Journal of Evolutionary Biology 19:1531–1544.

Jones, F. C., M. G. Grabherr, Y. F. Chan, P. Russell, E. Mauceli, J. Johnson, R. Swofford, M. Pirun, M. C. Zody, S. White, E. Birney, S. Searle, J. Schmutz, J. Grimwood, M. C. Dickson, R. M. Myers, C. T. Miller, B. R. Summers, A. K. Knecht, S. D. Brady, H. Zhang, A. A. Pollen, T. Howes, C. Amemiya, Broad Institute Genome Sequency Platform & Whole Genome Assemby Team, E. S. Lander, F. Di Palma, K. Lindblad-Toh, and D. M. Kingsley. 2012b. The genomic basis of adaptive evolution in threespine sticklebacks. Nature 484:55–61.

Jones, F. C., Y. F. Chan, J. Schmutz, J. Grimwood, S. D. Brady, A. M. Southwick, D. M. Absher, R. M. Myers, T. E. Reimchen, B. E. Deagle, D. Schluter, and D. M. Kingsley. 2012a. A genome-wide SNP genotyping array reveals patterns of global and repeated species-pair divergence in sticklebacks. Current Biology 22:83–90.

Jones, L. E., L. Becks, S. P. Ellner, N. G. Hairston Jr., T. Yoshida, and G. F. Fussmann. 2009. Rapid contemporary evolution and clonal food web dynamics. Philosophical Transactions of the Royal Society B. Biological Sciences 364:1579–1591.

Jones, R., D. C. Culver, and T. C. Kane. 1992. Are parallel morphologies of cave organisms the result of similar selection pressures? Evolution 46:353–365.

Jonsson, B., and N. Jonsson. 2001. Polymorphism and speciation in Arctic charr. Journal of Fish Biology 58:605–638.

Joron, M., and J. L. B. Mallet. 1998. Diversity in mimicry: paradox or paradigm? Trends in Ecology and Evolution 13:461–466.

Joy, J. B., and B. J. Crespi. 2007. Adaptive radiation of gall-inducing insects within a single host-plant species. Evolution 61:784–795.

Jump, A. S., and J. Penuelas. 2005. Running to stand still: adaptation and the response of plants to rapid climate change. Ecology Letters 8:1010–1020.

Kaeuffer, R., C. L. Peichel, D. I. Bolnick, and A. P. Hendry. 2012. Parallel and non-parallel aspects of ecological, phenotypic, and genetic divergence across replicate population pairs of lake and stream stickleback. Evolution 66:402–418.

Kaltz, O., and J. A. Shykoff. 1998. Local adaptation in host–parasite systems. Heredity 81:361–370.

Karasov, T., P. W. Messer, and D. A. Petrov. 2010. Evidence that adaptation in *Drosophila* is not limited by mutation at single sites. PLoS Genetics 6:e1000924.

Karell, P., K. Ahola, T. Karstinen, J. Valkama, and J. E. Brommer. 2011. Climate change drives microevolution in a wild bird. Nature Communications 2:208.

Karim, N., S. P. Gordon, A. K. Schwartz, and A. P. Hendry. 2007. This is not déjà vu all over again: male guppy colour in a new experimental introduction. Journal of Evolutionary Biology 20:1339–1350.

Kasada, M., M. Yamamichi, and T. Yoshida. 2014. Form of an evolutionary tradeoff affects eco-evolutionary dynamics in a predator-prey system. Proceedings of the National Academy of Sciences USA 111:16035–16040.

Kassen, R. 2014. Experimental evolution and the nature of biodiversity. Roberts and Company. Greenwood Village, CO.

Katano, O. 2011. Effects of individual differences in foraging of pale chub on algal biomass through trophic cascades. Environmental Biology of Fishes 92:101–112.

Kawecki, T. J. 2003. Sex-biased dispersal and adaptation to marginal habitats. American Naturalist 162:415–426.

Kawecki, T. J. 2008. Adaptation to marginal habitats. Annual Review of Ecology, Evolution, and Systematics 39:321–342.

Kawecki, T. J., and D. Ebert. 2004. Conceptual issues in local adaptation. Ecology Letters 7:1225–1241.

Kawecki, T. J., and R. D. Holt. 2002. Evolutionary consequences of asymmetric dispersal rates. American Naturalist 160:333–347.

Kawecki, T. J., R. E. Lenski, D. Ebert, B. Hollis, I. Olivieri, and M. C. Whitlock. 2012. Experimental evolution. Trends in Ecology and Evolution 27:547–560.

Keith, A. R., J. K. Bailey, and T. G. Whitham. 2010. A genetic basis to community repeatability and stability. Ecology 91:3398–3406.

Keller, L. F., and D. M. Waller. 2002. Inbreeding effects in wild populations. Trends in Ecology and Evolution 17:230–241.

Kellermann, V. M., B. van Heerwaarden, A. A. Hoffmann, and C. M. Sgrò. 2006. Very low additive genetic variance and evolutionary potential in multiple populations of two rainforest *Drosophila* species. Evolution 60:1104–1108.

Kellermann, V., B. van Heerwaarden, C. M. Sgrò, and A. A. Hoffmann. 2009. Fundamental evolutionary limits in ecological traits drive *Drosophila* species distributions. Science 325:1244–1246.

Kemp, D. J., D. N. Reznick, G. F. Grether, and J. A. Endler. 2009. Predicting the direction of ornament evolution in Trinidadian guppies (*Poecilia reticulata*). Proceedings of the Royal Society B. Biological Sciences 276:4335–4343.

Kendall, N. W., J. J. Hard, and T. P. Quinn. 2009. Quantifying six decades of fishery selection for size and age at maturity in sockeye salmon. Evolutionary Applications 2:523–536.

Kennedy, B. P., J. D. Blum, C. L. Folt, and K. H. Nislow. 2000. Using natural strontium isotopic signatures as fish markers: methodology and application. Canadian Journal of Fisheries and Aquatic Sciences 57:2280–2292.

Kerfoot, C. W., J. A. Robbins, and L. J. Weider. 1999. A new approach to historical reconstruction: Combining descriptive and experimental paleolimnology. Limnology and Oceanography 44:1232–1247.

Kessler, A., and I. T. Baldwin. 2001. Defensive function of herbivore-induced plant volatile emissions in nature. Science 291:2141–2144.

Kettlewell, B. 1973. The evolution of melanism: a study of a recurring necessity, with special reference to industrial melanism in Lepidoptera. Clarendon Press, Oxford, UK.

Kiflawi, M., J. Belmaker, E. Brokovich, S. Einbinder, and R. Holzman. 2007. Species diversity can drive speciation: comment. Ecology 88:2132–2135.

Kimmel, C. B., W. A. Cresko, P. C. Phillips, B. Ullmann, M. Currey, F. von Hippel, B. K. Kristjánsson, O. Gelmond, and K. McGuigan. 2012. Independent axes of genetic variation and parallel evolutionary divergence of opercle bone shape in threespine stickleback. Evolution 66:419–434.

Kingsolver, J. G., and D. W. Pfennig. 2004. Individual-level selection as a cause of Cope's rule of phyletic size increase. Evolution 58:1608–1612.

Kingsolver, J. G., and D. W. Schemske. 1991. Path analyses of selection. Trends in Ecology and Evolution 6:276–280.

Kingsolver, J. G., and S. E. Diamond. 2011. Phenotypic selection in natural populations: what limits directional selection? American Naturalist 177:346–357.

Kingsolver, J. G., and S. G. Smith. 1995. Estimating selection on quantitative traits using capture-recapture data. Evolution 49:384–388.

Kingsolver, J. G., H. E. Hoekstra, J. M. Hoekstra, D. Berrigan, S. N. Vignieri, C. E. Hill, A. Hoang, P. Gibert, and P. Beerli. 2001. The strength of phenotypic selection in natural populations. American Naturalist 157:245–261.

Kingsolver, J. G., R. Gomulkiewicz, and P. A. Carter. 2001. Variation, selection and evolution of function-valued traits. Genetica 112–113:87–104.

Kingsolver, J. G., S. E. Diamond, A. M. Siepielski, and S. M. Carlson. 2012. Synthetic analyses of phenotypic selection in natural populations: lessons, limitations and future directions. Evolutionary Ecology 26:1101–1118.

Kinnison, M. T., and A. P. Hendry. 2001. The pace of modern life II: from rates of contemporary microevolution to pattern and process. Genetica 112–113:145–164.

Kinnison, M. T., and N. G. Hairston Jr. 2007. Eco-evolutionary conservation biology: contemporary evolution and the dynamics of persistence. Functional Ecology 21:444–454.

Kinnison, M. T., M. J. Unwin, A. P. Hendry, and T. P. Quinn. 2001. Migratory costs and the evolution of egg size and number in introduced and indigenous salmon populations. Evolution 55:1656–1667.

Kinnison, M. T., M. J. Unwin, and T. P. Quinn. 2003. Migratory costs and contemporary evolution of reproductive allocation in male chinook salmon. Journal of Evolutionary Biology 16:1257–1269.

Kinnison, M. T., M. J. Unwin, and T. P. Quinn. 2008. Eco-evolutionary vs. habitat contributions to invasion in salmon: experimental evaluation in the wild. Molecular Ecology 17:405–414.

Kinnison, M. T., N. G. Hairston Jr., and A. P. Hendry. 2015. Cryptic eco-evolutionary dynamics. Annals of the New York Academy of Sciences 1360:120–144.

Kirkpatrick, M. 1982. Quantum evolution and punctuated equilibria in continuous genetic characters. American Naturalist 119:833–848.

Kirkpatrick, M. 2009. Patterns of quantitative genetic variation in multiple dimensions. Genetica 136:271–284.

Kirkpatrick, M., and N. Barton. 1997. Evolution of a species' range. American Naturalist 150:1–23.

Kirkpatrick, M., and S. L. Nuismer. 2004. Sexual selection can constrain sympatric speciation. Proceedings of the Royal Society B. Biological Sciences 271:687–693.

Kirkpatrick, M., and V. Ravigné. 2002. Speciation by natural and sexual selection: models and experiments. American Naturalist 159:S22–S35.

Kisel, Y., and T. G. Barraclough. 2010. Speciation has a spatial scale that depends on levels of gene flow. American Naturalist 175:316–334.

Kitano, J., D. I. Bolnick, D. A. Beauchamp, M. M. Mazur, S. Mori, T. Nakano, and C. L. Peichel. 2008. Reverse evolution of armor plates in the threespine stickleback. Current Biology 18:769–774.

Kitano, J., S. C. Lema, J. A. Luckenbach, S. Mori, Y. Kawagishi, M. Kusakabe, P. Swanson, and C. L. Peichel. 2010. Adaptive divergence in the thyroid hormone signaling pathway in the stickleback radiation. Current Biology 20:2124–2130.

Knapczyk, F. N., and J. K. Conner. 2007. Estimates of the average strength of natural selection are not inflated by sampling error or publication bias. American Naturalist 170:501–508.

Knight, T. M., M. Barfield, and R. D. Holt. 2008. Evolutionary dynamics as a component of stage-structured matrix models: an example using *Trillium grandiflorum*. American Naturalist 172:375–392.

Knoppien, P. 1985. Rare male mating advantage: a review. Biological Reviews 60:81–117.

Koenig, W. D., D. Van Vuren, and P. N. Hooge. 1996. Detectability, philopatry, and the distribution of dispersal distances in vertebrates. Trends in Ecology and Evolution 11:514–517.

Kolbe, J. J., L. J. Revell, B. Szekely, E. D. Brodie III, and J. B. Losos. 2011. Convergent evolution of phenotypic integration and its alignment with morphological diversification in Caribbean *Anolis* ecomorphs. Evolution 65:3608–3624.

Kondrashov, A. S., and F. A. Kondrashov. 1999. Interactions among quantitative traits in the course of sympatric speciation. Nature 400:351–354.

Konijnendijk, N., D. A. Joyce, H. D. J. Mrosso, M. Egas, and O. Seehausen. 2011. Community genetics reveal elevated levels of sympatric gene flow among morphologically similar but not among morphologically dissimilar species of Lake Victoria cichlid fish. International Journal of Evolutionary Biology 2011: Article ID 616320.

Kotowska, A. M., J. F. Cahill Jr, and B. A. Keddie. 2010. Plant genetic diversity yields increased plant productivity and herbivore performance. Journal of Ecology 98:237–245.

Kovach-Orr, C., and G. F. Fussmann. 2013. Evolutionary and plastic rescue in multitrophic model communities. Philosophical Transactions of the Royal Society B. Biological Sciences 368:20120084.

Krebs, C. J. 1978. A review of the Chitty Hypothesis of population regulation. Canadian Journal of Zoology 56:2463–2480.

Kruckeberg, A. R. 1969. The implications of ecology for plant systematics. Taxon 18:92–120.

Kruglyak, L. 2008. The road to genome-wide association studies. Nature 9:314–318.

Kruuk, L. E. B. 2004. Estimating genetic parameters in natural populations using the "animal model". Philosophical Transactions of the Royal Society B. Biological Sciences 359:873–890.

Kruuk, L. E. B., J. Slate, and A. J. Wilson. 2008. New answers for old questions: the evolutionary quantitative genetics of wild animal populations. Annual Review of Ecology, Evolution, and Systematics 39:525–548.

Kruuk, L. E. B., T. H. Clutton-Brock, J. Slate, J. M. Pemberton, S. Brotherstone, and F. E. Guinness. 2000. Heritability of fitness in a wild mammal population. Proceedings of the National Academy of Sciences USA 97:698–703.

Kwan, L., and H. D. Rundle. 2010. Adaptation to desiccation fails to generate pre- and postmating isolation in replicate *Drosophila melanogaster* laboratory populations. Evolution 64:710–723.

Kwiatkowski, M. A., and B. K. Sullivan. 2002. Geographic variation in sexual selection among populations of an iguanid lizard, *Sauromalus obesus* (=*ater*). Evolution 56:2039–2051.

Labonne, J., and A. P. Hendry. 2010. Natural and sexual selection giveth and taketh away reproductive barriers: models of population divergence in guppies. American Naturalist 176:26–39.

Lack, D. 1947. *Darwin's finches*. Cambridge University Press, Cambridge, UK.

Lackey, A. C. R., and J. W. Boughman. 2013. Loss of sexual isolation in a hybridizing stickleback species pair. Current Zoology 59: 591–603.

Lahti, D. C. 2005. Evolution of bird eggs in the absence of cuckoo parasitism. Proceedings of the National Academy of Sciences USA 102:18057–18062.

Lahti, D. C., N. A. Johnson, B. C. Ajie, S. P. Otto, A. P. Hendry, D. T. Blumstein, R. G. Coss, K. Donohue, and S. A. Foster. 2009. Relaxed selection in the wild. Trends in Ecology and Evolution 24:487–496.

Lande, R. 1976. Natural selection and random genetic drift in phenotypic evolution. Evolution 30:314–334.

Lande, R. 1979. Quantitative genetic analysis of multivariate evolution, applied to brain: body size allometry. Evolution 33:402–416.

Lande, R. 1981. Models of speciation by sexual selection on polygenic traits. Proceedings of the National Academy of Sciences USA 78:3721–3725.

Lande, R. 1992. Neutral theory of quantitative genetic variance in an island model with local extinction and colonization. Evolution 46:381–389.

Lande, R. 2009. Adaptation to an extraordinary environment by evolution of phenotypic plasticity and genetic assimilation. Journal of Evolutionary Biology 22:1435–1446.

Lande, R. 2014. Evolution of phenotypic plasticity and environmental tolerance of a labile quantitative character in a fluctuating environment. Journal of Evolutionary Biology 27:866–875.

Lande, R., and S. J. Arnold. 1983. The measurement of selection on correlated characters. Evolution 37:1210–1226.

Landry, L., and L. Bernatchez. 2010. Role of epibenthic resource opportunities in the parallel evolution of lake whitefish species pairs (*Coregonus* sp.). Journal of Evolutionary Biology 23:2602–2613.

Lane, J. E., L. E. B. Kruuk, A. Charmantier, J. O. Murie, and F. S. Dobson. 2012. Delayed phenology and reduced fitness associated with climate change in a wild hibernator. Nature 489:554–557.

Lange, R. T., and C. R. Graham. 1983. Rabbits and the failure of regeneration in Australian arid zone Acacia. Australian Journal of Ecology 8:377–381.

Langerhans, R. B., and T. J. DeWitt. 2004. Shared and unique features of evolutionary diversification. American Naturalist 164:335–349.

Langerhans, R. B., C. A. Layman, A. K. Langerhans, and T. J. Dewitt. 2003. Habitat-associated morphological divergence in two Neotropical fish species. Biological Journal of the Linnean Society 80:689–698.

Langerhans, R. B., C. A. Layman, and T. J. DeWitt. 2005. Male genital size reflects a tradeoff between attracting mates and avoiding predators in two live-bearing fish species. Proceedings of the National Academy of Sciences USA 102:7618–7623.

Langerhans, R. B., J. H. Knouft, and J. B. Losos. 2006. Shared and unique features of diversification in Greater Antillean *Anolis* ecomorphs. Evolution 60:362–369.

Langerhans, R. B., M. E. Gifford, and E. O. Joseph. 2007. Ecological speciation in *Gambusia* fishes. Evolution 61:2056–2074.

Lango Allen, H., K. Estrada, G. Lettre, S. I. Berndt, M. N. Weedon, et al. 2010. Hundreds of variants clustered in genomic loci and biological pathways affect human height. Nature 467:832–838.

Lankau, R. A. 2011. Rapid evolutionary change and the coexistence of species. Annual Review of Ecology, Evolution, and Systematics 42:335–354.

Lankau, R. A., and R. N. Nodurft. 2013. An exotic invader drives the evolution of plant traits that determine mycorrhizal fungal diversity in a native competitor. Molecular Ecology 22:5472–5485.

Lankau, R. A., and S. Y. Strauss. 2007. Mutual feedbacks maintain both genetic and species diversity in a plant community. Science 317:1561–1563.

Lankau, R. A., and S. Y. Strauss. 2011. Newly rare or newly common: evolutionary feedbacks through changes in population density and relative species abundance, and their management implications. Evolutionary Applications 4:338–353.

Lankau, R. A., V. Nuzzo, G. Spyreas, and A. S. Davis. 2009. Evolutionary limits ameliorate the negative impact of an invasive plant. Proceedings of the National Academy of Sciences USA 106:15362–15367.

Larsson, K., H. P. van der Jeugd, I. T. van der Veen, and P. Forslund. 1998. Body size declines despite positive directional selection on heritable size traits in a barnacle goose population. Evolution 52:1169–1184.

Lavergne, S., M. E. K. Evans, I. J. Burfield, F. Jiguet, and W. Thuiller. 2013. Are species' responses to global change predicted by past niche evolution? Philosophical Transactions of the Royal Society B. Biological Sciences 368:20120091.

Lawniczak, M. K. N., S. J. Emrich, A. K. Holloway, A. P. Regier, M. Olson, B. White, S. Redmond, L. Fulton, E. Appelbaum, J. Godfrey, C. Farmer, A. Chinwalla, S.-P. Yang, P. Minx, J. Nelson, K. Kyung, B. P. Walenz, E. Garcia-Hernandez, M. Aguiar, L. D. Viswanathan, Y.-H. Rogers, R. L. Strausberg, C. A. Saski, D. Lawson, F. H. Collins, F. C. Kafatos, G. K. Christophides, S. W. Clifton, E. F. Kirkness, and N. J. Besansky. 2010. Widespread divergence between incipient *Anopheles gambiae* species revealed by whole genome sequences. Science 330:512–514.

Lawrence, D., F. Fiegna, V. Behrends, J. G. Bundy, A. B. Phillimore, T. Bell, and T. G. Barraclough. 2012. Species interactions alter evolutionary responses to a novel environment. PLoS Biology 10:e1001330.

Lawson Handley, L. J., and N. Perrin. 2007. Advances in our understanding of mammalian sex-biased dispersal. Molecular Ecology 16:1559–1578.

LeRoy, C. J., D. G. Fischer, W. M. Andrews, L. Belleveau, C. H. Barlow, J. A. Schweitzer, J. K. Bailey, J. C. Marks, and J. C. Kallestad. 2016. Salmon carcasses influence genetic linkages between forests and streams. Canadian Journal of Fisheries and Aquatic Sciences 73:910–920.

Le Rouzic, A., K. Østbye, T. O. Klepaker, T. F. Hansen, L. Bernatchez, D. Schluter, and L. A. Vøllestad. 2011. Strong and consistent natural selection associated with armour reduction in sticklebacks. Molecular Ecology 20:2483–2493.

Le Rouzic, A., T. F. Hansen, T. P. Gosden, and E. I. Svensson. 2015. Evolutionary time-series analysis reveals the signature of frequency-dependent selection on a female mating polymorphism. American Naturalist 185:E182–E196.

Lescak, E. A., S. L. Bassham, J. Catchen, O. Gelmond, M. L. Sherbick, F. A. von Hippel, and W. A. Cresko. 2015. Evolution of stickleback in 50 years on earthquake-uplifted islands. Proceedings of the National Academy of Sciences USA 112:E7204–E7212.

Leal, M., and A. R. Gunderson. 2012. Rapid change in the thermal tolerance of a tropical lizard. American Naturalist 180:815–822.

Leblois, R., A. Estoup, and F. Rousset. 2003. Influence of mutational and sampling factors on the estimation of demographic parameters in a "continuous" population under isolation by distance. Molecular Biology and Evolution 20:491–502.

Lee, C.-R., and T. Mitchell-Olds. 2011. Quantifying effects of environmental and geographical factors on patterns of genetic differentiation. Molecular Ecology 20:4631–4642.

Lee-Yaw, J. A., H. M. Kharouba, M. Bontrager, C. Mahoney, A. M. Csergö, A. M. E. Noreen, Q. Li, R. Schuster, and A. L. Angert. 2016. A synthesis of transplant experiments and ecological niche models suggests that range limits are often niche limits. Ecology Letters 19:710–722.

Lefcheck, J. S., J. E. K. Byrnes, F. Isbell, L. Gamfeldt, J. N. Griffin, N. Eisenhauer, M. J. S. Hensel, A. Hector, B. J. Cardinale, and J. E. Duffy. 2015. Biodiversity enhances ecosystem multifunctionality across trophic levels and habitats. Nature Communications 6:6936.

Leibold, M. A., and M. A. McPeek. 2006. Coexistence of the niche and neutral perspectives in community ecology. Ecology 87:1399–1410.

Leimu, R., and M. Fischer. 2008. A meta-analysis of local adaptation in plants. PloS one 3:e4010.

Leinonen, T., J. M. Cano, and J. Merilä. 2011. Genetics of body shape and armour variation in threespine sticklebacks. Journal of Evolutionary Biology 24:206–218.

Leinonen, T., J. M. Cano, H. Mäkinen, and J. Merilä. 2006. Contrasting patterns of body shape and neutral genetic divergence in marine and lake populations of threespine sticklebacks. Journal of Evolutionary Biology 19:1803–1812.

Leinonen, T., R. B. O'Hara, J. M. Cano, and J. Merilä. 2008. Comparative studies of quantitative trait and neutral marker divergence: a meta-analysis. Journal of Evolutionary Biology 21:1–17.

Leinonen, T., R. J. S. McCairns, G. Herczeg, and J. Merilä. 2012. Multiple evolutionary pathways to decreased lateral plate coverage in freshwater threespine sticklebacks. Evolution 66:3866–3875.

Leinonen, T., R. J. S. McCairns, R. B. O'Hara, and J. Merilä. 2013. Q_{ST}–F_{ST} comparisons: evolutionary and ecological insights from genomic heterogeneity. Nature Reviews Genetics 14:179–190.

Lennon, J. T., and J. B. H. Martiny. 2008. Rapid evolution buffers ecosystem impacts of viruses in a microbial food web. Ecology Letters 11:1178–1188.

Lenoir, J., J. C. Gégout, P. A. Marquet, P. de Ruffray, and H. Brisse. 2008. A significant upward shift in plant species optimum elevation during the 20th century. Science 320:1768–1771.

Lenormand, T. 2002. Gene flow and the limits to natural selection. Trends in Ecology and Evolution 17:183–189.

León, J. A., and B. Charlesworth. 1978. Ecological versions of Fisher's fundamental theorem of natural selection. Ecology 59:457–464.

Lester, R. J. G. 1990. Reappraisal of the use of parasites for fish stock identification. Australian Journal of Marine and Freshwater Research 41:855–864.

Letcher, B. H., G. E. Horton, T. L. Dubreuil, and M. J. O'Donnell. 2005. A field test of the extent of bias in selection estimates after accounting for emigration. Evolutionary Ecology Research 7:643–650.

Levin, S. A. 1972. A mathematical analysis of the genetic feedback mechanism. American Naturalist 106:145–164.

Levine, J. M., and C. M. D'Antonio. 1999. Elton revisited: a review of evidence linking diversity and invasibility. Oikos 87:15–26.

Levins, R. 1968. *Evolution in changing environments: some theoretical explorations.* Princeton University Press, Princton, NJ.

Levinton, J. S., E. Suatoni, W. Wallace, R. Junkins, B. Kelaher, and B. J. Allen. 2003. Rapid loss of genetically based resistance to metals after the cleanup of a Superfund site. Proceedings of the National Academy of Sciences USA 100:9889–9891.

Lin, J., T. P. Quinn, R. Hilborn, and L. Hauser. 2008. Fine-scale differentiation between sockeye salmon ecotypes and the effect of phenotype on straying. Heredity 101:341–350.

Lind, M. I., and F. Johansson. 2007. The degree of adaptive phenotypic plasticity is correlated with the spatial environmental heterogeneity experienced by island populations of *Rana temporaria*. Journal of Evolutionary Biology 20:1288–1297.

Lind, M. I., and F. Johansson. 2009. Costs and limits of phenotypic plasticity in island populations of the common frog *Rana temporaria* under divergent selection pressures. Evolution 63:1508–1518.

Lind, M. I., P. K. Ingvarsson, H. Johansson, D. Hall, and F. Johansson. 2011. Gene flow and selection on phenotypic plasticity in an island system of *Rana temporaria*. Evolution 65:684–697.

Linhart, Y. B., and M. C. Grant. 1996. Evolutionary significance of local genetic differentiation in plants. Annual Review of Ecology and Systematics 27:237–277.

Linnen, C. R., Y.-P. Poh, B. K. Peterson, R. D. H. Barrett, J. G. Larson, J. D. Jensen, and H. E. Hoekstra. 2013. Adaptive evolution of multiple traits through multiple mutations at a single gene. Science 339:1312–1316.

Linsey, H. A., J. Gallie, S. Taylor, and B. Kerr. 2013. Evolutionary rescue from extinction is contingent on a lower rate of environmental change. Nature 494:463–467.

Liu, Y.-B., and B. E. Tabashnik. 1997. Experimental evidence that refuges delay insect adaptation to *Bacillus thuringiensis*. Proceedings of the Royal Society B. Biological Sciences 264:605–610.

Lively, C. M. 2012. Feedbacks between ecology and evolution: interactions between ΔN and Δp in a life-history model. Evolutionary Ecology Research 14:299–309.

Lively, C. M., and M. F. Dybdahl. 2000. Parasite adaptation to locally common host genotypes. Nature 405:679–681.

Loeuille, N. 2010. Influence of evolution on the stability of ecological communities. Ecology Letters 13:1536–1545.

Lojewski, N. R., D. G. Fischer, J. K. Bailey, J. A. Schweitzer, T. G. Whitham, and S. C. Hart. 2009. Genetic basis of aboveground productivity in two native *Populus* species and their hybrids. Tree Physiology 29:1133–1142.

Lopez, S., F. Rousset, F. H. Shaw, R. G. Shaw, and O. Ronce. 2008. Migration load in plants: role of pollen and seed dispersal in heterogeneous landscapes. Journal of Evolutionary Biology 21:294–309.

Lopez, S., F. Rousset, F. H. Shaw, R. G. Shaw, and O. Ronce. 2009. Joint effects of inbreeding and local adaptation on the evolution of genetic load after fragmentation. Conservation Biology 23:1618–1627.

Loreau, M. 2010. *From populations to ecosystems: theoretical foundations for a new ecological synthesis.* Princeton University Press, Princeton, NJ.

Loreau, M., S. Naeem, P. Inchausti, J. Bengtsson, J. P. Grime, A. Hector, D. U. Hooper, M. A. Huston, D. Raffaelli, B. Schmid, D. Tilman, and D. A. Wardle. 2001. Biodiversity and ecosystem functioning: current knowledge and future challenges. Science 294:804–808.

Losos, J. B. 2009. *Lizards in an evolutionary tree: ecology and adaptive radiation of Anoles.* University of California Press, Berkeley, CA.

Losos, J. B. 2011. Convergence, adaptation, and constraint. Evolution 65:1827–1840.

Losos, J. B., and D. Schluter. 2000. Analysis of an evolutionary species-area relationship. Nature 408:847–850.

Losos, J. B., and R. E. Glor. 2003. Phylogenetic comparative methods and the geography of speciation. Trends in Ecology and Evolution 18:220–227.

Losos, J. B., and R. E. Ricklefs. 2009. Adaptation and diversification on islands. Nature 457:830–836.

Losos, J. B., D. A. Creer, D. Glossip, R. Goellner, A. Hampton, G. Roberts, N. Haskell, P. Taylor, and J. Ettling. 2000. Evolutionary implications of phenotypic plasticity in the hindlimb of the lizard *Anolis sagrei*. Evolution 54:301–305.

Losos, J. B., D. J. Irschick, and T. W. Schoener. 1994. Adaptaptation and constraint in the evolution of specialization of bahamian *Anolis* lizards. Evolution 48:1786–1798.

Losos, J. B., K. I. Warheit, and T. W. Schoener. 1997. Adaptive differentiation following experimental island colonization in *Anolis* lizards. Nature 387:70–73.

Losos, J. B., T. W. Schoener, R. B. Langerhans, and D. A. Spiller. 2006. Rapid temporal reversal in predator-driven natural selection. Science 314:1111.

Lotterhos, K. E., and M. C. Whitlock. 2014. Evaluation of demographic history and neutral parameterization on the performance of F_{ST} outlier tests. Molecular Ecology 23:2178–2192.

Louda, S. M., R. W. Pemberton, M. T. Johnson, and P. A. Follett. 2003. Nontarget effects: the Achilles' heel of biological control? Retrospective analyses to reduce risk associated with biocontrol introductions. Annual Review of Entomology 48:365–396.

Lowry, D. B., J. L. Modliszewski, K. M. Wright, C. A. Wu, and J. H. Willis. 2008a. The strength and genetic basis of reproductive isolating barriers in flowering plants. Philosophical Transactions of the Royal Society B. Biological Sciences 363:3009–3021.

Lowry, D. B., R. C. Rockwood, and J. H. Willis. 2008b. Ecological reproductive isolation of coast and inland races of *Mimulus guttatus*. Evolution 62:2196–2214.

Lucek, K., M. P. Haesler, and A. Sivasundar. 2012. When phenotypes do not match genotypes:unexpected phenotypic diversity and potential environmental constraints in Icelandic stickleback. Journal of Heredity 103:579–584.

Luck, G. W., G. C. Daily, and P. R. Ehrlich. 2003. Population diversity and ecosystem services. Trends in Ecology and Evolution 18:331–336.

Luikart, G., N. Ryman, D. A. Tallmon, M. K. Schwartz, and F. W. Allendorf. 2010. Estimation of census and effective population sizes: the increasing usefulness of DNA-based approaches. Conservation Genetics 11:355–373.

Lundsgaard-Hansen, B., B. Matthews, P. Vonlanthen, A. Taverna, and O. Seehausen. 2013. Adaptive plasticity and genetic divergence in feeding efficiency during parallel adpative radiation of whitefish (*Coregonus* spp.). Journal of Evolutionary Biology 26:483–498.

Lundsgaard-Hansen, B., B. Matthews, and O. Seehausen. 2014. Ecological speciation and phenotypic plasticity affect ecosystems. Ecology 95:2723–2735.

Luo, S. S., and K. Koelle. 2013. Navigating the devious course of evolution: the importance of mechanistic models for indentifying eco-evolutionary dynamics in nature. American Naturalist 181:S58–S75.

Lynch, M., and B. Walsh. 1998. *Genetics and analysis of quantitative traits*. Sinauer Associates, Inc., Sunderland, MA.

Lynch, M., and W. Gabriel. 1987. Environmental tolerance. American Naturalist 129:283–303.

Maan, M. E., and M. E. Cummings. 2009. Sexual dimorphism and directional sexual selection on aposematic signals in a poison frog. Proceedings of the National Academy of Sciences USA 106:19072–19077.

Maan, M. E., and O. Seehausen. 2011. Ecology, sexual selection and speciation. Ecology Letters 14:591–602.

Macarthur, R., and R. Levins. 1967. Limiting similarity, convergence, and divergence of coexisting species. American Naturalist 101:377–385.

MacColl, A. D. C. 2011. The ecological causes of evolution. Trends in Ecology and Evolution 26:514–522.

Mackay, T. F. C., E. A. Stone, and J. F. Ayroles. 2009. The genetics of quantitative traits: challenges and prospects. Nature Reviews Genetics 10:565–577.

Madritch, M. D., and R. L. Lindroth. 2011. Soil microbial communities adapt to genetic variation in leaf litter inputs. Oikos 120:1696–1704.

Madritch, M. D., J. R. Donaldson, and R. L. Lindroth. 2006. Genetic identity of *Populus tremuloides* litter influences decomposition and nutrient release in a mixed forest stand. Ecosystems 9:528–537.

Magurran, A. E. 2005. *Evolutionary ecology: the Trinidadian guppy*. Oxford University Press, Oxford, UK.

Magurran, A. E., B. H. Seghers, P. W. Shaw, and G. R. Carvalho. 1995. The behavioral diversity and evolution of guppy, *Poecilia reticulata*, populations in trinidad. Advances in the Study of Behavior 24:155–202.

Mahler, D. L., T. Ingram, L. J. Revell, and J. B. Losos. 2013. Exceptional convergence on the macroevolutionary landscape in island lizard radiations. Science 341:292–295.

Majerus, M. E. N. 1998. *Melanism: evolution in action*. Oxford University Press, Oxford, UK.

Mäki-Tanila, A., and W. G. Hill. 2014. Influence of gene interaction on complex trait variation with multilocus models. Genetics 198:355–367.

Mäkinen, H., S. Papakostas, L. A. Vøllestad, E. H. Leder, and C. R. Primmer. 2016. Plastic and evolutionary gene expression responses are correlated in European grayling (*Thymallus thymallus*) subpopulations adapting to different thermal environments. Journal of Heredity 2016:82–89.

Mallet, J. 2007. Hybrid speciation. Nature 446:279–283.

Mallet, J. 2008. Hybridization, ecological races and the nature of species: empirical evidence for the ease of speciation. Philosophical Transactions of the Royal Society B. Biological Sciences 363:2971–2986.

Mallet, J. 2009. Rapid speciation, hybridization and adaptive radiation. Pages 177–194 *in* R. Butlin, J. Bridle, and D. Schluter, Editors. *Speciation and patterns of diversity*. Cambridge University Press, Cambridge, UK.

Mallet, J., A. Meyer, P. Nosil, and J. L. Feder. 2009. Space, sympatry and speciation. Journal of Evolutionary Biology 22:2332–2341.

Manceau, M., V. S. Domingues, C. R. Linnen, E. B. Rosenblum, and H. E. Hoekstra. 2010. Convergence in pigmentation at multiple levels: mutations, genes and function. Philosophical Transactions of the Royal Society B. Biological Sciences 365:2439–2450.

Manceau, M., V. S. Domingues, R. Mallarino, and H. E. Hoekstra. 2011. The developmental role of Agouti in color pattern evolution. Science 331:1062–1065.

Manel, S., O. E. Gaggiotti, and R. S. Waples. 2005. Assignment methods: matching biological questions with appropriate techniques. Trends in Ecology and Evolution 20:136–142.

Manolio, T. A., F. S. Collins, N. J. Cox, D. B. Goldstein, L. A. Hindorff, D. J. Hunter, M. I. McCarthy, E. M. Ramos, L. R. Cardon, A. Chakravarti, J. H. Cho, A. E. Guttmacher, A. Kong, L. Kruglyak, E. Mardis, C. N. Rotimi, M. Slatkin, D. Valle, A. S. Whittemore, M. Boehnke, A. G. Clark, E. E. Eichler, G. Gibson, J. L. Haines, T. F. C. Mackay, S. A. McCarroll, and P. M. Visscher. 2009. Finding the missing heritability of complex diseases. Nature 461:747–753.

Marr, A. B., L. F. Keller, and P. Arcese. 2002. Heterosis and outbreeding depression in descendants of natural immigrants to an inbred population of song sparrows (*Melospiza melodia*). Evolution 56:131–142.

Martin, C. H., and P. C. Wainwright. 2013. Multiple fitness peaks on the adaptive landscape drive adaptive radiation in the wild. Science 339:208–211.

Martin, G., R. Aguilée, J. Ramsayer, O. Kaltz, and O. Ronce. 2013. The probability of evolutionary rescue: towards a quantitative comparison between theory and evolution experiments. Philosophical Transactions of the Royal Society B. Biological Sciences 368:20120088.

Martin, G., S. F. Elena, and T. Lenormand. 2007. Distributions of epistasis in microbes fit predictions from a fitness landscape model. Nature Genetics 39:555–560.

Martin, R. A., and D. W. Pfennig. 2009. Disruptive selection in natural populations: the roles of ecological specialization and resource competition. American Naturalist 174:268–281.

Martínez-Fernández, M., M. P. de la Cadena, and E. Rolán-Alvarez. 2010. The role of phenotypic plasticity on the proteome differences between two sympatric marine snail ecotypes adapted to distinct micro-habitats. BMC Evolutionary Biology 10:65.

Matsubayashi, K. W., I. Ohshima, and P. Nosil. 2010. Ecological speciation in phytophagous insects. Entomologia Experimentalis et Applicata 134:1–27.

Matsuda, H., and P. A. Abrams. 1994. Runaway evolution to self-extinction under asymmetrical competition. Evolution 48:1764–1772.

Matsumura, S., R. Arlinghaus, and U. Dieckmann. 2012. Standardizing selection strengths to study selection in the wild: a critical comparison and suggestions for the future. BioScience 62:1039–1054.

Matthews, B., S. Hausch, C. Winter, C. A. Suttle, and J. B. Shurin. 2011a. Contrasting ecosystem-effects of morphologically similar copepods. PLoS ONE 6:e26700.

Matthews, B., A. Narwani, S. Hausch, E. Nonaka, H. Peter, M. Yamamichi, K. E. Sullam, K. C. Bird, M. K. Thomas, T. C. Hanley, and C. B. Turner. 2011b. Toward an integration of evolutionary biology and ecosystem science. Ecology Letters 14:690–701.

Matthews, B., L. De Meester, C. G. Jones, B. W. Ibelings, T. J. Bouma, V. Nuutinen, J. van de Koppel, and J. Odling-Smee. 2014. Under niche construction: an operational bridge between ecology, evolution, and ecosystem science. Ecological Monographs 84:245–263.

Matthews, B., T. Aebischer, K. Sullam, B. Lundsgaard-Hansen, and O. Seehausen. 2016. Experimental evidence of an eco-evolutionary feedback during adaptive divergence. Current Biology 26:483–489.

Mayr, E. 1947. Ecological factors in speciation. Evolution 1:263–288.

Mayr, E. 1963. *Animal species and evolution.* Belknap Press, Cambridge, MA.

McBride, C. S., and M. C. Singer. 2010. Field studies reveal strong postmating isolation between ecologically divergent butterfly populations. PLoS Biology 8:e1000529.

McCairns, R. J. S., and L. Bernatchez. 2010. Adaptive divergence between freshwater and marine sticklebacks: insights into the role of phenotypic plasticity from an integrated analysis of candidate gene expression. Evolution 64:1029–1047.

McGill, B. J., B. J. Enquist, E. Weiher, and M. Westoby. 2006. Rebuilding community ecology from functional traits. Trends in Ecology and Evolution 21:178–185.

McGuigan, K., and M. W. Blows. 2007. The phenotypic and genetic covariance structure of drosphilid wings. Evolution 61:902–911.

McGuigan, K., S. F. Chenoweth, and M. W. Blows. 2005. Phenotypic divergence along lines of genetic variance. American Naturalist 165:32–43.

McKay, J. K., and R. G. Latta. 2002. Adaptive population divergence: markers, QTL and traits. Trends in Ecology and Evolution 17:285–291.

McKellar, A. E., and A. P. Hendry. 2009. How humans differ from other animals in their levels of morphological variation. PLoS ONE 4:e6876.

McKinnon, J. S., and H. D. Rundle. 2002. Speciation in nature: the threespine stickleback model systems. Trends in Ecology and Evolution 17:480–488.

McKinnon, J. S., S. Mori, B. K. Blackman, L. David, D. M. Kingsley, L. Jamieson, J. Chou, and D. Schluter. 2004. Evidence for ecology's role in speciation. Nature 429:294–298.

McPeek, M. A., and J. M. Brown. 2000. Building a regional species pool: diversification of the *Enallagma* damselflies in Eastern North America. Ecology 81:904–920.

McPherson, A., P. A. Hohenlohe, and S. L. Nuismer. 2015. Trait dimensionality explains widespread variation in local adaptation. Proceedings of the Royal Society of London. B. Biological Sciences 282:20141570.

Medina, M. H., J. A. Correa, and C. Barata. 2007. Micro-evolution due to pollution: possible consequences for ecosystem responses to toxic stress. Chemosphere 67:2105–2114.

Meiri, S., D. Guy, T. Dayan, and D. Simberloff. 2009. Global change and carnivore body size: data are stasis. Global Ecology and Biogeography 18:240–247.

Mendelson, T. C. 2003. Sexual isolation evolves faster than hybrid inviability in a diverse and sexually dimorphic genus of fish (Percidae: Etheostoma). Evolution 57:317–327.

Mendelson, T. C., and K. L. Shaw. 2005. Sexual behaviour: rapid speciation in an arthropod. Nature 433:375–376.

Menzel, A., T. H. Sparks, N. Estrella, E. Koch, A. Aasa, R. Ahas, K. Alm-Kübler, P. Bissolli, O. Braslavská, A. Briede, F. M. Chmielewski, Z. Crepinsek, Y. Curnel, Å. Dahl, C. Defila, A. Donnelly, Y. Filella, K. Jatcza, F. Måge, A. Mestre, Ø. Nordli, J. Peñuelas, P. Pirinen, V. Remišová, H. Scheifinger, M. Striz, A. Susnik, A. J. H. van Vliet, F.-E. Wielgolaski, S. Zach, and A. Zust. 2006. European phenological response to climate change matches the warming pattern. Global Change Biology 12:1969–1976.

Merilä, J., A. Laurila, and B. Lindgren. 2004. Variation in the degree and costs of adaptive phenotypic plasticity among Rana temporaria populations. Journal of Evolutionary Biology 17:1132–1140.

Merilä, J., and A. P. Hendry. 2014. Climate change, adaptation, and phenotypic plasticity: the problem and the evidence. Evolutionary Applications 7:1–14.

Merilä, J., and B. C. Sheldon. 1999. Genetic architecture of fitness and nonfitness traits: empirical patterns and development of ideas. Heredity 83:103–109.

Merilä, J., and B. C. Sheldon. 2000. Lifetime reproductive success and heritability in nature. American Naturalist 155:301–310.

Merilä, J., and P. Crnokrak. 2001. Comparison of genetic differentiation at marker loci and quantitative traits. Journal of Evolutionary Biology 14:892–903.

Merilä, J., B. C. Sheldon, and L. E. B. Kruuk. 2001b. Explaining stasis: microevolutionary studies in natural populations. Genetica 112:199–222.

Merilä, J., L. E. B. Kruuk, and B. C. Sheldon. 2001a. Cryptic evolution in a wild bird population. Nature 412:76–79.

Merrill, R. M., Z. Gompert, L. M. Dembeck, M. R. Kronforst, W. O. McMillan, and C. D. Jiggins. 2011. Mate preference across the speciation continuum in a clade of mimetic butterflies. Evolution 65:1489–1500.

Mezey, J. G., and D. Houle. 2005. The dimensionality of genetic variation for wing shape in Drosophila melanogaster. Evolution 59:1027–1038.

Michel, A. P., S. Sim, T. H. Q. Powell, M. S. Taylor, P. Nosil, and J. L. Feder. 2010. Widespread genomic divergence during sympatric speciation. Proceedings of the National Academy of Sciences USA 107:9724–9729.

Micheli, F. 1999. Eutrophication, fisheries, and consumer-resource dynamics in marine pelagic ecosystems. Science 285:1396–1398.

Miehls, A. L. J., S. D. Peacor, L. Valliant, and A. G. McAdam. 2015. Evolutionary stasis despite selection on a heritable trait in an invasive zooplankton. Journal of Evolutionary Biology 28:1091–1102.

Milewski, A. V., T. P. Young, and D. Madden. 1991. Thorns as induced defences: experimental evidence. Oecologia 86:70–75.

Millar, N. P., and A. P. Hendry. 2012. Population divergence of private and non-private signals in wild guppies. Environmental Biology of Fishes 94:513–525.

Miller, C. T., S. Beleza, A. A. Pollen, D. Schluter, R. A. Kittles, M. D. Shriver, and D. M. Kingsley. 2007. cis-regulatory changes in Kit ligand expression and parallel evolution of pigmentation in sticklebacks and humans. Cell 131:1179–1189.

Mills, L. S., and F. W. Allendorf. 1996. The one-migrant-per-generation rule in conservation and management. Conservation Biology 10:1509–1518.

Mills, L. S., M. E. Soulé, and D. F. Doak. 1993. The keystone-species concept in ecology and conservation. BioScience 43:219–224.

Milot, E., F. M. Mayer, D. H. Nussey, M. Boisvert, F. Pelletier, and D. Réale. 2011. Evidence for evolution in response to natural selection in a contemporary human population. Proceedings of the National Academy of Sciences USA 108:17040–17045.

Miner, B. G., S. E. Sultan, S. G. Morgan, D. K. Padilla, and R. A. Relyea. 2005. Ecological consequences of phenotypic plasticity. Trends in Ecology and Evolution 20:685–692.

Mitchell-Olds, T., and R. G. Shaw. 1987. Regression analysis of natural selection: statistical inference and biological interpretation. Evolution 41:1149–1161.

Moczek, A. P., S. Sultan, S. Foster, C. Ledón-Rettig, I. Dworkin, H. F. Nijhout, E. Abouheif, and D. W. Pfennig. 2011. The role of developmental plasticity in evolutionary innovation. Proceedings of the Royal Society B. Biological Sciences 278:2705–2713.

Moeller, D. A., M. A. Geber, and P. Tiffin. 2011. Population genetics and the evolution of geographic range limits in an annual plant. American Naturalist 178:S44–S61.

Mojica, J. P., and J. K. Kelly. 2010. Viability selection prior to trait expression is an essential component of natural selection. Proceedings of the Royal Society B. Biological Sciences 277:2945–2950.

Møller, A. P., D. Rubolini, and E. Lehikoinen. 2008. Populations of migratory bird species that did not show a phenological response to climate change are declining. Proceedings of the National Academy of Sciences USA 105:16195–16200.

Molofsky, J., and J.-B. Ferdy. 2005. Extinction dynamics in experimental metapopulations. Proceedings of the National Academy of Sciences USA 102:3726–3731.

Montesinos, D., G. Santiago, and R. M. Callaway. 2012. Neo-allopatry and rapid reproductive isolation. American Naturalist 180:529–533.

Moore, A. J., E. D. Brodie III, and J. B. Wolf. 1997. Interacting phenotypes and the evolutionary process: I. direct and indirect genetic effects of social interactions. Evolution 51:1352–1362.

Moore, J.-S., and A. P. Hendry. 2009. Can gene flow have negative demographic consequences? Mixed evidence from stream threespine stickleback. Philosophical Transactions of the Royal Society B. Biological Sciences 364:1533–1542.

Moore, J.-S., J. L. Gow, E. B. Taylor, and A. P. Hendry. 2007. Quantifying the constraining influence of gene flow on adaptive divergence in the lake-stream threespine stickleback system. Evolution 61:2015–2026.

Moose, S. P., J. W. Dudley, and T. R. Rocheford. 2004. Maize selection passes the century mark: a unique resource for 21st century genomics. Trends in Plant Science 9:358–364.

Mopper, S., P. Stiling, K. Landau, D. Simberloff, and P. Van Zandt. 2000. Spatiotemporal variation in leafminer population structure and adaptation to individual oak trees. Ecology 81:1577–1587.

Moran, E. V., and J. M. Alexander. 2014. Evolutionary responses to global change: lessons from invasive species. Ecology Letters 17:637–649.

Morbey, Y. E., and A. P. Hendry. 2008. Adaptation of salmonids to spawning habitat. Pages 15–36 *in* D. A. Sear and P. DeVries, Editors. *Salmon spawning habitat in rivers: physical controls, biological responses, and approaches to remediation.* American Fisheries Society Symposium 65, Quebec City, Canada.

Morgan, A. D., S. Gandon, and A. Buckling. 2005. The effect of migration on local adaptation in a coevolving host-parasite system. Nature 437:253–256.

Morgan, D. C., and H. Smith. 1979. A systematic relationship between phytochrome-controlled development and species habitat, for plants grown in simulated natural radiation. Planta 145:253–258.

Morjan, C. L., and L. H. Rieseberg. 2004. How species evolve collectively: implications of gene flow and selection for the spread of advantageous alleles. Molecular Ecology 13:1341–1356.

Morris, D. W., and P. Lundberg. 2011. *The pillars of evolution: fundamental principles of the eco-evolutionary process.* Oxford University Press, Oxford, UK.

Morrissey, M. B., and J. D. Hadfield. 2012. Directional selection in temporally replicated studies is remarkably consistent. Evolution 66:435–442.

Morrissey, M. B., D. J. Parker, P. Korsten, J. M. Pemberton, L. E. B. Kruuk, and A. J. Wilson. 2012. The prediction of adaptive evolution: empirical application of the secondary theorem of selection and comparison to the breeder's equation. Evolution 66:2399–2410.

Morrissey, M. B., L. E. B. Kruuk, and A. J. Wilson. 2010. The danger of applying the breeder's equation in observational studies of natural populations. Journal of Evolutionary Biology 23:2277–2288.

Morrissey, M. B., and K. Sakrejda. 2013. Unification of regression-based methods for the analysis of natural selection. Evolution 67:2094–2100.

Moser, D., A. Frey, and D. Berner. 2016. Fitness differences between parapatric lake and stream stickleback revealed by a field transplant. Journal of Evolutionary Biology 29:711–719.

Mousseau, T. A., and D. A. Roff. 1987. Natural selection and the heritability of fitness components. Heredity 59:181–197.

Moya-Laraño, J. 2011. Genetic variation, predator-prey interactions and food web structure. Philosophical Transactions of the Royal Society B. Biological Sciences 366:1425–1437.

Moya-Laraño, J. O. Verdeny-Vilalta, J. Rowntree, N. Melguizo-Ruiz, M. Montserrat, and P. Laiolo. 2014. Climate change and eco-evolutionary dynamics in food webs. Advances in Ecological Research 47:1–80.

Murren, C. J., J. R. Auld, H. Callahan, C. K. Ghalambor, C. A. Handelsman, M. A. Heskel, J. G. Kingsolver, H. J. Maclean, J. Masel, H. Maughan, D. W. Pfennig, R. A. Relyea, S. Seiter, E. Snell-Rood, U. K. Steiner, and C. D. Schlichting. 2015. Constraints on the evolution of phenotypic plasticity: limits and costs of phenotype and plasticity. Heredity 115:293–301.

Myles, S., N. Bouzekri, E. Haverfield, M. Cherkaoui, J.-M. Dugoujon, and R. Ward. 2005. Genetic evidence in support of a shared Eurasian-North African dairying origin. Human Genetics 117:34–42.

Nagel, L., and D. Schluter. 1998. Body size, natural selection, and speciation in sticklebacks. Evolution 52:209–218.

Nagy, E. S. 1997. Selection for native characters in hybrids between two locally adapted plant subspecies. Evolution 51:1469–1480.

Nagy, E. S., and K. J. Rice. 1997. Local adaptation in two subspecies of an annual plant: implications for migration and gene flow. Evolution 51:1079–1089.

Nei, M., and A. Chakravarti. 1977. Drift variances of F_{ST} and G_{ST} statistics obtained from a finite number of isolated populations. Theoretical Population Biology 11:307–325.

Nei, M., T. Maruyama, and C.-I. Wu. 1983. Models of evolution of reproductive isolation. Genetics 103:557–579.

Neigel, J. E. 1997. A comparison of alternative strategies for estimating gene flow from genetic markers. Annual Review of Ecology and Systematics 28:105–128.

Newton, I. 1994. The role of nest sites in limiting the numbers of hole-nesting birds: a review. Biological Conservation 70:265–276.

Nicholson, A. J. 1957. The self-adjustment of populations to change. Cold Spring Harbor Symposia on Quantitative Biology 22:153–173.

Nicolakakis, N., D. Sol, and L. Lefebvre. 2003. Behavioural flexibility predicts species richness in birds, but not extinction risk. Animal Behaviour 65:445–452.

Niemiller, M. L., B. M. Fitzpatrick, and B. T. Miller. 2008. Recent divergence with gene flow in Tennessee cave salamanders (Plethodontidae: *Gyrinophilus*) inferred from gene genealogies. Molecular Ecology 17:2258–2275.

Noor, M. A. 1995. Speciation driven by natural selection in *Drosophila*. Nature 375:674–675.

Noor, M. A., and S. M. Bennett. 2009. Islands of speciation or mirages in the desert? Examining the role of restricted recombination in maintaining species. Heredity 103:439–444.

Norberg, J., M. C. Urban, M. Vellend, C. A. Klausmeier, and N. Loeuille. 2012. Eco-evolutionary responses of biodiversity to climate change. Nature Climate Change 2:747–751.

Nosil, P. 2007. Divergent host plant adaptation and reproductive isolation between ecotypes of *Timema cristinae* walking sticks. American Naturalist 169:151–162.

Nosil, P. 2009. Adaptive population divergence in cryptic color-pattern following a reduction in gene flow. Evolution 63:1902–1912.

Nosil, P. 2012. *Ecological speciation*. Oxford University Press, Oxford, UK.

Nosil, P. 2013. Degree of sympatry affects reinforcement in *Drosophila*. Evolution 67:868–872.

Nosil, P., and B. J. Crespi. 2004. Does gene flow constrain adaptive divergence or vice versa? A test using ecomorphology and sexual isolation in *Timema cristinae* walking-sticks. Evolution 58:102–112.

Nosil, P., and B. J. Crespi. 2006. Experimental evidence that predation promotes divergence in adaptive radiation. Proceedings of the National Academy of Sciences USA 103:9090–9095.

Nosil, P., and C. P. Sandoval. 2008. Ecological niche dimensionality and the evolutionary diversification of stick insects. PloS one 3:e1907.

Nosil, P., and D. Schluter. 2011. The genes underlying the process of speciation. Trends in Ecology and Evolution 26:160–167.

Nosil, P., and S. M. Flaxman. 2011. Conditions for mutation-order speciation. Proceedings of the Royal Society B. Biological Sciences 278:399–407.

Nosil, P., B. J. Crespi, and C. P. Sandoval. 2002. Host-plant adaptation drives the parallel evolution of reproductive isolation. Nature 417:440–443.

Nosil, P., B. J. Crespi, and C. P. Sandoval. 2003. Reproductive isolation driven by the combined effects of ecological adaptation and reinforcement. Proceedings of the Royal Society B. Biological Sciences 270:1911–1918.

Nosil, P., D. J. Funk, and D. Ortiz-Barrientos. 2009a. Divergent selection and heterogeneous genomic divergence. Molecular Ecology 18:375–402.

Nosil, P., L. J. Harmon, and O. Seehausen. 2009b. Ecological explanations for (incomplete) speciation. Trends in Ecology and Evolution 24:145–156.

Nosil, P., S. P. Egan, and D. J. Funk. 2008. Heterogeneous genomic differentiation between walking-stick ecotypes: "isolation by adaptation" and multiple roles for divergent selection. Evolution 62:316–336.

Nosil, P., T. H. Vines, and D. J. Funk. 2005. Reproductive isolation caused by natural selection against immigrants from divergent habitats. Evolution 59:705–719.

Nuismer, S. L. 2006. Parasite local adaptation in a geographic mosaic. Evolution 60:24–30.

Nuismer, S. L., and S. Gandon. 2008. Moving beyond common-garden and transplant designs: insight into the causes of local adaptation in species interactions. American Naturalist 171:658–668.

Nussey, D. H., A. J. Wilson, and J. E. Brommer. 2007. The evolutionary ecology of individual phenotypic plasticity in wild populations. Journal of Evolutionary Biology 20:831–844.

Nussey, D. H., E. Postma, P. Gienapp, and M. E. Visser. 2005. Selection on heritable phenotypic plasticity in a wild bird population. Science 310:304–306.

Oke, K. B., M. Bukhari, R. Kaeuffer, G. Rolshausen, K. Räsänen, D. I. Bolnick, C. L. Peichel, and A. P. Hendry. Does plasticity enhance or dampen phenotypic parallelism? A test with three lake-stream stickleback pairs. Journal of Evolutionary Biology 29:126–143.

O'Neil, P. 1999. Selection on flowering time: an adaptive fitness surface for nonexistent character combinations. Ecology 80:806–820.

O'Steen, S., A. J. Cullum, and A. F. Bennett. 2002. Rapid evolution of escape ability in Trinidadian guppies (Poecilia reticulata). Evolution 56:776–784.

Odling-Smee, J. F., K. N. Laland, and M. W. Feldman. 2003. Niche construction: the neglected process in evolution. Princeton University Press, Princeton, NJ.

Odling-Smee, J., D. H. Erwin, E. P. Palkovacs, M. W. Feldman, and K. N. Laland. 2013. Niche construction theory: a practical guide for ecologists. Quarterly Review of Biology 88:3–28.

Ogden, R., and R. S. Thorpe. 2002. Molecular evidence for ecological speciation in tropical habitats. Proceedings of the National Academy of Sciences USA 99:13612–13615.

Ohlberger, J., Ø. Langangen, E. Edeline, D. Claessen, I. J. Winfield, N. C. Stenseth, and A. Vøllestad. 2011. Stage-specific biomass overcompensation by juveniles in response to increased adult mortality in a wild fish population. Ecology 92:2175–2182.

Olendorf, R., F. H. Rodd, D. Punzalan, A. E. Houde, C. Hurt, D. N. Reznick, and K. A. Hughes. 2006. Frequency-dependent survival in natural guppy populations. Nature 441:633–636.

Oloriz, F., B. Marques, and F. J. Rodriguez-Tovar. 1991. Eustatism and faunal associations: examples from the South Iberian margin during the late Jurassic (Oxfordian-Kimmeridgian). Eclogae Geologicae Helvetiae 84:83–106.

Olson, E. C., and R. L. Miller. 1958. Morphological integration. Chicago University Press, Chicago, IL.

Olson-Manning, C. F., M. R. Wagner, and T. Mitchell-Olds. 2012. Adaptive evolution: evaluating empirical support for theoretical predictions. Nature Reviews Genetics 13:867–877.

O'Reilly-Wapstra, J. M., M. Hamilton, B. Gosney, C. Whiteley, J. K. Bailey, D. Williams, T. Wardlaw, R. E. Vaillancourt, and B. M. Potts. 2014. Genetic correlations in multi-species plant/herbivore interactions at multiple genetic scales: implications for eco-evolutionary dynamics. Advances in Ecological Research 50:267–295.

Ord, T. J., and J. A. Stamps. 2009. Species identity cues in animal communication. American Naturalist 174:585–593.

Orr, H. A. 1998. The population genetics of adaptation: the distribution of factors fixed during adaptive evolution. Evolution 52:935–949.

Orr, H. A. 1999. The evolutionary genetics of adaptation: a simulation study. Genetical Research 74:207–214.

Orr, H. A. 2000. Adaptation and the cost of complexity. Evolution 54:13–20.

Orr, H. A. 2005. The genetic theory of adaptation: a brief history. Nature Reviews Genetics 6:119–127.

Orr, H. A. 2009. Fitness and its role in evolutionary genetics. Nature Reviews Genetics 10:531–539.

Orr, H. A., and R. L. Unckless. 2008. Population extinction and the genetics of adaptation. American Naturalist 172:160–169.

Orsini, L., J. Vanoverbeke, I. Swillen, J. Mergeay, and L. De Meester. 2013. Drivers of population genetic differentiation in the wild: isolation by dispersal limitation, isolation by adaptation, and isolation by colonization. Molecular Ecology 22:5983–5999.

Ortiz-Barrientos, D., A. Grealy, and P. Nosil. 2009. The genetics and ecology of reinforcement: implications for the evolution of prezygotic isolation in sympatry and beyond. Annals of the New York Academy of Sciences 1168:156–182.

Ostevik, K. L., B. T. Moyers, G. L. Owens, and L. H. Rieseberg. 2012. Parallel ecological speciation in plants? International Journal of Ecology 2012:Article ID 939862.

Otto, S. P., and C. D. Jones. 2000. Detecting the undetected: estimating the total number of loci underlying a quantitative trait. Genetics 156:2093–2107.

Otto, S. P., and J. Whitton. 2000. Polyploid incidence and evolution. Annual Review of Genetics 34:401–437.

Otto, S. P., M. R. Servedio, and S. L. Nuismer. 2008. Frequency-dependent selection and the evolution of assortative mating. Genetics 179:2091–2112.

Ouborg, N. J., Y. Piquot, and J. M. van Groenendael. 1999. Population genetics, molecular markers, and the study of dispersal in plants. Journal of Ecology 87:551–568.

Ovaskainen, O., M. Karhunen, C. Zheng, J. M. C. Arias, and J. Merilä. 2011. A new method to uncover signatures of divergent, and stabilizing selection in quantitative traits. Genetics 189:621–632.

Ożgo, M. 2011. Rapid evolution in unstable habitats: a success story of the polymorphic land snail Cepaea nemoralis (Gastropoda: Pulmonata). Biological Journal of the Linnean Society 102:251–262.

Ozgul, A., D. Z. Childs, M. K. Oli, K. B. Armitage, D. T. Blumstein, L. E. Olson, S. Tuljapurkar, and T. Coulson. 2010. Coupled dynamics of body mass and population growth in response to environmental change. Nature 466:482–485.

Ozgul, A., S. Tuljapurkar, T. G. Benton, J. M. Pemberton, T. H. Clutton-Brock, and T. Coulson. 2009. The dynamics of phenotypic change and the shrinking sheep of St. Kilda. Science 325:464–467.

Paenke, I., B. Sendhoff, and T. J. Kawecki. 2007. Influence of plasticity and learning on evolution under directional selection. American Naturalist 170:E47–E58.

Paine, R. T. 1980. Food webs: linkage, interaction strength, and community infrastructure: the third Tansley lecture. Journal of Animal Ecology 49:667–685.

Pakanen, V.-M., O. Hildén, A. Rönkä, E. J. Belda, A. Luukkonen, L. Kvist, and K. Koivula. 2011. Breeding dispersal strategies following reproductive failure explain low apparent survival of immigrant Temminck's stints. Oikos 120:615–622.

Palacio-López, K., B. Beckage, S. Scheiner, and J. Molofsky. 2015. The ubiquity of phenotypic plasticity in plants: a synthesis. Ecology and Evolution 5:3389–3400.

Palkovacs, E. P., and D. M. Post. 2008. Eco-evolutionary interactions between predators and prey: can predator-induced changes to prey communities feed back to shape predator foraging traits? Evolutionary Ecology Research 10:699–720.

Palkovacs, E. P., and D. M. Post. 2009. Experimental evidence that phenotypic divergence in predators drives community divergence in prey. Ecology 90:300–305.

Palkovacs, E. P., and C. M. Dalton. 2012. Ecosystem consequences of behavioural plasticity and contemporary evolution. Pages 175–189 in U. Candolin, and B. B. M. Wong, Editors. Behavioural responses to a changing world: mechanisms and consequences. Oxford University Press, Oxford, UK.

Palkovacs, E. P., K. B. Dion, D. M. Post, and A. Caccone. 2008. Independent evolutionary origins of landlocked alewife populations and rapid parallel evolution of phenotypic traits. Molecular Ecology 17:582–597.

Palkovacs, E. P., M. C. Marshall, B. A. Lamphere, B. R. Lynch, D. J. Weese, D. F. Fraser, D. N. Reznick, C. M. Pringle, and M. T. Kinnison. 2009. Experimental evaluation of evolution and coevolution as agents of ecosystem change in Trinidadian streams. Philosophical Transactions of the Royal Society B. Biological Sciences 364:1617–1628.

Palkovacs, E. P., B. A. Wasserman, and M. T. Kinnison. 2011. Eco-evolutionary trophic dynamics: loss of top predators drives trophic evolution and ecology of prey. PLoS ONE 6:e18879.

Palkovacs, E. P., E. G. Mandeville, and D. M. Post. 2014. Contemporary trait change in a classic ecological experiment: rapid decrease in alewife gill-raker spacing following introduction to an inland lake. Freshwater Biology 59:1897–1901.

Palkovacs, E. P., D. C. Fryxell, N. E. Turley, and D. M. Post. 2015. Ecological effects of intraspecific consumer biodiversity for aquatic communities and ecosystems. Pages 37–52 in A. Belgrano, G. Woodward, and U. Jacob, Editors. *Aquatic functional biodiversity: an eco-evolutionary approach.* Academic Press, Elsevier, London, UK.

Palumbi, S. R. 2001. Humans as the world's greatest evolutionary force. Science 293:1786–1790.

Panhuis, T. M., R. Butlin, M. Zuk, and T. Tregenza. 2001. Sexual selection and speciation. Trends in Ecology and Evolution 16:364–371.

Pantel, J. H., C. Duvivier, and L. De Meester. 2015. Rapid local adaptation mediates zooplankton community assembly in experimental mesocosms. Ecology Letters 18:992–1000.

Papadopulos, A. S. T., W. J. Baker, D. Crayn, R. K. Butlin, R. G. Kynast, I. Hutton, and V. Savolainen. 2011. Speciation with gene flow on Lord Howe Island. Proceedings of the National Academy of Sciences USA 108:13188–13193.

Parmesan, C., and G. Yohe. 2003. A globally coherent fingerprint of climate change impacts across natural systems. Nature 421:37–42.

Partridge, L. 1988. The rare-male effect: what is its evolutionary significance? Philosophical Transactions of the Royal Society of London B. Biological Sciences 319:525–539.

Paul, J. R., S. N. Sheth, and A. L. Angert. 2011. Quantifying the impact of gene flow on phenotype-environment mismatch: a demonstration with the scarlet monkeyflower *Mimulus cardinalis.* American Naturalist 178:S62–S79.

Pavey, S. A., H. Collin, P. Nosil, and S. M. Rogers. 2010. The role of gene expression in ecological speciation. Annals of the New York Academy of Sciences 1206:110–129.

Pavey, S. A., J. L. Nielsen, and T. R. Hamon. 2010. Recent ecological divergence despite migration in sockeye salmon (*Oncorhynchus nerka*). Evolution 64:1773–1783.

Payne, R. B., L. L. Payne, J. L. Woods, and M. D. Sorenson. 2000. Imprinting and the origin of parasite-host species associations in brood-parasitic indigobirds, *Vidua chalybeata.* Animal Behaviour 59:69–81.

Pearse, D. E., S. A. Hayes, M. H. Bond, C. V. Hanson, E. C. Anderson, R. B. Macfarlane, and J. C. Garza. 2009. Over the falls? Rapid evolution of ecotypic differentiation in steelhead/rainbow trout (*Oncorhynchus mykiss*). Journal of Heredity 100:515–525.

Pease, C. M., R. Lande, and J. J. Bull. 1989. A model of population growth, dispersal and evolution in a changing environment. Ecology 70:1657–1664.

Peay, K. G., M. Belisle, and T. Fukami. 2012. Phylogenetic relatedness predicts priority effects in nectar yeast communities. Proceedings of the Royal Society B. Biological Sciences 279:749–758.

Peccoud, J., A. Ollivier, M. Plantegenest, and J.-C. Simon. 2009. A continuum of genetic divergence from sympatric host races to species in the pea aphid complex. Proceedings of the National Academy of Sciences USA 106:7495–7500.

Peek, M. S., A. J. Leffler, S. D. Flint, and R. J. Ryel. 2003. How much variance is explained by ecologists? Additional perspectives. Oecologia 137:161–170.

Peichel, C. L., K. S. Nereng, K. A. Ohgi, B. L. E. Cole, P. F. Colosimo, C. A. Buerkle, D. Schluter, and D. M. Kingsley. 2001. The genetic architecture of divergence between threespine stickleback species. Nature 414:901–905.

Pelletier, F., D. Garant, and A. P. Hendry. 2009. Eco-evolutionary dynamics. Philosophical Transactions of the Royal Society B. Biological Sciences 364:1483–1489.

Pelletier, F., T. Clutton-Brock, J. Pemberton, S. Tuljapurkar, and T. Coulson. 2007. The evolutionary demography of ecological change: linking trait variation and population growth. Science 315:1571–1574.

Pellmyr, O. 2003. Yuccas, yucca moths, and coevolution: a review. Annals of the Missouri Botanical Garden 90:35–55.

Pelz, H.-J., S. Rost, M. Hünerberg, A. Fregin, A.-C. Heiberg, K. Baert, A. D. MacNicoll, C. V. Prescott, A.-S. Walker, J. Oldenburg, and C. R. Müller. 2005. The genetic basis of resistance to anticoagulants in rodents. Genetics 170:1839–1847.

Penn, D. J., and W. K. Potts. 1999. The evolution of mating preferences and major histocompatibility complex genes. American Naturalist 153: 145–164.

Perez-Jvostov, F., A. P. Hendry, G. F. Fussmann, and M. E. Scott. 2015. Testing for local host-parasite adaptation: an experiment with Gyrodactylus ecotoparasites and guppy hosts. International Journal of Parasitology 45:409–417.

Perkins, T. A. 2012. Evolutionarily labile species interactions and spatial spread of invasive species. American Naturalist 179:E37–E54.

Perron, G. G., A. Gonzalez, and A. Buckling. 2008. The rate of environmental change drives adaptation to an antibiotic sink. Journal of Evolutionary Biology 21:1724–1731.

Persson, L., and A. M. de Roos. 2013. Symmetry breaking in ecological systems through different energy efficiencies of juveniles and adults. Ecology 94:1487–1498.

Persson, L., P.-A. Amundsen, A. M. de Roos, A. Klemetsen, R. Knudsen, and R. Primicerio. 2007. Culling prey promotes predator recovery—alternative states in a whole-lake experiment. Science 316:1743–1746.

Peterson, D. A., R. Hilborn, and L. Hauser. 2014. Local adaptation limits lifetime reproductive success of dispersers in a wild salmon metapopulation. Nature Communications 5:3696.

Petraitis, P. S., A. E. Dunham, and P. H. Niewiarowski. 1996. Inferring multiple causality: the limitations of path analysis. Functional Ecology 10:421–431.

Petren, K., P. R. Grant, B. R. Grant, and L. F. Keller. 2005. Comparative landscape genetics and the adaptive radiation of Darwin's finches: the role of peripheral isolation. Molecular Ecology 14:2943–2957.

Pfennig, D. W., A. M. Rice, and R. A. Martin. 2007. Field and experimental evidence for competition's role in phenotypic divergence. Evolution 61:257–271.

Pfennig, D. W., and M. McGee. 2010. Resource polyphenism increases species richness: a test of the hypothesis. Philosophical Transactions of the Royal Society B. Biological Sciences 365:577–591.

Pfennig, D. W., and P. J. Murphy. 2002. How fluctuating competition and phenotypic plasticity mediate species divergence. Evolution 56:1217–1228.

Pfennig, D. W., M. A. Wund, E. C. Snell-Rood, T. Cruickshank, C. D. Schlichting, and A. P. Moczek. 2010. Phenotypic plasticity's impacts on diversification and speciation. Trends in Ecology and Evolution 25:459–467.

Pfennig, K. S. 1998. The evolution of mate choice and the potential for conflict between species and mate-quality recognition. Proceedings of the Royal Society B. Biological Sciences 265:1743–1748.

Pfennig, K. S., and D. W. Pfennig. 2009. Character displacement: ecological and reproductive responses to a common evolutionary problem. Quaterly Review of Biology 84:253–276.

Pfennig, D. W., and K. S. Pfennig. 2012. *Evolution's wedge: competition and the origins of diversity*. University of California Press, Berkeley, CA.

Pfennig, D. W., C. K. Akcali, and D. W. Kikuchi. 2015. Batesian mimicry promotes pre- and postmating isolation in a snake mimicry complex. Evolution 69:1085–1090.

Phillimore, A. B., and T. D. Price. 2008. Density-dependent cladogenesis in birds. PLoS Biology 6:483–489.

Phillimore, A. B., J. D. Hadfield, O. R. Jones, and R. J. Smithers. 2010. Differences in spawning date between populations of common frog reveal local adaptation. Proceedings of the National Academy of Sciences USA 107:8292–8297.

Phillips, B. L., and R. Shine. 2004. Adapting to an invasive species: toxic cane toads induce morphological change in Australian snakes. Proceedings of the National Academy of Sciences USA 101:17150–17155.

Phillips, B. L., and R. Shine. 2005. The morphology, and hence impact, of an invasive species (the cane toad, *Bufo marinus*): changes with time since colonisation. Animal Conservation 8:407–413.

Phillips, B. L., and R. Shine. 2006. An invasive species induces rapid adaptive change in a native predator: cane toads and black snakes in Australia. Proceedings of the Royal Society B. Biological Sciences 273:1545–1550.

Phillips, B. L., G. P. Brown, and R. Shine. 2003. Assessing the potential impact of cane toads on Australian snakes. Conservation Biology 17:1738–1747.

Phillips, B. L., G. P. Brown, and R. Shine. 2010. Evolutionarily accelerated invasions: the rate of dispersal evolves upwards during the range advance of cane toads. Journal of Evolutionary Biology 23:2595–2601.

Phillips, B. L., G. P. Brown, J. K. Webb, and R. Shine. 2006. Invasion and the evolution of speed in toads. Nature 439:803.

Phillips, P. C. 2008. Epistasis—the essential role of gene interactions in the structure and evolution of genetic systems. Nature Reviews Genetics 9:855–867.

Phillips, P. C., and S. J. Arnold. 1989. Visualizing multivariate selection. Evolution 43:1209–1222.

Phillips, P. C., and S. J. Arnold. 1999. Hierarchical comparison of genetic variance-covariance matrices. I. using the flury hierarchy. Evolution 53:1506–1515.

Phillips, P. C., M. C. Whitlock, and K. Fowler. 2001. Inbreeding changes the shape of the genetic covariance matrix in *Drosophila melanogaster*. Genetics 158:1137–1145.

Phillis, C. C., J. W. Moore, M. Buoro, S. A. Hayes, J. C. Garza, and D. E. Pearse. 2016. Shifting thresholds: rapid evolution of migratory life histories in steelhead/rainbow trout, *Oncorhynchus mykiss*. Journal of Heredity 2016:51–60.

Pianka, E. R. 1970. On r- and K-selection. American Naturalist 104:592–597.

Pigliucci, M. 2003. Phenotypic integration: studying the ecology and evolution of complex phenotypes. Ecology Letters 6:265–272.

Pigliucci, M. 2006. Genetic variance-covariance matrices: a critique of the evolutionary quantitative genetics research program. Biology and Philosophy 21:1–23.

Pimentel, D. 1968. Population regulation and genetic feedback. Science 159:1432–1437.

Pimentel, D., E. H. Feinberg, P. W. Wood, and J. T. Hayes. 1965. Selection, spatial distribution, and the coexistence of competing fly species. American Naturalist 99:97–109.

Pimentel, D., W. P. Nagel, and J. L. Madden. 1963. Space-time structure of environment and survival of parasite-host systems. American Naturalist 97:141–167.

Pimm, S. L., L. Dollar, and O. L. Bass Jr. 2006. The genetic rescue of the Florida panther. Animal Conservation 9:115–122.

Pimm, S., P. Raven, A. Peterson, Ç. H. Şekercioğlu, and P. R. Ehrlich. 2006. Human impacts on the rates of recent, present, and future bird extinctions. Proceedings of the National Academy of Sciences USA 103:10941–10946.

Pinho, C., and J. Hey. 2010. Divergence with gene flow: models and data. Annual Review of Ecology, Evolution, and Systematics 41:215–230.

Plaistow, S. J., C. T. Lapsley, and T. G. Benton. 2006. Context-dependent intergenerational effects: the interaction between past and present environments and its effect on population dynamics. American Naturalist 167:206–215.

Podos, J. 2001. Correlated evolution of morphology and vocal signal structure in Darwin's finches. Nature 409:185–188.

Podos, J. 2010. Acoustic discrimination of sympatric morphs in Darwin's finches: a behavioural mechanism for assortative mating? Philosophical Transactions of the Royal Society B. Biological Sciences 365:1031–1039.

Polechová, J., N. Barton, and G. Marion. 2009. Species' range: adaptation in space and time. American Naturalist 174:E186–E204.

Porlier, M., A. Charmantier, P. Bourgault, P. Perret, J. Blondel, and D. Garant. 2012. Variation in phenotypic plasticity and selection patterns in blue tit breeding time: between- and within-population comparisons. Journal of Animal Ecology 81:1041–1051.

Porter, A. H., and N. A. Johnson. 2002. Speciation despite gene flow when developmental pathways evolve. Evolution 56:2103–2111.

Pörtner, H. O., and R. Knust. 2007. Climate change affects marine fishes through the oxygen limitation of thermal tolerance. Science 315:95–97.

Post, D. M., and E. P. Palkovacs. 2009. Eco-evolutionary feedbacks in community and ecosystem ecology: interactions between the ecological theatre and the evolutionary play. Philosophical Transactions of the Royal Society B. Biological Sciences 364:1629–1640.

Post, D. M., E. P. Palkovacs, E. G. Schielke, and S. I. Dodson. 2008. Intraspecific variation in a predator affects community structure and cascading trophic interactions. Ecology 89:2019–2032.

Post, D. M., N. E. Turley, E. P. Palkovacs, J. K. Bailey, S. des Roches, A. P. Hendry, M. T. Kinnison, and J. A. Schweitzer. Unpublished data. The ecological importance of biodiversity within species.

Post, E., and M. C. Forchhammer. 2008. Climate change reduces reproductive success of an Arctic herbivore through trophic mismatch. Philosophical Transactions of the Royal Society B. Biological Sciences 363:2369–2375.

Posthuma, L., and N. M. Van Straalen. 1993. Heavy-metal adaptation in terrestrial invertebrates: a review of occurrence, genetics, physiology and ecological consequences. Comparative Biochemistry and Physiology C. Pharmacology, Toxicology, and Endocrinology 106:11–38.

Postma, E., J. Visser, and A. J. van Noordwijk. 2007. Strong artificial selection in the wild results in predicted small evolutionary change. Journal of Evolutionary Biology 20:1823–1832.

Pregitzer, C. C., J. K. Bailey, S. C. Hart, and J. A. Schweitzer. 2010. Soils as agents of selection: feedbacks between plants and soils alter seedling survival and performance. Evolutionary Ecology 24:1045–1059.

Pregitzer, C. C., J. K. Bailey, and J. A. Schweitzer. 2013. Genetic by environment interactions affect plant-soil linkages. Ecology and Evolution 3:2322–2333.

Present, T. M. C., and D. O. Conover. 1992. Physiological basis of latitudinal growth differences in *Menidia*: variation in consumption or efficiency? Functional Ecology 6:23–31.

Price, G. R. 1970. Selection and covariance. Nature 227:520–521.

Price, T. 1998. Sexual selection and natural selection in bird speciation. Philosophical Transactions of the Royal Society of London Series B. Biological Sciences 353:251– 260.

Price, T. 2008. *Speciation in birds*. Roberts and Company, Geenwood Village, CO.

Price, T., and D. Schluter. 1991. On the low heritablity of life-history traits. Evolution 45:853–861.

Price, T. D., A. Qvarnström, and D. E. Irwin. 2003. The role of phenotypic plasticity in driving genetic evolution. Proceedings of the Royal Society of London B. Biological Sciences 270:1433–1440.

Price, T., M. Kirkpatrick, and S. J. Arnold. 1988. Directional selection and the evolution of breeding date in birds. Science 240:798–799.

Price, T., M. Turelli, and M. Slatkin. 1993. Peak shifts produced by correlated response to selection. Evolution 47:280–290.

Price, T. D., D. M. Hooper, C. D. Buchanan, U. S. Johansson, D. T. Tietze, P. Alström, U. Olsson, M. Ghosh-Harihar, F. Ishtiaq, S. K. Gupta, J. Martens, B. Harr, P. Singh, and D. Mohan. 2014. Niche filling slows the diversification of Himalayan songbirds. Nature 509:222–225.

Pringle, J. M., A. M. H. Blakeslee, J. E. Byers, and J. Roman. 2011. Asymmetric dispersal allows an upstream region to control population structure throughout a species' range. Proceedings of the National Academy of Sciences USA 108:15288–15293.

Prokop, Z. M., Ł. Michalczyk, S. M. Drobniak, M. Herdegen, and J. Radwan. 2012. Meta-analysis suggests choosy females get sexy sons more than "good genes". Evolution 66:2665–2673.

Provine, W. B. 1971. *The origins of theoretical population genetics*. University of Chicago Press, Chicago, IL.

Przybylo, R., B. C. Sheldon, and J. Merilä. 2000. Climatic effects on breeding and morphology: evidence for phenotypic plasticity. Journal of Animal Ecology 69:395–403.

Ptacek, M. B. 2000. The role of mating preferences in shaping interspecific divergence in mating signals in vertebrates. Behavioural Processes 51:111–134.

Puebla, O. 2009. Ecological speciation in marine v. freshwater fishes. Journal of Fish Biology 75:960–996.

Puebla, O., E. Bermingham, and F. Guichard. 2012. Pairing dynamics and the origin of species. Proceedings of the Royal Society B. Biological Sciences 279:1085–1092.

Puebla, O., E. Bermingham, F. Guichard, and E. Whiteman. 2007. Colour pattern as a single trait driving speciation in *Hypoplectrus* coral reef fishes? Proceedings of the Royal Society B. Biological Sciences 274:1265–1271.

Purvis, A., C. D. L. Orme, N. H. Toomey, and P. N. Pearson. 2009. Temporal patterns in diversification rates. Pages 278–300 *in* R. Butlin, J. Bridle, and D. Schluter, Editors. *Speciation and patterns of diversity*. Cambridge University Press, Cambridge, UK.

Pybus, O. G., and P. H. Harvey. 2000. Testing macro-evolutionary models using incomplete molecular phylogenies. Proceedings of the Royal Society B. Biological Sciences 267:2267–2272.

Quinn, T. P., and D. J. Adams. 1996. Environmental changes affecting the migratory timing of American shad and sockeye salmon. Ecology 77:1151–1162.

Quinn, T. P., E. C. Volk, and A. P. Hendry. 1999. Natural otolith microstructure patterns reveal precise homing to natal incubation sites by sockeye salmon (*Oncorhynchus nerka*). Canadian Journal of Zoology 77:766–775.

Quinn, T. P., M. T. Kinnison, and M. J. Unwin. 2001. Evolution of chinook salmon (*Oncorhynchus tshawytscha*) populations in New Zealand: pattern, rate, and process. Genetica 112–113:493–513.

Quinn, T. P., S. Hodgson, L. Flynn, R. Hilborn, and D. E. Rogers. 2007. Directional selection by fisheries and the timing of sockeye salmon (*Oncorhynchus nerka*) migrations. Ecological applications 17:731–739.

Quinn, T. P., M. J. Unwin, and M. T. Kinnison. 2000. Evolution of temporal isolation in the wild: genetic divergence in timing of migration and breeding by introduced chinook salmon populations. Evolution 54: 1372–1385.

Råberg, L., D. Sim, and A. F. Read. 2007. Disentangling genetic variation for resistance and tolerance to infectious diseases in animals. Science 318:812–814.

Ralph, P. L., and G. Coop. 2015. The role of standing variatoni in geographic convergent adaptation. American Naturalist 186:S5–S23.

Ramsey, J., H. D. Bradshaw, and D. W. Schemske. 2003. Components of reproductive isolation between the monkeyflowers *Mimulus lewisii* and *M. cardinalis* (Phrymaceae). Evolution 57:1520–1534.

Räsänen, K., A. Laurila, and J. Merilä. 2003. Geographic variation in acid stress tolerance of the moor frog, *Rana arvalis*. I. Local adaptation. Evolution 57:352–362.

Räsänen, K., and A. P. Hendry. 2008. Disentangling interactions between adaptive divergence and gene flow when ecology drives diversification. Ecology Letters 11:624–636.

Räsänen, K., and A. P. Hendry. 2014. Asymmetric reproductive barriers and mosaic reproductive isolation: insights from Misty lake-stream stickleback. Ecology and Evolution 4:1166–1175.

Ratcliffe, L. M., and P. R. Grant. 1983. Species recognition in Darwin's finches (*Geospiza*, Gould). I. discrimination by morphological clues. Animal Behaviour 31:1139–1153.

Rausher, M. D. 1992. The measurement of selection on quantitative traits: biases due to environmental covariances between traits and fitness. Evolution 46:616–626.

Raymond, M., C. Berticat, M. Weill, N. Pasteur, and C. Chevillon. 2001. Insecticide resistance in the mosquito *Culex pipiens*: what have we learned about adaptation? Genetica 112:287–296.

Rebar, D., and R. L. Rodríguez. 2014. Trees to treehoppers: genetic variation in host plants contributes to variation in the mating signals of a plant-feeding insect. Ecology Letters 17:203–210.

Reed, D. H., and R. Frankham. 2001. How closely correlated are molecular and quantitative measures of genetic variation? A meta-analysis. Evolution 55:1095–1103.

Reed, T. E., V. Grøtan, S. Jenouvrier, B.-E. Sæther, and M. E. Visser. 2013. Population growth in a wild bird is buffered against phenological mismatch. Science 340:488–491.

Reed, T. E., D. E. Schindler, M. J. Hague, D. A. Patterson, E. Meir, R. S. Waples, and S. G. Hinch. 2011. Time to evolve? Potential evolutionary responses of Fraser River sockeye salmon to climate change and effects on persistence. PLoS ONE 6:e20380.

Reimchen, T. E., and P. Nosil. 2002. Temporal variation in divergent selection on spine number in threespine stickleback. Evolution 56:2472–2483.

Renaut, S., A. W. Nolte, and L. Bernatchez. 2009. Gene expression divergence and hybrid misexpression between lake whitefish species pairs (*Coregonus* spp. Salmonidae). Molecular Biology and Evolution 26:925–936.

Renaut, S., A. W. Nolte, S. M. Rogers, N. Derome, and L. Bernatchez. 2011. SNP signatures of selection on standing genetic variation and their association with adaptive phenotypes along gradients of ecological speciation in lake whitefish species pairs (*Coregonus* spp.). Molecular Ecology 20:545–559.

Renaut, S., G. L. Owens, and L. H. Rieseberg. 2014. Shared selective pressure and local genomic landscape lead to repeatable patterns of genomic divergence in sunflowers. Molecular Ecology 23: 311–324.

Reusch, T. B. H. 2014. Climate change in the oceans: evolutionary versus phenotypically plastic responses of marine animals and plants. Evolutionary Applications 7:104–122.

Reusch, T. B. H., A. Ehlers, A. Hämmerli, and B. Worm. 2005. Ecosystem recovery after climatic extremes enhanced by genotypic diversity. Proceedings of the National Academy of Sciences USA 102:2826–2831.

Rettelbach, A., M. Kopp, U. Dieckmann, and J. Hermisson. 2013. Three modes of adaptive speciation in spatially structured populations. American Naturalist 182:E215–E234.

Reznick, D. N., and C. K. Ghalambor. 2001. The population ecology of contemporary adaptations: what empirical studies reveal about the conditions that promote adaptive evolution. Genetica 112–113:183–198.

Reznick, D. N., and C. K. Ghalambor. 2005. Selection in nature: experimental manipulations of natural populations. Integrative and Comparative Biology 45:456–462.

Reznick, D. N., and H. A. Bryga. 1996. Life-history evolution in guppies (*Poecilia reticulata*: Poeciliidae). V. Genetic basis of parallelism in life histories. American Naturalist 147:339–359.

Reznick, D. N., and H. Bryga. 1987. Life-history evolution in guppies (*Poecilia reticulata*). 1. Phenotypic and genetic changes in an introduction experiment. Evolution 41:1370–1385.

Reznick, D. N., F. H. Rodd, and M. Cardenas. 1996b. Life-history evolution in guppies (*Poecilia reticulata*: Poeciliidae). IV. Parallelism in life-history phenotypes. American Naturalist 147:319–338.

Reznick, D. N., F. H. Shaw, F. H. Rodd, and R. G. Shaw. 1997. Evaluation of the rate of evolution in natural populations of guppies (*Poecilia reticulata*). Science 275:1934–1937.

Reznick, D. N., M. J. Butler, F. H. Rodd, and P. Ross. 1996a. Life-history evolution in guppies (*Poecilia reticulata*). 6. Differential mortality as a mechanism for natural selection. Evolution 50:1651–1660.

Reznick, D., H. Rodd, and L. Nunney. 2004. Empirical evidence for rapid evolution. Pages 101–118 *in* R. Ferrière, U. Dieckmann, and D. Couvet, Editors. *Evolutionary conservation biology*. Cambridge University Press, Cambridge, UK.

Reznick, D., M. J. Bryant, and F. Bashey. 2002. r- and K-selection revisited: the role of population regulation in life-history evolution. Ecology 83:1509–1520.

Reznick, D., M. J. Butler Iv, and H. Rodd. 2001. Life-history evolution in guppies. VII. The comparative ecology of high- and low-predation environments. American Naturalist 157:126–140.

Rhode, J. M., and M. B. Cruzan. 2005. Contributions of heterosis and epistasis to hybrid fitness. American Naturalist 166:E124–E139.

Rice, W. R., and E. E. Hostert. 1993. Laboratory experiments on speciation: what have we learned in 40 years? Evolution 47:1637–1653.

Rice, W. R., and G. W. Salt. 1990. The evolution of reproductive isolation as a correlated character under sympatric conditions: experimental evidence. Evolution 44:1140–1152.

Richards, C. L., O. Bossdorf, N. Z. Muth, J. Gurevitch, and M. Pigliucci. 2006. Jack of all trades, master of some? On the role of phenotypic plasticity in plant invasions. Ecology Letters 9:981–993.

Richards, C. M. 2000. Inbreeding depression and genetic rescue in a plant metapopulation. American Naturalist 155:383–394.

Richmond, J. Q., E. L. Jockusch, and A. M. Latimer. 2011. Mechanical reproductive isolation facilitates parallel speciation in western North American scincid lizards. American Naturalist 178:320–332.

Ricker, W. E. 1954. Stock and recruitment. Journal of the Fisheries Research Board of Canada 11:559–623.

Ricklefs, R. E. 1987. Community diversity: relative roles of local and regional processes. Science 235:167–171.

Rico, C., and G. F. Turner. 2002. Extreme microallopatric divergence in a cichlid species from Lake Malawi. Molecular Ecology 11:1585–1590.

Ridenour, W. M., J. M. Vivanco, Y. Feng, J.-I. Horiuchi, and R. M. Callaway. 2008. No evidence for trade-offs: *Centaurea* plants from America are better competitors and defeners. Ecological Monographs 78:369–386.

Riechert, S. E. 1993. Investigation of potential gene flow limitation of behavioral adaptation in an aridlands spider. Behavioral Ecology and Sociobiology 32:355–363.

Rieseberg, L. H. 2001. Chromosomal rearrangements and speciation. Trends in Ecology and Evolution 16:351–358.

Rieseberg, L. H., and N. C. Ellstrand. 1993. What can molecular and morphological markers tell us about plant hybridization? Critical Reviews in Plant Sciences 12:213–241.

Rieseberg, L. H., O. Raymond, D. M. Rosenthal, Z. Lai, K. Livingstone, T. Nakazato, J. L. Durphy, A. E. Schwarzbach, L. A. Donovan, and C. Lexer. 2003. Major ecological transitions in wild sunflowers facilitated by hybridization. Science 301:1211–1216.

Ritchie, M. G. 2007. Sexual selection and speciation. Annual Review of Ecology, Evolution, and Systematics 38:79–102.

Robertson, A. 1966. A mathematical model of the culling process in dairy cattle. Animal Production 8:95–108.

Robinson, B. W., and D. S. Wilson. 1994. Character release and displacement in fishes: a neglected literature. American Naturalist 144:596–627.

Robinson, B. W., and K. J. Parsons. 2002. Changing times, spaces, and faces: tests and implications of adaptive morphological plasticity in the fishes of northern postglacial lakes. Canadian Journal of Fisheries and Aquatic Sciences 59:1819–1833.

Rockman, M. V. 2012. The QTN program and the alleles that matter for evolution: all that's gold does not glitter. Evolution 66:1–17.

Rodríguez, R. L., J. W. Boughman, D. A. Gray, E. A. Hebets, G. Hobel, and L. B. Symes. 2013. Diversification under sexual selection: the relative roles of mate preference strength and the degree of divergence in mate preferences. Ecology Letters 16:964–974.

Roesti, M., A. P. Hendry, W. Salzburger, and D. Berner. 2012. Genome divergence during evolutionary diversification as revealed in replicate lake-stream stickleback population pairs. Molecular Ecology 21:2852–2862.

Roesti, M., S. Gavrilets, A.P. Hendry, W. Salzburger, and D. Berner. 2014. The genomic signature of parallel adaptation from shared genetic variation. Molecular Ecology 23:3944–3956.

Roesti, M., B. Kueng, D. Moser, and D. Berner. 2015. The genomics of ecological vicariance in threespine stickleback fish. Nature Communications 6:8767.

Roff, D. A. 1992. *Evolution of life histories: theory and analysis*. Chapman and Hall, New York.

Roff, D. A. 1996. The evolution of genetic correlations: an analysis of patterns. Evolution 50:1392–1403.

Roff, D. A. 1997. *Evolutionary quantitative genetics*. Chapman and Hall, New York.

Roff, D. A. 2002. *Life history evolution*. Sinauer Associates, Inc., Sunderland.

Roff, D. A. 2007. A centennial celebration for quantitative genetics. Evolution 61:1017–1032.

Roff, D. A., and K. Emerson. 2006. Epistasis and dominance: evidence for differential effects in life-history versus morphological traits. Evolution 60:1981–1990.

Roff, D. A., and T. A. Mousseau. 1987. Quantitative genetics and fitness: lessons from *Drosophila*. Heredity 58:103–118.

Rogers, S. M., P. Tamkee, B. Summers, S. Balabahadra, M. Marks, D. M. Kingsley, and D. Schluter. 2012. Genetic signature of adaptive peak shift in threespine stickleback. Evolution 66:2439–2450.

Rokyta, D. R., Z. Abdo, and H. A. Wichman. 2009. The genetics of adaptation for eight microvirid bacteriophages. Journal of Molecular Evolution 69:229–239.

Rollinson, N., and L. Rowe. 2015. Persistent directional selection on body size and a resolution to the paradox of stasis. Evolution 69:2441–2451.

Rolshausen, G., G. Segelbacher, K. A. Hobson, and H. M. Schaefer. 2009. Contemporary evolution of reproductive isolation and phenotypic divergence in sympatry along a migratory divide. Current Biology 19:2097–2101.

Rolshausen, G., S. Muttalib, R. Kaeuffer, K. B. Oke, D. Hanson, and A. P. Hendry. 2015a. When maladaptive gene flow does not increase selection. Evolution 69:2289–2302.

Rolshausen, G., D. A. T. Phillip, D. M. Beckles, A. Akbari, S. Ghoshal, P. B. Hamilton, C. R. Tyler, A. G. Scarlett, I. Ramnarine, P. Bentzen, and A. P. Hendry. 2015b. Do stressful conditions make adaptation difficult? Guppies in the oil-polluted environments of southern Trinidad. Evolutionary Applications 8:854–870.

Ronce, O. 2007. How does it feel to be like a rolling stone? Ten questions about dispersal evolution. Annual Review of Ecology, Evolution, and Systematics 38:231–253.

Ronce, O., and M. Kirkpatrick. 2001. When sources become sinks: migrational meltdown in heterogeneous habitats. Evolution 55:1520–1531.

Root, R. B. 1967. Niche exploitation pattern of the Blue-Gray Gnatcatcher. Ecological Monographs 37:317–350.

Root, T. L., J. T. Price, K. R. Hall, S. H. Schneider, C. Rosenzweig, and J. A. Pounds. 2003. Fingerprints of global warming on wild animals and plants. Nature 421:57–60.

Rosenblum, E. B., and L. J. Harmon. 2011. "Same same but different": replicated ecological speciation at white sands. Evolution 65:946–960.

Rosenblum, E. B., C. E. Parent, and E. E. Brandt. 2014. The molecular basis of phenotypic convergence. Annual Review of Ecology, Evolution and Systematics. 45:203–226.

Rosenthal, G. G., T. Y. F. Martinez, F. J. G. de León, and M. J. Ryan. 2001. Shared preferences by predators and females for male ornaments in swordtails. American Naturalist 158:146–154.

Rosenzweig, M. L. 1978. Competitive speciation. Biological Journal of the Linnean Society 10:275–289.

Rosenzweig, M. L. 2001. The four questions: what does the introduction of exotic species do to diversity? Evolutionary Ecology Research 3:361–367.

Rosindell, J., S. P. Hubbell, and R. S. Etienne. 2011. The unified neutral theory of biodiversity and biogeography at age ten. Trends in Ecology and Evolution 26:340–348.

Roughgarden, J. 1979. *Theory of population genetics and evolutionary ecology: an introduction*. Macmillan, New York.

Rousset, F. 2000. Genetic differentiation between individuals. Journal of Evolutionary Biology 13:58–62.

Rousset, F. 2004. *Genetic structure and selection in subdivided populations*. Princeton University Press, Princeton, NJ.

Roy, D., K. Lucek, R. P. Walter, and O. Seehausen. 2015. Hybrid 'superswarm' leads to rapid diverence and establishiment of populations during a biological invasion. Molecular Ecology 24:5394–5411.

Royte, E. 2001. *The tapir's morning bath: mysteries of the tropical rain forest and the scientists who are trying to solve them*. First Mariner Books, New York.

Rudman, S. M., and D. Schluter. 2016. Ecological impacts of reverse speciation in threespine stickleback. Current Biology 26:490–495.

Rudman, S. M., M. A. Rodriguez-Cabal, A. Stier, T. Sato, J. Heavyside, R. W. El-Sabaawi, and G. M. Crutsinger. 2015. Adaptive genetic variation mediates bottom-up and top-down control in an aquatic ecosytem. Proceedings of the Royal Society B. Biological Sciences 282:20151234.

Rueffler, C., T. J. M. Van Dooren, O. Leimar, and P. A. Abrams. 2006. Disruptive selection and then what? Trends in Ecology and Evolution 21:238–245.

Rumpho, M. E., K. N. Pelletreau, A. Moustafa, and D. Bhattacharya. 2011. The making of a photosynthetic animal. Journal of Experimental Biology 214:303–311.

Rundell, R. J., and T. D. Price. 2009. Adaptive radiation, nonadaptive radiation, ecological speciation, and nonecological speciation. Trends in Ecology and Evolution 24:394–399.

Rundle, H. D. 2002. A test of ecologically dependent postmating isolation between sympatric sticklebacks. Evolution 56:322–329.

Rundle, H. D. 2003. Divergent environments and population bottlenecks fail to generate premating isolation in *Drosophila pseudoobscura*. Evolution 57:2557–2565.

Rundle, H. D., and M. C. Whitlock. 2001. A genetic interpretation of ecologically dependent isolation. Evolution 55:198–201.

Rundle, H. D., and P. Nosil. 2005. Ecological speciation. Ecology Letters 8:336–352.

Rundle, H. D., L. Nagel, J. W. Boughman, and D. Schluter. 2000. Natural selection and parallel speciation in sympatric sticklebacks. Science 287:306–308.

Ryan, M. J., and A. Keddy-Hector. 1992. Directional patterns of female mate choice and the role of sensory biases. American Naturalist 139:S4–S35.

Ryan, M., J. H. Fox, and W. Wilczynski. 1990. Sexual selection for sensory exploitation in the frog *Physalaemus pustulosus*. Nature 343:66–67.

Ryan, P. G., P. Bloomer, C. L. Moloney, T. J. Grant, and W. Delport. 2007. Ecological speciation in South Atlantic island finches. Science 315:1420–1423.

Saccheri, I., and I. Hanski. 2006. Natural selection and population dynamics. Trends in Ecology and Evolution 21:341–347.

Samani, P., and G. Bell. 2016. The ghosts of selection past reduces the probability of plastic rescue but increase the likelihood of evolutionary rescue to novel stressors in experimental populations of wild yeast. Ecology Letters 19:289–298.

Samis, K. E., and C. G. Eckert. 2009. Ecological correlates of fitness across the northern geographic range limit of a Pacific Coast dune plant. Ecology 90:3051–3061.

Sanchez, A., and J. Gore. 2013. Feedback between population and evolutionary dynamics determines the fate of social microbial populations. PLoS Biology 11:e1001547.

Sasa, M. M., P. T. Chippindale, and N. A. Johnson. 1998. Patterns of postzygotic isolation in frogs. Evolution 52:1811–1820.

Sasaki, A., and S. Ellner. 1997. Quantitative genetic variance maintained by fluctuating selection with overlapping generations: variance components and covariances. Evolution 51:682–696.

Savolainen, O., T. Pyhäjärvi, and T. Knürr. 2007. Gene flow and local adaptation in trees. Annual Review of Ecology, Evolution, and Systematics 38:595–619.

Sax, D. F., and S. D. Gaines. 2008. Species invasions and extinction: the future of native biodiversity on islands. Proceedings of the National Academy of Sciences USA 105:11490– 11497.

Sax, D. F., S. D. Gaines, and J. H. Brown. 2002. Species invasions exceed extinctions on islands world-wide: a comparative study of plants and birds. American Naturalist 160:766–783.

Schaum, C. E., and S. Collins. 2014. Plasticity predicts evolution in a marine alga. Proceedings of the Royal Society B. Biological Sciences 281:20141486.

Scheiner, S. M. 1993. Genetics and evolution of phenotypic plasticity. Annual Review of Ecology and Systematics 24:35–68.

Scheiner, S. M., and D. Berrigan. 1998. The genetics of phenotypic plasticity. VIII. The cost of plasticity in *Daphnia pulex*. Evolution 52:368–378.

Scheiner, S. M., and R. D. Holt. 2012. The genetics of phenotypic plasticity. X. Variation versus uncertainty. Ecology and Evolution 2:751–767.

Scheiner, S. M., R. J. Mitchell, and H. S. Callahan. 2000. Using path analysis to measure natural selection. Journal of Evolutionary Biology 13:423–433.

Schemske, D. W. 2010. Adaptation and the origin of species. American Naturalist 176:S4–S25.

Schemske, D. W., and H. D. Bradshaw Jr. 1999. Pollinator preference and the evolution of floral traits in monkeyflowers (*Mimulus*). Proceedings of the National Academy of Sciences USA 96:11910–11915.

Schiestl, F. P., and P. M. Schlüter. 2009. Floral isolation, specialized pollination, and pollinator behavior in orchids. Annual Review of Entomology 54:425–446.

Schieving, F., and H. Poorter. 1999. Carbon gain in a multispecies canopy: the role of specific leaf area and photosynthetic nitrogen-use efficiency in the tragedy of the commons. New Phytologist 143:201–211.

Schiffers, K., E. C. Bourne, S. Lavergne, W. Thuiller, and J. M. J. Travis. 2013. Limited evolutionary rescue of locally adapted populations facing climate change. Philosophical Transactions of the Royal Society B. Biological Sciences 368:20120083.

Schindler, D. E., R. Hilborn, B. Chasco, C. P. Boatright, T. P. Quinn, L. A. Rogers, and M. S. Webster. 2010. Population diversity and the portfolio effect in an exploited species. Nature 465:609–612.

Schlaepfer, M. A., M. C. Runge, and P. W. Sherman. 2002. Ecological and evolutionary traps. Trends in Ecology and Evolution 17:474–480.

Schlichting, C. D., and M. Pigliucci. 1998. *Phenotypic evolution: a reaction norm perspective*. Sinauer Associates, Inc., Sunderland, MA.

Schlichting, C. D., and M. A. Wund. 2014. Phenotypic plasticity and epigenetic marking: an assessment of evidence for genetic accommodation. Evolution 68:656–672.

Schliewen, U. K., and B. Klee. 2004. Reticulate sympatric speciation in Cameroonian crater lake cichlids. Frontiers in Zoology 1:5.

Schluter, D. 1984. Morphological and phylogenetic relations among the Darwin's finches. Evolution 38:921–930.

Schluter, D. 1988. Estimating the form of natural selection on a quantitative trait. Evolution 42:849–861.

Schluter, D. 1993. Adaptive radiation in sticklebacks: size, shape, and habitat use efficiency. Ecology 74:699–709.

Schluter, D. 1995. Adaptive radiation in sticklebacks: trade-offs in feeding performance and growth. Ecology 76:82–90.

Schluter, D. 1996a. Adaptive radiation along genetic lines of least resistance. Evolution 50:1766–1774.

Schluter, D. 1996b. Ecological speciation in postglacial fishes. Philosophical Transactions of the Royal Society of London B. Biological Sciences 351:807–814.

Schluter, D. 2000a. *The ecology of adaptive radiation*. Oxford University Press, UK.

Schluter, D. 2000b. Ecological character displacement in adaptive radiation. American Naturalist 156:S4–S16.

Schluter, D. 2003. Frequency dependent natural selection during character displacement in sticklebacks. Evolution 57:1142–1150.

Schluter, D. 2009. Evidence for ecological speciation and its alternative. Science 323:737–741.

Schluter, D., and D. Nychka. 1994. Exploring fitness surfaces. American Naturalist 143:597–616.

Schluter, D., and J. D. McPhail. 1992. Ecological character displacement and speciation in sticklebacks. American Naturalist 140:85–108.

Schluter, D., and P. R. Grant. 1984. Determinants of morphological patterns in communities of Darwin finches. American Naturalist 123:175–196.

Schluter, D., T. D. Price, and L. Rowe. 1991. Conflicting selection pressures and life history trade-offs. Proceedings of the Royal Society B. Biological Sciences 246:11–17.

Schmitz, O. J., and K. B. Suttle. 2001. Effects of top predator species on direct and indirect interactions in a food web. Ecology 82:2072–2081.

Schmitz, O. J. 2013. Global climate change and the evolutionary ecology of ecosystem functioning. Annals of the New York Academy of Sciences 1297:61–72.

Schoener, T. W. 2011. The newest synthesis: understanding the interplay of evolutionary and ecological dynamics. Science 331:426–429.

Schoustra, S. E., T. Bataillon, D. R. Gifford, and R. Kassen. 2009. The properties of adaptive walks in evolving populations of fungus. PLoS Biology 7:e1000250.

Schröder, A., L. Persson, and A. M. de Roos. 2009. Culling experiments demonstrate size-class specific biomass increases with mortality. Proceedings of the National Academy of Sciences USA 106:2671–2676.

Schtickzelle, N., G. Mennechez, and M. Baguette. 2006. Dispersal depression with habitat fragmentation in the bog fritillary butterfly. Ecology 87:1057–1065.

Schwander, T., and O. Leimar. 2011. Genes as leaders and followers in evolution. Trends in Ecology and Evolution 26:143–151.

Schwartz, A. K., and A. P. Hendry. 2006. Sexual selection and the detection of ecological speciation. Evolutionary Ecology Research 8:399–413.

Schwartz, A. K., D. J. Weese, P. Bentzen, M. T. Kinnison, and A. P. Hendry. 2010. Both geography and ecology contribute to mating isolation in guppies. PLoS ONE 5:e15659.

Schweitzer, J. A., M. D. Madritch, J. K. Bailey, C. J. LeRoy, D. G. Fischer, R. J. Rehill, R. L. Lindroth, A. E. Hagerman, S. C. Wooley, S. C. Hart, and T. G. Whitham. 2008. From genes to ecosystems: the genetic basis of condensed tannins and their role in nutrient regulation in a *Populus* model system. Ecosystems 11:1005–1020.

Schweitzer, J. A., D. G. Fischer, B. J. Rehill, S. C. Wooley, S. A. Woolbright, R. L. Lindroth, T. G. Whitham, D. R. Zak, and S. C. Hart. 2011. Forest gene diversity is correlated with the composition and function of soil microbial communities. Population Ecology 53:35–46.

Schweitzer, J. A., J. K. Bailey, B. J. Rehill, G. D. Martinsen, S. C. Hart, R. L. Lindroth, P. Keim, and T. G. Whitham. 2004. Genetically based trait in a dominant tree affects ecosystem processes. Ecology Letters 7:127–134.

Schweitzer, J. A., J. K. Bailey, S. C. Hart, and T. G. Whitham. 2005. Nonadditive effects of mixing cottonwood genotypes on litter decomposition and nutrient dynamics. Ecology 86:2834–2840.

Schweitzer, J. A., I. Juric, T. F. J. van de Voorde, K. Clay, W. H. van der Putten, and J. K. Bailey. 2014. Are there evolutionary consequences of plant-soil feedbacks along soil gradients. Functional Ecology 28:55–64.

Seddon, N., C. A. Botero, J. A. Tobias, P. O. Dunn, H. E. A. MacGregor, D. R. Rubenstein, J. A. C. Uy, J. T. Weir, L. A. Whittingham, and R. J. Safran. 2013. Sexual selection accelerates signal evolution during speciation in birds. Proceedings of the Royal Society B. Biological Sciences 280: 20131065.

Seehausen, O. 2004. Hybridization and adaptive radiation. Trends in Ecology and Evolution 19:198–207.

Seehausen, O. 2006. African cichlid fish: a model system in adaptive radiation research. Proceedings of the Royal Society B. Biological Sciences 273:1987–1998.

Seehausen, O., G. Takimoto, D. Roy, and J. Jokela. 2008a. Speciation reversal and biodiversity dynamics with hybridization in changing environments. Molecular Ecology 17:30–44.

Seehausen, O., J. J. M. van Alphen, and F. Witte. 1997. Cichlid fish diversity threatened by eutrophication that curbs sexual selection. Science 277:1808–1811.

Seehausen, O., Y. Terai, I. S. Magalhaes, K. L. Carleton, H. D. J. Mrosso, R. Miyagi, I. van der Sluijs, M. V. Schneider, M. E. Maan, H. Tachida, H. Imai, and N. Okada. 2008b. Speciation through sensory drive in cichlid fish. Nature 455:620–627.

Servedio, M. R. 2001. Beyond reinforcement: the evolution of premating isolation by direct selection on preferences and postmating, prezygotic incompatibilities. Evolution 55:1909–1920.

Servedio, M. R. 2015. Geography, assortative mating, and the effects of sexual selection on speciation with gene flow. Evolutionary Applications 9:91–102.

Servedio, M. R., and M. A. F. Noor. 2003. The role of reinforcement in speciation: theory and data. Annual Review of Ecology, Evolution, and Systematics 34:339–364.

Servedio, M. R., G. S. Van Doorn, M. Kopp, A. M. Frame, and P. Nosil. 2011. Magic traits in speciation: "magic" but not rare? Trends in Ecology and Evolution 26:389–397.

Servedio, M. R., and R. Bürger. 2014. The counterintuitive role of sexual selection in species maintenance and specation. Proceedings of the National Academy of Sciences USA 111: 8113–8118.

Sexton, J. P., S. B. Hangartner, and A. A. Hoffmann. 2014. Genetic isolation by environment or distance: which pattern of gene flow is most common? Evolution 68: 1–15.

Sexton, J. P., P. J. McIntyre, A. L. Angert, and K. J. Rice. 2009. Evolution and ecology of species range limits. Annual Review of Ecology, Evolution, and Systematics 40:415–436.

Sexton, J. P., S. Y. Strauss, and K. J. Rice. 2011. Gene flow increases fitness at the warm edge of a species' range. Proceedings of the National Academy of Sciences USA 108:11704–11709.

Shafer, A. B. A., and J. B. W. Wolf. 2013. Widespread evidence for incipient ecological speciation: a meta-analysis of isolation-by-ecology. Ecology Letters 16:940–950.

Sharon, G., D. Segal, J. M. Ringo, A. Hefetz, I. Zilber-Rosenberg, and E. Rosenberg. 2010. Commensal bacteria play a role in mating preference of *Drosophila melanogaster*. Proceedings of the National Academy of Sciences USA 107:20051–20056.

Sharpe, D. M. T., and A. P. Hendry. 2009. Life history change in commercially exploited fish stocks: an analysis of trends across studies. Evolutionary Applications 2:260–275.

Sharpe, D. M. T., K. Räsänen, D. Berner, and A. P. Hendry. 2008. Genetic and environmental contributions to the morphology of lake and stream stickleback: implications for gene flow and reproductive isolation. Evolutionary Ecology Research 10:849–866.

Shaw, K. L., and S. P. Mullen. 2011. Genes versus phenotypes in the study of speciation. Genetica 139:649–661.

Shaw, R. G., and C. J. Geyer. 2010. Inferring fitness landscapes. Evolution 64:2510–2520.

Shea, K., and P. Chesson. 2002. Community ecology theory as a framework for biological invasions. Trends in Ecology and Evolution 17:170–176.

Sheets, H. D., and C. E. Mitchell. 2001. Why the null matters: statistical tests, random walks and evolution. Genetica 112–113:105–125.

Sheldon, S. P., and K. N. Jones. 2001. Restricted gene flow according to host plant in an herbivore feeding on native and exotic watermilfoils (Myriophyllum: Haloragaceae). International Journal of Plant Sciences 162:793–799.

Shelton, A. M., J. D. Tang, R. T. Roush, T. D. Metz, and E. D. Earle. 2000. Field tests on managing resistance to Bt-engineered plants. Nature Biotechnology 18:339–342.

Shimada, Y., T. Shikano, and J. Merilä. 2011. A high incidence of selection on physiologically important genes in the three-spined stickleback, *Gasterosteus aculeatus*. Molecular Biology and Evolution 28:181–193.

Shindo, C., M. J. Aranzana, C. Lister, C. Baxter, C. Nicholls, M. Nordborg, and C. Dean. 2005. Role of FRIGIDA and FLOWERING LOCUS C in determining variation in flowering time of *Arabidopsis*. Plant Physiology 138:1163–1173.

Shuster, S. M., E. V. Lonsdorf, G. M. Wimp, J. K. Bailey, and T. G. Whitham. 2006. Community heritability measures the evolutionary consequences of indirect genetic effects on community structure. Evolution 60:991–1003.

Siefert, A., C. Violle, L. Chalmandrier, C. H. Albert, A. Taudiere, A. Fajardo, L. W. Aarssen, C. Baraloto, M. B. Carlucci, M. V. Cianciaruso, V. de L. Dantas, F. de Bello, L. D. S. Duarte, C. R. Fonseco, G. T. Freschet, S. Gaucherand, N. Gross, K. Hikosaka, B. Jackson, V. Jung, C. Kamiyama, M. Katabuchi, S. W. Kembel, E. Kichenin, N. J. B. Kraft, A. Lagerström, Y. Le Bagousse-Pinguet, Y. Li, N.

Mason, J. Messier, T. Nakashizuka, J. McC. Overton, D. A. Peltzer, I. M. Pérez-Ramos, V. D. Pillar, H. C. Prentice, S. Richardson, T. Sasaki, B. S. Schamp, C. Schöb, B. Shipley, M. Sundqvist, M. T. Sykes, M. Vandewalle, and D. A. Wardle. 2015. A global meta-analysis of the relative extent of intraspecific trait variation in plant communities. Ecology Letters 18:1406–1419.

Siepielski, A. M., J. D. DiBattista, and S. M. Carlson. 2009. It's about time: the temporal dynamics of phenotypic selection in the wild. Ecology Letters 12:1261–1276.

Siepielski, A. M., J. D. DiBattista, J. A. Evans, and S. M. Carlson. 2011. Differences in the temporal dynamics of phenotypic selection among fitness components in the wild. Proceedings of the Royal Society B. Biological Sciences 278:1572–1580.

Siepielski, A. M., K. M. Gotanda, M. B. Morrissey, S. E. Diamond, J. D. DiBattista, and S. M. Carlson. 2013. The spatial patterns of directional phenotypic selection. Ecology Letters 16:1382–1392.

Siepielski, A. M., A. Nemirov, M. Cattivera, and A. Nickerson. 2016. Experimental evidence for an eco-evolutionary coupling between local adaptation and intraspecific competition. American Naturalist 187: 447–456.

Sih, A., A. M. Bell, J. C. Johnson, and R. E. Ziemba. 2004. Behavioral syndromes: an integrative overview. Quaterly Review of Biology 79:241–277.

Sih, A., D. I. Bolnick, B. Luttbeg, J. L. Orrock, S. D. Peacor, L. M. Pintor, E. Preisser, J. S. Rehage, and J. R. Vonesh. 2010. Predator-prey naïveté, antipredator behavior, and the ecology of predator invasions. Oikos 119:610–621.

Silvertown, J. 2004. Plant coexistence and the niche. Trends in Ecology and Evolution 19:605–611.

Silvertown, J., C. Servaes, P. Biss, and D. Macleod. 2005. Reinforcement of reproductive isolation between adjacent populations in the Park Grass Experiment. Heredity 95:198–205.

Simberloff, D., and T. Dayan. 1991. The guild concept and the structure of ecological communities. Annual Review of Ecology and Systematics 22:115–143.

Simms, E. L., and J. Triplett. 1994. Costs and benefits of plant responses to disease: resistance and tolerance. Evolution 48:1973–1985.

Simpson, G. G. 1944. *Tempo and mode in evolution*. Columbia University Press, New York.

Simpson, G. G. 1953a. The Baldwin effect. Evolution 7:110–117.

Simpson, G. G. 1953b. *The major features of evolution*. Columbia University Press, New York.

Sinervo, B., and C. M. Lively. 1996. The rock-paper-scissors game and the evolution of alternative male strategies. Nature 380:240–243.

Sinervo, B., E. Svensson, and T. Comendant. 2000. Density cycles and an offspring quantity and quality game driven by natural selection. Nature 406:985–988.

Sinervo, B., P. Doughty, R. B. Huey, and K. Zamudio. 1992. Allometric engineering: a causal analysis of natural selection on offspring size. Science 258:1927–1930.

Singer, M. C. 2015. Adaptive and maladaptive consequences of "matching habitat choice:" lessons from a rapidly-evolving butterfly metapopulation. Evolutionary Ecology 29:905–925.

Singer, M. C., and C. S. McBride. 2010. Multitrait, host-associated divergence among sets of butterfly populations: implications for reproductive isolation and ecological speciation. Evolution 64:921–933.

Siwertsson, A., R. Knudsen, K. K. Kahilainen, K. Præbel, R. Primicerio, and P.-A. Amundsen. 2010. Sympatric diversification as influenced by ecological opportunity and historical contingency in a young species lineage of whitefish. Evolutionary Ecology Research 12:929–947.

Skelly, D. K., L. N. Joseph, H. P. Possingham, L. K. Freidenburg, T. J. Farrugia, M. T. Kinnison, and A. P. Hendry. 2007. Evolutionary responses to climate change. Conservation Biology 21:1353–1355.

Skúlason, S., and T. B. Smith. 1995. Resource polymorphisms in vertebrates. Trends in Ecology and Evolution 10:366–370.

Slate, J. 2005. Quantitative trait locus mapping in natural populations: progress, caveats and future directions. Molecular Ecology 14:363–379.

Slatkin, M. 1973. Gene flow and selection in a cline. Genetics 75:733–756.

Slatkin, M. 1976. The rate of spread of an advantageous allele in a subdivided population. Pages 767–780 *in* S. Karlin and E. Nevo, Editors. *Population genetics and ecology*. Academic Press, New York.

Slatkin, M. 1987. Gene flow and the geographic structure of natural populations. Science 236:787–792.

Slatkin, M. 1993. Isolation by distance in equilibrium and non-equilibrium populations. Evolution 47:264–279.

Slatkin, M. 2005. Seeing ghosts: the effect of unsampled populations on migration rates estimated for sampled populations. Molecular Ecology 14:67–73.

Slobodkin, L. B. 1961. *Growth and regulation of animal populations*. Holt, Rinehart and Winston, New York.

Smallegange, I. M., and T. Coulson. 2013. Towards a general, population-level understanding of eco-evolutionary change. Trends in Ecology and Evolution 28:143–148.

Smith, B. R., and D. T. Blumstein. 2008. Fitness consequences of personality: a meta-analysis. Behavioral Ecology 19:448–455.

Smith, D. S., M. K. Lau, R. Jacobs, J. A. Monroy, S. M. Shuster, and T. G. Whitham. 2015. Rapid plant evolution in the presence of an introduced species alters community composition. Oecologia 179:563–572.

Smith, G. R., C. Badgley, T. P. Eiting, and P. S. Larson. 2010. Species diversity gradients in relation to geological history in North American freshwater fishes. Evolutionary Ecology Research 12:693–726.

Smith, J. W., and C. W. Benkman. 2007. A coevolutionary arms race causes ecological speciation in crossbills. American Naturalist 169:455–465.

Smith, S. A., G. Bell, and E. Bermingham. 2004. Cross-Cordillera exchange mediated by the Panama Canal increased the species richness of local freshwater fish assemblages. Proceedings of the Royal Society B. Biological Sciences 271:1889–1896.

Smith, T. B. 1993. Disruptive selection and the genetic basis of bill size polymorphism in the African finch Pyrenestes. Nature 363:618–620.

Smith, T. B., and S. Skúlason. 1996. Evolutionary significance of resource polymorphisms in fishes, amphibians, and birds. Annual Review of Ecology and Systematics 27:111–133.

Smith, T. B., R. K. Wayne, D. J. Girman, and M. W. Bruford. 1997. A role for ecotones in generating rainforest biodiversity. Science 276:1855–1857.

Smouse, P. E., and V. L. Sork. 2004. Measuring pollen flow in forest trees: an exposition of alternative approaches. Forest Ecology and Management 197:21–38.

Snowberg, L. K., and D. I. Bolnick. 2008. Assortative mating by diet in a phenotypically unimodal but ecologically variable population of stickleback. American Naturalist 172:733–739.

Sobel, J. M., and G. F. Chen. 2014. Unification of methods for estimating the strength of reproductive isolation. Evolution 68: 1511–1522.

Sobel, J. M., G. F. Chen, L. R. Watt, and D. W. Schemske. 2010. The biology of speciation. Evolution 64:295–315.

Sol, D., R. P. Duncan, T. M. Blackburn, P. Cassey, and L. Lefebvre. 2005a. Big brains, enhanced cognition, and response of birds to novel environments. Proceedings of the National Academy of Sciences USA 102:5460–5465.

Sol, D., D. G. Stirling, and L. Lefebvre. 2005b. Behavioral drive or behavioral inhibition in evolution: subspecific diversification in Holarctic passerines. Evolution 59:2669–2677.

Sol, D., S. Bacher, S. M. Reader, and L. Lefebvre. 2008. Brain size predicts the success of mammal species introduced into novel environments. American Naturalist 172:S63–S71.

Sol, D., S. Timmermans, and L. Lefebvre. 2002. Behavioural flexibility and invasion success in birds. Animal Behaviour 63:495–502.

Soria-Carrasco, V., Z. Gompert, A. A. Comeault, T. E. Farkas, T. L. Parchman, J. S. Johnston, C. A. Buerkle, J. L. Feder, J. Bast, T. Schwander, S. P. Egan, B. J. Crespi, and P. Nosil. 2014. Stick insect genomes reveal natural selection's role in parallel speciation. Science 344:738–742.

Spalding, D. 1873. Instinct, with original observations on young animals. Macmillan's Magazine 27:282–293.

Spitze, K. 1993. Population structure in *Daphnia obtusa*: quantitative genetic and allozymic variation. Genetics 135:367–374.

Standen, E. M., T. Y. Du, and H. C. E. Larsson. 2014. Developmental plasticity and the origin of tetrapods. Nature 513: 54–58.

Stapley, J., J. Reger, P. G. D. Feulner, C. Smadja, J. Galindo, R. Ekblom, C. Bennison, A. D. Ball, A. P. Beckerman, and J. Slate. 2010. Adaptation genomics: the next generation. Trends in Ecology and Evolution 25:705–712.

Stearns, S. C. 1983. The genetic basis of differences in life-history traits among six populations of mosquitofish (*Gambusia affinis*) that shared ancestors in 1905. Evolution 37:618–627.

Stearns, S. C. 1992. *The evolution of life histories*. Oxford University Press, Oxford, UK.

Stearns, S. C., and R. D. Sage. 1980. Maladaptation in a marginal population of the mosquito fish, *Gambusia affinis*. Evolution 34:65–75.

Steiner, U. K., and J. Van Buskirk. 2008. Environmental stress and the costs of whole-organism phenotypic plasticity in tadpoles. Journal of Evolutionary Biology 21:97–103.

Steppan, S. J., P. C. Phillip, and D. Houle. 2002. Comparative quantitative genetics: evolution of the G matrix. Trends in Ecology and Evolution 17:320–327.

Stern, D. L., and V. Orgogozo. 2008. The loci of evolution: how predictable is genetic evolution? Evolution 62:2155–2177.

Stinchcombe, J. R., A. F. Agrawal, P. A. Hohenlohe, S. J. Arnold, and M. W. Blows. 2008. Estimating nonlinear selection gradients using quadratic regression coefficients: double or nothing? Evolution 62:2435–2440.

Stinchcombe, J. R., and H. E. Hoekstra. 2008. Combining population genomics and quantitative genetics: finding the genes underlying ecologically important traits. Heredity 100:158–170.

Stinchcombe, J. R., F. T. W. Group, and M. Kirkpatrick. 2012. Genetics and evolution of function-valued traits: understanding environmentally responsive phenotypes. Trends in Ecology and Evolution 27:637–647.

Stinchcombe, J. R., M. T. Rutter, D. S. Burdick, P. Tiffin, M. D. Rausher, and R. Mauricio. 2002. Testing for environmentally induced bias in phenotypic estimates of natural selection: theory and practice. American Naturalist 160:511–523.

Stirling, D. G., D. Réale, and D. A. Roff. 2002. Selection, structure and the heritability of behaviour. Journal of Evolutionary Biology 15:277–289.

Stockwell, C. A., A. P. Hendry, and M. T. Kinnison. 2003. Contemporary evolution meets conservation biology. Trends in Ecology and Evolution 18:94–101.

Stoks, R., L. Govaert, K. Pauwels, B. Jansen, and L. De Meester. 2016. Resurrecting complexity: the interplay of plasticity and rapid evolution in the multiple trait response to strong changes in predation pressure in the waterflea *Daphnia magna*. Ecology Letters 19:180–190.

Stoddard, P. K. 1999. Predation enhances complexity in the evolution of electric fish signals. Nature 400:254–256.

Stomp, M., M. A. van Dijk, H. M. J. van Overzee, M. T. Wortel, C. A. M. Sigon, M. Egas, H. Hoogveld, H. J. Gons, and J. Huisman. 2008. The timescale of phenotypic plasticity and its impact on competition in fluctuating environments. American Naturalist 172:169–185.

Storfer, A., and A. Sih. 1998. Gene flow and ineffective antipredator behavior in a stream-breeding salamander. Evolution 52:558–565.

Stoyan, D., and S. Wagner. 2001. Estimating the fruit dispersion of anemochorous forest trees. Ecological Modelling 145:35–47.

Strasburg, J. L., and L. H. Rieseberg. 2010. How robust are "isolation with migration" analyses to violations of the im model? A simulation study. Molecular Biology and Evolution 27:297–310.

Strasburg, J. L., C. Scotti-Saintagne, I. Scotti, Z. Lai, and L. H. Rieseberg. 2009. Genomic patterns of adaptive divergence between chromosomally differentiated sunflower species. Molecular Biology and Evolution 26:1341–1355.

Strauss, S. Y. 2014. Ecological and evolutionary responses in complex communities: implications for invasions and eco-evolutionary feedbacks. Oikos 123:257–266.

Strauss, S. Y., J. A. Lau, and S. P. Carroll. 2006. Evolutionary responses of natives to introduced species: what do introductions tell us about natural communities? Ecology Letters 9:357–374.

Strauss, S. Y., J. A. Lau, T. W. Schoener, and P. Tiffin. 2008. Evolution in ecological field experiments: implications for effect size. Ecology Letters 11:199–207.

Strong, D. R., L. A. Szyska, and D. S. Simberloff. 1979. Tests of community-wide character displacement against null hypotheses. Evolution 33:897–913.

Stuart, S. N., J. S. Chanson, N. A. Cox, B. E. Young, A. S. L. Rodrigues, D. L. Fischman, and R. W. Waller. 2004. Status and trends of amphibian declines and extinctions worldwide. Science 306:1783–1786.

Stuart, Y. E., and J. B. Losos. 2013. Ecological character displacement: glass half full or half empty? Trends in Ecology and Evolution 28:402–408.

Stuart, Y. E., T. S. Campbell, P. A. Hohenlohe, R. G. Reynolds, L. J. Revell, and J. B. Losos. 2014. Rapid evolution of a native species following invasion by a congener. Science 346:463–466.

Stutz, W. E., M. Schmerer, J. L. Coates, and D. I. Bolnick. 2015. Among-lake reciprocal transplants induce convergent expression of immune genes in threespine stickleback. Molecular Ecology 24:4629–4646.

Sullam, K. E., B. E. R. Rubin, C. M. Dalton, S. S. Kilham, A. S. Flecker, and J. A. Russell. 2015. Divergence across diet, time and populations rules out parallel evolution in the gut microbiomes of Trinidadian guppies. The ISME Journal 9:1508–1522.

Sultan, S. E., and H. G. Spencer. 2002. Metapopulation structure favors plasticity over local adaptation. American Naturalist 160:271–283.

Sultan, S. E., T. Horgan-Kobelski, L. M. Nichols, C. E. Riggs, and R. K. Waples. 2013. A resurrection study reveals rapid adaptive evolution within populations of an invasive plant. Evolutionary Applications 6:266–278.

Suzuki, Y., and H. F. Nijhout. 2006. Evolution of a polyphenism by genetic accommodation. Science 311:650–652.

Svanbäck, R., M. Pineda-Krch, and M. Doebeli. 2009. Fluctuating population dynamics promotes the evolution of phenotypic plasticity. American Naturalist 174:176–189.

Svardal, H., C. Rueffler, and M. Doebeli. 2014. Organismal complexity and the potential for evolutionary diversification. Evolution 68: 3248–3259.

Svensson, E. I. 2012. Non-ecological speciation, niche conservatism and thermal adaptation: how are they connected? Organisms Diversity and Evolution 12:229–240.

Svensson, E. I., F. Eroukhmanoff, and M. Friberg. 2006. Effects of natural and sexual selection on adaptive population divergence and premating isolation in a damselfly. Evolution 60:1242–1253.

Svensson, E. I., J. Abbott, and R. Hardling. 2005. Female polymorphism, frequency dependence, and rapid evolutionary dynamics in natural populations. American Naturalist 165:567–576.

Swindell, W. R., and J. L. Bouzat. 2006. Gene flow and adaptive potential in *Drosophila melanogaster*. Conservation Genetics 7:79–89.

Swindell, W. R., M. Huebner, and A. P. Weber. 2007. Plastic and adaptive gene expression patterns associated with temperature stress in *Arabidopsis thaliana*. Heredity 99:143–150.

Tabashnik, B. E., A. J. Gassmann, D. W. Crowder, and Y. Carrière. 2008. Insect resistance to Bt crops: evidence versus theory. Nature Biotechnology 26:199–202.

Tabashnik, B. E., J. B. J. Van Rensburg, and Y. Carrière. 2009. Field-evolved insect resistance to Bt crops: definition, theory, and data. Journal of Economic Entomology 102:2011–2025.

Tabashnik, B. E., T. J. Dennehy, and Y. Carrière. 2005. Delayed resistance to transgenic cotton in pink bollworm. Proceedings of the National Academy of Sciences USA 102:15389–15393.

Tack, A. J. M., and T. Roslin. 2010. Overrun by the neighbors: landscape context affects strength and sign of local adaptation. Ecology 91:2253–2260.

Tack, A. J. M., M. T. J. Johnson, and T. Roslin. 2012. Sizing up community genetics: it's a matter of scale. Oikos 121:481–488.

Tack, A. J. M., O. Ovaskainen, P. Pulkkinen, and T. Roslin. 2010. Spatial location dominates over host plant genotype in structuring an herbivore community. Ecology 91:2660–2672.

Takahata, N. 1983. Gene identity and genetic differentiation of populations in the finite island model. Genetics 104:497–512.

Tallmon, D. A., G. Luikart, and R. S. Waples. 2004. The alluring simplicity and complex reality of genetic rescue. Trends in Ecology and Evolution 19:489–496.

Tammi, J., M. Appelberg, U. Beier, T. Hesthagen, A. Lappalainen, and M. Rask. 2003. Fish status survey of Nordic lakes: effects of acidification, eutrophication and stocking activity on present fish species composition. Ambio 32:98–105.

Tang, J. D., H. L. Collins, T. D. Metz, E. D. Earle, J. Z. Zhao, R. T. Roush, and A. M. Shelton. 2001. Greenhouse tests on resistance management of Bt transgenic plants using refuge strategies. Journal of Economic Entomology 94:240–247.

Taylor, E. B. 1999. Species pairs of north temperate freshwater fishes: evolution, taxonomy, and conservation. Reviews in Fish Biology and Fisheries 9:299–324.

Taylor, E. B., J. W. Boughman, M. Groenenboom, M. Sniatynski, D. Schluter, and J. L. Gow. 2006. Speciation in reverse: morphological and genetic evidence of the collapse of a three-spined stickleback (*Gasterosteus aculeatus*) species pair. Molecular Ecology 15:343–355.

Tebbich, S., K. Sterelny, and I. Teschke. 2010. The tale of the finch: adaptive radiation and behavioural flexibility. Philosophical Transactions of the Royal Society of London Series B. Biological Sciences 365:1099–1109.

Templeton, A. R. 2008. The reality and importance of founder speciation in evolution. BioEssays 30:470–479.

Teplitsky, C., J. A. Mills, J. W. Yarrall, and J. Merilä. 2009. Heritability of fitness components in a wild bird population. Evolution 63:716–726.

Teplitsky, C., M. Tarka, A. P. Møller, S. Nakagawa, J. Balbontín, T. A. Burke, C. Doutrelant, A. Gregoire, B. Hansson, D. Hasselquist, L. Gustafsson, F. de Lope, A. Marzal, J. A. Mills, N. T. Wheelwright, J. W. Yarrall, and A. Charmantier. 2014. Assessing multivariate constraints to evolution across ten long-term avian studies. PLoS ONE 9:e90444.

Terekhanova, N. V., M. D. Logacheva, A. A. Penin, T. V. Neretina, A. E. Barmintseva, G. A. Bazykin, A. S. Kondrashov, and N. S. Mugue. 2015. Fast evolution from precast bricks: genomics of young freshwater populations of threespine stickleback *Gasterosteus aculeatus*. PLoS Genetics 10:e1004696.

terHorst, C. P., T. E. Miller, and D. R. Levitan. 2010a. Evolution of prey in ecological time reduces the effect size of predators in experimental microcosms. Ecology 91:629–636.

terHorst, C. P., T. E. Miller, and E. Powell. 2010b. When can competition for resources lead to ecological equivalence? Evolutionary Ecology Research 12:843–854.

Tétard-Jones, C., M. A. Kertesz, P. Gallois, and R. F. Preziosi. 2007. Genotype-by-genotype interactions modified by a third species in a plant-insect system. American Naturalist 170:492–499.

Thibert-Plante, X., and S. Gavrilets. 2013. Evolution of mate choice and the so-called magic traits in ecological speciation. Ecology Letters 16:1004–1013.

Thibert-Plante, X., and A. P. Hendry. 2009. Five questions on ecological speciation addressed with individual-based simulations. Journal of Evolutionary Biology 22:109–123.

Thibert-Plante, X., and A. P. Hendry. 2010. When can ecological speciation be detected with neutral loci? Molecular Ecology 19:2301–2314.

Thibert-Plante, X., and A. P. Hendry. 2011a. Factors influencing progress toward sympatric speciation. Journal of Evolutionary Biology 24:2186–2196.

Thibert-Plante, X., and A. P. Hendry. 2011b. The consequences of phenotypic plasticity for ecological speciation. Journal of Evolutionary Biology 24:326–342.

Thomas, C. D., A. Cameron, R. E. Green, M. Bakkenes, L. J. Beaumont, Y. C. Collingham, B. F. N. Erasmus, M. F. de Siqueira, A. Grainger, L. Hannah, L. Hughes, B. Huntley, A. S. van Jaarsveld, G. F. Midgley, L. Miles, M. A. Ortega-Huerta, A. T. Peterson, O. L. Phillips, and S. E. Williams. 2004. Extinction risk from climate change. Nature 427:145–148.

Thomas, C. D., and J. J. Lennon. 1999. Birds extend their ranges northwards. Nature 399:213.

Thomas, C. D., E. J. Bodsworth, R. J. Wilson, A. D. Simmons, Z. G. Davies, M. Musche, and L. Conradt. 2001. Ecological and evolutionary processes at expanding range margins. Nature 411:577–581.

Thomas, Y., M.-T. Bethenod, L. Pelozuelo, B. Frérot, and D. Bourguet. 2003. Genetic isolation between two sympatric host-plant races of the European corn borer, *Ostrinia nubilalis* Hübner. I. Sex pheromone, moth emergence timing, and parasitism. Evolution 57:261–273.

Thompson, J. N. 1994. *The coevolutionary process.* Chicago University Press, Chicago, USA.

Thompson, J. N. 1998. Rapid evolution as an ecological process. Trends in Ecology and Evolution 13:329–332.

Thompson, J. N. 2005. *The geographic mosaic of coevolution.* Chicago University Press, Chicago, USA.

Thorogood, R., and N. B. Davies. 2013. Reed warbler hosts fine-tune their defenses to track three decades of cuckoo decline. Evolution 67:3545–3555.

Thorpe, R. S., J. T. Reardon, and A. Malhotra. 2005. Common garden and natural selection experiments support ecotypic differentiation in the Dominican anole (*Anolis oculatus*). American Naturalist 165:495–504.

Thorpe, R. S., Y. Surget-Groba, and H. Johansson. 2010. Genetic tests for ecological and allopatric speciation in anoles on an island archipelago. PLoS Genetics 6:e1000929.

Thuiller, W., S. Lavorel, M. B. Araújo, M. T. Sykes, and I. C. Prentice. 2005. Climate change threats to plant diversity in Europe. Proceedings of the National Academy of Sciences USA 102:8245–8250.

Thurman, T. J., and R. D. H. Barrett. 2016. The genetic consequences of selection in natural populations. Molecular Ecology 25: 1429–1448.

van Tienderen, P. H. 1997. Generalists, specialists, and the evolution of phenotypic plasticity in sympatric populations of distinct species. Evolution 51:1372–1380.

Tilman, D. 1982. *Resource competition and community structure.* Princeton University Press, Princeton, NJ.

Tilman, D. 1994. Competition and biodiversity in spatially structured habitats. Ecology 75:2–16.

Tilman, D. 2004. Niche tradeoffs, neutrality, and community structure: a stochastic theory of resource competition, invasion, and community assembly. Proceedings of the National Academy of Sciences USA 101:10854–10861.

Tilman, D., D. Wedin, and J. Knops. 1996. Productivity and sustainability influenced by biodiversity in grassland ecosystems. Nature 379:718–720.

Tilman, D., S. S. Kilham, and P. Kilham. 1982. Phytoplankton community ecology: the role of limiting nutrients. Annual Review of Ecology and Systematics 13:349–372.

Tilmon, K. J. (Editor). 2008. *Specialization, speciation, and radiation: the evolutionary biology of herbivorous insects.* University of California Press, Berkeley, CA.

Tinghitella, R. M. 2008. Rapid evolutionary change in a sexual signal: genetic control of the mutation 'flatwing' that renders male field crickets (*Teleogryllus oceanicus*) mute. Heredity 100: 261–267.

Tishkoff, S. A., F. A. Reed, A. Ranciaro, B. F. Voight, C. C. Babbitt, J. S. Silverman, K. Powell, H. M. Mortensen, J. B. Hirbo, M. Osman, M. Ibrahim, S. A. Omar, G. Lema, T. B. Nyambo, J. Ghori, S. Bumpstead, J. K. Pritchard, G. A. Wray, and P. Deloukas. 2007. Convergent adaptation of human lactase persistence in Africa and Europe. Nature Genetics 39:31–40.

Tokeshi, M. 1999. *Species coexistence: ecological and evolutionary perspectives.* Blackwell Science, John Wiley & Sons, Oxford, UK.

Tomanek, L. 2008. The importance of physiological limits in determining biogeographical range shifts due to global climate change: the heat-shock response. Physiological and Biochemical Zoology 81:709–717.

Tooker, J. F., and S. D. Frank. 2012. Genotypically diverse cultivar mixtures for insect pest management and increased crop yields. Journal of Applied Ecology 49:974–985.

Torres-Dowdall, J., C. A. Handelsman, D. N. Reznick, and C. K. Ghalambor. 2012. Local adaptation and the evolution of phenotypic plasticity in Trinidadian guppies (*Poecilia reticulata*). Evolution 66:3432–3443.

Travis, J., J. Leips, and F. H. Rodd. 2013. Evolution in population parameters: density-dependent selection or density-dependent fitness? American Naturalist 181:S9–S20.

Tuda, M., and Y. Iwasa. 1998. Evolution of contest competition and its effect on host-parasitoid dynamics. Evolutionary Ecology 12:855–870.

Tufto, J. 2001. Effects of releasing maladapted individuals: a demographic-evolutionary model. American Naturalist 158:331–340.

Tufto, J. 2010. Gene flow from domesticated species to wild relatives: migration load in a model of multivariate selection. Evolution 64:180–192.

Tufto, J. 2015. Genetic evolution, plasticity, and bet-hedging as adaptive responses to temporally autocorrelated fluctuating selection: A quantitative genetic model. Evolution 69:2034–2049.

Turcotte, M. M., D. N. Reznick, and J. D. Hare. 2011a. Experimental assessment of the impact of rapid evolution on population dynamics. Evolutionary Ecology Research 13:113–131.

Turcotte, M. M., D. N. Reznick, and J. D. Hare. 2011b. The impact of rapid evolution on population dynamics in the wild: experimental test of eco-evolutionary dynamics. Ecology Letters 14:1084–1092.

Turcotte, M. M., D. N. Reznick, and J. D. Hare. 2013. Experimental test of an eco-evolutionary dynamic feedback loop between evolution and population density in the green peach aphid. American Naturalist 181:S46–S57.

Turcotte, M. M., A. K. Lochab, N. E. Turley, and M. T. J. Johnson. 2015. Plant domestication slows pest evolution. Ecology Letters 18:907–915.

Turelli, M., J. R. Lipkowitz, and Y. Brandvain. 2013. On the Coyne and Orr-igin of species: effects of intrinsic postzygotic isolation, ecological differentiation, X chromosome size, and sympatry on *Drosophila* speciation. Evolution 68:1176–1187.

Turley, N. E., and M. T. J. Johnson. 2015. Ecological effects of aphid abundance, genotypic variation, and contemporary evolution on plants. Oecologia 178:747–759.

Turley, N. E., W. C. Odell, H. Schaefer, G. Everwand, M. J. Crawley, and M. T. J. Johnson. 2013. Contemporary evolution of plant growth rate following experimental removal of herbivores. American Naturalist 181:S21–S34.

Turner, T. L., M. W. Hahn, and S. V. Nuzhdin. 2005. Genomic islands of speciation in *Anopheles gambiae*. PLoS Biology 3:1572–1578.

Tutt, J. W. 1896. *British moths*. Routledge, London.

Tyerman, J. G., M. Bertrand, C. C. Spencer, and M. Doebeli. 2008. Experimental demonstration of ecological character displacement. BMC Evolutionary Biology 8:34.

Urban, M. C. 2011. The evolution of species interactions across natural landscapes. Ecology Letters 14:723–732.

Urban, M. C., and D. K. Skelly. 2006. Evolving metacommunities: toward an evolutionary perspective on metacommunities. Ecology 87:1616–1626.

Urban, M. C., and L. De Meester. 2009. Community monopolization: local adaptation enhances priority effects in an evolving metacommunity. Proceedings of the Royal Society B. Biological Sciences 276:4129–4138.

Urban, M. C., B. L. Phillips, D. K. Skelly, and R. Shine. 2007. The cane toad's (*Chaunus* [*Bufo*] *marinus*) increasing ability to invade Australia is revealed by a dynamically updated range model. Proceedings of the Royal Society B. Biological Sciences 274:1413–1419.

Urban, M. C., M. A. Leibold, P. Amarasekare, L. De Meester, R. Gomulkiewicz, M. E. Hochberg, C. A. Klausmeier, N. Loeuille, C. de Mazancourt, J. Norberg, J. H. Pantel, S. Y. Strauss, M. Vellend, and M. J. Wade. 2008. The evolutionary ecology of metacommunities. Trends in Ecology and Evolution 23:311–317.

Urban, M. C., J. L. Richardson, and N. A. Freidenfelds. 2014. Plasticity and genetic adaptation mediate amphibian and reptile responses to climate change. Evolutionary Applications 7: 88–103.

Urbanelli, S., D. Porretta, V. Mastrantonio, R. Bellini, G. Pieraccini, R. Romoli, G. Crasta, and G. Nascetti. 2014. Hybridization, natural selection, and evolution of reproductive isolation: a 25-years survey of an artificial sympatric area between two mosquito sibling species of the *Aedes mariae* complex. Evolution 68:3030–3038.

Uyeda, J. C., T. F. Hansen, S. J. Arnold, and J. Pienaar. 2011. The million-year wait for macroevolutionary bursts. Proceedings of the National Academy of Sciences USA 108:15908–15913.

Valiente-Banuet, A., A. V. Rumebe, M. Verdú, and R. M. Callaway. 2006. Modern quaternary plant lineages promote diversity through facilitation of ancient tertiary lineages. Proceedings of the National Academy of Sciences USA 103:16812–16817.

Vamosi, S. M., S. B. Heard, J. C. Vamosi, and C. O. Webb. 2009. Emerging patterns in the comparative analysis of phylogenetic community structure. Molecular Ecology 18:572–592.

Van Buskirk, J., and U. K. Steiner. 2009. The fitness costs of developmental canalization and plasticity. Journal of Evolutionary Biology 22:852–860.

Van Buskirk, J., R. S. Mulvihill, and R. C. Leberman. 2012. Phenotypic plasticity alone cannot explain climate-induced change in avain migration timing. Ecology and Evolution 2:2430–2437.

van der Putten, W. H., R. D. Bardgett, J. D. Bever, T. M. Bezemer, B. B. Casper, T. Fukami, P. Kardol, J. N. Klironomos, A. Kulmatiski, J. A. Schweitzer, K. N. Suding, T. F. J. Van de Voorde, and D. A. Wardle. 2013. Plant-soil feedbacks: the past, the present and future challenges. Journal of Ecology 101:265–276.

van der Sluijs, I., T. J. M. Van Dooren, O. Seehausen, and J. J. M. Van Alphen. 2008. A test of fitness consequences of hybridization in sibling species of Lake Victoria cichlid fish. Journal of Evolutionary Biology 21:480–491.

van der Sluijs, I., S. M. Gray, M. C. P. Amorim, I. Barber, U. Candolin, A. P. Hendry, R. Krahe, M. E. Maan, A. C. Utne-Palm, H.-J. Wagner, and B. B. M. Wong. 2011. Communication in troubled waters: responses of fish communication systems to changing environments. Evolutionary Ecology 25:623–640.

van Doorn, G. S., P. Edelaar, and F. J. Weissing. 2009. On the origin of species by natural and sexual selection. Science 326:1704–1707.

Van Doorslaer, W., J. Vanoverbeke, C. Duvivier, S. Rousseaux, M. Jansen, B. Jansen, H. Feuchtmayr, D. Atkinson, B. Moss, R. Stoks, and L. De Meester. 2009. Local adaptation to higher temperatures reduces immigration success of genotypes from a warmer region in the water flea *Daphnia*. Global Change Biology 15:3046–3055.

van Heerwaarden, B., and C. M. Sgró. 2014. Is adaptation to climate change really constrained in niche specialists? Proceedings of the Royal Society B. Biological Sciences 281:20140396.

van Kleunen, M., and M. Fischer. 2005. Constraints on the evolution of adaptive phenotypic plasticity in plants. The New Phytologist 166:49–60.

van Klinken, R. D., and O. R. Edwards. 2002. Is host-specificity of weed biological control agents likely to evolve rapidly following establishment? Ecology Letters 5:590–596.

van Oppen, M. J. H., G. F. Turner, C. Rico, J. C. Deutsch, K. M. Ibrahim, R. L. Robinson, and G. M. Hewitt. 1997. Unusually fine-scale genetic structuring found in rapidly speciating Malawi cichlid fishes. Proceedings of the Royal Society B. Biological Sciences 264:1803–1812.

Van Rossum, F., and L. Triest. 2010. Pollen dispersal in an insect-pollinated wet meadow herb along an urban river. Landscape and Urban Planning 95:201–208.

van Tienderen, P. H. 1991. Evolution of generalists and specialists in spatially heterogeneous environments. Evolution 45:1317–1331.

van Tienderen, P. H. 2000. Elasticities and the link between demographic and evolutionary dynamics. Ecology 81:666–679.

van Tienderen, P. H., and G. de Jong. 1994. A general model of the relation between phenotypic selection and genetic response. Journal of Evolutionary Biology 7:1–12.

Van Valen, L. 1973. A new evolutionary law. Evolutionary Theory 1:1–30.

Vander Wal, E., D. Garant, M. Festa-Bianchet, and F. Pelletier. 2013. Evolutionary rescue in vertebrates: evidence, applications and uncertainty. Philosophical Transactions of the Royal Society B. Biological Sciences 368:20120090.

Vasseur, D. A., and J. W. Fox. 2011. Adaptive dynamics of competition for nutritionally complementary resources: character convergence, displacement, and parallelism. American Naturalist 178:501–514.

Vasseur, D. A., P. Amarasekare, V. H. W. Rudolf, and J. M. Levine. 2011. Eco-evolutionary dynamics enable coexistence via neighbor-dependent selection. American Naturalist 178:E96–E109.

Vedder, O., S. Bouwhuis, and B. C. Sheldon. 2013. Quantitative assessment of the importance of phenotypic plasticity in adaptation to climate change in wild bird populations. PLoS Biology 11:e1001605.

Veen, T., T. Borge, S. C. Griffith, G.-P. Saetre, S. Bures, L. Gustafsson, and B. C. Sheldon. 2001. Hybridization and adaptive mate choice in flycatchers. Nature 411:45–50.

Velema, G. J., J. S. Rosenfeld, and E. B. Taylor. 2012. Effects of the invasive American signal crayfish (*Pacifastacus leniusculus*) on the reproductive behaviour of threespine stickleback (*Gasterosteus aculeatus*) sympatric species pairs. Canadian Journal of Zoology 90: 1328–1338.

Vellend, M. 2006. The consequences of genetic diversity in competitive communities. Ecology 87:304–311.

Vellend, M. 2008. Effects of diversity on diversity: consequences of competition and facilitation. Oikos 117:1075–1085.

Vellend, M. 2010. Conceptual synthesis in community ecology. Quarterly Review of Biology 85:183–206.

Vellend, M., G. Lajoie, A. Bourret, C. Múrria, S. W. Kembel, and D. Garant. 2014. Drawing ecological inferences from coincident patterns of population- and community-level biodiversity. Molecular Ecology 23:2890–2901.

Vermeij, G. J. 1987. *Evolution and escalation: an ecological history of life*. Princeton University Press, Princeton, NJ.

Via, S. 1999. Reproductive isolation between sympatric races of pea aphids. I. Gene flow restriction and habitat choice. Evolution 53:1446–1457.

Via, S. 2001. Sympatric speciation in animals: the ugly duckling grows up. Trends in Ecology and Evolution 16:381–390.

Via, S., A. C. Bouck, and S. Skillman. 2000. Reproductive isolation between divergent races of pea aphids on two hosts. II. Selection against migrants and hybrids in the parental environments. Evolution 54:1626–1637.

Via, S., and J. West. 2008. The genetic mosaic suggests a new role for hitchhiking in ecological speciation. Molecular Ecology 17:4334–4345.

Via, S., and R. Lande. 1985. Genotype-environment interaction and the evolution of phenotypic plasticity. Evolution 39:505–522.

Vincenzi, S. 2014. Extinction risk and eco-evolutionary dynamics in a variable environment with
increasing frequency of extreme events. Journal of the Royal Society Interface 11:20140441.

Vindenes, Y., and Ø. Langangen. 2015. Individual heterogeneity in life histories and eco-evolutionary
dynamics. Ecology Letters 18:417–432.

Vines, T. H., and D. Schluter. 2006. Strong assortative mating between allopatric sticklebacks as a
by-product of adaptation to different environments. Proceedings of the Royal Society B. Biological
Sciences 273:911–916.

Violle, C., D. R. Nemergut, Z. C. Pu, and L. Jiang. 2011. Phylogenetic limiting similarity and competitive
exclusion. Ecology Letters 14:782–787.

Visser, M. E., P. Gienapp, A. Husby, M. Morrisey, I. de la Hera, F. Pullido, and C. Both. 2015. Effects
of spring temperatures on the strength of selection on timing of reproduction in a long-distance
migratory bird. PLoS Biology 13:e1002120.

Waddington, C. 1953. Genetic assimilation of an acquired character. Evolution 7:118–126.

Waddington, C. H. 1961. Genetic assimilation. Advances in Genetics 10:257–293.

Wade, M. J. 2002. A gene's eye view of epistasis, selection and speciation. Journal of Evolutionary
Biology 15:337–346.

Wade, M. J., and S. Kalisz. 1990. The causes of natural selection. Evolution 44:1947–1955.

Wade, M. J., and C. J. Goodnight. 1998. The theories of Fisher and Wright in the context of metapopulations: when nature does many small experiments. Evolution 52:1537–1553.

Wade, M. J., and S. Kalisz. 1989. The additive partitioning of selection gradients. Evolution 43:1567–1569.

Wagner, G. P., and L. Altenberg. 1996. Complex adaptations and the evolution of evolvability. Evolution
50:967–976.

Walker, J. A. 2007. A general model of functional constraints on phenotypic evolution. American
Naturalist 170:681–689.

Walker, J. A., and M. A. Bell. 2000. Net evolutionary trajectories of body shape evolution within a
microgeographic radiation of threespine sticklebacks (Gasterosteus aculeatus). Journal of Zoology
252:293–302.

Walker, J. A. 2014. The effect of unmeasured confounders on the ability to estimate a true performance
or selection gradient (and other partial regression coefficients). Evolution 68:2128–2136.

Wallace, B. 1975. Hard and soft selection revisited. Evolution 29:465–473.

Walsh, B., and M. W. Blows. 2009. Abundant genetic variation plus strong selection = multivariate
genetic constraints: a geometric view of adaptation. Annual Review of Ecology, Evolution, and
Systematics 40:41–59.

Walsh, M. R., and D. M. Post. 2011. Interpopulation variation in a fish predator drives evolutionary
divergence in prey in lakes. Proceedings of the Royal Society of London B. Biological Sciences
278:2628–2637.

Walsh, M. R., D. F. Fraser, R. D. Bassar, and D. N. Reznick. 2011. The direct and indirect effects
of guppies: implications for life-history evolution in Rivulus hartii. Functional Ecology
25:227–237.

Walters, C. J., and F. Juanes. 1993. Recruitment limitation as a consequence of natural selection for
use of restricted feeding habitats and predation risk-taking by juvenile fishes. Canadian Journal of
Fisheries and Aquatic Sciences 50:2058–2070.

Wang, J. L. 2005. Estimation of effective population sizes from data on genetic markers. Philosophical
Transactions of the Royal Society B. Biological Sciences 360:1395–1409.

Wang, I. J., and G. S. Bradburd. 2014. Isolation by environment. Molecular Ecology 23: 5649–5662.

Waples, R. S. 1998. Separating the wheat from the chaff: patterns of genetic differentiation in high gene
flow species. Journal of Heredity 89:438–450.

Waples, R. S., and O. Gaggiotti. 2006. What is a population? An empirical evaluation of some genetic
methods for identifying the number of gene pools and their degree of connectivity. Molecular
Ecology 15:1419–1439.

Wardle, D. A., R. D. Bardgett, L. R. Walker, and K. I. Bonner. 2009. Among- and within-species variation in plant litter decomposition in contrasting long-term chronosequences. Functional Ecology 23:442–453.

Webb, C. 2003. A complete classification of Darwinian extinction in ecological interactions. American Naturalist 161:181–205.

Webb, C. O., D. D. Ackerly, M. A. McPeek, and M. J. Donoghue. 2002. Phylogenies and community ecology. Annual Review of Ecology and Systematics 33:475–505.

Weeks, A. R., C. M. Sgro, A. G. Young, R. Frankham, N. J. Mitchell, K. A. Miller, M. Byrne, D. J. Coates, M. D. B. Eldridge, P. Sunnucks, M. F. Breed, E. A. James, and A. A. Hoffmann. 2011. Assessing the benefits and risks of translocations in changing environments: a genetic perspective. Evolutionary Applications 4:709–725.

Weersing, K., and R. J. Toonen. 2009. Population genetics, larval dispersal, and connectivity in marine systems. Marine Ecology Progress Series 393:1–12.

Weese, D. J., A. K. Schwartz, P. Bentzen, A. P. Hendry, and M. T. Kinnison. 2011. Eco-evolutionary effects on population recovery following catastrophic disturbance. Evolutionary Applications 4:354–366.

Weese, D. J., S. P. Gordon, A. P. Hendry, and M. T. Kinnison. 2010. Spatiotemporal variation in linear natural selection on body color in wild guppies (*Poecilia reticulata*). Evolution 64:1802–1815.

Weiblen, G. D. 2002. How to be a fig wasp. Annual Review of Entomology 47:299–330.

Weis, J. J., and D. M. Post. 2013. Intraspecific variation in a predator drives cascading variation in primary producer community composition. Oikos 122:1343–1349.

Welch, J. J., and D. Waxman. 2003. Modularity and the cost of complexity. Evolution 57:1723–1734.

Weldon, W. F. R. 1901. A first study of natural selection in *Clausilia laminata* (Montagu). Biometrika 1:109–124.

Werner, E. E., and S. D. Peacor. 2003. A review of trait-mediated indirect interactions in ecological communities. Ecology 84:1083–1100.

West-Eberhard, M. J. 2003. *Developmental plasticity and evolution*. Oxford University Press, Oxford, UK.

West-Eberhard, M. J. 1983. Sexual selection, social competition, and speciation. Quaterly Review of Biology 58:155–183.

Westemeier, R. L., J. D. Brawn, S. A. Simpson, T. L. Esker, R. W. Jansen, J. W. Walk, E. L. Kershner, J. L. Bouzat, and K. N. Paige. 1998. Tracking the long-term decline and recovery of an isolated population. Science 282:1695–1698.

Westley, P. A. H. 2011. What invasive species reveal about the rate and form of contemporary phenotypic change in nature. American Naturalist 177:496–509.

Whalon, M. E., D. Mota-Sanchez, and R. M. Hollingworth. 2008. *Global Pesticide Resistance in Arthropods*. CABI, Oxfordshire

Whiteley, A. R., S. W. Fitzpatrick, W. C. Funk, and D. A. Tallmon. 2015. Genetic rescue to the rescue. Trends in Ecology and Evolution 30:42–49.

Whitham, T. G., J. K. Bailey, J. A. Schweitzer, S. M. Shuster, R. K. Bangert, C. J. LeRoy, E. V. Lonsdorf, G. J. Allan, S. P. DiFazio, B. M. Potts, D. G. Fischer, C. A. Gehring, R. L. Lindroth, J. C. Marks, S. C. Hart, G. M. Wimp, and S. C. Wooley. 2006. A framework for community and ecosystem genetics: from genes to ecosystems. Nature Reviews Genetics 7:510–523.

Whitham, T. G., W. P. Young, G. D. Martinsen, C. A. Gehring, J. A. Schweitzer, S. M. Shuster, G. M. Wimp, D. G. Fischer, J. K. Bailey, R. L. Lindroth, S. Woolbright, and C. R. Kuske. 2003. Community and ecosystem genetics: a consequence of the extended phenotype. Ecology 84:559–573.

Whitlock, M. C. 1992. Temporal fluctuations in demographic parameters and the genetic variance among populations. Evolution 46:608–615.

Whitlock, M. C. 1995. Variance-induced peak shifts. Evolution 49:252–259.

Whitlock, M. C. 2008. Evolutionary inference from Q_{ST}. Molecular Ecology 17:1885–1896.

Whitlock, M. C., and D. E. McCauley. 1999. Indirect measures of gene flow and migration: $F_{ST} \neq 1/(4Nm+1)$. Heredity 82:117–125.

Whitlock, M. C., and K. E. Lotterhos. 2015. Reliable detection of loci responsible for local adaptation: inference of a null model through trimming the distribution of F_{ST}. American Naturalist 186:S24–S36.

Whitlock, M. C., P. K. Ingvarsson, and T. Hatfield. 2000. Local drift load and the heterosis of interconnected populations. Heredity 84:452–457.

Whittaker, R. H. 1977. Evolution of species diversity in land communities. Evolutionary Biology 10:1–67.

Wiens, J. J., and C. H. Graham. 2005. Niche conservatism: integrating evolution, ecology, and conservation biology. Annual Review of Ecology Evolution and Systematics 36:519–539.

Wiens, K. E., E. Crispo, and L. J. Chapman. 2014. Phenotypic plasticity is maintained despite geographical isolation in an African cichlid fish, *Pseudocrenilabrus multicolor*. Integrative Zoology 9:85–96.

Wikelski, M., and L. M. Romero. 2003. Body size, performance and fitness in Galapagos marine iguanas. Integrative and Comparative Biology 43:376–386.

Wilbur, H. M. 1997. Experimental ecology of food webs: complex systems in temporary ponds. Ecology 78:2279–2302.

Willi, Y., and A. A. Hoffmann. 2009. Demographic factors and genetic variation influence population persistence under environmental change. Journal of Evolutionary Biology 22:124–133.

Willi, Y., M. van Kleunen, S. Dietrich, and M. Fischer. 2007. Genetic rescue persists beyond first-generation outbreeding in small populations of a rare plant. Proceedings of the Royal Society B. Biological Sciences 274:2357–2364.

Williams, E. E. 1972. The origin of faunas. Evolution of lizard congeners in a complex island fauna: a trial analysis. Evolutionary Biology 6:47–89.

Williamson, M., and A. Fitter. 1996. The varying success of invaders. Ecology 77:1661–1666.

Willis, C. G., B. Ruhfel, R. B. Primack, A. J. Miller-Rushing, and C. C. Davis. 2008. Phylogenetic patterns of species loss in Thoreau's woods are driven by climate change. Proceedings of the National Academy of Sciences USA 105:17029–17033.

Wilson, A. J., D. Réale, M. N. Clements, M. M. Morrissey, E. Postma, C. A. Walling, L. E. B. Kruuk, and D. H. Nussey. 2010. An ecologist's guide to the animal model. Journal of Animal Ecology 79:13–26.

Wimp, G. M., G. D. Martinsen, K. D. Floate, R. K. Bangert, and T. G. Whitham. 2005. Plant genetic determinants of arthropod community structure and diversity. Evolution 59:61–69.

Wimp, G. M., W. P. Young, S. A. Woolbright, G. D. Martinsen, P. Keim, and T. G. Whitham. 2004. Conserving plant genetic diversity for dependent animal communities. Ecology Letters 7:776–780.

Winkelmann, K., M. J. Genner, T. Takahashi, and L. Rüber. 2014. Competition-driven speciation in a cichlid fish. Nature Communications 5:3412.

Wolf, J. B. W., J. Lindell, and N. Backström. 2010. Speciation genetics: current status and evolving approaches. Philosophical Transactions of the Royal Society B. Biological Sciences 365:1717–1733.

Wolf, J. B., and M. J. Wade. 2009. What are maternal effects (and what are they not)? Philosophical Transactions of the Royal Society B. Biological Sciences 364:1107–1115.

Wolf, J. B., E. D. Brodie III, and M. J. Wade. 2000. *Epistasis and the evolutionary process*. Oxford University Press, Oxford, UK.

Wolf, J. B., E. D. Brodie III, J. M. Cheverud, A. J. Moore, and M. J. Wade. 1998. Evolutionary consequences of indirect genetic effects. Trends in Ecology and Evolution 13:64–69.

Wolf, M., and F. J. Weissing. 2012. Animal personalities: consequences for ecology and evolution. Trends in Ecology and Evolution 27:452–461.

Wolkovich, E. M., B. I. Cook, J. M. Allen, T. M. Crimmins, J. L. Betancourt, S. E. Travers, S. Pau, J. Regetz, T. J. Davies, N. J. B. Kraft, T. R. Ault, K. Bolmgren, S. J. Mazer, G. J. McCabe, B. J. McGill, C. Parmesan, N. Salamin, M. D. Schwartz, and E. E. Cleland. 2012. Warming experiments underpredict plant phenological responses to climate change. Nature 485:494–497.

Wray, G. A. 2007. The evolutionary significance of cis-regulatory mutations. Nature Reviews Genetics 8:206–216.

Wright, S. 1931. Evolution in Mendelian populations. Genetics 16:97–159.

Wright, T. F., J. R. Eberhard, E. A. Hobson, M. L. Avery, and M. A. Russello. 2010. Behavioral flexibility and species invasions: the adaptive flexibility hypothesis. Ethology Ecology and Evolution 22:393–404.

Wu, C.-I. 2001. The genic view of the process of speciation. Journal of Evolutionary Biology 14:851–865.

Wund, M. A., J. A. Baker, B. Clancy, J. L. Golub, and S. A. Foster. 2008. A test of the "flexible stem" model of evolution: ancestral plasticity, genetic accommodation, and morphological divergence in the threespine stickleback radiation. American Naturalist 172:449–462.

Wyles, J. S., J. G. Kunkel, and A. C. Wilson. 1983. Birds, behavior, and anatomical evolution. Proceedings of the National Academy of Sciences USA 80:4394–4397.

Wymore, A. S., A. T. H. Keeley, K. M. Yturralde, M. L. Schroer, C. R. Propper, and T. G. Whitham. 2011. Genes to ecosystems: exploring the frontiers of ecology with one of the smallest biological units. The New Phytologist 191:19–36.

Xie, X. F., J. Rull, A. P. Michel, S. Velez, A. A. Forbes, N. F. Lobo, M. Aluja, and J. L. Feder. 2007. Hawthorn-infesting populations of *Rhagoletis pomonella* in Mexico and speciation mode plurality. Evolution 61:1091–1105.

Yamamichi, M., T. Yoshida, and A. Sasaki. 2011. Comparing the effects of rapid evolution and phenotypic plasticity on predator-prey dynamics. American Naturalist 178:287–304.

Yamamoto, S., and T. Sota. 2012. Parallel allochronic divergence in a winter moth due to disruption of reproductive period by winter harshness. Molecular Ecology 21:174–183.

Yang, J. A., B. Benyamin, B. P. McEvoy, S. Gordon, A. K. Henders, D. R. Nyholt, P. A. Madden, A. C. Heath, N. G. Martin, G. W. Montgomery, M. E. Goddard, and P. M. Visscher. 2010. Common SNPs explain a large proportion of the heritability for human height. Nature Genetics 42:565–569.

Yang, J., A. Bakshi, Z. Zhu, G. Hemani, A. A. E. Vinkhuyzen, S. H. Lee, M. R. Robinson, J. R. B. Perry, I. M. Nolte, J. V. van Vliet-Ostaptchouk, H. Snieder, The LifeLines Cohort Study, T. Esko, L. Milani, R. Mägi, A. Metspalu, A. Hamsten, P. K. E. Magnusson, N. L. Pedersen, E. Ingelsson, N. Soranzo, M. C. Keller, N. R. Wray, M. E. Goddard, and P. M. Visscher. 2015. Genetic variance estimation with imputed variants finds negligible missing heritability for human height and body mass index. Nature Genetics 47:1114–1120.

Yeaman, S. 2015. Local adaptation by alleles of small effect. American Naturalist 186:S74–S89.

Yeaman, S., and F. Guillaume. 2009. Predicting adaptation under migration load: the role of genetic skew. Evolution 63:2926–2938.

Yoder, J. B., E. Clancey, S. Des Roches, J. M. Eastman, L. Gentry, W. Godsoe, T. J. Hagey, D. Jochimsen, B. P. Oswald, J. Robertson, B. A. J. Sarver, J. J. Schenk, S. F. Spear, and L. J. Harmon. 2010. Ecological opportunity and the origin of adaptive radiations. Journal of Evolutionary Biology 23:1581–1596.

Yoshida, T., L. E. Jones, S. P. Ellner, G. F. Fussmann, and N. G. Hairston Jr. 2003. Rapid evolution drives ecological dynamics in a predator-prey system. Nature 424:303–306.

Yoshida, T., S. P. Ellner, L. E. Jones, B. J. M. Bohannan, R. E. Lenski, and N. G. Hairston Jr. 2007. Cryptic population dynamics: rapid evolution masks trophic interactions. PLoS Biology 5:1868–1879.

Young, A., T. Boyle, and T. Brown. 1996. The population genetic consequences of habitat fragmentation for plants. Trends in Ecology and Evolution 11:413–418.

Yukilevich, R. 2012. Asymmetrical patterns of speciation uniquely support reinforcement in *Drosophila*. Evolution 66:1430–1446.

Zangerl, A. R., and M. R. Berenbaum. 2005. Increase in toxicity of an invasive weed after reassociation with its coevolved herbivore. Proceedings of the National Academy of Sciences USA 102:15529–15532.

Zhang, D. Y., G. J. Sun, and X. H. Jiang. 1999. Donald's ideotype and growth redundancy: a game theoretical analysis. Field Crops Research 61:179–187.

Zheng, C.Z., O. Ovaskainen, and I. Hanski. 2009. Modelling single nucleotide effects in phosphoglucose isomerase on dispersal in the Glanville fritillary butterfly: coupling of ecological and evolutionary dynamics. Philosophical Transactions of the Royal Society. B. Biological Sciences 364:1519–1532.

Zhu, Y. Y., H. R. Chen, J. H. Fan, Y. Y. Wang, Y. Li, J. B. Chen, J.X. Fan, S. S. Yang, L. P. Hu, H. Leung, T. W. Mew, P. S. Teng, Z. H. Wang, and C. C. Mundt. 2000. Genetic diversity and disease control in rice. Nature 406:718–722.

Zuk, M., and G. R. Kolluru. 1998. Exploitation of sexual signals by predators and parasitoids. Quarterly Review of Biology 73:415–438.

Zuk, M., J. T. Rotenberry, and R. M. Tinghitella. 2006. Silent night: adaptive disappearance of a sexual signal in a parasitized population of field crickets. Biology Letters 2:521–524.

Zuppinger-Dingley, D., B. Schmid, J. S. Petermann, V. Yadav, G. B. De Deyn, and D. F. B. Flynn. 2014. Selection for niche differentiation in plant communities increases biodiversity effects. Nature 515:108–111.

Zytynska, S. E., S. Fleming, C. Tétard-Jones, M. A. Kertesz, and R. F. Preziosi. 2010. Community genetic interactions mediate indirect ecological effects between a parasitoid wasp and rhizobacteria. Ecology 91:1563–1568.

Index